IUTAM SYMPOSIUM ON WAVES IN LIQUID/GAS AND LIQUID/VAPOUR TWO-PHASE SYSTEMS

FLUID MECHANICS AND ITS APPLICATIONS
Volume 31

Series Editor: R. MOREAU
MADYLAM
Ecole Nationale Supérieure d'Hydraulique de Grenoble
Boîte Postale 95
38402 Saint Martin d'Hères Cedex, France

Aims and Scope of the Series

The purpose of this series is to focus on subjects in which fluid mechanics plays a fundamental role.

As well as the more traditional applications of aeronautics, hydraulics, heat and mass transfer etc., books will be published dealing with topics which are currently in a state of rapid development, such as turbulence, suspensions and multiphase fluids, super and hypersonic flows and numerical modelling techniques.

It is a widely held view that it is the interdisciplinary subjects that will receive intense scientific attention, bringing them to the forefront of technological advancement. Fluids have the ability to transport matter and its properties as well as transmit force, therefore fluid mechanics is a subject that is particulary open to cross fertilisation with other sciences and disciplines of engineering. The subject of fluid mechanics will be highly relevant in domains such as chemical, metallurgical, biological and ecological engineering. This series is particularly open to such new multidisciplinary domains.

The median level of presentation is the first year graduate student. Some texts are monographs defining the current state of a field; others are accessible to final year undergraduates; but essentially the emphasis is on readability and clarity.

For a list of related mechanics titles, see final pages.

IUTAM Symposium on

Waves in Liquid/Gas and Liquid/Vapour Two-Phase Systems

Proceedings of the IUTAM Symposium
held in Kyoto, Japan,
9–13 May 1994

Edited by

SHIGEKI MORIOKA

Department of Aeronautical Engineering,
Kyoto University, Japan

and

LEEN VAN WIJNGAARDEN

Department of Mechanical Engineering,
University of Twente, Enschede,
The Netherlands

SPRINGER-SCIENCE+BUSINESS MEDIA, B.V.

A C.I.P. Catalogue record for this book is available from the Library of Congress.

ISBN 978-94-010-4030-3 ISBN 978-94-011-0057-1 (eBook)
DOI 10.1007/978-94-011-0057-1

Printed on acid-free paper

CONTENTS

(4) Waves in liquid-vapor transition

(5) Waves with vapor-liquid transition

(6) Wave propagation near critical point

International Scientific Committee

D.G. Crighton (UK)
J.M. Delhaye (France)
V.K. Kedrinskii (Russia)
G.E.A. Meier (FRG)
S. Morioka (Japan), Co-Chairman
A. Prosperetti (USA)
B. Sturtevant (USA)
L. van Wijngaarden (The Netherlands), Co-Chairman

Local Organizing Committee

S. Fujikawa
T. Fukano
F. Higashino
T. Kawahara
M. Kiya
Y. Kobayashi
S. Kubo
Y. Matsumoto
S. Morioka
Y. Onishi
A. Sakurai
Y. Sone
N. Sugimoto
K. Takayama

Supporting Organizations

International Union of Theoretical and Applied Mechanics (IUTAM)
The Japan Society of Aeronautical and Space Sciences
The Japan Society of Fluid Mechanics
Kyoto University, Faculty of Engineering

Organizations and Companies for Financial Aid

Japan World Exposition Commemorative Association
International Union of Theoretical and Applied Mechanics

Daihatsu Motor Co. Ltd.
Hitachi Ltd.
Kansai Electric Power Co. Inc.
Kawasaki Heavy Industries Ltd.
Mitsubishi Electric Corp.
Mitsubishi Heavy Industries Ltd.
Toyota Motor Corp.

Preface

A IUTAM symposium on 'Waves in Liquid/Gas and Liquid/Vapor Two-Phase Systems' was held in Kyoto, Japan, 9-13 May 1994. Sixty-three scientists participated coming from ten countries, and forty-two lectures were presented. The list of participants and the program are included in this volume.

The symposium was held in response to the request of the participants in the IUTAM symposium 'Adiabatic Waves in Liquid-Vapor System' held at Göttingen in 1989. At that time, the need for another symposium in about five years had been indicated by all the participants. This symposium intends to develop the subject of wave properties in more general liquid-gas two-phase systems.

Topics in this symposium may be classified as (1) waves in liquid-gas bubble systems including interfacial effects, (2) waves in gas(vapor)-droplets systems, (3) waves in films or stratified systems, (4) waves with liquid-vapor transition, (5) waves with vapor-liquid transition, (6) wave propagation near the critical point and (7) waves with low pressure effect.

As for topic (1), experiments, numerical simulations and analytical approaches to waves in bubly liquids were discussed. The importance of interbubble interactions through the liquid-field is now well established at least in terms of potential theory. There was also a progress concerning the well-posedness of governing equations for void waves. For pressure waves there were some new phenomena, such as bubble-cluster formation and the occurrence of three-dimensional structures, in addition to a progress from more qualitative studies to quantitative ones.

The researches reported in (2) reach from statistically homogeneous and dilute droplet distributions to statistically inhomogeneous and dense ones. This tendency can be seen in the researches of (1) too. Numerical simulation seems to be the most promising research tool here. It is important to understand the elementary processes, such as collisions between droplet and droplet, liquid layer or solid wall. In this symposium, various wave phenomena were discussed in connection with such collision processes.

There were many contributions for topic (3). The equations describing such waves can be reduced to Benny's equation, which has chaotic solutions. In fact, one of the results in this symposium is that this type of equation appears in a variety of flow patterns in two-phase flow.

As for (4), (5), and (7), possible waves are jumps between two different states of single-/two-phases for ordinary and retrograde fluids. Now interest focuses on the spatial and temporal structure of these waves. In this symposium, some nice experimental results regarding this topic were presented as well as attempts at modelling.

As for (6), interest is now in modelling of the flow pattern and in possible wave modes associated with singular behaviour of the fluid state near the critical point.

The scientific committee, the members of which are listed on the first page, contributed to this symposium by recommending invited lecturers, serving as chairmen during the symposium and by selecting abstracts and by reviewing of papers for the proceedings. The local organizing committee the members of which are also listed on the first page contributed in the preparation and organization of the symposium. In particular Prof. T. Kawahara, Prof. N. Sugimoto, Dr. H. Monji, Dr. K. Ueno and Mr. A. Tsukamoto made substantial contributions. We also thank the participants for their cooperations and lively discussions during the symposium.

Very helpful financial support for the symposium was provided by the Japan World Exposition Commemorative Association. Throughout the preparation period Japan was in a state of heavy depression and it was difficult to expect any contributions from companies. Without the support from the Japan World Exposition Commemorative Association, the symposium would not have succeeded. Here we express our appreciation to this Association for their contribution which enabled the participation of many foreign participants. Financial support was also provided by the International Union of Theoretical and Applied Mechanics and by several companies listed on the following page. We thank those companies for their understanding of the significance of this symposium and their contribution under difficult circumstances.

Kyoto was the metropolis in Japan from 794 to 1868 and this year is just 1200 years after the transfer to Kyoto capital. We were very happy that this symposium could be held in Kyoto at this time.

Kyoto, May 1994

Shigeki Morioka

Leen van Wijngaarden

List of Participants

Prof. Teruaki Akamatsu
Dept. of Mechanical Engineering,
Kyoto Univ., Yoshida-Honmachi,
Sakyo-ku, Kyoto 606-01, Japan

Prof. Kazuo Aoki
Dept. of Aeronautical Engineering,
Kyoto Univ., Yoshida-Honmachi,
Sakyo-ku, Kyoto 606-01, Japan

Prof. Alfred E. Beylich
Stoßwellenlabor, Technische Hochschule,
Templergraben 55, 52056 Aachen, Germany

Dr. Arie Biesheuvel
J.M. Burgerscentre for Fluid Mechanics,
Univ. of Twente, P.O. Box 217,
7500 AE Enschede, The Netherlands

Ir. Gerrit Jan de Bruin
Dept. of Applied Physics,
Univ. of Twente, P.O. Box 217
7500 AE Enschede, The Netherlands

Prof. David G. Crighton
Dept. of Applied Mathematics and
Theoretical Physics, Univ. of Cambridge,
Silver Street, Cambridge CB3 9EW, U.K.

Prof. Luis Antonio Dávalos-Orozco
Inst. de Investigaciones en Materiales,
Univ. Nacional Autonoma de México,
Apartado Postal 70-360 Delegación de
Coyoacan, 04510 México, D.F., México

Prof. Can F. Delale
Dept. of Mechanical Engineering,
City College of New York,
Convent Avenue and 138th Street,
New York, NY 10031, USA

Prof. Jean-Marc Delhaye
Centre d'Etudes Nucléaires de Grenoble,
Service de Thermohydraulique pour les
Applications Industrielles, 38054
Grenoble Cedex 9, France

Mr. Paul C. Duineveld
Dept. of Applied Physics,
Univ. of Twente, P.O. Box 217
7500 AE Enschede, The Netherlands

Prof. Jean Fabre
Inst. de Mécanique des Fluides,
Avenue du Professeur Camille Soula,
31400 Toulouse, France

Prof. David L. Frost
Dept. of Mechanical Engineering,
McGill Univ., 817 Sherbrooke St. W.,
Montreal, Quebec H3A 2K6, Canada

Dr. Shigeo Fujikawa
Dept. of Mechanical System Engineering,
Toyama Prefectural Univ.,
Kosugi-machi, Toyama 939-03, Japan

Mr. Hitoshi Fujimoto
Dept. of Mineral Science and Technology,
Kyoto Univ., Yoshida-Honmachi,
Sakyo-ku, Kyoto 606-01, Japan

Prof. Toru Fukano
Dept. of Mechanical Engineering,
Kyushu Univ., 6-10-1 Hakozaki,
Higashi-ku, Fukuoka 812, Japan

Dr. Sergey L. Gavrilyuk
Lavrentyev Inst. of Hydrodynamics,
The Russian Academy of Sciences,
Novosibirsk, 630090, Russia

Dr. Nail A. Gumerov
Inst. of Energy Resources Transportation
pr. Oktyabrya 144/3, Ufa, 450055, Russia

Prof. Yoshimichi Hagiwara
Kyoto Inst. of Technology
Matsugasaki, Sakyo-ku, Kyoto 606, Japan

Dr. Hisao Hayakawa
Dept. of Physics, Tohoku Univ.
Aramaki, Aoba-ku, Sendai 980, Japan

Prof. Fumio Higashino
Dept. of Mechanical Systems Engineering,
Tokyo Univ. of Agriculture and Technology
2-22-16 Nakano, Koganei, Tokyo 184, Japan

Dr. Ryuji Ishii
Dept. of Aeronautical Engineering,
Kyoto Univ., Yoshida-Honmachi,
Sakyo-ku, kyoto 606-01, Japan

Dr. Sang W. Joo
Dept. of Mechanical Engineering,
Wayne State Univ., Detroit,
MI 48202, USA

Dr. Masaharu Kameda
Dept. of Mechanical Engineering,
The Univ. of Tokyo, 7-3-1, Hongo,
Bunkyo-ku, Tokyo 113, Japan

Dr. Isao Kataoka
Inst. of Atomic Energy, Kyoto Univ.,
Gokasyo, Uji, Kyoto 611, Japan

Dr. Takuji Kawahara
Dept. of Physics, Kyoto Univ.,
Kitashirakawa-Oiwakecho, Sakyo-ku,
Kyoto 606-01, Japan

Prof. Valery K. Kedrinskii
Lavrentyev Inst. of Hydrodynamics,
The Russian Academy of Sciences,
Novosibirsk 630090, Russia

Prof. Alfred Kluwick
Inst. für Strömungslehre u. Wärme-
übertrangung, Technische Univ. Wien,
Wiedner Hauptstraße 7/322,
A-1040 Wien, Austria

Prof. Yasunori Kobayashi
Inst. of Engineering Mechanics,
Univ. of Tsukuba, 1-1-1 Tennodai,
Tsukuba, Ibaraki 305, Japan

Prof. Syozo Kubo
Dept. of Applied Mathematics and Physics,
Tottori Univ., Tottori 680, Japan

Dr. Heidy M. Mader
Inst. of Environmental and Biological
Sciences, Lancaster Univ., Lancaster,
LA1 4YQ, U.K.

Dr. Kazuo Maeno
Dept. of Mechanical Engineering,
Chiba Univ., 1-33 Yayoi-cho,
Chiba 260, Japan

Prof. Goichi Matsui
Inst. of Engineering Mechanics,
Univ. of Tsukuba, 1-1-1 Tennodai,
Tsukuba, Ibaraki 305, Japan

Dr. Mitsuhiro Matsumoto
Dept. of Applied Physics, Nagoya Univ.,
Furo-cho, Tikusa-ku, Nagoya 464, Japan

Prof. Yoichiro Matsumoto
Dept. of Mechanical Engineering,
The Univ. of Tokyo, 7-3-1 Hongo,
Bunkyo-ku, Tokyo 113, Japan

Prof. Gerd E.A. Meier
Inst. für Strömungsmechanik,
Deutschen Forschungsanstalt für Luft-
und Raumfahrt e.V., Bunsenstraße 10,
37073 Göttingen, Germany

Dr. Rob Miesen
Koninklijke/Shell-Laboratorium,
Amsterdam, Postbus 38000,
1030 BN Amsterdam, The Netherlands

Dr. Hiroyuki Miura
Dept. of Mechanical Engineering,
Kinki Univ., 3-4-1 Kowakae,
Higashi-Osaka, Osaka 577, Japan

Dr. Hideaki Monji
Inst. of Engineering Mechanics,
Univ. of Tsukuba, 1-1-1 Tennodai,
Tsukuba, Ibaraki 305, Japan

Prof. Shigeki Morioka
Dept. of Aeronautical Engineering,
Kyoto Univ., Yoshida-Honmachi,
Sakyo-ku, Kyoto 606-01, Japan

Prof. Yoshimoto Onishi
Dept. of Applied Mathematics and Physics,
Tottori Univ., Tottori 680, Japan

Mr. Takeshi Ooshida
Dept. of Physics, Kyoto Univ.
Kitashirakawa-Oiwakeco, Sakyo-ku,
Kyoto 606-01, Japan

Prof. Boris G. Pokusaev
Moscow State Academy of Chemical Engineering, Staraya Basmannaya 21/4
Moscow 107884, Russia

Prof. Andrea Prosperetti
Dept. of Mechanical Engineering,
The Johns Hopkins Univ.,
122 Latrobe Hall, Baltimore
Maryland 21218, USA

Dr. Martin Rein
Max-Planck-Inst. für Strömungsforschung,
Bunsenstraße 10, D-37073 Göttingen,
Germany

Prof. Akira Sakurai
Tokyo Denki Univ.,
1-8-22 Hongo-Komagome, Bunkyo-ku,
Tokyo 113, Japan

Prof. Ashok S. Sangani
Dept. of Chemical Engineering and
Materials Sciences, Syracuse Univ.,
Syracuse, NY 13244-1190, USA

Prof. Günter H. Schnerr
Inst. für Strömungslehre und Strömungsmaschinen, Univ. Karlsruhe (TH)
Kaiserstraße 12, D-76128 Karlsruhe,
Germany

Prof. Akimi Serizawa
Dept. of Nuclear Engineering,
Kyoto Univ., Yoshida-Honmachi,
Sakyo-ku, Kyoto 606-01, Japan

Prof. Vladimir P. Skripov
Inst. of Thermophysics, Ural Branch
of Russian Academy of Sciences,
Pervomaiskaya st. 91, GSP-828,
Ekaterinburg, 620219, Russia

Prof. Yoshio Sone
Dept. of Aeronautical Engineering,
Kyoto Univ., Yoshida-Honmachi,
Sakyo-ku, Kyoto 606-01, Japan

Dr. Nobumasa Sugimoto
Dept. of Mechanical Engineering,
Faculty of Engineering Science,
Univ. of Osaka, Toyonaka,
Osaka 560, Japan

Prof. Kenjiro Suzuki
Dept. of Mechanical Engineering,
Kyoto Univ., Yoshida-Honmachi,
Sakyo-ku, Kyoto 606-01, Japan

Dr. Hiroyuki Takahira
Dept. of Mechanical Engineering,
Kyoto Univ., Yoshida-Honmachi,
Sakyo-ku, Kyoto 606-01, Japan

Prof. Kazuyoshi Takayama
Shock Wave Research Center,
Inst. of Fluid Science, Tohoku Univ.,
2-1-1 Katahira, Aoba-ku, Sendai 980
Japan.

Prof. Tomomasa Tatsumi
Kyoto Inst. of Technology,
Matsugasaki, Sakyo-ku, Kyoto 606, Japan

Prof. Vladimir M. Teshukov
Lavrentyev Inst. of Hydrodynamics,
The Russian Academy of Sciences,
Novosibirsk, 630090, Russia

Mr. Akimasa Tsukamoto
Dept. of Aeronautical Engineering,
Kyoto Univ., Yoshida-Honmachi,
Sakyo-ku, Kyoto 606-01, Japan

Mr. Hidekazu Tsuji
Dept. of Applied Mathematics and Physics,
Tottori Univ., Tottori 680, Japan

Dr. Kazuyuki Ueno
Inst. of Fluid Science, Tohoku Univ.,
2-1-1 Katahira, Aoba-ku, Sendai 980,
Japan

Mr. Hisashi Umekawa
Dept. of Mechanical Engineering,
Kansai Univ., 3-3-35 Yamate-cho,
Suita, Osaka 564, Japan

Prof. Leen van Wijngaarden
Laboratorium voor Warmteen Stromingsleer,
Univ. Twente, Postbus 217
7500 AE Enschede, The Netherlands

Mr. Kiyoshi Yamada
Shock Wave Research Center,
Inst. of Fluid Science, Tohoku Univ.
2-1-1 Katahira, Aoba-ku, Sendai 980,
Japan

Prof. Stéphane Zaleski
Laboratoire de Modélisation en Mécanique,
Univ. Pierre et Marie Curie,
Tour 66, 4 Place Jussieu, 75252 Paris
Cedex 05, France

Structure and stability of void waves in bubbly flows

A. BIESHEUVEL
J.M.Burgerscentre for Fluid Mechanics, University of Twente,
P.O.Box 217, 7500 AE Enschede, The Netherlands

1 Introduction

Observations of bubbly flows in a vertical pipe show that upon increasing the gas flow rate, at some point along the pipe a uniform bubbly flow changes into a kind of agitated, 'turbulent' bubbly flow. This turbulent flow is characterized by large collections of bubbles that violently move about in a kind of zig-zag motion. A further increase in the gas flux moves the point of transition downwards. An increase in the gas flux raises the gas volume concentration in the pipe, and since this volume concentration increases with height along the tube, due to the loss of hydrostatic head, this suggests that there is a critical value of the gas volume concentration associated with the flow transition. For still higher gas fluxes the gas bubbles rapidly coalesce within the region of turbulent flow, and large gas slugs are formed that eventually fill the entire cross-section of the pipe. The flow then has changed into a regime in which gas slugs move upwards with high speed, separated by regions with an approximately uniform bubbly fluid.

In the past 10 years the work of Leen van Wijngaarden and myself has been concerned with hydrodynamic interactions between gas bubbles with the aim of obtaining a description of the behaviour of bubbly flows. The above-mentioned flow transition serves as a 'reference-problem', that ultimately should be understood. It just so happened that Leen concentrated on the dynamics of interacting pairs of bubbles and the properties of dilute dispersions, while I looked at the behaviour of non-dilute bubbly flows. Leen van Wijngaarden (1992) has recently reviewed his work and that of his graduate students Jim Kok and Cees Kapteyn, here my work with Henk Bulthuis and Jeroen Lammers will be described.

The magnificent work of G.K.Batchelor always has been an important source of inspiration for us at Twente. I have the impression that Batchelor's work on the stability of fluidized beds has been largely overlooked in the literature on bubbly flows, so this article starts with a description of his ideas as applied to bubbly flows. In §3 the work of Bulthuis (1992) on bubble interactions is described, followed by a brief discussion of recent papers by Ashok Sangani, Andrea Prosperetti and co-

S. Morioka and L. van Wijngaarden (eds.),
IUTAM Symposium on Waves in Liquid/Gas and Liquid/Vapour Two-Phase Systems, 1-24.

workers in §4. Finally the work of Lammers (1994) on the structure of void waves and instabilities will be presented in §5 and §6.

2 Batchelor's theory

In a brilliant analysis of the conditions under which uniform fluidized beds are unstable to planar voidage disturbances Batchelor (1988) showed that when spatial gradients are sufficiently small, the mass and momentum conservation equations of the dispersed phase, in a reference frame in which there is no net volume flux of material, are given by

$$\frac{\partial \phi}{\partial t} + V\frac{\partial \phi}{\partial x} + \phi\frac{\partial V}{\partial x} = 0, \tag{1}$$

$$\phi(1+\theta)\left(\frac{\partial V}{\partial t} + V\frac{\partial V}{\partial x}\right) - \phi\zeta V\frac{\partial V}{\partial x} + Q\frac{\partial \phi}{\partial x} - \frac{\partial}{\partial x}\left(\phi\eta\frac{\partial V}{\partial x}\right)$$
$$= -\frac{\gamma\tilde{g}}{U}\phi(V-U). \tag{2}$$

The arguments that lead Batchelor to formulate these conservation equations apply equally well to non-uniform bubbly flows. Then ϕ and V denote the volume concentration and the mean velocity of the bubbles, $U(\phi)$ is the mean bubble velocity in a homogeneous dispersion with volume concentration ϕ when the bubbles move only under the action of gravity. The constant γ, takes a value between 1 and 2, depending on the functional relationship between the mean frictional force and the mean bubble velocity under uniform conditions (see Batchelor's paper for a precise definition). The reduced gravity is defined by

$$\tilde{g} = g\frac{\rho_l - \rho_g}{\rho_g},$$

in which ρ_g and ρ_l denote the density of the bubbles and the fluid.

The first two terms on the left-hand side of (2) represent the acceleration reaction of the bubbles. The functions $\theta(\phi)$ and $\zeta(\phi)$ are related to the added mass coefficient $C(\phi)$ through

$$\theta(\phi) = \frac{\rho_l}{\rho_g}C(\phi), \quad \text{and} \quad \zeta(\phi) = \frac{\rho_l}{\rho_g}\phi\left(\frac{\mathrm{d}C}{\mathrm{d}\phi}\right).$$

The other two terms in the momentum conservation equation represent the mean force per unit mass of the bubbles which arises in a non-uniform dispersion because of momentum transport by velocity fluctuations of the bubbles, and by forces which the bubbles exert on each other hydrodynamically; $\phi\rho_g Q(\phi)$ is the bulk modulus of elasticity of the bubbles (see §4) and $\phi\rho_g\eta(\phi)$ is an effective bubble viscosity.

To analyse the conditions for instability one linearizes the conservation equations. If it is assumed that the disturbances are proportional to $\exp\{\mathrm{i}\kappa(x-ct)\}$, with complex phase speed c and real wave number κ, it follows that c has to satisfy the relation

$$(1+\theta)(c-U)^2 + (c-U)\left(\zeta U + \mathrm{i}\kappa\eta + \frac{\mathrm{i}\gamma\tilde{g}}{\kappa U}\right) - \left(Q + \frac{\gamma\tilde{g}}{\kappa U}\mathrm{i}W\right) = 0,$$

where the functions of the volume concentration have to be evaluated for the value that characterizes the uniform state. The velocity W is defined as

$$W(\phi) = \phi\left(\frac{\mathrm{d}U}{\mathrm{d}\phi}\right).$$

From the solution of this equation Batchelor deduces that the condition for instability of the flow reads

$$\left\{\frac{\gamma\tilde{g}W}{\kappa^2\eta U + \gamma\tilde{g}} + \frac{\frac{1}{2}\zeta U}{(1+\theta)}\right\}^2 > \frac{Q}{1+\theta} + \frac{\frac{1}{4}\zeta^2 U^2}{(1+\theta)^2}. \tag{3}$$

Taking the limit $\kappa \to \infty$ it becomes clear that a sufficient condition for instability is that $Q < 0$. This is not unexpected, because then the dispersed phase has negative elasticity. To consider the case $Q > 0$, note first that the left-hand side of the above expression obtains a maximum for some wavenumber $\kappa = \kappa_m$, which in general will depend on the values of the functions U, θ and η. Provided that $W+\frac{1}{2}\zeta U/(1+\theta) < 0$, a condition that holds in general for bubbly flows, this maximum is reached for $\kappa_m = 0$. Thus a bubbly flow becomes unstable at a value of the volume concentration ϕ such that

$$Q - W\{(1+\theta)W + \zeta U\} < 0.$$

Using the definitions of W and ζ, and noting that $\rho_l \gg \rho_g$, this condition can be rewritten as

$$Q - \frac{\rho_l}{\rho_g}\phi^2\left\{\frac{\mathrm{d}U}{\mathrm{d}\phi}\right\}\left\{\frac{\mathrm{d}(CU)}{\mathrm{d}\phi}\right\} < 0. \tag{4}$$

Since in general the mean velocity of rise of the bubbles becomes less with increasing void fraction, it appears that when the bulk elasticity of the bubbles is absent in a description of the bubbly flow, the instability condition has the very simple form $\mathrm{d}(CU)/\mathrm{d}\phi < 0$. This expression can for instance be found in van Wijngaarden & Biesheuvel (1988). Note that when $\mathrm{d}(CU)/\mathrm{d}\phi > 0$ the flow can only be unstable when the bulk elasticity is negative.

Information on the correct expression for the bulk elasticity is absent, but Batchelor suggests by an elegant argument that it can be written as

$$\rho_g\phi Q(\phi) = \rho_g\phi\left(\frac{\gamma\tilde{g}}{U}D + \frac{\mathrm{d}(\phi H U^2)}{\mathrm{d}\phi}\right). \tag{5}$$

The first contribution arises from the hydrodynamic interaction forces between the bubbles which cause a diffusive flux of bubbles in the presence of a concentration gradient, with diffusivity $D(\phi)$. The second contribution represents transport of mean bubble momentum due to velocity fluctuations, with $H(\phi)$ a function that vanishes as $\phi \to 0$ and at closest packing. The latter contribution is negligibly small of course, and one therefore expects that the condition $Q > 0$ is not violated.

We see then that what is required to obtain a numerical estimate for the instability condition of a uniform bubbly flow is a knowledge of the functions U, C, D, which all can be determined from experiments, in principle. The mean velocity of rise of the bubbles is usually expressed in the form

$$U(\phi) = U_\infty(1-\phi)^p, \tag{6}$$

where U_∞ is the velocity of a single bubble rising under the action of gravity in an infinite liquid, and p is a constant which depends on the physical properties of the two phases. Not much is known about the function D, so Batchelor proposes to write

$$D(\phi) = \alpha a|U|,$$

where α is a constant of order 1, and a denotes the radius of the bubbles. The condition for instability (4) now reduces to

$$N \equiv \xi p^2\phi^2(1-\phi)^{2(p-1)}\left(C - \frac{1-\phi}{p}\frac{\mathrm{d}C}{\mathrm{d}\phi}\right) > \alpha, \tag{7}$$

with ξ the bubble Froude number, defined as

$$\xi = \frac{U_\infty{}^2}{\gamma a g}.$$

Thus, the value at which the flow first becomes unstable upon increasing the gas volume flux, ϕ_c say, is essentially determined by the size of the bubbles and the liquid properties. This is exemplified for dispersions of equally-sized gas bubbles with equivalent radius in the range 0.5 mm to 0.9 mm rising in pure water. Single bubbles of this size rise in a straight line and have oblate ellipsoidal shape. An approximate expression for their velocity has been given by Moore (1965),

$$U_\infty = \frac{a^2 g}{9\nu G(\chi)}, \tag{8}$$

where χ is the ratio between the major axis and the minor axis, and

$$G(\chi) = \frac{1}{3}\chi^{4/3}(\chi^2-1)^{3/2}\frac{\sqrt{\chi^2-1}-(2-\chi^2)\cos^{-1}\chi^{-1}}{(\chi^2\cos^{-1}\chi^{-1}-\sqrt{\chi^2-1})^2}.$$

The deformation follows from (8) together with a relation for the Weber number

$$W(\chi) = \frac{2\rho_l U_\infty{}^2 a}{\sigma} = 4\chi^{-4/3}(\chi^3+\chi-2)\{\chi^2\sec^{-1}\chi - \sqrt{\chi^2-1}\}^2(\chi^2-1), \tag{9}$$

where σ denotes the coefficient of surface tension. A good approximation for the function $C(\phi)$ is

$$C(\phi) = \frac{1}{2}Q(\chi)\frac{1+(2/Q(\chi))\phi}{1-\phi} \tag{10}$$

with

$$Q(\chi) = 2\chi^2\frac{\sqrt{\chi^2-1}-\cos^{-1}\chi^{-1}}{\chi^2\cos^{-1}\chi^{-1}-\sqrt{\chi^2-1}}.$$

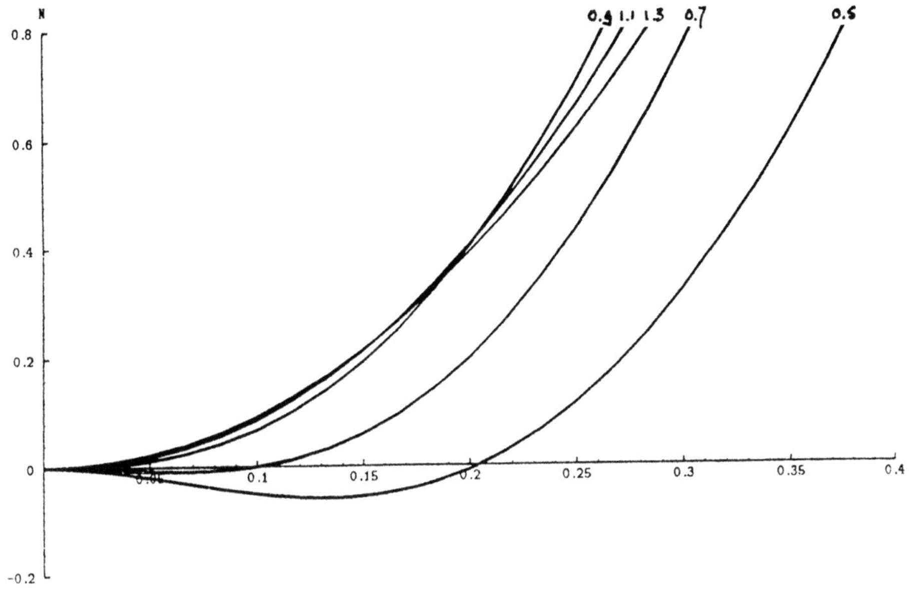

Figure 1: Stability criterion for dispersions of equally-sized bubbles in water. Bubble diameters 0.5, 0.7, 0.9, 1.1 and 1.3 mm. The flow is unstable to the right of the curves, above a horizontal line $N = \alpha$.

(The standard notation $W(\chi)$ and $Q(\chi)$ will hopefully cause no confusion with the functions $W(\phi)$ and $Q(\phi)$ introduced by Batchelor.)

Bubbles with an equivalent radius in the range 0.9 mm to 1.4 mm perform a spiralling motion. The deformation can also be calculated to a good approximation from (8) and (9), but now expression (8) gives the (approximately constant) speed along the bubble trajectory; the mean speed of rise is about 10 % less (Duineveld 1994). In figure 1 the function N has been plotted for bubble radii between 0.5 mm and 1.3 mm, with the values for p taken equal to 2. If α is known one can draw a horizontal line at $N = \alpha$ to find the critical value ϕ_c. We see that the latter value strongly depends on the bubble size, and the value of α. In pipe flow experiments ϕ increases along the pipe because the bubbles grow in size due to the loss of hydrostatic head. Thus even though the flow may be stable in the bottom section, it may eventually become unstable higher up. There are indications (see §6) that the velocity fluctuations of the bubbles diminish as they move downstream, even when their size does not change. The value of α then depends on the distance from the bubble generator, and so does ϕ_c. There is clearly a need for more information regarding α before definite predictions can be made.

3 Hydrodynamic interaction between bubbles

Experiments performed at Twente by Kok (1993) have shown that the motion of pairs of bubbles with a diameter of the order of 1 mm to 1.4 mm is accurately described by irrotational flow theory. This suggests to perform numerical simulations of large numbers of bubbles on the basis of this theory, for instance to obtain expressions for the flow properties mentioned in §3 without having to take recourse to simple, but arbitrary probability distributions of the bubble positions and velocities. Bulthuis (1992) has devised a simple approximate scheme for these simulations. If the bubbles are represented by dipoles the fluid velocity potential induced by each of of these dipoles is given by

$$\phi(\mathbf{r}_p) = \frac{\mathbf{D}_p \cdot \mathbf{r}_p}{r_p^3} + \mathbf{U}_p \cdot \mathbf{r}_p.$$

Here $\mathbf{r}_p$ denotes the position vector with centre at dipole p, and $\mathbf{U}_p$ is the velocity at dipole p induced by the other singularities. The dipole strength is

$$\mathbf{D}_p = -\frac{1}{2}a_p^3(\mathbf{V}_p - \mathbf{U}_p),$$

where $\mathbf{V}_p$ and a_p are the velocity and radius of bubble p. Note that the bubbles do not need to have equal size. The velocity induced at p by the other dipoles is given by

$$\mathbf{U}_p = \sum_{q \neq p} \frac{\mathbf{D}_q}{d_{pq}^3} - \frac{3\mathbf{D}_q \cdot \mathbf{d}_{pq}}{d_{pq}^5}\mathbf{d}_{pq} = \sum_{q \neq p} \mathbf{C}_{pq} \cdot \mathbf{D}_q,$$

with $\mathbf{d}_{pq}$ the seperation-vector of the dipoles p and q. It is convenient to use the matrix notation

$$\underline{\mathbf{U}} = \underline{\mathbf{C}} \cdot \underline{\mathbf{D}}. \tag{11}$$

The dipole strength can then be written as

$$\underline{\mathbf{D}} = -\underline{\mathbf{I}} \cdot (\underline{\mathbf{V}} - \underline{\mathbf{U}}), \tag{12}$$

where $[\underline{I}]_{pq} = \frac{1}{2}a_p^3\mathbf{I}$, for $p = q$, and $\mathbf{0}$ otherwise; $\mathbf{I}$ is the 3×3 identity-matrix and $\mathbf{0}$ is the 3×3 null-matrix. Combining (11) and (12) it follows that

$$\underline{\mathbf{D}} = -(\underline{\mathbf{I}}^{-1} - \underline{\mathbf{C}})^{-1} \cdot \underline{\mathbf{V}}. \tag{13}$$

Next, the fluid impulse associated with bubble p is defined by

$$\mathbf{J}_p = -\rho_l \int \phi_p \mathrm{d}\mathbf{A}_p = -\rho_l(4\pi\mathbf{D}_p + \frac{4}{3}\pi a_p^3\mathbf{V}_p),$$

or in matrix notation, using (13),

$$\underline{\mathbf{J}} = 4\pi\rho_l\left((\underline{\mathbf{I}}^{-1} - \underline{\mathbf{C}})^{-1} - \frac{2}{3}\underline{\mathbf{I}}\right) \cdot \underline{\mathbf{V}} = \rho_l\underline{\mathbf{M}} \cdot \underline{\mathbf{V}}. \tag{14}$$

The kinetic energy of the fluid is then given by

$$T = \frac{1}{2}\rho_l\underline{\mathbf{V}} \cdot \underline{\mathbf{M}} \cdot \underline{\mathbf{V}},$$

and the rate of energy dissipation, within the dipole approximation, by

$$\Phi = 48\pi\mu_l \sum_p \frac{\mathbf{D}_p \cdot \mathbf{D}_p}{a_p^5} = 48\pi\mu_l(2\underline{\mathbf{I}})^{-\frac{5}{3}} \cdot \underline{\mathbf{D}} \cdot \underline{\mathbf{D}}.$$

The buoyancy force on the bubbles can be represented by

$$\underline{\mathbf{F}}_g = \rho_l \frac{8\pi}{3} \underline{\mathbf{I}} \cdot \underline{\mathbf{G}},$$

where $[\underline{G}]_q = \mathbf{g}$.

The equations of motion now follow from Lagrange's equations:

$$\frac{\mathrm{d}}{dt}(\underline{\mathbf{M}} \cdot \underline{\mathbf{V}}) = \frac{1}{2}\underline{\mathbf{V}} \cdot \frac{\partial \underline{\mathbf{M}}}{\partial \underline{\mathbf{x}}} \cdot \underline{\mathbf{V}} + \frac{8\pi}{3}\underline{\mathbf{I}} \cdot \underline{\mathbf{G}} - 48\pi\nu_l(2\underline{\mathbf{I}})^{-\frac{5}{3}} \cdot (\underline{\mathbf{I}}^{-1} - \underline{\mathbf{C}})^{-2} \cdot \underline{\mathbf{V}}, \qquad (15)$$

where $[\underline{x}]_p = \mathbf{x}_p$ refers to the position vector of bubble p, and $\underline{\mathbf{M}}$ is defined by (14).

Finally, the crucial step is to recognize that within the dipole approximation it is consistent to write

$$(\underline{\mathbf{I}}^{-1} - \underline{\mathbf{C}})^{-1} \simeq \underline{\mathbf{I}} + \underline{\mathbf{I}} \cdot (\underline{\mathbf{C}} \cdot \underline{\mathbf{I}}) + \underline{\mathbf{I}} \cdot (\underline{\mathbf{C}} \cdot \underline{\mathbf{I}})^2.$$

One may interpret this as follows: The first term represents the case when no hydrodynamic interactions are taken into account. The second term corrects each dipole strength for the presence of a velocity induced by the other dipoles, each with a strength as if there where no other dipoles around. The third term accounts for pairwise reflections. In principle the method can be extended to include higher-order multipoles, and the method can be easily implemented on a 80486-processor based PC.

We have simulated the motion of two bubbles in this manner and the bubble trajectories agreed well with what Kok (1993) found in his experiments; in fact small discrepancies between his analytical and experimental results could be explained by small differences in the size of the bubbles. We did not bother to investigate the motion of large numbers of bubbles, as soon as we heard from similar work done by Sangani & Didwania (1993*b*) for dispersions of equally-sized bubbles, but using a different method (and an IBM 3090).

These authors find that due to inertial forces the bubbles rapidly arrange themselves in planes, and that due to viscous forces the velocity fluctuations diminish and finally the bubbles end up in one big lump. This behaviour is not observed in practice. The precise reasons are not clear at present, but an explanation may perhaps be found in experiments on the motion of two bubbles performed by Paul Duineveld (1994). He observes that bubbles with diameters up to 1.4 mm in general coalesce after they have arranged themselves in a plane normal to direction of gravity, whereas larger bubbles with diameters of the order of 1.6 mm bounce off after a collision and then repel each other while performing a spiralling motion. This behaviour of the larger bubbles is presumably due to the presence of the wakes. The bubbles that are normally used in bubbly flow experiments are of still larger size, and also show this repelling behaviour. It seems that numerical simulations like those mentioned above cannot be expected to mimic the behaviour of these flows. Research aimed at improving this situation seems of vital importance.

4 Elasticity of a bubbly fluid

Mention should be made of the interesting work of Sangani & Didwania (1993*a*) and Bulthuis, Prosperetti & Sangani (1993) in which an expression for the bulk modulus of elasticity of a bubbly fluid is derived, on the assumptions that the bubbles are equally-sized spheres and the surrounding flow field can be described by irrotational flow theory. These authors show that the stress arising from interaction forces and collisions between the bubbles can be split into two contributions: a 'kinetic' stress S_k and a 'potential' stress S_p. The names are borrowed from the kinetic theory of gases, which can be used to obtain definitions of these stresses in terms of the probability distribution functions of the bubble positions and velocities.

In the presence of a concentration gradient the kinetic stress gives a contribution to the mean force on the bubbles

$$\frac{\partial S_k}{\partial x} = -\frac{\partial(n\langle vi\rangle)}{\partial x}, \tag{16}$$

where n is the mean number density. Just as the velocity fluctuations of the bubbles cause a transport of bubble momentum, and as a consequence, under non-uniform conditions will give rise to a force, they will also lead to transport of fluid impulse, and consequently to an additional force in the presence of gradients. In (16) i denotes the fluctuation in the 'fluid impulse of a bubble', the latter defined in the usual manner as the evaluation of the integral $-\rho_l \int \phi \mathbf{n} \mathrm{d}A$ over the surface of a bubble. There is nothing known about the functional form of the kinetic stress S_k, for ease of discussion we propose to write it as

$$n\langle vi\rangle = \phi\rho_l G(\phi)\langle v^2\rangle, \quad \text{with} \quad \langle v^2\rangle = H(\phi)U^2.$$

G an unknown function of the void fraction, which for spheres is equal to $\frac{1}{2}$ in the limit of vanishing volume concentration, and the mean square velocity fluctuations are written in the manner of Batchelor (1988).

With regard to the potential stress S_p, Sangani & Didwania (1993*a*) and Bulthuis *et al.* (1993) find for dilute dispersions consisting of cubic arrays of bubbles moving with equal velocity U, that S_p is given by

$$S_p = -\lambda_1 \rho_l \frac{\phi^2 U^2}{(1-\phi)^2} + \mathcal{O}(\phi^{\frac{13}{3}}).$$

The constant λ_1 is equal to 0.386 for a simple cubic array, -1.26 for a body-centred array, and -1.17 for a face-centred array. In addition, Bulthuis *et al.* (1993) show that for a random array

$$S_p = \frac{1}{5}\rho_l \phi^2 \{1 + C(\phi)\}^2 U^2,$$

correct to the same order. Here $C(\phi)$ is given by (10), with $Q(\chi) = 1$. Combining the above expressions it can be deduced that the bulk modulus of elasticity of the bubbles is given by

$$\rho_g \phi Q(\phi) = \rho_g \phi \left(\frac{\rho_l}{\rho_g} \frac{\mathrm{d}(\phi^2 PU^2)}{\mathrm{d}\phi} + \frac{\rho_l}{\rho_g} \frac{\mathrm{d}(\phi GHU^2)}{\mathrm{d}\phi} + \frac{\mathrm{d}(\phi HU^2)}{\mathrm{d}\phi} \right), \tag{17}$$

where the functions $\lambda_1/(1-\phi)^2$ and $-\frac{1}{5}\{1+C(\phi)\}^2$ have been replaced by $P(\phi)$.

It is not clear how to reconcile expression (17) with expression (5) given by Batchelor. Of course, the model of a bubbly fluid as a collection of equally-sized spheres moving about in a wholly irrotational flow is not a good representation of reality, but it is nevertheless possible that (17) is a reasonable approximation of the elasticity when inertia forces dominate the motion of the bubbles.

Batchelor does not mention a mean force on the particles arising from transport of impulse through particle velocity fluctuations. Either the second contribution to the elasticity in (17) is in some way represented in the term $\gamma\tilde{g}D/U$ appearing in (5), together with the first contribution, or it has been overlooked and only the first term in (17) has some relation with the diffusivity of the particles. In either case the association with a particle diffusivity is still to be made. What is remarkable about expression (17) is that it points at the possibility that for certain arrangements of the bubbles, under conditions in which the velocity fluctuations are small, Q becomes negative. Small-scale variations in the volume concentration will then be amplified. Observations indicate that these conditions prevail in the agitated, 'turbulent' flow regime, and it may well be that the formation of gas slugs can be understood in terms of the elasticity becoming negative locally.

Evidently, further research towards the correct representation of the bulk modulus of elasticity is necessary, and the estimates of the critical values of the volume concentration given above may need modification.

5 Void waves

Suppose that a uniform bubbly fluid is characterized by a gas volume concentration ϕ_0 and a mean bubble velocity $U(\phi_0) \equiv U_0$, the latter defined with respect to a zero-volume flux reference frame. To obtain a description of the disturbed flow we can use the time-scale of the imposed disturbances T and the velocity U_0 to render the conservation equations (1) and (2) dimensionless. They become

$$\frac{\partial \phi}{\partial \bar{t}} + \bar{V}\frac{\partial \phi}{\partial \bar{x}} + \frac{\partial \bar{V}}{\partial \bar{x}} = 0. \tag{18}$$

$$\phi(1+\theta)\left(\frac{\partial \bar{V}}{\partial \bar{t}} + \bar{V}\frac{\partial \bar{V}}{\partial \bar{x}}\right) - \phi\zeta\bar{V}\frac{\partial \bar{V}}{\partial \bar{x}} + \bar{Q}\frac{\partial \phi}{\partial \bar{x}} - \frac{1+\theta_0}{\mathrm{Re}}\frac{\partial}{\partial \bar{x}}\left(\frac{\phi\bar{\eta}}{\eta_0}\frac{\partial \bar{V}}{\partial \bar{x}}\right)$$
$$= -\frac{1+\theta_0}{\mathrm{Fr}}\phi\left(\frac{\bar{V}-\bar{U}}{\bar{U}}\right). \tag{19}$$

Here we have introduced the non-dimensional variables

$$\bar{t} = t/T, \quad \bar{x} = x/U_0T, \quad \bar{V} = V/U_0, \quad \bar{U} = U/U_0,$$
$$\bar{Q} = Q/U_0^2, \quad \bar{\eta} = \eta/U_0^2T.$$

The suffix 0 means that the functions are to be evaluated at the volume fraction of the undisturbed flow. The dimensionless numbers Fr and Re are defined as

$$\mathrm{Fr} = \frac{(1+\theta_0)U_0}{\gamma\tilde{g}T}, \quad \mathrm{Re} = \frac{(1+\theta_0)U_0^2T}{\eta_0}. \tag{20}$$

The first number can be interpreted as the ratio between the relaxation time of the bubbles, $(1+\theta)U/\gamma\tilde{g}$, and the time-scale of the disturbances. It will be assumed that T is so large that $\mathrm{Fr} \ll 1$ and $\mathrm{Re} \gg 1$, the precise conditions for this to be true will not be considered, but they hold in the experiments that are described below.

An approximate solution is obtained by neglecting the left-hand side of (19), which yields $\bar{V} = \bar{U}$, and upon substitution in (18),

$$\frac{\partial \phi}{\partial x} + \frac{1}{c(\phi)}\frac{\partial \phi}{\partial t} = 0, \qquad c(\phi) = U(\phi) + \phi\frac{\mathrm{d}U}{\mathrm{d}\phi}, \tag{21}$$

written again in physical variables. This represents a simple wave propagating with velocity $\phi(\mathrm{d}U/\mathrm{d}\phi)$ with respect to the bubbles. The solution is obtained by Riemann's method, fitting in 'concentration shocks' with Whitham's rule, where it becomes multi-valued. Since $\mathrm{d}c/\mathrm{d}\phi < 0$ for bubbly flows, the shocks break backwards, i.e. the volume fraction in front of the shocks is larger than at the rear. This description of the wave motion is precisely that of the Kynch theory (1952) of sedimentation waves. The relation with the theory of nonlinear acoustics will be clear; Kynch writes on page 166 "*Changes in particle density are propagated through a dispersion just as sound is propagated through air ...*".

An approximate equation describing weakly nonlinear waves can be derived by substituting

$$\phi = \phi_0 + \epsilon\phi', \qquad \bar{V} = 1 + \epsilon\bar{V}'; \qquad \epsilon \ll 1,$$

in (18) and (19), and retaining only term of $\mathcal{O}(\epsilon)$ and $\mathcal{O}(\mathrm{Fr})$. This yields, in physical variables,

$$\frac{\partial \phi'}{\partial x} + \left(\frac{1}{c_0} - \frac{c_1}{c_0^2}\phi'\right)\frac{\partial \phi'}{\partial t} = \frac{\mathcal{D}}{c_0^3}\frac{\partial^2 \phi'}{\partial t^2} \tag{22}$$

with velocities c_0 and c_1 given by $c_0 = c(\phi_0)$ and $c_1 = \{\mathrm{d}c/\mathrm{d}\phi\}_{\phi=\phi_0}$. The diffusivity $\mathcal{D}$ is defined as

$$\mathcal{D} = \frac{U}{\gamma\tilde{g}}\left(Q - \frac{\rho_l}{\rho_g}\phi^2\left\{\frac{\mathrm{d}U}{\mathrm{d}\phi}\right\}\left\{\frac{\mathrm{d}(CU)}{\mathrm{d}\phi}\right\}\right).$$

in which the functions of the volume concentration are evaluated for $\phi = \phi_0$. Comparing the expression for $\mathcal{D}$ with (7) we see that the instability criterion comes down to the diffusivity of long waves becoming negative. Equation (22) is Burgers' equation in the form appropriate to signalling problems. That this equation would be obtained could have been anticipated, of course, by recognizing that the assumption $\mathrm{Re} \gg 1$ puts the conservation equations in a form that is equivalent to the wave-hierarchy problems studied by Lighthill and Whitham during the fifties (see Whitham 1974). They showed that, in the order of approximation considered here, the 'dynamical' effects represented by the first three terms on the left-hand side of (19) have a diffusive effect on the 'kinematic' waves described by (21). Combination of weak nonlinearity in the simple wave description with linear diffusive effects yields Burgers' equation. It was not until 1956 that Lighthill rigorously established the relation between the Burgers equation and the propagation of finite-amplitude sound waves, which makes the above-cited observation by Kynch the more remarkable.

We now summarize features of solutions to the Burgers equation that were obtained in the context of nonlinear acoustics. The solutions concern signalling problems with sinusoidal variations in the boundary condition. There exists an extensive literature on the subject; very readable accounts are Lighthill (1956) and Crighton (1992), on which much of what is written below is based.

For an analysis of the propagation of sinusoidal disturbances it pays to rewrite the Burgers equation in variables made dimensionless with the frequency ω and amplitude ϕ_s of the source-signal. Define

$$\theta = \omega\left(t - \frac{x}{c_0}\right), \qquad z = \frac{\omega\phi_s c_1}{c_0^2}x, \qquad \phi^* = \frac{\phi'}{\phi_s}, \qquad \nu = \frac{\omega\mathcal{D}}{c_0 c_1 \phi_s}.$$

The Burgers equation now becomes

$$\frac{\partial\phi^*}{\partial z} - \phi^*\frac{\partial\phi^*}{\partial\theta} = \nu\frac{\partial^2\phi^*}{\partial\theta^2}. \tag{23}$$

The use of the symbol θ here as well will cause no confusion. This formulation shows how solutions to (23) can be used to study the features of the concentration wave motion when in experiments the frequency and the amplitude of the source-signal is varied, and *vice versa*.

An exact solution to (23) with sinusoidal boundary condition has been given by Cole (1951), it reads

$$\phi^* = 2\nu\phi_s\frac{\partial}{\partial\theta}\ln\left(\sum_{n=0}^{\infty}(-1)^n\epsilon_n I_n(1/2\nu)\cos(n\theta)\,\mathrm{e}^{-n^2\nu z}\right). \tag{24}$$

Here $I_n(x)$ is the modified Bessel function and ϵ_n denotes the Neumann symbol: $\epsilon_0 = 1$ and $\epsilon_n = 2$ when $n \neq 0$. As it stands the Cole-solution is not very informative. As far as I know there are only two limits in which this solution converges after a few terms. One limit is for $\nu \gg 1$, which is for waves in which nonlinear effects are small. The expansion

$$\phi^* = \sin\theta\,\mathrm{e}^{-\nu z}\left(1 + \frac{1}{32\nu^2}\left(-1 + 2\mathrm{e}^{-2\nu z} - \mathrm{e}^{-4\nu z}\right)\right) + \frac{\sin 2\theta}{4\nu}\left(\mathrm{e}^{-2\nu z} - \mathrm{e}^{-4\nu z}\right),$$

correct to $\mathcal{O}(\nu^{-3})$, is useful for assessing the effects of nonlinearity on the motion of small-amplitude waves. The solution of the linearized Burgers equation, $\nu \to \infty$ for constant νz, is seen to be given by

$$\phi' = \phi_s\,\mathrm{e}^{-\nu z}\sin\theta = \phi_s\,\mathrm{e}^{-(\omega^2\mathcal{D}/c_0^3)x}\sin\omega(t - x/c_0). \tag{25}$$

The other limit is for $\nu z \gg 1$, in which case (24) can be approximated by

$$\phi^* \simeq 4\nu\frac{I_1(1/2\nu)}{I_0(1/2\nu)}\,\mathrm{e}^{-\nu z}\sin\theta.$$

For $\nu \gg 1$, constant νz, the linear wave (25) is found again; if ν is small, i.e. for relatively large source-amplitudes, the behaviour at large distances is given by

$$\phi' \sim 4\nu\phi_s\mathrm{e}^{-\nu z}\sin\theta = \frac{4\mathcal{D}\omega}{c_0 c_1}\mathrm{e}^{-(\omega^2\mathcal{D}/c_0^3)x}\sin\omega(t - x/c_0), \tag{26}$$

a linear wave damped by diffusive effects, but now with an amplitude that has become independent of the amplitude of the source-signal. This is the phenomenon of *amplitude saturation*. In fact this is the 'old-age' stage of any nonlinear periodic wave. The structure results because the higher harmonics are attenuated more strongly than the fundamental.

The case of small non-dimensional diffusivity is particularly interesting. An informative picture can then be obtained by studying approximate solutions of the problem. In the limit $\nu z \ll 1$ weak-shock theory applies. The initial evolution of the wave, neglecting diffusive effects altogether, is given by the Fubini-solution

$$\phi^* = \sum_{n=1}^{\infty} \frac{2J_n(nz)}{nz} \sin(n\theta),$$

where $J_n(x)$ is the nth-order Bessel function. This may be viewed as a Fourier series representation of the Riemann-solution

$$\phi^* = \sin\xi, \quad \text{along the curve} \quad \theta = \xi - z\sin\xi,$$

where $\phi^*(\theta, 0) = \sin\xi$. The expansion of J_n for small argument reveals that the initial development of the amplitude of nth harmonic is equal to

$$\hat{\phi}_n \sim \frac{(z/2)^{n-1}}{nn!}.$$

The solution breaks down at $z = 1$, $\theta = 2N\pi$ $(N = 0, \pm1, \pm2, \ldots)$. A discontinuity has to be inserted for $z > 1$ with amplitude $2\sin(X)$; here $X = \operatorname{sinc}^{-1}(1/z)$, where $\operatorname{sinc}^{-1}(x)$ denotes the inverse function of $\operatorname{sinc}(x) = \sin(x)/x$. For $z < 1$ the choice $X = 0$ is appropriate.

To exemplify small diffusive effects one can turn to an exact solution of (23) found by Khokhlov (Rudenko & Soluyan 1977),

$$\phi^* \simeq \frac{1}{z}\left(\pi \tanh\frac{\pi\theta}{2\nu z} - \theta\right), \quad \theta \in [-\pi, \pi] \quad (\operatorname{mod} 2\pi). \tag{27}$$

This is not an exact solution for the Burgers equation with sinusoidal boundary coundition, but it is the asymptotic form of the Cole-solution for $z > 1$ in the limit $\nu \to 0$. If $z = \mathcal{O}(1)$ and θ not close to 0, $\tanh(\pi\theta/2\nu z)$ is -1 for $\theta > 0$ and $+1$ for $\theta < 0$, so that in this region (27) represents the sawtooth wave

$$\phi' \sim \phi_s \frac{\pi - \theta}{z} = \frac{c_0^2}{\omega c_1 x}(\pi - \omega(t - x/c_0)).$$

This is amplitude saturation again. The behaviour is familiar, of course, for the lossless equation: any periodic finite-amplitude signal eventually becomes a sawtooth wave with amplitude depending on the frequency of the lowest harmonic. On the other hand, for $z = \mathcal{O}(1)$, $\theta = \mathcal{O}(\nu)$ the second term may be neglected and

$$\phi' \simeq \phi_s \frac{\pi}{z} \tanh\left(\frac{\pi\theta}{2\nu z}\right) = \frac{\pi c_0^2}{\omega c_1 x} \tanh\frac{\pi c_0^3(t - x/c_0)}{\omega \mathcal{D}}.$$

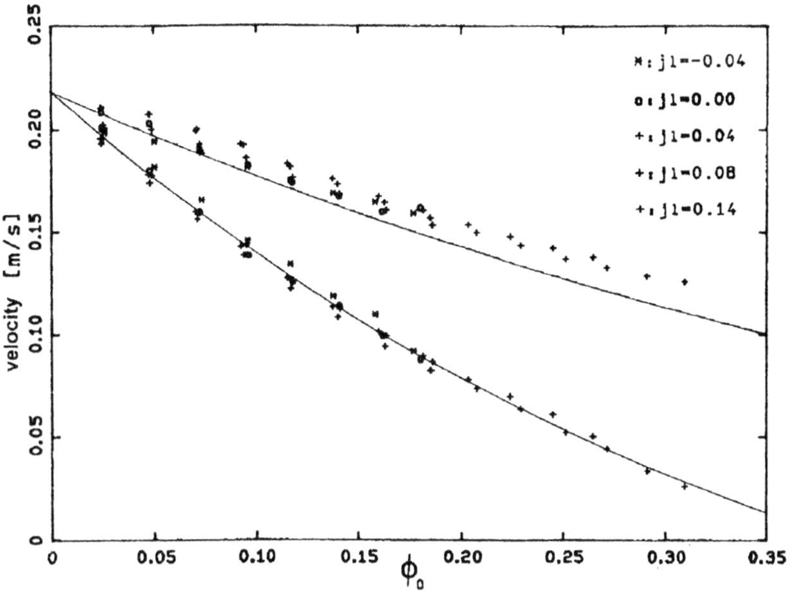

Figure 2: Data for the mean bubble velocity (top) and linear void wave speed (bottom), with respect to a zero volume-flux reference frame. The upper curve is calculated from a fit through the wave speed data.

This describes a weak shock wave with strength $2\pi/z$. In physical variables the shock strength is $4\pi c_0^2/\omega c_1 x$ and the shock thickness is $4\omega\mathcal{D}x/\pi c_0^2$. As z increases the shock thickens and becomes weaker, and weak shock theory becomes less appropriate. At $z = \mathcal{O}(1/\nu)$ the shock thickness has become $\mathcal{O}(1)$, i.e. comparable to the wavelength. For the behaviour at large distances one can look again at the Cole-solution (26).

Experiments

Lammers (1994) has performed experiments on periodic waves in a vertical pipe with inner diameter 8.0 cm and 5.5 m length, run with countercurrent liquid flow ($J_l \simeq -0.04$ m/s), stagnant fluid, and cocurrent liquid flow (J_l between 0.03 m/s and 0.16 m/s); J_l is the superficial liquid velocity. The liquid used was tap water, and the bubble diameter varied with the mean volume concentration from 2.8 mm to 4.2 mm, with a variance of about 0.6 mm. When generating the concentration waves care was taken to maintain a constant total volume flux of the two phases. The waves were recorded with 4 impedance volume concentration meters mounted at the wall at 0.87 m, 2.04 m, 3.22 m, and 4.47 m from the inlet to the pipe.

The mean bubble velocity was determined from measurements of the volume concentration and the prescribed mass flow rates of gas. As shown in figure 2 the results for the various liquid flow rates fit nicely into a single curve when expressed relative to the total volume flux. A correlation of the form (6) yields $U_\infty = 21.4$ cm/s and $p = 1.70$. For single bubbles with an equivalent diameter of about 3.5 mm, rising

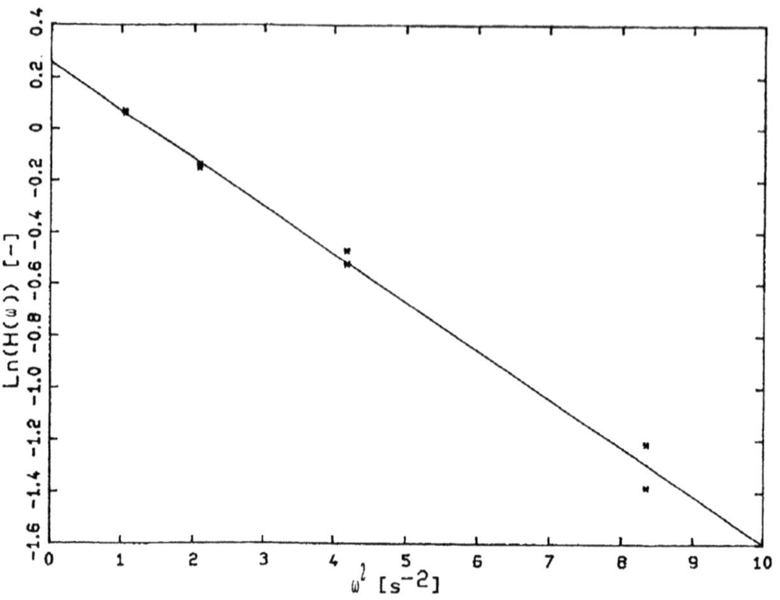

Figure 3: Attenuation of linear void waves over a distance of 4.46 m, represented by the transmission factor H as function of the square of the frequency ω. The mean volume concentration is 10 %.

in tap water, Haberman & Morton (1953) measured velocities of about 23 cm/s, and values of p between 1.5 and 2.3 are common for bubbly flows in pipes. Data for the linear wave speeds were obtained from measurements of the fundamental mode of imposed small amplitude disturbances. To confirm the relationship between the kinematic wave speed and the velocity of the bubbles first a polynomial fit of the wave speed measurements was made,

$$c_0(\phi) = 0.219(1-\phi)(1-3.08\phi+2.77\phi^2-0.55\phi^3), \tag{28}$$

and then relation (21) was integrated, which gave

$$U(\phi) = 0.219(1-\phi)(1-1.0\phi+0.55\phi^2).$$

The fit and the result of the integration are also shown in figure 2.

Data for the damping of the small-amplitude waves over a distance of 3.6 m are shown in figure 3; $\phi_0 = 10\%$. The relation between the attenuation and the angular frequency ω is expressed by a transmission factor

$$H(\omega) = \exp\left(\lambda - \frac{\mathcal{D}\omega^2}{c_0^3}\right)x.$$

The value on the abscissa corresponds to $\lambda = 0.07\text{m}^{-1}$, and from the slope of the curve an effective diffusivity of $\mathcal{D} = 2.7\ 10^{-4}\ \text{m}^2/\text{s}$ can be inferred. The growth coefficient λ can be explained from the increase in bubble volume along the pipe.

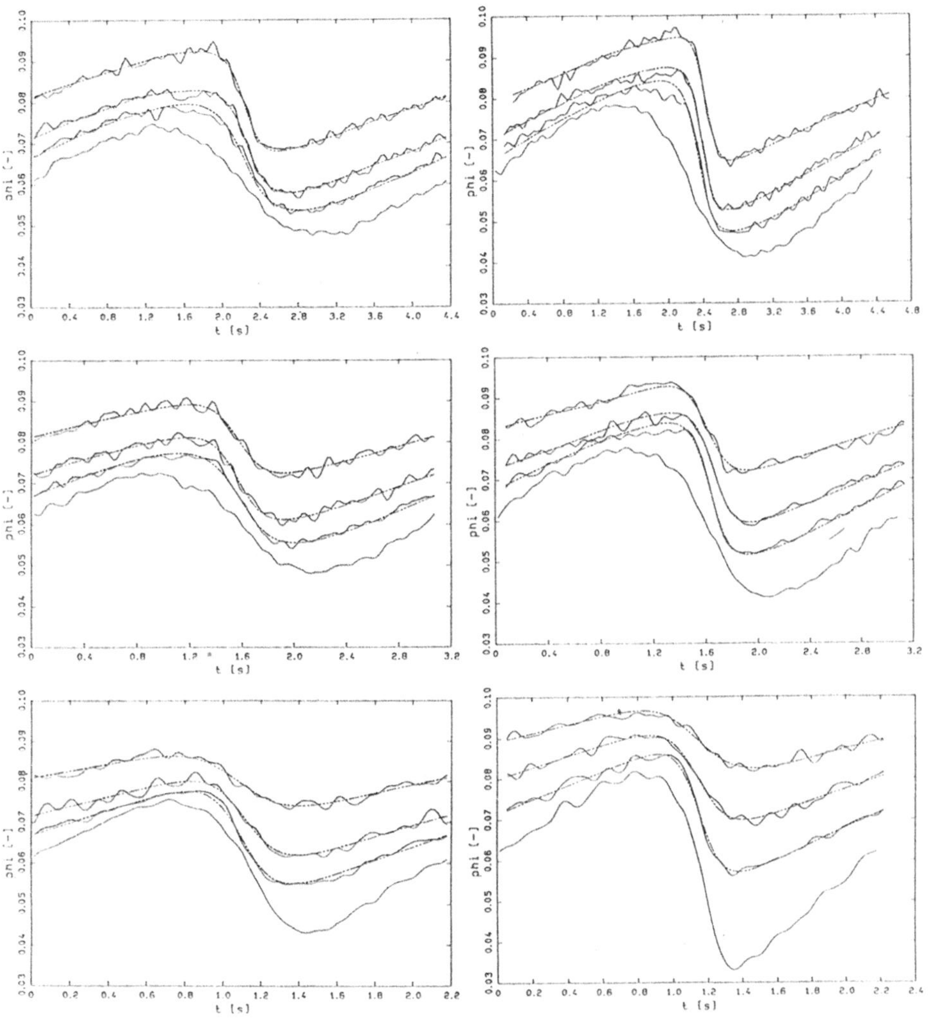

Figure 4: Comparisons between the recorded propagation of sinusoidal concentration disturbances and calculations based on a generalized Burgers equation. Superficial liquid velocity 0.03 m/s, mean volume concentration 6 %. Wave periods 2.176 s (bottom), 3.072 s (middle), 4.352 s (top); signal amplitudes 1.5 % (left), 2.0 % (right).

This effect can be accounted for by replacing the right-hand side of (1) by a term $\lambda\phi V$, with λ defined as $\lambda = -(1/\rho_g)\mathrm{d}\rho_g/\mathrm{d}x$. The weakly nonlinear wave equation is then modified into a generalized Burgers equation

$$\frac{\partial\phi'}{\partial x} + \left(\frac{1}{c_0} - \frac{c_1}{c_0^2}\phi'\right)\frac{\partial\phi'}{\partial t} = \frac{\mathcal{D}}{c_0^3}\frac{\partial^2\phi'}{\partial t^2} + \lambda\phi'. \tag{29}$$

The propagation of sinusoidal finite-amplitude disturbances was studied for the water flow rates mentioned above and for various mean volume concentrations. For each of these, data analysis was performed on time series from nine experimental runs with four different frequencies and amplitudes each; 64 wave cycles were taken, and 128 points per cycle. The recorded signals of each electrode were taken as a boundary condition in a numerical calculation of the generalized Burgers equation (29) using a pseudo-spectral method with 1024 harmonics, and the calculated result compared with the spectral decomposition of the recorded signal of the next electrode. A Levenberg-Marquardt procedure then gave values for the diffusivity $\mathcal{D}$ and the nonlinear speed c_1, where it was assumed that these parameters had the same values in each of the nine runs and between each measuring station. For the growth factor λ and the velocity c_0 the values obtained from the experiments on small amplitude waves were used. Examples are shown in figure 4, for a volume concentration $\phi_0 = 0.06$ at the inlet of the pipe and superficial liquid velocity $J_l = 0.03$ m/s; the broken line is the calculated wave shape. The wave periods are 2.176 s (bottom), 3.072 s (middle), and 4.352 s (top); the signal amplitudes are $\phi_s = 0.015$ (left) and $\phi_s = 0.02$ (right). Note that in order to be able to represent the signals in a single figure a phase shift and a shift in the average concentration has been applied. In these cases, where strictly speaking the signal-amplitude was too large for the theory to apply, the values $\mathcal{D} = 2.1\,10^{-4}$ m^2/s and $c_1 = -0.96$ m/s were obtained.

The stages in which the waves have developed into a decaying sawtooth and an 'old-age sinoid' are illustrated in figure 5 and figure 7. Here $J_l = -0.03$ m/s, the mean void fraction, signal-strength and period are $\phi_0 = 0.06$, $\phi_s = 0.054$, $T = 15.36$ s for the sawtooth wave, and $\phi_0 = 0.07$, $\phi_s = 0.050$, $T = 2.56$ s for the old-age sinoid. The wave velocities and diffusivities were found to be $c_0 = 0.140$ m/s, $c_1 = -0.80$ m/s, $\mathcal{D} = 2.5\,10^{-4}$ m^2/s (sawtooth), and $c_0 = 0.136$ m/s, $c_1 = -0.75$ m/s, $\mathcal{D} = 2.1\,10^{-4}$ m^2/s (sinoid).

The phenomenon of amplitude saturation is exemplified in figure 6 and figure 8, in which the encircled points correspond to the cases shown in figures 5 and 7, respectively. Figure 6 presents data for the amplitudes of the first three harmonics as a function of the source amplitude, each point recorded at 4.47 m above the pipe inlet. It was estimated that $\nu z = 0.065$ and $z = 83\,\phi_s$ in this case. The broken curves were calculated from a Fourier series representation of the solution for the 'embryo-shock' region given by Lighthill (1956, §8.3), corrected for the effect of bubble growth. In figure 8 data for the signal amplitude are given, recorded at the same point, but with $\nu z = 2.0$ and $z = 388\,\phi_s$. The drawn curve is calculated from the Cole-solution (26), corrected again for bubble growth effects.

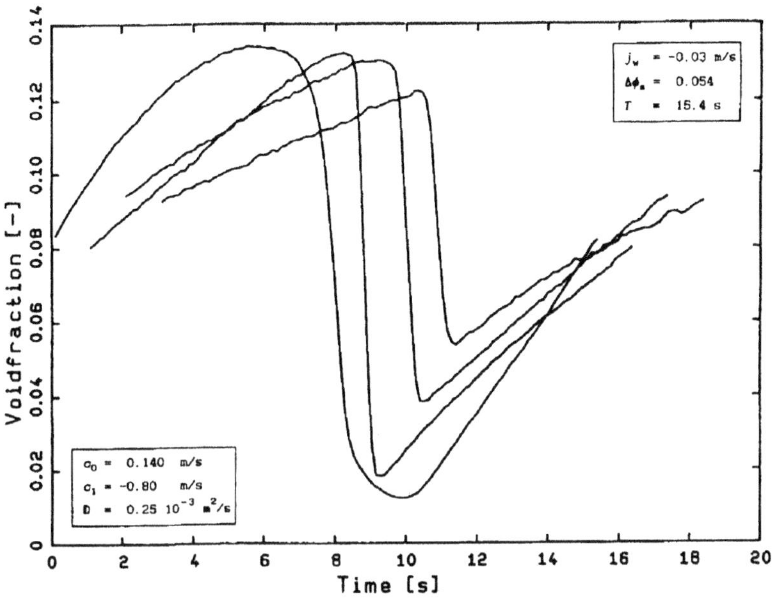

Figure 5: Registration of a finite-amplitude sinusoidal concentration disturbance which has developed into a sawtooth wave. Flow parameters are indicated in the boxes.

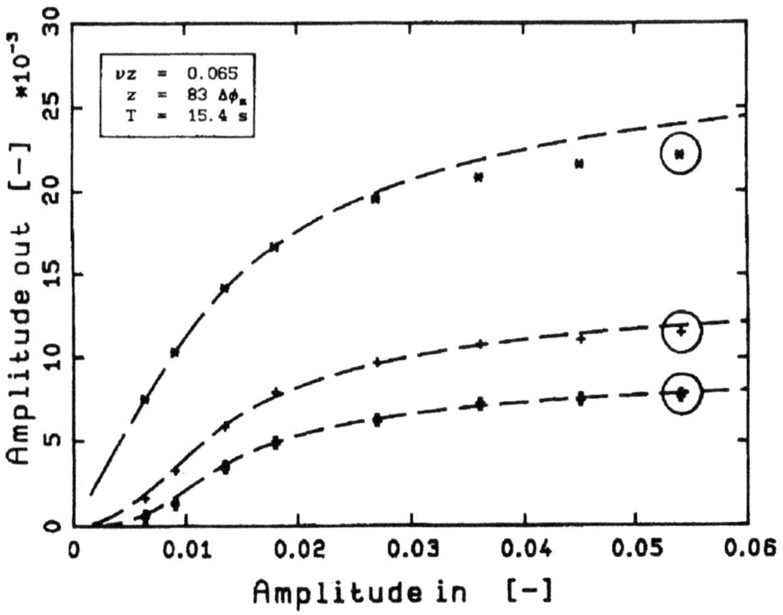

Figure 6: Amplitude saturation in the sawtooth wave-regime; illustrated for the first three harmonics of a Fourier series representation of the wave shape. The encircled data correspond to the waves shown in figure 5. The broken curves are based on an analytical solution of Burgers equation given by Lighthill (1956).

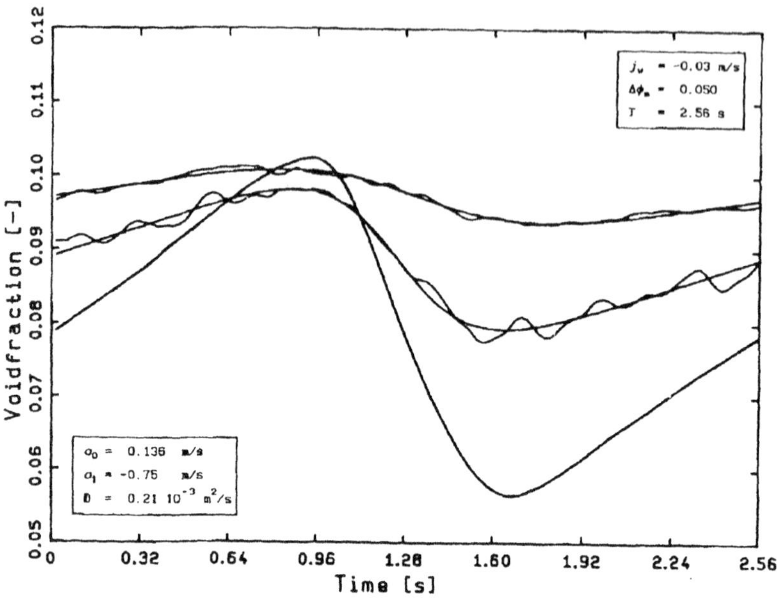

Figure 7: Registration of a finite-amplitude sinusoidal concentration disturbance in the final stage of decay. Flow parameters are indicated in the boxes.

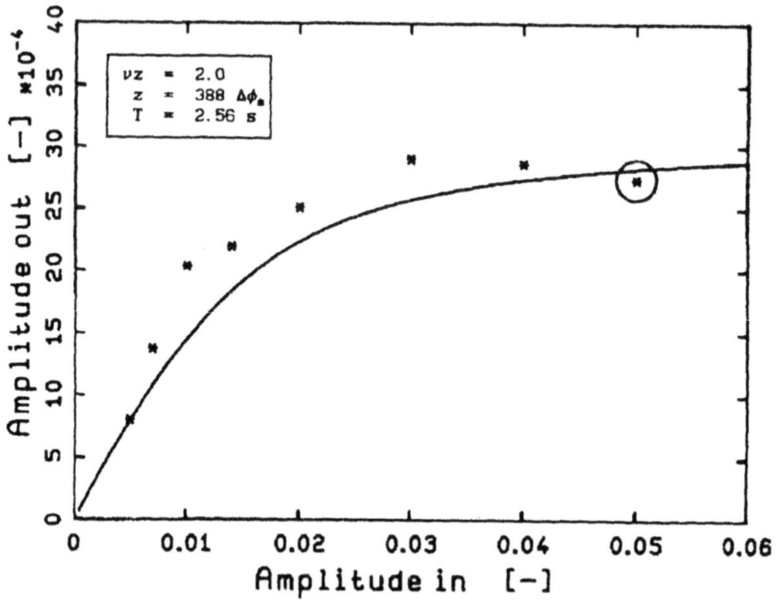

Figure 8: Amplitude saturation in the 'old age sinoid' stage of the waves. The encircled point corresponds to the waves shown in figure 7. The broken curve is based on an analytical solution of the Burgers equation given by Cole (1951).

6 Instability of bubbly flows

The experiments described in the previous section establish without doubt that the propagation of long weakly nonlinear void waves is governed by Burgers' equation. This agrees with Batchelor's theory, but can hardly be judged as a corroboration. In this section three observations of what happens near the transition to 'turbulent' bubbly flow are presented, which all at first sight seem to conflict with this theory. But arguments can be given that show that it is too early to jump to the conclusion that Batchelor's theory is not valid.

The experiments on concentration waves were all performed under stable flow conditions. Figure 9 shows for two different liquid flow rates how the mean velocity of the bubbles, relative to a zero volume flux frame, changes with the volume concentration as one goes beyond the point of transition. In the region of turbulent flow bubble clusters can be seen to move with high speed preferably near the centre of the pipe, which explains why the volume concentration may become less above the point of transition even though the gas volume flux increases. But the thing to notice is the strong dependency of the critical value of the concentration on the liquid flow rate (compare upper figure with lower figure), and on the distance from the bubble distributor (compare left and right figures). These phenomena do not seem to follow from Batchelor's theory. Now, in the experiments the bubble size varies somewhat with the liquid flow rate, with smaller bubbles when the flow rate becomes less, and looking again at figure 1 for the case when α is sufficiently large, this might explain what is observed. That for a given liquid flow rate ϕ_c is less when the point of transition is higher-up in the tube seems to confirm that the bubble velocity fluctuations diminish as the bubbles rise; indications for this behaviour are found in the numerical simulations discussed in §3. We are presently investigating the first possibility, however drawing definite conclusions is hampered by a lack of knowledge of the function $C(\phi)$, and of precise values for p, α, and the Froude number ξ, for bubbles with diameters in the range between 2.8 mm to 4.2 mm.

The second observation can be made from figure 10, in which values of the diffusivity $\mathcal{D}$ obtained from wave propagation experiments for volume concentrations up to the critical value, have been collected. These values were calculated from recordings of only two measuring stations; at the largest volume concentrations the point of transition was right at the upper station. The number 0 corresponds to case of stagnant water, the numbers 4 and 8 to superficial liquid velocities of 0.04 m/s and 0.08 m/s, and + indicates a superficial liquid velocity of 0.16 m/s. The diffusivity is markedly higher in the last case, for which various explanations can be given, but which do not concern us here. The remarkable thing is that there is no indication of a downward trend in the diffusivity as the volume concentration approaches the critical value. A possible explanation may be found by looking at the weakly nonlinear stability of the flow. To analyse this problem one can introduce again dimensionless numbers Fr and Re, defined in a similar manner as in (20), but with T replaced by h/U_0, where h is a characteristic length-scale of a disturbance. To obtain an equation for the evolution of weakly nonlinear waves when $\phi_0 \simeq \phi_c$, assume that in the

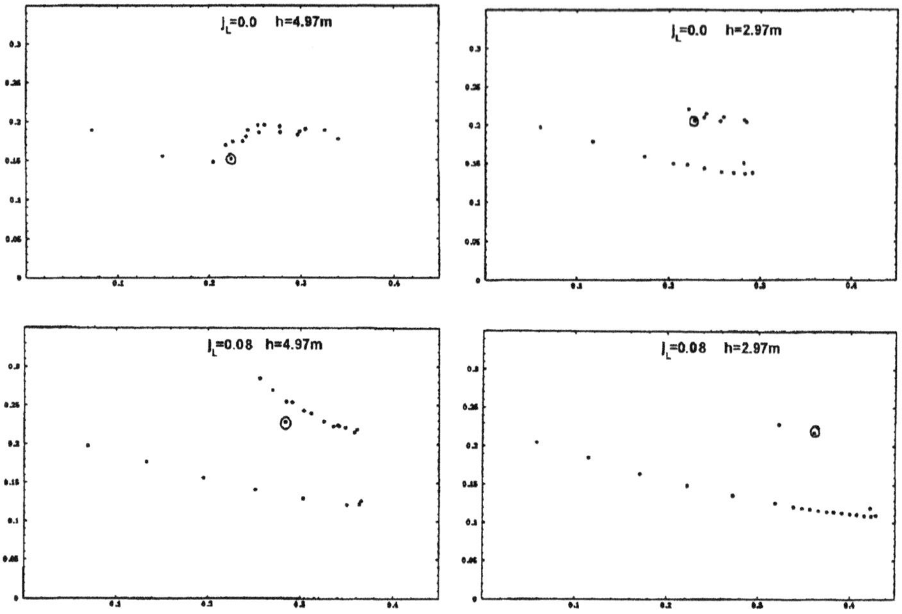

Figure 9: Bubble velocity as a function of the volume concentration, measured at 4.97 m (left) and 2.97 m (right) above the pipe inlet, for two different liquid flow rates: stagnant water (top), superficial liquid velocity 0.08 m/s (bottom). The encircled points indicate the measured values when the instability was seen to occur right at the measuring station.

disturbed state

$$\phi = \phi_0 + \epsilon\phi', \qquad \bar{V} = 1 + \epsilon V', \qquad \epsilon \ll 1,$$

and use the scalings

$$\mathrm{Fr} = \mathcal{O}(\epsilon^{\frac{1}{2}}), \quad \mathrm{Re} = \mathcal{O}(\epsilon^{\frac{1}{2}}), \quad |\mathcal{D}/(U_0 h)| = \mathcal{O}(\epsilon^{\frac{3}{2}}).$$

Linear stability analysis shows that the maximum growth rate occurs near $\kappa \simeq \kappa_n$, where κ_n determines the range of wavenumbers for which the flow is unstable. The above-mentioned scaling amounts to $h \simeq 1/\kappa_n$. It is now straightforward to show from (1) and (2) that to first order the disturbances are described by

$$\frac{\partial \phi'}{\partial t} + (U+W)\frac{\partial \phi'}{\partial x} = 0,$$

and to $\mathcal{O}(\epsilon)$ by

$$\frac{\partial \phi'}{\partial t} + (U+W)\frac{\partial \phi'}{\partial x} + \left\{\frac{\mathrm{d}(U+W)}{\mathrm{d}\phi}\right\}\phi'\frac{\partial \phi'}{\partial x} + \frac{\eta U}{\gamma \tilde{g}} W \frac{\partial^3 \phi'}{\partial x^3} = 0,$$

where of course it is understood that the coefficients are to be evaluated for the value of the volume concentration of the undisturbed flow. Initially the disturbances propagate with the kinematic wave speed without change of form, but after a dimensionless

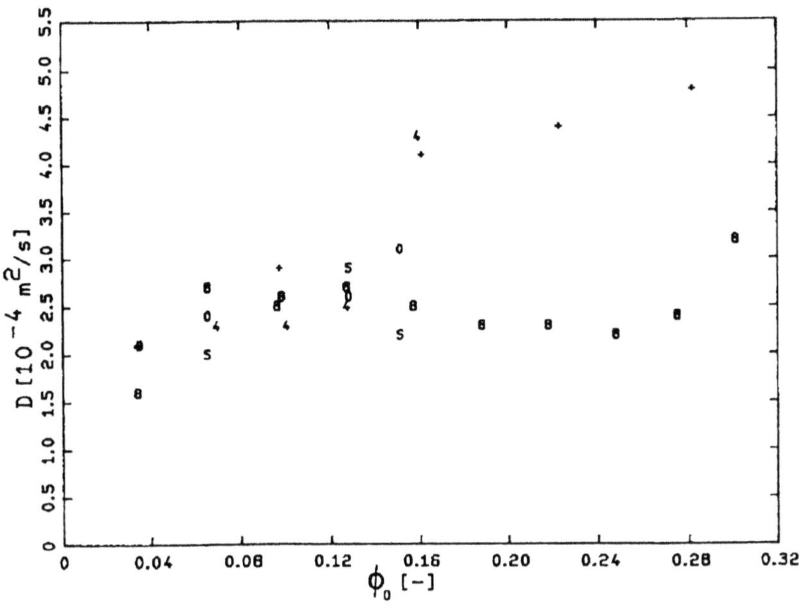

Figure 10: Values of the diffusivity $\mathcal{D}$ determined from experiments on periodic finite-amplitude concentration waves; data indicated by 0, 4, 8, + correspond to superficial liquid velocities of 0, 4, 8, and 16 m/s. S are from shock waves experiments not discussed here.

time of $\mathcal{O}(\epsilon^{-1})$ nonlinear and dispersive effects become important. At this stage the waves are described by the Korteweg-de Vries equation.

To $\mathcal{O}(\epsilon^{\frac{3}{2}})$ the pertinent equation becomes

$$\frac{\partial\phi'}{\partial t} + (U+W)\frac{\partial\phi'}{\partial x} + \left\{\frac{\mathrm{d}(U+W)}{\mathrm{d}\phi}\right\}\phi'\frac{\partial\phi'}{\partial x} \tag{30}$$

$$= \mathcal{D}\frac{\partial^2\phi'}{\partial x^2} + \left\{\frac{\mathrm{d}\mathcal{D}}{\mathrm{d}\phi}\right\}\frac{\partial}{\partial x}\left(\phi'\frac{\partial\phi'}{\partial x}\right) - \frac{\eta U}{\gamma\tilde{g}}W\left(1 - \frac{(1+\theta)U}{\gamma\tilde{g}}\mathcal{C}\frac{\partial}{\partial x}\right)\frac{\partial^3\phi'}{\partial x^3},$$

where we have written for conveniency $\mathcal{C} = 2W + \zeta/(1+\theta)$. It appears that the effects of a negative diffusivity only become essential at a dimensionless time of $\mathcal{O}(\epsilon^{-\frac{3}{2}})$. The behaviour near the point of instability critically depends on the precise values of the coefficients that appear in (30), about which unfortunately not much is known. Nonetheless, it is found that a bifurcation occurs at ϕ_c, and that for this to be a subcritical or a supercritical instability depends on whether the value of the parameter $(\gamma\tilde{g}/(1+\theta)U\mathcal{C})(\{\mathrm{d}\mathcal{D}/\mathrm{d}\phi\}/\{\mathrm{d}(U+W)/\mathrm{d}\phi\})$ is larger or less than 1, respectively. A rather rough estimate indicates that the former is the case for bubbly flows. If this is true, then it may be possible that small, but finite fluctuations induced by the imposed disturbances cause an instability to set in for values less than ϕ_c; then the behaviour of $\mathcal{D}$ when ϕ_c is reached is difficult to detect.

Finally, figure 11 shows the change in the radial profiles of the local volume fraction and local mean bubble velocity as the flow becomes unstable, for $J_l = 0.08$

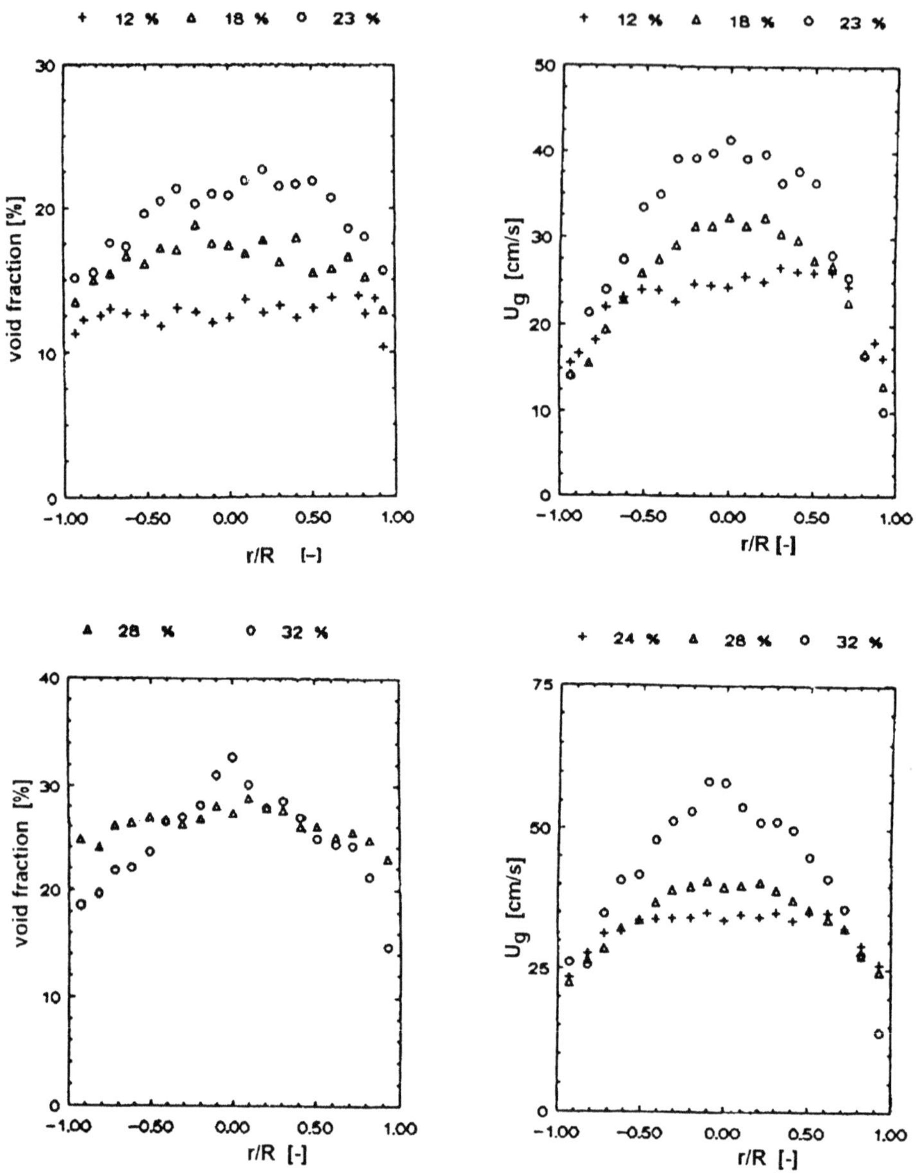

Figure 11: Radial profiles of the volume concentration and the bubble velocity (in a laboratory reference frame) near the point of instability of a uniform bubbly flow; stagnant flow (top), superficial liquid velocity 0.08 m/s (below). Measurement point 3.35 m above the inlet, indicated values of the mean volume concentration as measured just above the inlet. (+) 1 m below, (Δ) just below, and (O) 1 m above the point of transition, as judged by eye.

m/s (above) and $J_l = 0.16$ m/s (below). These data were obtained with an optical probe, introduced into the flow at a height of 3.35 m above the inlet. The indicated values of the mean volume concentration are those at the inlet of the pipe, and in each of the figures the point of transition, as seen by eye, was 1 m above, just above, and 1 m below the probe. What is observed is that under stable conditions the bubbly flow is fairly uniform, but that just below the point of transition the radial distribution has become of parabolic shape. This suggests that the initial stage of the transition is associated with an instability to radial disturbances, rather than planar disturbances as is assumed in Batchelor's theory. However, what is judged by eye to be the point of transition, may in fact be located just above the point where the actual instability occurs. The parabolic profile may be the result of a second stage of an instability to planar disturbances (see Batchelor 1993). At this moment there does not exist a satisfactory three-dimensional description of the dynamics of bubbly flows, which can be used to settle this question.

Acknowledgments

This work is part of the research program of the "Stichting voor Fundamenteel Onderzoek der Materie (FOM)", which is financially supported by the "Nederlandse organisatie voor wetenschappelijk onderzoek (NWO)". I gratefully acknowledge Paul Duineveld for the preparation of figure 1, and Jeroen Lammers for many valuable discussions and providing the rest of the figures.

References

[1] BATCHELOR, G. K. 1988 A new theory of the instability of a uniform fluidized bed. *J. Fluid Mech.* **184**, 75–110.

[2] BATCHELOR, G. K. 1993 Secondary instability of a gas-fluidized bed. *J. Fluid Mech.* **257**, 359–371.

[3] BULTHUIS, H. F. 1992 Effectieve eigenschappen van water-bellen mengsels. M.Sc.thesis, University of Twente.

[4] BULTHUIS, H. F., PROSPERETTI, A. & SANGANI, A. S. 1993 "Particle stress" in disperse two-phase flow. Submitted to *J. Fluid Mech.*

[5] COLE, J. D. 1951 On a quasi-linear parabolic equation occurring in aerodynamics. *Quart. Appl. Math.* **9**, 225–236.

[6] CRIGHTON, D. G. 1992 Nonlinear acoustics. In *Modern Methods in Analytical Acoustics* (ed. D.G.Crighton). pp.684–670. Springer-Verlag

[7] DUINEVELD, P. C. 1994 Bouncing and coalescence of two bubbles in water. Ph.D.thesis, University of Twente.

[8] HABERMAN, W. L. & MORTON, R. K. 1953 An experimental investigation of the drag and shape of air bubbles rising in various liquids. *David Taylor Model Basin Rep.* No. 802.

[9] KOK, J. B. W. 1993 Dynamics of a pair of gas bubbles moving through liquid. *Eur. J. Mech.*,B/*Fluids* **4**, 515–560.

[10] KYNCH, G. J. 1952 A theory of sedimentation *Trans. Faraday Soc.* **48**, 166–176.

[11] LAMMERS, J. H. 1994 The stability of bubbly flows. Ph.D.thesis, University of Twente.

[12] LIGHTHILL, M. J. 1956 Viscosity effects in sound waves of finite amplitude. In *Surveys in Mechanics* (eds G. K. Batchelor & R. M. Davies), pp. 250–351. Cambridge University Press.

[13] MOORE, D. W. 1965 The velocity of rise of distorted gas bubbles in a liquid of small viscosity. *J. Fluid Mech.* **23**, 749–766.

[14] RUDENKO, O. V. & SOLUYAN, S. I. 1977 *Theoretical Foundations of Nonlinear Acoustics.* Consultants Bureau (English translation), Plenum.

[15] SANGANI, A. S. & DIDWANIA, A. K. 1993*a* Dispersed-phase stress tensor in flows of bubbly liquids at large Reynolds numbers. *J. Fluid Mech.* **248**, 27–54.

[16] SANGANI, A. S. & DIDWANIA, A. K. 1993*b* Dynamic simulations of flows of bubbly liquids at large Reynolds numbers. *J. Fluid Mech.* **250**, 307–337.

[17] WHITHAM, G. B. 1974 *Linear and Nonlinear Waves.* Wiley & Sons.

[18] VAN WIJNGAARDEN, L. 1992 Dispersed flows. *Proc. 3rd Intl Workshop on Two-phase Flow Fundamentals*, London. To be published in *Multiphase Science and Technology* (eds G. F. Hewitt & J.-M. Delhaye). Hemisphere.

[19] VAN WIJNGAARDEN, L. & BIESHEUVEL, A. 1988 Voidage waves in mixtures of liquid and gas bubbles. In *Transient Phenomena in Multiphase Flow* (ed. N. Afgan), pp. 275–289. Hemisphere.

Void and Pressure Waves in Slug Flow

J. Fabre, A. Liné, E. Gadoin
Institut de Mécanique des Fluides,
URA CNRS 005
Allées Camille Soula, 31400 Toulouse, France

Introduction

When gas and liquid flow in a pipe the interface between the phases may take a variety of different patterns, the most fascinating and misunderstood being probably slug flow. As this kind of flow occurs over a very wide range of intermediate flow rates of gas and liquid, it is of interest for many practical problems involving flow instabilities and transient behaviour.

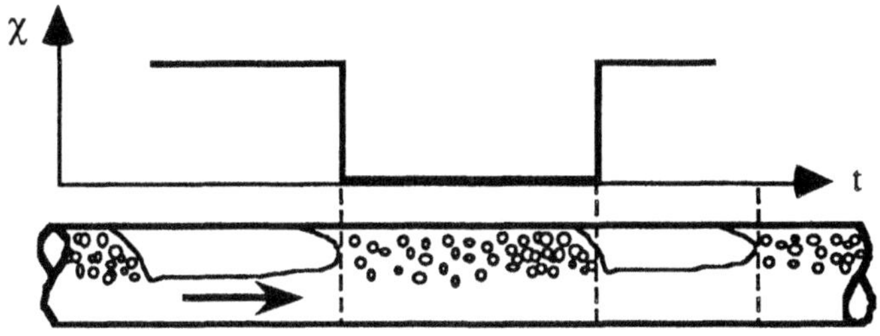

Figure 1. Schematic description of slug flow

The primary characteristic of slug flow is its inherent unsteadiness. An observer looking at a fixed position along the axis would see the passage of a sequence of slugs of liquid containing dispersed bubbles, each looking somewhat like a length of bubbly pipe flow, alternating with sections of separated flow which take on a stratified configuration in horizontal pipes (Fig. 1). These two states follow in a random-like manner, inducing pressure, velocity and phase fraction fluctuations which are not predictable. In contrast, several statistical averaged quantities can be predicted. For flows with steady inlet conditions, the methods leading to the determination of pressure, phase fractions and velocities have been firmly established in the past twenty years (see for example the overviews of Taitel and Barnea, 1990; Fabre and Liné, 1992); their limitations have been clearly identified as well (Dukler and Fabre, 1992).

S. Morioka and L. van Wijngaarden (eds.),
IUTAM Symposium on Waves in Liquid/Gas and Liquid/Vapour Two-Phase Systems, 25-44.

In opposition to this, slug flow with transient inlet or outlet conditions has not been extensively studied although it is of primary importance in many practical situations: this is due partly to the complexity of the flow itself and partly to the lack of theoretical framework.

In two-phase flows, transient states are controlled by the behaviour of the perturbations of pressure, phase fractions and velocities. These perturbations form a set of different nonlinear waves that may propagate upstream or downstream. The ideal method to study the behaviour of these waves would be to separate the linear and nonlinear effects and to start with small amplitude harmonic waves (Tournaire, 1987; Matsui and Arimoto, 1978). Slug flow being unsteady, random void and pressure fluctuations of large amplitude occur even at constant flow conditions. Small perturbations imposed at the inlet or outlet result in void and pressure perturbations that are hardly ever visible. Thus the problem has to be, at least experimentally, tackled in a different way. A possible method already used for studying hyperbolic waves is to investigate the response of the flow to finite steps of inlet or outlet conditions, these conditions being either flow rate or pressure or both. Such a method allows to identify the different propagative modes existing in the system and to understand part of their behaviour. The physical interpretation from two-phase flow models is less obvious: however we will attempt to throw light on the capabilities and weaknesses of drift flux model, generalised to slug flows.

The available experiments covering both steady and transient slug flows being limited to horizontal flow, we will focus the discussion to this particular case, even if the theory was built in a more general framework. The analysis of the flow development is out of the scope of this paper. However the length of the long bubbles evolves from the inlet to the outlet: it is mainly due to gas expansion. In addition, the coalescence of long bubbles occurs near the inlet.

In a first part a general description of the main characteristics of slug flow will be given in order to point out the limitations imposed by the flow structure. In a second part the experimental occurrence of three different waves will be given from the investigations of Théron (1989) and Gadoin (1993): the behaviour of these waves will be discussed. In a third part, we will use the general method of Fabre et al (1989) to develop the averaged equations for slug flow, leading to different hyperbolic wave models which can be viewed as generalisations of the drift flux model. Finally we will discuss the capabilities of this model to capture the physics.

STEADY SLUG FLOW: DESCRIPTION

As already stated slug flow consists of a space-time sequence of long bubbles flowing over a liquid film and liquid slugs carrying small bubbles. The gas is mainly transported in the large bubbles whose velocity is typically greater than the velocity of the liquid. Thus there exists a negative net flux of liquid with respect to the long bubbles which causes the film to travel at a much lower velocity than the gas. This provokes a flow separation at the rear of the long bubble from which small bubbles are shed to form a bubbly wake. They escape by turbulent dispersion into the slugs: as a consequence there is a negative net flux of gas with respect to the long bubbles. Typically the mean slug length is about 10 to 40 pipe diameters, whereas the long bubbles are more sensitive to the ratio of the gas to liquid rate, their length ranging from a few to several tenth of diameters.

Slug flow possesses three very distinctive features which deserve to be recalled: it is intermittent, chaotic and wavy. Some experimental results may help to illustrate these points.

The intermittence can be seen by an eulerian observer looking at a fixed position of the pipe. The flow switches from a stratified pattern in which the phases are separated by an almost flat interface to a dispersed pattern where small bubbles are conveyed by the liquid: an example is given in Fig. 2 from the time evolution of void fraction R_G for slug flow with non bubbly liquid. At low mixture velocity j the liquid fraction R_L in liquid slugs is close to unity indicating that little gas is entrained from the long bubbles (Fig. 3); when j increases R_L decreases. However the same trends is observed for the liquid fraction in the long bubbles so that the flow is always characterised by a significant change in void fraction when the liquid slugs and the long bubbles are passing.

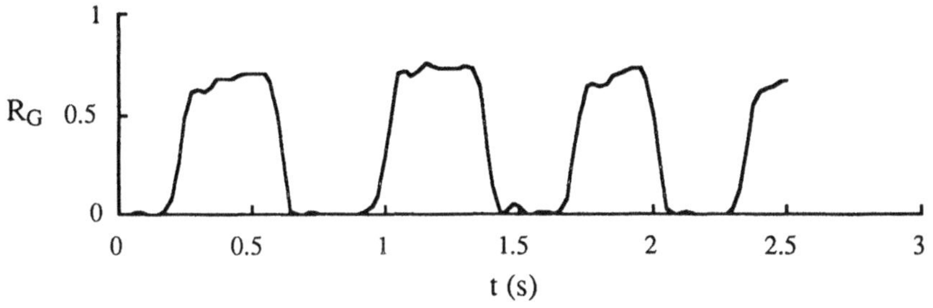

Figure 2. Time trace of void fraction for slug flow with non bubbly liquid

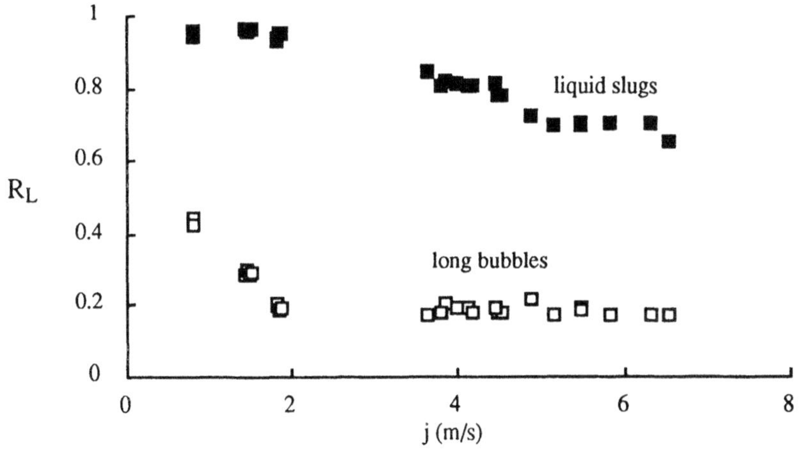

Figure 3. Distribution of liquid fraction versus mixture velocity (GADOIN, 1993)

The chaotic nature of slug flow may be proved by the fractal analysis of the successive lengths of long bubbles and liquid slugs. From an investigation carried out at IMF over a very large number of successive bubbles and slugs, no order could be put in evidence.

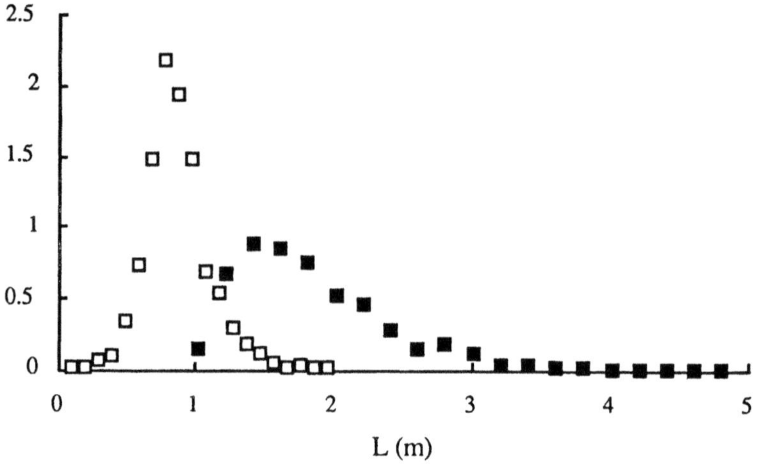

Figure 4. Probability density distribution of lengths of:
■ long bubbles, □ liquid slugs at j_G=1.25 m/s and j_L=0.97 m/s (Fabre et al, 1993)

Another less conclusive argument consists in analysing the probability distribution of these lengths. For most flow conditions the bubble and slug lengths are widely distributed about their average. This property is clearly shown in Fig. 4 in which the long bubbles are found within the range 1–5 m and the liquid slugs within 0.5–2 m.

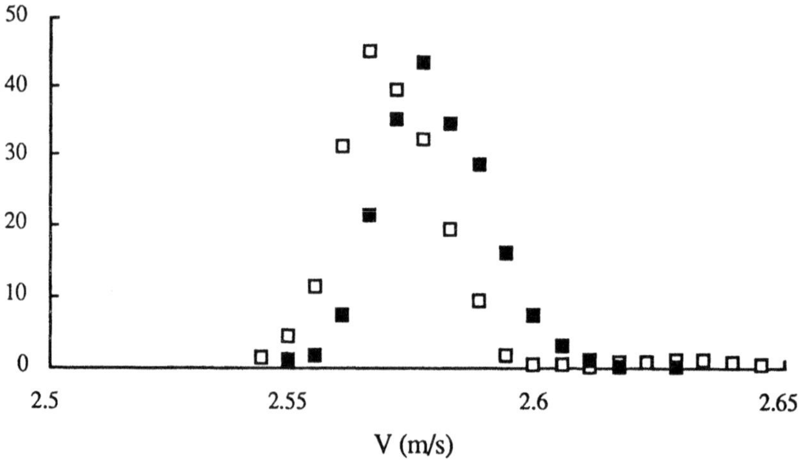

Figure 5. Probability density distribution of velocities of:
■ long bubbles, □ liquid slugs at j_G=1.25 m/s and j_L=0.97 m/s (Fabre et al, 1993)

The bubble and slug velocities do not display the same properties as their lengths. Fig. 5 shows clearly the following feature: the velocities are narrowly distributed about their average over a range of nearly 2%. As a consequence, an observer moving at the average velocity would see the whole structure almost frozen. Although this property becomes less evident at high gas or liquid flow rate, this quasi-steady behaviour in a moving frame is the key of the success of the unit cell model (Wallis, 1969). Thus we must arrive to the conclusion that there exists a signal (the characteristic function of the structure) which moves at a particular and recognisable velocity. In the wide sense given by Whitham (1974), slug flow may be seen as a wave. The fact that this wavy structure is shape conservative in a certain reference frame has a noticeable influence upon the propagation of void fraction wave, as discussed further. This property leads to a simplification in flow modelling since a change of frame transforms an unsteady flow problem into a steady one. Another simplification still arises from the bubble velocity V which can be expressed linearly with respect to the mixture velocity j:

$$V = C_0 j + C_\infty \sqrt{gD} \qquad (1)$$

where C_0 and C_∞ may be considered constant for wide ranges of inlet conditions. This model due to Nicklin et al (1962) applies for horizontal flow as well (Fig. 6). The long bubble velocity V must be seen as the velocity of the flow structure. In some sense it has to be interpreted as the wave velocity.

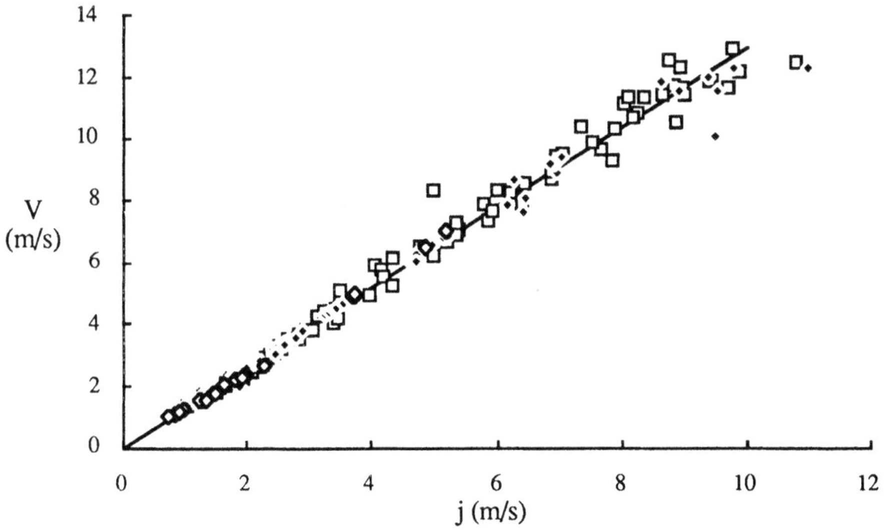

Figure 6. Velocity of long bubbles: □ Hubbard (1965), ◆ Ferré (1979), ◇ Théron (1989)

TRANSIENT SLUG FLOW: EXPERIMENTAL EVIDENCE OF WAVES

Keeping in mind the properties of steady slug flow, we come back to our initial problem concerning the transient effects.

The transient experiments were carried out by Théron (1989) and Gadoin (1993) in the same experimental facility. The two-phase flow loop is made of a long horizontal pipe, of 5 cm inside diameter and 90 m length in which air and water were injected.

The time evolution of pressure and void fraction was obtained in 4 measuring test sections located at 25 m, 45 m, 65 m and 89 m from the mixer: they will be further referred to as sections 2, 3, 4, 5. Each test section was equipped with a pressure transducer and a conductance probe. At the outlet, the mixture was separated by gravity in a free surface reservoir. The instant liquid rate could be determined at the outlet from the instantaneous water level in the reservoir, measured with a pressure transducer placed at the bottom. The conductance probes used for the measurement of void fraction were similar to these of Andreussi et al (1988): each probe was made up of 2 stainless steel rings flush mounted at the wall, 5 cm away from each other. The variation of the conductivity of the liquid with temperature was compensated by using a reference probe fully immersed in the liquid upstream the mixer. The probe conductance was converted to a DC voltage through a differential home made amplifier. In order to identify the bubble and slug front (Fig. 7), a numerical procedure was applied to determine for each run the void fraction threshold above which a long bubble is present: this was made from the probability distribution of void fraction which is clearly bimodal for the cases which were investigated. Thus a preliminary experiment was done for each run, in order to determine this distribution.

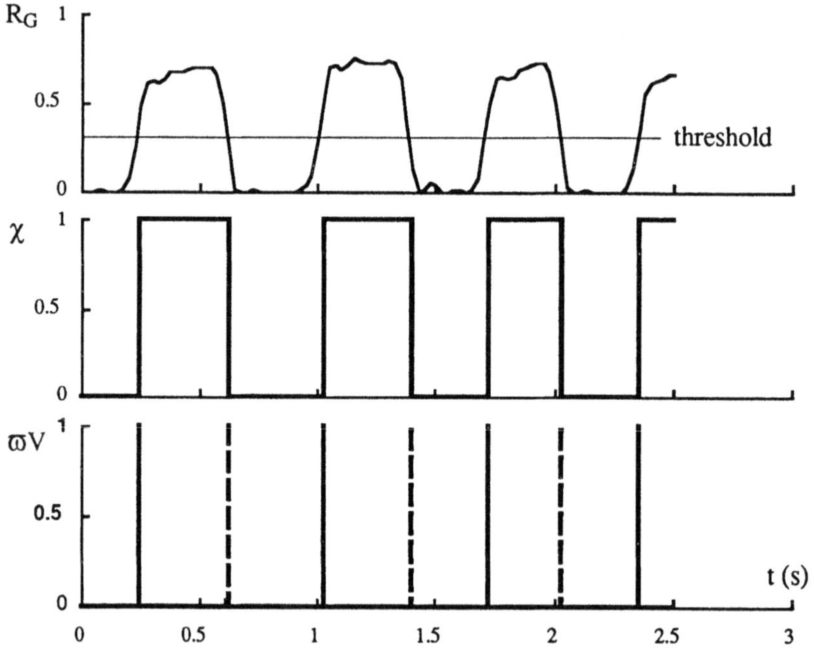

Figure 7. Time evolution of the flow structure.

Let us discuss briefly the different signals of Fig. 7. By applying a threshold to the time variation of the void fraction we obtain a bolean function which is defined as follows:

$$\chi(x,t)=\frac{1}{2}\left\{\operatorname{sign}\left[R_G(x,t)-R_{G,th}\right]+1\right\} \tag{2}$$

where $R_{G,th}$ is the threshold value. The function χ may be understood as a characteristic function of occurrence of the long bubbles (Fig. 1). Its value is unity when these bubbles are present at x,t.. It will be called intermittence function and may be expressed as a difference of two series of Heaviside functions:

$$\chi(x,t)=\sum_{j=1,n}\left\{\mathrm{H}\left[x-x_{2j+1}(t)\right]-\mathrm{H}\left[x-x_{2j}(t)\right]\right\} \tag{3}$$

where j is the cell number that the observer see at point x, the even subscript $2j$ referring to bubble front and odd subscript $2j+1$ to slug front. By taking the time derivative of the intermittence function one obtains:

$$\frac{\partial\chi}{\partial t}=\varpi_S V_S-\varpi_D V_D \tag{4}$$

where V_S and V_D are the velocity function of bubble and slug fronts (S for separated flow and D for dispersed flow), ϖ_S and ϖ_D are the space derivative of Heaviside functions which are expressed as Dirac delta function series:

$$\varpi_S(x,t)=\sum_{j=1,n}\delta\left[x-x_{2j}(t)\right] \qquad \text{and} \qquad \varpi_D(x,t)=\sum_{j=1,n}\delta\left[x-x_{2j+1}(t)\right] \tag{5}$$

As they have the dimension of the inverse of a length, each product ϖV may be viewed as the local frequency at which the fronts pass at a given point. Each of these functions may be easily deduced from the experimental recording of the void fraction by a numerical signal processing.

Transient phenomena were obtained during short periods following a perturbation generated at the inlet or the outlet. For inlet perturbation, sudden changes in gas or liquid flow rate were created by operating fast closing valves, disposed for each fluid, in parallel to the main line. For outlet perturbation, sudden changes of aperture of a guillotine valve were provoked. The steps of flow rate or valve aperture were chosen (i) weak enough to maintain the same flow pattern, (ii) high enough to be able to measure significant variations of the void fraction. In each case the resulting changes of the mean void fraction were typically less than 0.1. In order to extract the coherent perturbations from the "noise" due to the random fluctuations linked to the slug flow structure, ensemble averages were carried out over a large number of positive and negative steps. A hundredth of samples was recorded for each flow condition. Each run extended over up to 8 hours.

In order to put in evidence the void fraction wave we have to take the conditional ensemble average of the void fraction defined by:

$$\overline{R}_G(x,\tau)=\frac{1}{N}\sum_{i=1,N}R_G(x,t_i+\tau) \tag{6}$$

where t_i are the starting times of the positive or negative perturbation.

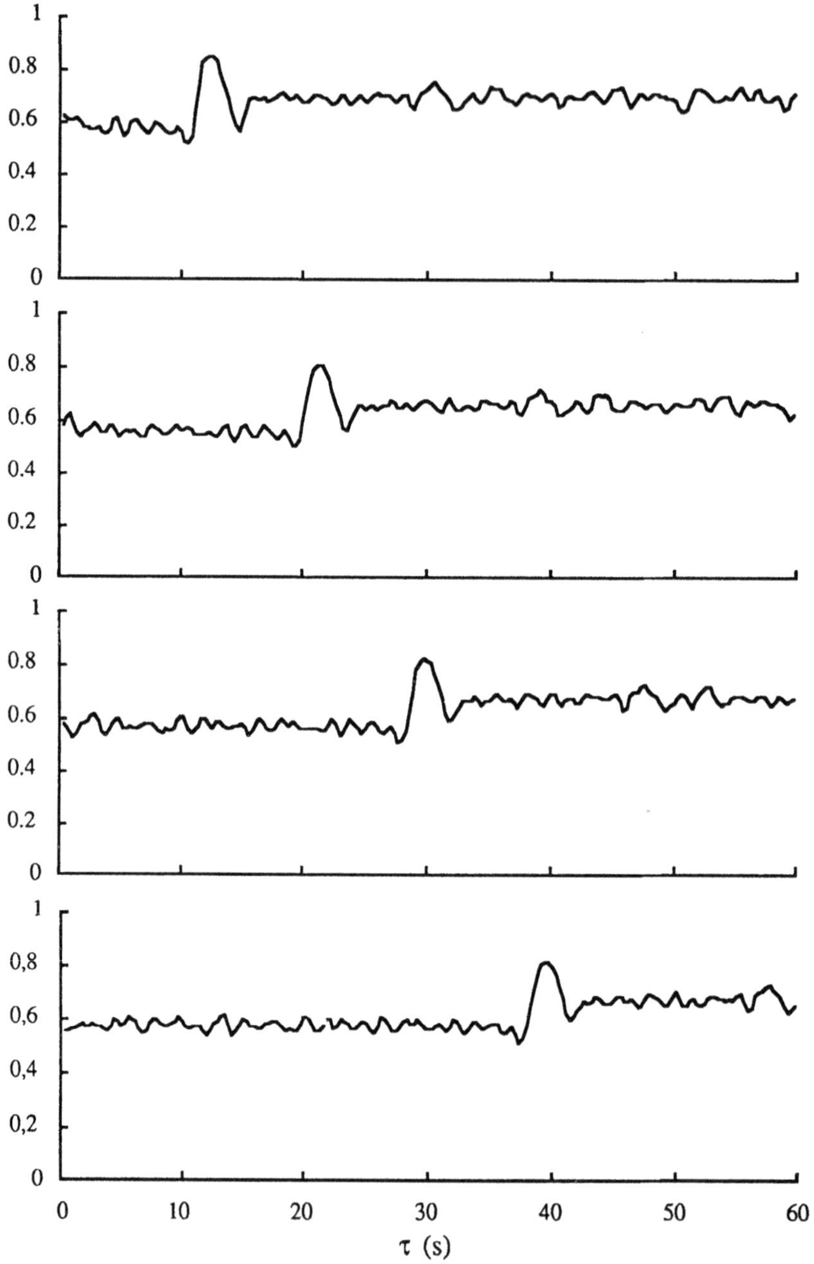

Figure 8. Time evolution of gas fraction at different locations: from top to bottom, sections 2, 3, 4 & 5 (j_L=0.5 m/s, j_G=1 m/s, Δj_G=+0.55 m/s)

The existence of the void fraction wave is illustrated in Fig. 8 in which the time trace of the mean void fraction is plotted at different measuring test sections. In the present example, the wave was created by a sudden increase of gas flow rate at the inlet, the liquid flow rate being kept constant. As a result the void fraction increases with a clearly recognisable shape. As the time is passing the wave reaches the different test sections. It must be stressed that this wave cannot be identified on a single shot. Some interesting features were observed more or less clearly for each flow conditions: these void fraction waves travelled at constant velocity and were found almost shape conservative. These properties are shown in Fig. 9 for the same flow conditions as these of Fig. 8. The exact wave locations were calculated to ensure the best superposition of their shape after a translation with time. The results are quite impressive. The wave is seen to travel with a uniform motion, its velocity being close to the long bubble velocity. The shape similarity at different locations is seen in the rippled structure of the wave. This result was observed whatever the sign of the perturbation indicating that the wave is weakly dispersive.

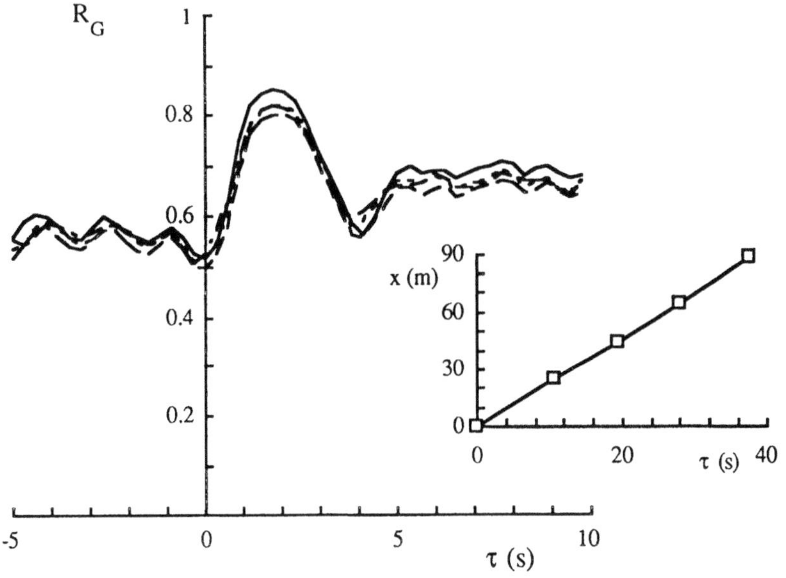

Figure 9. Location and superposition of void fraction waves observed at different measuring sections

The mean pressure is defined in a similar way as we did for void fraction. It is thus given like in Eq. (6) by:

$$\overline{P}(x,\tau) = \frac{1}{N} \sum_{i=1,N} P(x, t_i + \tau) \qquad (7)$$

However, in contrast to void fraction, the pressure perturbation is clearly distinguishable in a single shot. As expected, a pressure disturbance following the initial variation of flow rate travels downstream at a larger velocity than the void fraction wave (Fig. 10). The pressure suddenly increases at the inlet, the signal presenting an overshot due to the strong increase in momentum that the pressure force has to balance. This overshot was seen to damp as it propagates. As the outlet was maintained at atmospheric pressure, the pressure perturbation almost vanishes as it reaches the end of the pipe. Contrary to void fraction the pressure wave experiences a shape modification as it travels.

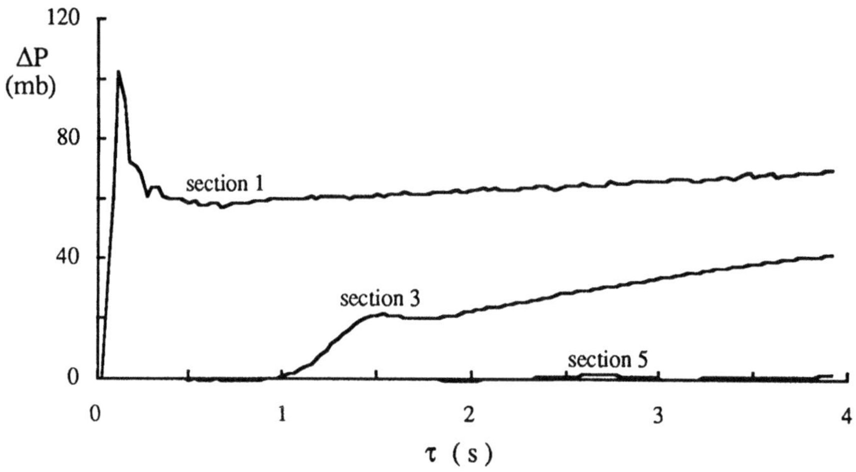

Figure 10. Time evolution of pressure at different locations
(j_L=0.5 m/s, j_G=1 m/s, Δj_G=+0.55 m/s)

Concerning the pressure variation an additional remark has to be done. In each cross section, it varies with time according to the following sequence (Fig. 11). When the disturbance reaches a given location, the pressure changes suddenly as previously mentioned; then it remains nearly constant until the void fraction wave reaches this point; finally, the pressure slowly evolves toward the final steady conditions which are obtained when the void fraction wave has reached the outlet. In fact the pressure results from a twofold mechanism: the fast change which propagates rapidly is controlled by inertia; the slow variation is mainly due to wall friction which depends upon the fluid velocity downstream.

In principle all the physical quantities are coupled through the mass and momentum conservation: when pressure is perturbed so is void fraction. According to this remark, what we called "void fraction wave" and "pressure wave" would be a matter of convenience without the following observations. On the one hand, when the fast pressure change was created we could not put in evidence a void fraction perturbation which emerges from noise. On the other hand, when the void fraction perturbation arrives, a weak perturbation of pressure is visible in the time traces (see Fig. 11).

If the coupling between pressure and void fraction is weak, what about between void fraction and intermittence. Indeed it is helpful to understand what causes the void fraction to change. Let us define the intermittence factor β in the same way as we did for pressure and void fraction:

$$\beta(x,\tau) = \frac{1}{N} \sum_{i=1,N} \chi(x,t_i + \tau) \qquad (8)$$

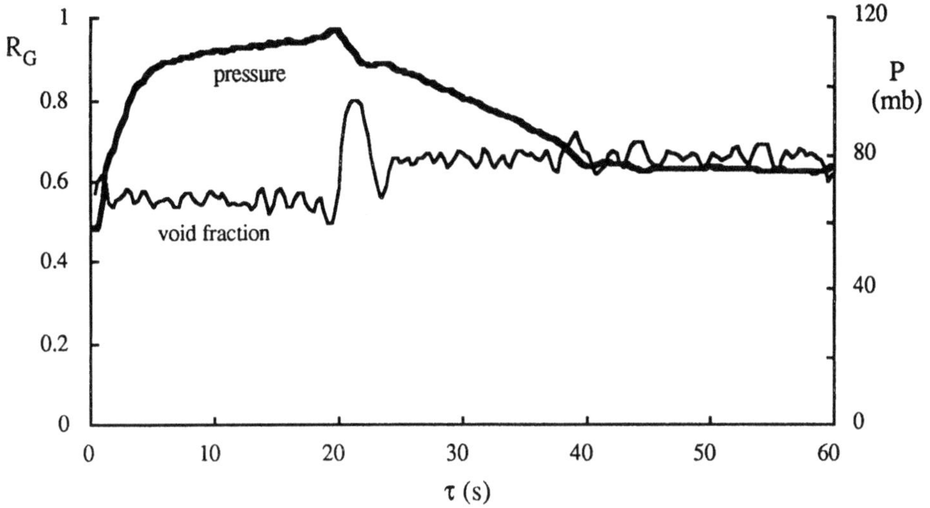

Figure 11. Time traces of void fraction and pressure at section 3; (j_L=0.5 m/s, j_G=1 m/s, Δj_G=+0.55 m/s)

For completeness, we define the rate of occurrence of the front of bubbles and slugs by taking the ensemble average of the function defined in Eq. (4). This demands a careful mathematical definition since the ensemble average of a set of Dirac delta functions is not a continuous function of time. This question has been discussed by Fabre et al. (1989) who overcame the problem by putting:

$$v_q(x,\tau) = \frac{1}{NT_h} \sum_{i=1,N} \left[\int_{\tau}^{\tau+T_h} \varpi_q(x,t_i+\theta) V_q(x,t_i+\theta) d\theta \right] \tag{9}$$

where q stands for S or D. In the above definition, a time averaging is followed by a statistical averaging. The first is defined over the interval T_h. If T_c is the time scale of the fluctuations to be filtered, and T_p the time scale of transient phenomena, N and T_h must satisfy the following conditions:

$$NT_h \gg T_c \qquad \text{and} \qquad T_h \ll T_p \tag{10}$$

In Fig. 12 the time traces of void fraction, intermittence factor and frequency of bubble front are plotted for comparison. It appears clearly that the three physical quantities are strongly coupled. When the gas flux suddenly increases at the inlet, both R_G and β increase and β goes to unity for a few seconds. During this time the eulerian observer see a long bubble, this fact being confirmed by the value of the frequency which goes to zero. Therefore the void perturbation comes essentially from a structure perturbation.

Last but not least, the propagation of void fraction and pressure waves displayed some interesting consequences at the outlet. In this view, the case which was treated along the paper is very illustrative. Let us consider the time evolution of the liquid flow rate at the pipe outlet, keeping in mind that it was maintained constant at the inlet (Fig. 13). A few seconds after the gas flow rate was increased, the liquid rate was observed to increase at the outlet. It happens that the liquid rate began to change when the pressure wave reached the outlet. Then the liquid rate remained almost constant up to the arrival time of the void fraction wave when it suddenly dropped to its lowest value: this is consequence of the modification of the flow structure. After the void fraction perturbation was evacuated the liquid rate rebuilt to its initial value. Thus the transient period resulting from sudden inlet change in flow rates is the period within the dates of arrival of the two perturbations travelling downstream: the faster corresponds to pressure perturbation, the slower corresponds to void fraction and intermittence factor perturbations. During the transient period a quasi steady state is observed. This is a typical behaviour of a physical mechanism described by hyperbolic equations.

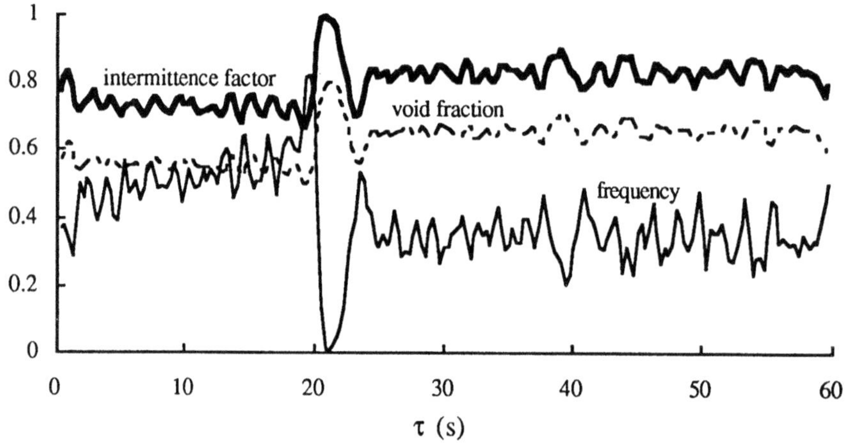

Figure 12. Time traces of intermittence factor ß, void fraction R_G and frequency ν at section 3; (j_L=0.5 m/s, j_G=1 m/s, Δj_G=+0.55 m/s)

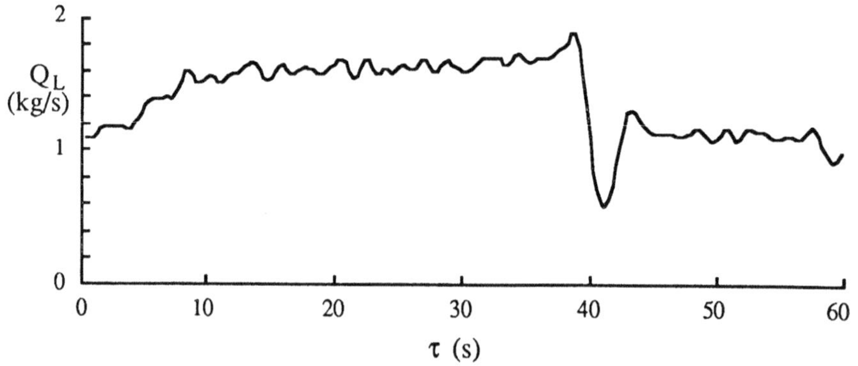

Figure 13. Time evolution of liquid flow rate at the outlet (j_L=0.5 m/s, j_G=1 m/s, Δj_G=+0.55 m/s)

Transients induced by outlet perturbation were investigated in the same way than transients induced by inlet perturbations. Unlike inlet perturbations the time traces display neither a void fraction nor a structure wave. Like inlet perturbations, a pressure wave is observed (Fig. 14): it propagates upward with a weak amplitude; it is not shape conservative and presents a tendency to damp.

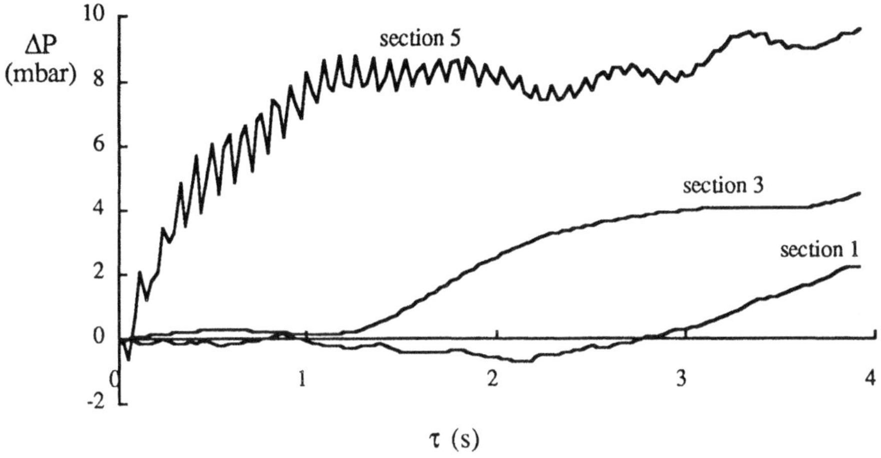

Figure 14. Time traces of pressure for a perturbation created at the outlet; (j_L=0.5 m/s, j_G=1 m/s, decrease of aperture of guillotine valve)

GENERALISED DRIFT FLUX MODEL FOR SLUG FLOW

A mathematical description of transient slug flow has been proposed by Fabre et al. (1989). We will recall here the main steps of this method. Conceptually, this approach is similar to the conditioned averaging applied to intermittent turbulence in single-phase flows by Libby (1975) and Dopazo (1977). The complexity of slug flow arises from its particular structure, which is neither periodic in space nor in time. This suggests that one should formulate balance equations, without ignoring the chaotic nature of the flow. The method closely follows the experimental procedure previously discussed.

The first step consists in the assumption that slug flow is a miscellaneous flow pattern, inhomogeneous with respect to space. It has thus to be viewed as a space-time sequence of two homogeneous flow patterns: separated flow and dispersed flow whose identification may be done by using the intermittence function already defined. That we recognise the existence of two different patterns leads to treat two different sets of equations, or more precisely the same set of equations with different closure laws.

The second step consists in writing the equations of mass, momentum and energy for each homogeneous flow pattern and to render these equations uniformly valid at any point and any time by using the intermittence as a weighting function. After little algebra one obtain the local-instant 1D equations for slug flow, written in conservative form.

The third step ends up the procedure by taking the ensemble average of the mass, momentum and energy equations. Again we must stress that the method follows closely the statistical experimental analysis.

The results may be summarised in limiting our attention to the equations of mass applied to each phase and the equation of momentum applied to the mixture (Fabre et al, 1989).

$$\frac{\partial}{\partial t}(\rho_k R_k) + \frac{\partial}{\partial t}(\rho_k R_k U_k) = 0 \qquad k=L, G \qquad (11)$$

$$R_k^S(V - U_k^S) = R_k^D(V - U_k^D) \qquad k=L, G \qquad (12)$$

$$\sum_{k=L,G}\left[\frac{\partial}{\partial t}(\rho_k R_k U_k) + \frac{\partial}{\partial x}(\rho_k R_k U_k^2) + \frac{\partial}{\partial x}\left(\rho_k \beta^S \beta^D \frac{R_k^S R_k^D}{R_k}(U_k^S - U_k^D)^2\right)\right] = -\frac{\partial P}{\partial x} + \sum_{k=L,G}(T_k - \rho_k R_k g \sin\theta) \qquad (13)$$

Note that Eq. (11) is the bulk mass conservation whereas Eq. (12) expresses the mass conservation across the discontinuity between the long bubbles and the liquid slugs. For momentum it is unnecessary to use the conservation across the discontinuity since we do not distinguish between the pressure in the long bubble and the pressure in the liquid slugs.

In the above equations we used the following ensemble averages of the physical quantities:

$$\beta^q = \overline{\chi_q} \qquad\qquad \beta^q F^q = \overline{\chi_q F_q} \qquad (14)$$

where q refers to separated or dispersed flow (q=S, D); θ is the pipe inclination; R_k, U_k, T_k are the statistical averages of phase fraction, velocity and wall friction defined by:

$$R_k = \beta_k^S R_k^S + \beta_k^D R_k^D \qquad k=L, G \qquad (15)$$

$$R_k U_k = \beta_k^S R_k^S U_k^S + \beta_k^D R_k^D U_k^D \qquad k=L, G \qquad (16)$$

$$T_k = \beta_k^S T_k^S + \beta_k^D T_k^D \qquad k=L, G \qquad (17)$$

In order to understand clearly the closure problem let us consider in a first step that both gas and liquid have constant densities. Then the mass equations appear uncoupled with the momentum equation which is only needed to determine the pressure as soon as the velocities and phase fractions are deduced from Eqs. (11), (12), (15) and (16). These 8 equations contain 11 unknown quantities, so that the missing information has to be restored by three closure laws which constitute the corner stone of slug flow modelling (Fabre and Liné, 1992). The first one has been discussed since Eq. (1) relates the long bubble velocity to the mixture velocity j given by:

$$j = R_G U_G + R_L U_L \tag{18}$$

The two others express the void fraction in the liquid slugs and in the long bubbles. This closure problem will not be discussed as it would complicate the discussion without any significant physical benefit.

Let us now come back to the general case by taking into account the gas compressibility. Then the mass and momentum equations become coupled through the relation between pressure and density. In fact the equation of energy would be needed together with the state equation for the gas in order to relate these two quantities. However we may replace this equation by a suitable assumption. For convenience we will consider that, due to the large heat capacity of the liquid phase, the mixture is isothermal. It will be seen that this assumption as a minor influence. Thus at given temperature the gas density and the pressure are linked by the state equation of perfect gas:

$$P = \rho_G r T \tag{19}$$

On comparing to the classical drift flux model the present set of equations displays some difference: indeed the momentum equation contains an extra term, the 3rd term of the l.h.s. of Eq. (13). This term accounts for the fact that each phase moves at different velocity in long bubbles and in liquid slugs. It may be demonstrated that, if we neglect the gas inertia, this term can be put under the following form

$$\sum_{k=L,G} \left[\rho_k \beta^S \beta^D \frac{R_k^S R_k^D}{R_k} \left(U_k^S - U_k^D \right)^2 \right] \cong C \rho_L R_L (V - U_L)^2 \tag{20}$$

where C may be interpreted as the virtual inertia coefficient given by:

$$C = \beta^S \beta^D \frac{\left(R_L^D - R_L^S \right)^2}{R_L^D R_L^S} \tag{21}$$

PROPERTIES OF THE GENERALISED DRIFT FLUX MODEL

In the limit of zero void fraction in liquid slugs, the foregoing model possesses some interesting features which illuminates the physics of wave propagation. In this case the gas is conveyed in the long bubbles yielding:

$$U_G^S = U_G = V \tag{22}$$

If we neglect the induced inertia, the set of partial differential equations is fully hyperbolic for certain range of void fraction. It has three real characteristics. In the physical plane (x,t) the slope V_1 of the first characteristics C_1 is equal to the gas velocity:

$$\frac{dx}{dt} = U_G = V_1 \qquad \text{on } C_1 \qquad (23)$$
$$\frac{d}{dt}\frac{PR_G}{1-C_0R_G} = 0$$

Interesting enough is the fact that the solution admits a Riemann invariant since the quantity $PR_G/(1-C_0R_G)$ is constant along this characteristic. This point has been already put in evidence by Fabre et al (1990). It may be shown that the eigenvalue V_1 corresponds to the velocity of the void fraction wave. On considering the perturbation of Fig. 8 as a shock it is possible to relate the jumps of pressure and void fraction (Théron, 1989):

$$[P]_1 \approx -\rho_L (U_G - U_L)^2 [R_G]_1 \qquad (24)$$

where $[F]_1$ is the jump of F across the characteristic 1. According to Eq. (24) the jump in pressure remains weak even for a not too small jump in void fraction: this proves that across the first characteristic only void fraction is significantly perturbed. The comparison between the experimental values of V_1 and U_G is shown in Fig.15. At low gas velocity, there is a very good agreement between the theory and the experiments. At higher gas velocity the void fraction wave was found to travel faster than expected. However some assumptions may be questionable: the void fraction in liquid slugs is not negligible so that the expression of V_1 involves a complementary term; the induced inertia may play a role.

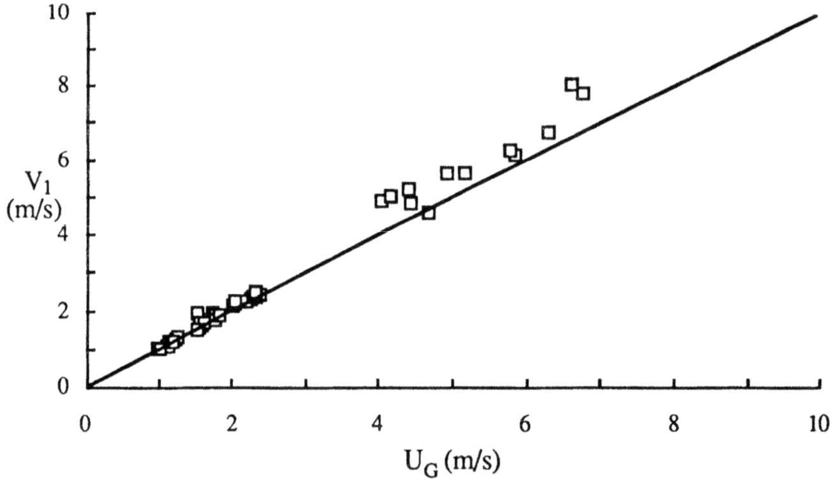

Figure 15. Velocity of void fraction wave versus gas velocity

The two other characteristics C_2 and C_3 are real if $R_G<1/C_0$. According to the recommended value of 1.2–1.3 for C_0 this leads to $R_G<0.75–0.85$ which corresponds more or less to the critical void fraction at which a transition between slug flow and annular flow is observed. On these characteristics, the following ordinary differential equations hold:

$$\begin{aligned}&\frac{dx}{dt}=U_L\pm W=V_{2,3}\\&\frac{dP}{dt}\pm\rho_L W(U_G-U_L)\frac{dR_G}{dt}+(V_{2,3}-U_G)\left(\rho_L R_L\frac{dU_L}{dt}-Q\right)=0\end{aligned}\qquad \text{on } C_2 \text{ and } C_3 \quad (25)$$

where $Q=\sum_{k=L,G}(T_k-\rho_k R_k g\sin\theta)$

In the two-phase medium, the pressure perturbations travel upstream and downstream at a velocity W with respect to the liquid. This velocity depends on both pressure and void fraction through the following relationship:

$$W=\sqrt{\frac{P}{\rho_L R_G(1-C_0R_G)}}\qquad(26)$$

As for void fraction waves, the jump in pressure and void fraction across a shock are related by:

$$[P]_{2,3}\approx-\rho_L W^2[R_G]_{2,3}\qquad(27)$$

As W is large, the jump in void fraction is negligible compared to the jump in pressure.

The velocity W has to be compared to the expression of pressure wave velocity W_h from homogeneous model of two-phase mixture:

$$W_h=\sqrt{\frac{P}{\rho_L R_G(1-R_G)}}\qquad(28)$$

It turns out that Eqs. (26) and (28) are very similar, the single difference is that the drift between phases is not allowed in the homogeneous model. Nevertheless it can be seen that the drift flux model for slug flow does not display a significant difference since we found the same pressure dependence. The two models do not differ in fact at low enough void fraction (Fig. 16).

The experimental data were obtained at different pipe locations, that is at different pressure conditions. In order to compare the experimental data to the theory, a modified velocity has been introduced:

$$W_0=W\sqrt{\frac{P}{P_0}}\qquad(29)$$

where P_0 is the atmospheric pressure. Thus W_0 is the velocity that pressure perturbations would experience if the medium would have been at atmospheric pressure. The results have been plotted in Fig. 16. The experimental data combine the results from both inlet and outlet perturbations. They show the same trend for the waves travelling upstream and downstream. The agreement with Eq. (26) is better than with the homogeneous result of Eq. (28). However it is not still satisfactory and there is a tendency to underestimate the pressure wave velocity with the present theory. In contrast to the present results Matsui and Arimoto (1978) found that their data compared well with the homogeneous model. But their results were obtained in a vertical pipe of 5 mm diameter in which the structure was a sequence of liquid and gas plugs. The difference between plug and slug flow is significant: contrary to plug flow, there exists in slug flow a flux of liquid across the cell. This means that there is a drift between phases, this drift depending *a priori* on the transient behaviour of the phases. This is the major weakness of the drift flux model to be inefficient in predicting the pressure wave behaviour.

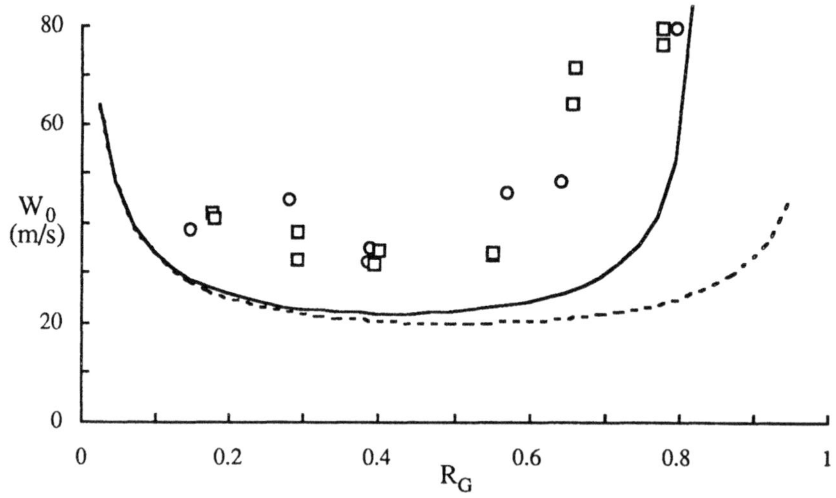

Figure 16. Velocity of pressure waves propagating downstream (○), upstream (□) versus void fraction; —— Eq. (26) with C_0=1.2, - - - Eq. (28)

CONCLUSION

An experimental investigation of transient phenomena in two-phase slug flow has been carried out. The objective was to put in evidence the existence of waves induced by perturbations of finite amplitude created at the inlet or the outlet of a long horizontal pipe:

- Sudden changes of gas or liquid flow rate at the inlet generated two distinct waves propagating downstream, a fast pressure wave significantly damped and a slow void fraction wave with a conservative shape. It was observed that the void fraction wave results primarily from a perturbation of the flow structure.
- Sudden changes of aperture of a guillotine valve placed at the outlet induced a pressure perturbation travelling upstream.

Although the experimental method used in this study does not lead to the same level of information as the analysis of harmonic waves of small amplitude, it provides useful information for the evaluation of two-phase transient models. The experimental results are well explained by a generalised drift-flux model. This model based on three partial differential equations possesses three eigenvalues, each of these corresponding to one of the waves which were experimentally observed. The model, even in its simplified form, predicts reasonably well the velocity of the void fraction wave, whereas it underestimates the velocity of the pressure waves. However it does better than the homogeneous model, especially at large void fraction. Indeed the weakness of both drift flux and homogeneous models has the same origin: the prediction of the pressure is based on the momentum equation of the mixture, the drift between phases being imposed by an analytical closure law. The momentum transfer between phases is thus poorly modelled. The two-fluid model, generalised to slug flow should improve the results.

ACKNOWLEDGEMENTS

The authors are greatly indebted to Institut Français du Pétrole, Elf and Total for supporting this work.

REFERENCES

Andreussi, P., Di Donfrancesco, A., Messia, M. An impedance method for the measurement of liquid hold-up in two phase flow. *Int. J. Multiphase Flow* 14, pp 777–87, 1988

Dopazo, C. On conditioned averages for intermittent turbulent flows. *J. Fluid Mech.* 81, pp 433–38, 1977

Dukler, A. E., Fabre, J. Gas liquid slug flow : knots and loose ends. *Third International Workshop on Two-Phase Flow Fundamentals,* June 15-19, London, UK, 1992

Dukler, A. E., Hubbard, M. G. A model for gas liquid slug flow in horizontal and near horizontal tubes. *Ind. Eng. Chem. Fundam.* 14, pp 337–47, 1975

Fabre, J., Grenier, P., Gadoin, E., Evolution of Slug Flow in Long Pipe, *6th International Conference on Multi Phase Production*, Cannes, France, pp 165-177, Ed. A. Wilson, BHRG Group Publication, 1993

Fabre, J., Liné, A. Modelling of two-phase slug flow. *Annu. Rev. Fluid Mech.* 24, pp 21–46, 1992

Fabre, J., Liné, A., Péresson, L. Two fluid/ Two flow pattern model for transient gas liquid flow in pipes. *Proc. 4th Int. Conf. Multi-Phase Flow*, Nice, France, pp. 269–84. Cranfield, BHRA, 1989

Fabre, J., Péresson, L., Corteville, J., Bourgeois, T., Odello, R., Severe slugging in Pipeline/Riser Systems, *SPE Production Engineering*, 299-305, August, 1990

Ferré, D. Ecoulements diphasiques à poches en conduite horizontale. *Rev. Inst. Fr. Pét.* 34, pp 113–42, 1979

Gadoin, E., *Régime intermittent "gaz-liquide" en conduite horizontale : écoulements non établis et transitoires.* Thèse, Inst. Natl. Polytech. de Toulouse, France, 1993

Libby, P. A. On the prediction of intermittent turbulent flows. *J. Fluid Mech.* 68, pp 273–95, 1975

Matsui, G., Arimoto, S. Propagation characteristics of pressure wave through a gas-liquid plug-train system in a pipe. *Proceedings of joint symposium on design and operation of fluid machinery,* June 12-14, Colorado State University, Colorado, USA, 1978

Nicklin, D.J., Wilkes, J.O., Davidson, J.F. Two phase flow in vertical tubes. *Trans. Inst. Chem. Engs.* 40, pp 61–68, 1962

Taitel, Y., Barnea, D. Two-phase slug flow. *Adv. Heat Transfer,* 20, pp 83–132, 1990

Théron, B. *Ecoulements instationnaires intermittents de gaz et de liquide en conduite horizontale.* Thèse, Inst. Natl. Polytech. de Toulouse, 1989

Tournaire, A. *Détection et étude des ondes de taux de vide en écoulement diphasique à bulles jusqu'à la transition bulle-bouchon.* Thèse, Inst. Natl. Polytech. de Grenoble, France, 1987

Wallis, G.B. *One-Dimensional Two-Phase Flow.* New-York, McGraw-Hill, 1969

Whitham, G. B. *Linear and Nonlinear Waves.* New-York, Interscience, Wiley, 1974

Behavior of void wave on the basis of turbulent liquid-bubbles two-phase flow model

S. Morioka
Dept. of Aeronautical Engineering
Kyoto University, 606-01, Japan

1 Introduction

Twenty years have gone since D. Gidaspow et al[1]. pointed out that the basic equations for liquid gas-bubbles two-phase flow were ill-posed as an initial value problem. In the meantime, many works have been made by many researchers and many facts have been disclosed. Nevertheless the problem does not yet seem to have arrived at the final conclusion.

As is well known, the ill-posedness of the initial value problem means that the characteristics become complex in spite of an initial value problem. In other wards, this means that an unstable wave exists and its growth rate increases infinitely with increasing wavenumber. Therefore we cannot ask for any solution developing from a given initial state continuously.

Of course, mathematically, we can consider a variety of terms required to make the basic equations well-posed, but there remains the essential problem which physical mechanism corresponds to them in liquid and gas- bubbles two-phase flow. At present, we can say explicitly that there exist two kinds of characteristics in the equations governing a bubbly liquid flow, and one corresponds to the void wave and the other to the pressure wave. The characteristic curve which becomes complex is that corresponding to the void wave. The origin of the trouble is the terms involved in the equation of motion of the bubbles, and the ill-posedness appears only if there is a relative velocity between the bubbles and the surrounding liquid, that is, a velocity slip exists. Hence, such a problem does not occure as long as we consider a one velocity model.

The purpose of this paper is to show that, if we assume that the bubbly liquid flow is essentially in a fluctuating and turbulent state as long as a velocity slip exists, not only the problem of ill-posedness is settled but also useful informations are available on the stability and the characteristics of fluctuation and turbulence in bubbly liquid flow.

2 Basic equations

The description should be simplified if we remove the interaction between the void wave and the pressure wave, since the questionable point of the present problem is in the void wave. For this purpose, we assume incompressibility for both gas and liquid. In fact,

S. Morioka and L. van Wijngaarden (eds.),
IUTAM Symposium on Waves in Liquid/Gas and Liquid/Vapour Two-Phase Systems, 45-53.

this assumption may be valid for the void wave, because of its relatively slow motion.

Here we discuss the case where the number density of bubbles is small, the so-called case of rarefied bubbles, because we can expect unambiguous equations and clear velocity slip for this case[2]. Further, we assume one dimensional motion for the simplicity of description.

Under the above assumptions, the basic equations are closed only by the phase averaged continuity equations of liquid and gas and the equation of motion of bubbles. Here the phase average means averaging over a space involving many bubbles. The variables are the void fraction α, the velocity of liquid phase u, and the velocity of gas phase v.

The continuity equations are

$$\frac{\partial}{\partial t}(1-\alpha)+\frac{\partial}{\partial x}(1-\alpha)u=0, \tag{1}$$

for liquid phase, and

$$\frac{\partial \alpha}{\partial t}+\frac{\partial}{\partial x}\alpha v=0, \tag{2}$$

for gas phase. The equation of motion for bubbles is given by

$$\frac{1}{2}\left[\left(\frac{\partial v}{\partial t}+v\frac{\partial v}{\partial x}\right)-\left(\frac{\partial u}{\partial t}+u\frac{\partial u}{\partial x}\right)\right]=\left(\frac{\partial u}{\partial t}+u\frac{\partial u}{\partial x}\right)+g(1-\alpha)-C_D(v-u)+\kappa_G\frac{\partial^2 v}{\partial x^2}, \tag{3}$$

where the left hand side denotes a virtual inertia force and the terms on the right hand side are the virtual force due to accelerating liquid, the buoyancy, the viscous drag, and the interbubble bouncing force, respectively. According to Batchelor's explanation[3], when the averaged velocity gradient of bubbles is negative, the bubbles approach each other and resist to push out the liquid between bubbles. This resistance appears as interbubble bouncing force, and it is reasonably assumed to be proportional to the gradient of the averaged velocity gradient of bubbles.

At a glance, the terms in the left hand side of the equation of motion of bubbles, that is the virtual inertia of bubbles, seems to be leading actor, and the last term in the right hand side opposes the bubble motion and controls the runaway of bubbles, and hence the problem of ill-posedness would be settled. However, the fact is not so.

In contrast to solid particles[4], gas bubbles have negligible mass, and the fluctuation of bubbles is governed by surrounding liquid motion (the flow field of liquid) principally. In bubbly liquid, the leading actor is the balance between the buoyance and the viscous drag, and the interbubble bouncing force gives only a dispersive effect.

On the other hand, the virtual inertia force and the virtual force due to the accelerating liquid act as a negative damping for bubble motion arising from an unbalance between the buoyance and the viscous drag. The instability of the void wave is induced by this negative damping, and the ill-posedness in the initial value problem is also attributed to this negative damping.

This fact is understood clearly by linearizing the above equations. A steady state is given as $\alpha=\alpha_0$, $u=u_0$, $v=v_0$, where $g(1-\alpha_0)=C_D(v_0-u_0)$ must be satisfied. Taking the perturbation from the above steady state; $\alpha=\alpha_0+\alpha'$, $u=u_0+u'$, $v=v_0+v'$, and linearizing (1) to (3) with respect to the perturbations α', u', v', and further eliminating the variables u' and v', then we obtain

$$\frac{\partial \alpha'}{\partial t} + [v_0 - 2\alpha_0(v_0 - u_0)]\frac{\partial \alpha'}{\partial x}$$
$$= -\frac{1}{2C_D}\left[(1-\alpha_0)\left(\frac{\partial}{\partial t} + v_0\frac{\partial}{\partial x}\right)^2 + 3\alpha_0\left(\frac{\partial}{\partial t} + u_0\frac{\partial}{\partial x}\right)^2\right]\alpha' - \frac{1}{2C_D}\kappa_G(1-\alpha_0)\frac{\partial^3 \alpha'}{\partial x^3}.$$

For a sufficiently larger time than the characteristic time of the void wave, $(1-\alpha_0)/2C_D$, the equation satisfied by the asymptotic solution may be provided by replacing $\partial/\partial t$ to $-[v_0 - 2\alpha_0(v_0 - u_0)]\partial/\partial x$ in the above equation[5]. Then we have

$$\frac{\partial \alpha'}{\partial t} + [v_0 - 2\alpha_0(v_0 - u_0)]\frac{\partial \alpha'}{\partial x}$$
$$= -\frac{1}{2C_D}(3 - 8\alpha_0 + 8\alpha_0^2)\alpha_0(v_0 - u_0)^2\frac{\partial^2 \alpha'}{\partial x^2} - \frac{\kappa_G}{C_D}(1-\alpha_0)\alpha_0(v_0 - u_0)\frac{\partial^3 \alpha'}{\partial x^3}. \quad (4)$$

Evidently we find that the virtual inertia force and virtual force due to the accelerating liquid act as negative damping in the motion of the bubbles if there is a velocity slip in the main flow, and the interbubble bouncing force acts only as dispersive effect.

3 Fluctuating/turbulent model

From a mathematical point of view, in order that the void wave is stable, a term of the second order derivative is necessary to cancel such a negative damping. Otherwise, in order that the void wave is unstable but the growth rate is finite over all wavenumber, a term of the fourth order derivative with a positive coefficient is necessary. Expressing these terms as $A\partial^2\alpha'/\partial x^2$ and $B\partial^4\alpha'/\partial x^4$ and adding them to (4), we have

$$\frac{\partial \alpha'}{\partial t} + [v_0 - 2\alpha_0(v_0 - u_0)]\frac{\partial \alpha'}{\partial x}$$
$$= A\frac{\partial^2 \alpha'}{\partial x^2} - \frac{1}{2C_D}(3 - 8\alpha_0 + 8\alpha_0^2)\alpha_0(v_0 - u_0)^2\frac{\partial^2 \alpha'}{\partial x^2}$$
$$- \frac{\kappa_G}{C_D}(1-\alpha_0)\alpha_0(v_0 - u_0)\frac{\partial^3 \alpha'}{\partial x^3} + B\frac{\partial^4 \alpha'}{\partial x^4}$$

Now the problem is whether there exists a physical explanation of such terms. The answer is that these terms appear simultaneously if we consider the fluctuating/turbulent diffusion of bubbles.

If we add a term expressing the fluctuating/turbulent diffusion on the right hand side of the continuity equation of bubbles, it is modified to

$$\frac{\partial \alpha'}{\partial t} + v_0\frac{\partial \alpha'}{\partial x} + \alpha_0\frac{\partial v'}{\partial x} = D_G\frac{\partial^2 \alpha'}{\partial x^2} \quad (5)$$

and then the linearized equation of the void fraction becomes

$$\begin{aligned}&\frac{\partial\alpha'}{\partial t}+[v_0-2\alpha_0(v_0-u_0)]\frac{\partial\alpha'}{\partial x}\\&=\left[(1-\alpha_0)D_G-\frac{1}{2C_D}(3-8\alpha_0+8\alpha_0^2)\alpha_0(v_0-u_0)^2\right]\frac{\partial^2\alpha'}{\partial x^2}\\&\quad-\frac{1}{C_D}(1-\alpha_0)\alpha_0(v_0-u_0)(D_G+\kappa_G)\frac{\partial^3\alpha'}{\partial x^3}+\frac{1}{2C_D}(1-\alpha_0)\kappa_G D_G\frac{\partial^4\alpha'}{\partial x^4}.\end{aligned}\tag{6}$$

In the averaging process of fluctuating/turbulent flow, Reynolds stress terms may appear in the equation of motion of the bubbles. It is difficult to deduce the functional form. Here we have neglected it, but a possibility for it to take the same form as the interbubble bouncing force may be discussed. If the fluctuation of liquid phase u' increases in the flow direction x, $\partial(\overline{u'u'})/\partial x$ should be positive. Then the averaged liquid velocity $\overline{u}$ will decrease if there is no contribution from the pressure, since the energy of the fluctuation should be supplied by the main flow. Then, a proportionality may be assumed between $\partial(\overline{u'u'})/\partial x$ and $\partial^2\overline{u}/\partial x^2$, since the gradient of the averaged flow velocity seems to increase in the flow direction. Such a proportionality may be assumed also between $\overline{v'v'}$ and $\overline{v}$, since $\overline{v'v'}$ and $\overline{v}$ will be proportional to $\overline{u'u'}$ and $\overline{u}$, respectively. If this is true, the Reynolds stress can be taken into account by the effectively increased value of κ_G.

The presence of a term of the fourth order derivative with a positive coefficient assures that the void wave of short wavelength is always damped out. Physically, this means that fluctuating/turbulent motion of the liquid phase cancels out the void wave of short wavelength. The coefficient in front of the fourth order derivative shows that both the interbubble bouncing force and the fluctuating/turulent diffusion force are necessary for this purpose.

When there is the fourth order derivative with a positive coefficient, the growth rate of the void wave remains finite even if it is unstable, and then we can follow the developmemt from a given initial state continuously. That is the basic equation becomes well-posed as an initial value problem.

On the other hand, the coefficient of the second order derivative modified by fluctuating/turbulent diffusion is given as

$$(1-\alpha_0)D_G-\frac{1}{2C_D}(3-8\alpha_0+8\alpha_0^2)\alpha_0\left(\frac{v_0}{u_0}-1\right)^2u_0^2.\tag{7}$$

If the fluctuation/turbulence in the liquid is relatively large, while the void fraction α_0, the velocity slip ratio v_0/u_0 and the flow velocity of liquid u_0 are relatively small, this coefficient has a positive value, and then the void wave is stable over all wavenumbers.

The value D_G of the fluctuating/turbulent diffusion coefficient is taken to increase with increasing fluctuating/turbulent intensity in liquid. Then it is reasonable to consider that the void wave instability appears and hence the fluctuating/turbulent diffusion coefficient increases as the void fraction, the velocity slip ratio and the flow velocity of liquid increase. However, it reacts to suppress the void wave instability. Thus, we can consider that a liquid-bubbles two-phase flow has both mechanisms generating and suppressing flutuation/turbulence concurrently.

4 Evolution equation of void wave

Now the initial value problem has become well-posed due to the existence of the fourth order derivative, but if the instability remains in a range of wavenumer, which kind of growing wave is brought about ? It is necessary to include the effect of nonlinearity for this purpose. If the effect of nonlinearity is weak, this can be investigated by using a perturbation method. In fact, the basic equation, that is the evolution equation, obtained by the reductive perturbation method[6] is reduced to the Benny equation. The Benny equation will be explained in detail by Kawahara in his lecture in this symposium, and it will be shown that an initial periodic variance developes to a state between a soliton-like puls train and a chaotic puls train through various transition process, depending on the magnitude of the coefficient in the dispersion term, that is the third order derivative term.

First we expand the dependent variable as follows, in the basic equations (1)–(3),

$$\left.\begin{aligned} \alpha &= \alpha_0 + \varepsilon\alpha_1 + \varepsilon^2\alpha_2 + \cdots, \\ u &= u_0 + \varepsilon u_1 + \varepsilon^2 u_2 + \cdots, \\ v &= v_0 + \varepsilon v_1 + \varepsilon^2 v_2 + \cdots, \end{aligned}\right\} \tag{8}$$

where ε denotes a parameter indicating the order of nonlinearity. On the other hand, taking into account that the nonlinear convection term balances with the highest order dissipation term (the fourth order term) after the lapse of time on the basis of the previous linearlized equation, the time and space coordinates are contracted and shifted as follows,

$$\tau = \varepsilon^{4/3}t, \qquad \xi = \varepsilon^{1/3}\left(\frac{x}{c} - t\right), \tag{9}$$

where c is an unknown constant representing the propagation velocity of void wave.

Substituting the expansions of the depending variables (8) and the transformation of time and space coordinate (9) into the basic equation (1) to (3), we have the relation $g(1-\alpha_0) = C_D(v_0 - u_0)$ from the lowest order terms with respect to ε, and $c = v_0 - 2\alpha_0(v_0 - u_0)$ from the next lower order terms. As expected, the latter coincides with the propagation velocity of the void wave in the linearized theory.

From the further lower order terms, we have

$$\begin{aligned} &\frac{\partial\alpha_1}{\partial\tau} - \frac{(2-3\alpha_0)(v_0-u_0)}{(1-\alpha_0)c}\frac{\partial}{\partial\xi}\alpha_1^2 \\ &\quad - \left[\frac{(1-\alpha_0)D_G}{c^2} - \frac{(3-8\alpha_0+8\alpha_0^2)\alpha_0(v_0-u_0)^2}{2C_Dc^2}\right]\varepsilon^{-2/3}\frac{\partial^2\alpha_1}{\partial\xi^2} \\ &\quad - \frac{(1-\alpha_0)\alpha_0(v_0-u_0)\kappa_G\varepsilon^{-1/3}}{C_Dc^3}\frac{\partial^3\alpha_1}{\partial\xi^3} + \frac{(1-\alpha_0)\kappa_G D_G}{2C_Dc^4}\frac{\partial^4\alpha_1}{\partial\xi^4} = 0. \end{aligned} \tag{10}$$

This is a Benny type equation. In order that the coefficient of the second derivative is of $O(1)$, $(1-\alpha_0)D_G/c^2 - (3-8\alpha_0+8\alpha_0^2)\alpha_0(v_0-u_0)^2/2C_Dc^3$ must be small as the order of $\varepsilon^{2/3}$.

Benny equation can be represented in the standard form as[7]

$$\frac{\partial\phi}{\partial t} + \phi\frac{\partial\phi}{\partial x} - \frac{\partial^2\phi}{\partial x^2} + \delta\frac{\partial^3\phi}{\partial x^3} + \frac{\partial^4\phi}{\partial x^4} = 0.$$

It is known that, if the coefficient δ of the dispersion term is of a large value, an initial periodic wave train develops to a soliton-like puls train, but if δ is of a small value, it develops to a chaotic puls train. By comparing the above nonlinear evolution equation given by the reductive perturbation method with the Benny equation, we can predict a variety of growing process of void wave and various fluctuating/turbulent characteristics in bubbly liquid flow, since the coefficient in each term takes various values depending on the main flow condition. However it will not bring about a large change such as the whole aspect will be over turned.

5 Fluctuating/turburent characteristics in bubbly liquid nozzle flow

The above theoretical results may be examined by a bubbly liquid flow in an accelerating nozzle flow, at least qualitatively. Figure 1 shows a facility used in this experiment. The facility is of a vertical blow down type to generate bubbly liquid flow and a diverging-converging nozzle or a diverging-straight nozzle is set in the test section. The measurements have been performed for the bubble velocity by double exposure stroboscopic photography and for pressure by piezo transducers set on the nozzle wall, as well as measurements of the injecting gas flow rate and the blow down time for a given volume of liquid. From these data, we calculate the averaged void fraction, the averaged flow velocity of liquid, and hence the velocity slip ratio, by assuming quasi-one dimensional flow.

We can deduce the fluctuation in the void fraction from the fluctuation in the pressure, because it has been shown that a pressure fluctuation can be induced by a fluctuation in void fraction if there is a velocity slip[8].

Figures 2 and 3 show an example of the experimental results on bubbly liquid flow in a diverging-converging nozzle, measured by our students S. Murata and H. Takase. In view of the distributions of the bubble velocity, the liquid flow velocity and the pressure along the nozzle, the flow is considered to be in a supersonic regime. Figure 4 shows the power spectrum density function of the pressure fluctuation measured at four points along a nozzle (22mm upstream of the throat, the throat, and 22mm and 50mm downstream of the throat).

As expected theoretically, the intensity of the pressure fluctuation becomes larger with increasing velocity slip ratio, increasing void fraction and increasing flow velocity of the liquid phase. The strong fluctuation of lower frequency may be relation to the relative motion between the bubbles and the surrounding liquid, while the intensified turbulence of higher frequency to the void wave instability. The strong fluctuation of lower frequency and the intensified turbulence detected at throat and 22mm downstream of the throat, in which the velocity slip is largest, suggest the existence of the void wave instability. On the other hand, the intensified fluctuation of middle frequency and reduced fluctuation and turbulence of lower and higher frequencies suggest the stabilized fluctuating/turbulent state. The growing and decaying region of the void wave instability is restricted in a blow down type facility, because the bubbles are always subjected to upward buoyancy force, which diminishes the velocity slip.

References

1. Lyczkowski, R.W., Solbrig, C.W., Gidaspow, D. & Hughes, E.D. 1975 ASME paper presented at Winter Annual Meeting (Houston, USA).
2. Auton, T.R., Hunt, J.C.R. & Prud'Homme, M. 1988 J. Fluid Mech, 197, 341.
3. Batchelor, G.K. 1988 J. Fluid Mech, 193, 75.
4. Morioka, S. & Nakajima, T. 1987 J. de Méc. Théor. et Appl. 6, 77.
5. Witham, G.B. 1974 Linear and nonlinear waves (Wiley-Interscience), 74.
6. Taniuti, T. & Wei, C.C. 1968 J. Phys. Soc. Jpn. 24, 941.
7. Kawahara, T. & Toh, S. 1988 Phys, Fluids, 31, 2103.
8. Morioka, S. 1989 Proc. of IUTAM Symposium on Adiabatic waves in liquid-vapor systems (Springer-Verlag), 261.

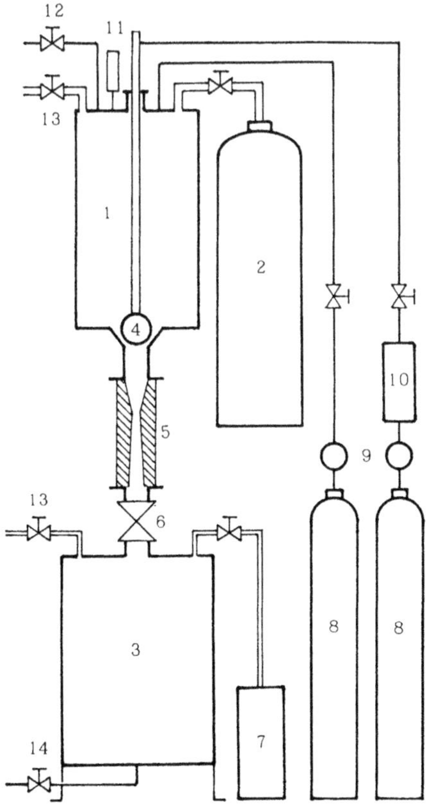

Fig. 1. Schematic diagram of experimantal facility

(1) upstream tank
(2) auxiliary tank
(3) downstream dump tank
(4) gas injector
(5) test section
(6) shutter valve
(7) vacuum pump
(8) nitrogen bumb
(9) pressure regulator
(10) flowmeter
(11) overflow valve
(12) feed pipe
(13) leak valve
(14) drain valve

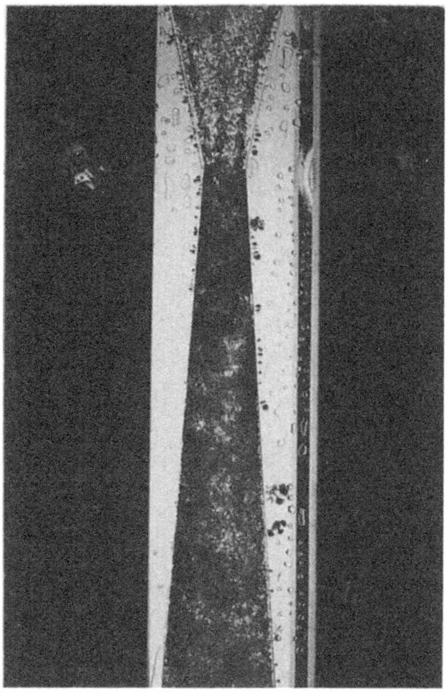

Fig. 2. Photograph of nozzle flow of bubbly liquid : upstream tank pressure p_u=1.5 atm, downstream dump tank pressure p_d=1.0 atm, nitrogen flow rate V_G=5.0 Nℓ/min.

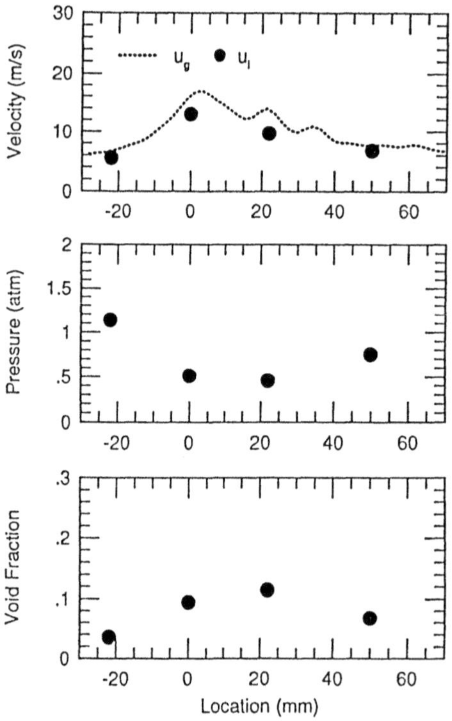

Fig. 3. Distributions of averaged velocities, pressure and void fraction along nozzle : p_u=1.5 atm, p_d=1.0 atm, V_G=5.0 Nℓ/min.

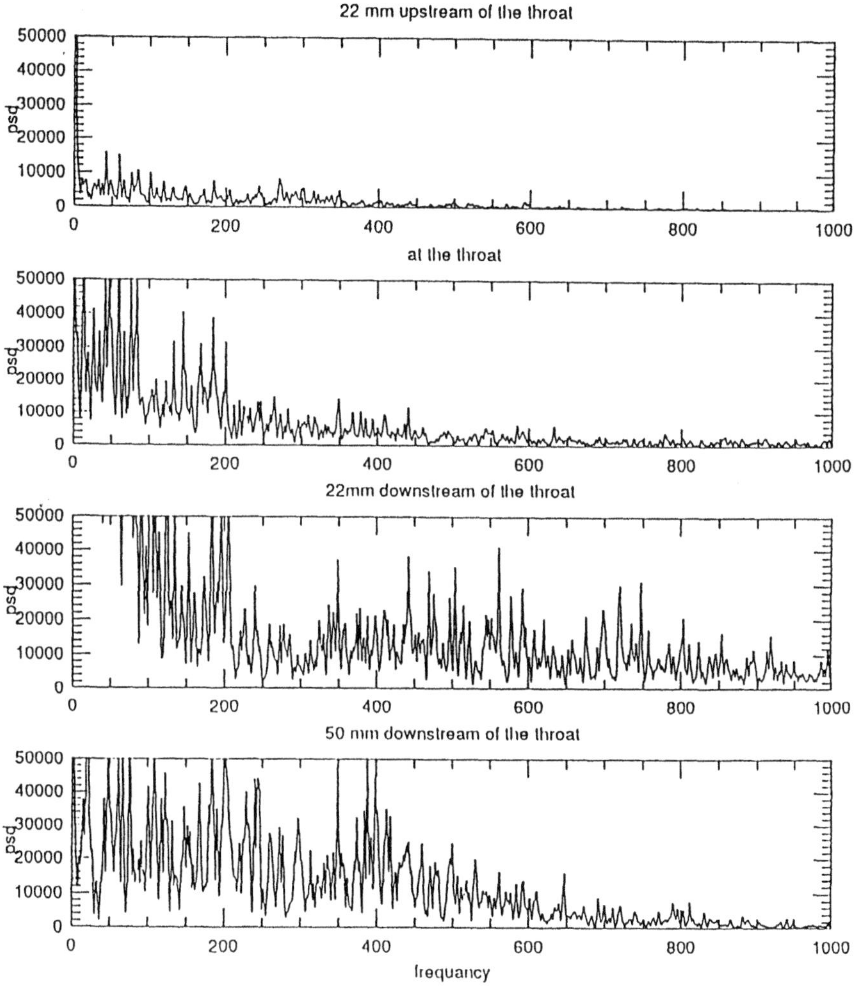

Fig. 4. Power spectrum density function of pressure fluctuation at four measuring points : p_u=1.5 atm, p_d=1.0 atm, V_G=5.0 Nl/min.

LINEAR WAVES IN BUBBLY LIQUIDS

Andrea Prosperetti
Department of Mechanical Engineering, The Johns Hopkins University
Baltimore MD 21218 USA

Abstract. A summary of the current status of the modeling of the propagation of pressure waves in bubbly liquid is presented. An apparent failure of the theory near and above the resonance frequency of the bubbles is found by comparison with experimental data. In the second part of the paper applications to oceanic noise, laboratory bubble clouds, and resonance-parametric generation of low frequency underwater sound are described.

Key words: Bubbly liquids, Two-phase flow, Oceanic noise

1 Foldy's theory

The lowest-order theory of wave propagation in bubbly liquid was essentially worked out by Foldy in a well-known 1945 paper [3], and later applied by him and Carstensen to the analysis of data [4]. The following, although by no means rigorous, is a very simple derivation of Foldy's result.

In the context of potential flow theory, to leading order, bubbles behave as monopoles with strength given by $\dot{v}$, the time derivative of their volume. Consider then N bubbles immersed in an "incident" flow given by a potential ϕ_∞. The total flow due to the incident flow plus the effect of the bubbles has then the potential

$$\phi = \phi_\infty + \sum_{j=1}^{N} \frac{\dot{v}_j}{4\pi|\mathbf{x} - \mathbf{x}_j|}, \tag{1}$$

where $\mathbf{x}_j$ is the position of the j-th bubble. If N is large, the preceding relation may be approximated as

$$\phi \simeq \phi_\infty + \int \frac{\dot{v}}{4\pi|\mathbf{x} - \mathbf{x}'|}\, n(\mathbf{x}')\, d^3x', \tag{2}$$

where n is the bubble number density. Upon taking the Laplacian of this expression, since ϕ_∞ is regular in the region of interest, we have

$$\nabla^2\phi = n\dot{v}, \tag{3}$$

which is essentially Foldy's result for an incompressible liquid. This can be written in a somewhat different form by noting that, for the linear problem,

$$\mathbf{u} = \nabla\phi \qquad P = -\rho\frac{\partial\phi}{\partial t}, \tag{4}$$

S. Morioka and L. van Wijngaarden (eds.),
IUTAM Symposium on Waves in Liquid/Gas and Liquid/Vapour Two-Phase Systems, 55-65.

where $\mathbf{u}$ and P are the mixture velocity and pressure fields. By eliminating ϕ we thus have

$$\nabla \cdot \mathbf{u} = n\dot{v}\,, \tag{5}$$

$$\rho\frac{\partial \mathbf{u}}{\partial t} + \nabla P = 0\,. \tag{6}$$

The seeming simplicity of this derivation actually hides many subtleties. As pointed out by Caflisch et al. [5], the step leading from the left-hand to the right-hand side of (2) is actually a limit process in which $n \to \infty$ and the bubble radius $a \to 0$ in such a way that $na \to$ const. As a consequence, in this limit the gas volume fraction

$$\beta = nv = \frac{4}{3}\pi a^3 n\,, \tag{7}$$

is extremely small, $\beta \sim a^2 \to 0$. This is the reason why in the previous derivation there was no need to worry whether the point $\mathbf{x}$ was in the liquid phase, where ϕ is defined, or inside the bubbles, where it is not. Indeed, in the limit where the previous results applies, the probability that $\mathbf{x}$ is in the gas is vanishingly small.

An equation of motion for the bubble radius is necessary to close the system of equations (5), (6). For this purpose we use the linearized form of the Rayleigh-Plesset equation,

$$\rho a\frac{\partial^2 R}{\partial t^2} = p - P - \frac{2\sigma}{R} - \frac{4\mu}{a}\frac{\partial R}{\partial t}\,. \tag{8}$$

Here σ is the surface tension coefficient, μ the liquid viscosity, and p the pressure in the bubble. We also use a to denote the equilibrium value of the radius, and R to denote the time-varying instantaneous value. Upon writing

$$R = a(1+X)\,, \tag{9}$$

we may express the pressure p as

$$p = p_e(1 - \Phi X)\,, \tag{10}$$

where p_e is the equilibrium pressure for a constant ambient pressure P_∞,

$$p_e = \frac{2\sigma}{a} + P_\infty\,. \tag{11}$$

In general Φ is an integral operator involving a convolution [12]. For the special case of time harmonic motion with frequency ω, however, one can show that

$$\frac{3\gamma}{\Phi} = 1 - 3(\gamma - 1)i\chi\left[\sqrt{\frac{i}{\chi}}\coth\frac{i}{\chi} - 1\right], \qquad \chi = \frac{D}{\omega a^2}\,, \tag{12}$$

where γ is the adiabatic index of the gas and D its thermal diffusivity. Upon substituting (9) and (12) into (8), and separating real and imaginary part of Φ, one finds an equation formally coincident with that of a driven damped oscillator,

$$\frac{\partial^2 X}{\partial t^2} + 2b\frac{\partial X}{\partial t} + \omega_0^2 X = -\frac{P - P_\infty}{\rho a^2}, \tag{13}$$

where

$$b = \frac{p_e}{2\rho a^2 \omega} \operatorname{Im} \Phi + \frac{2\mu}{\rho a^2}, \qquad \omega_0^2 \frac{p_e}{\rho a^2}\left(\operatorname{Re} \Phi - \frac{2\sigma}{a p_e}\right). \tag{14}$$

The similarity with a simple oscillator is however deceptive, as this relation is only valid for *steady* motion when $P - P_\infty \propto \exp i\omega t$. In particular, the "parameters" of the oscillator, b and ω_0, depend on the driving frequency ω.

For this case of harmonic motion (i.e., monochromatic pressure waves), the following well-known dispersion relation is readily obtained from (5), (6), and (13):

$$\frac{\kappa^2}{\omega^2} = \frac{3\beta}{a^2} \frac{1}{\omega_0^2 - \omega^2 + 2ib\omega}, \tag{15}$$

where κ is the wave number in the bubbly liquid. This relation can be generalized to include the liquid compressibility and a distribution of bubble sizes (see e.g. [19]),

$$\frac{\kappa^2}{\omega^2} = \frac{1}{c_L^2} + \int da \frac{4\pi a f(a)}{\omega_0^2(a) - \omega^2 + 2ib(a)\omega}, \tag{16}$$

where c_L is the speed of sound in the liquid and $f(a)$ is the radius size distribution normalized to 1.

The relation (15) (with the inclusion of liquid compressibility) is plotted in Fig. 1 for $a = 0.1$, 1, and 3 mm. Pressure and temperature have the standard values and the gas volume fraction is 1%. Starting from the low-frequency side, the curves exhibit at first a slight increase, which is the more marked the bigger the bubbles. This feature is a consequence of the fact that, as $\omega \to 0$, $\Phi \to 3$ and the bubbles oscillate isothermally. However, as ω increases, the flow of heat is impeded, the gas temperature changes, and the bubbles get stiffer. As the frequency is further incresed, the "softness" of the bubbles near resonance makes itself felt, the compressibility of the mixture increases, and the speed of sound falls. Just above resonance, the bubbles tend to *expand* upon *compression* and no propagation would be possible in an incompressible liquid. Liquid compressibility renders propagation possible, but with a much greater speed of sound. Finally, as $\omega \to \infty$, the amplitude of the bubble oscillations vanishes and the speed of sound reverts to that in the pure liquid.

These predictions for the phase velocity are in good agreement with data. An example is shown in Fig. 2, where the data are taken from the well-know paper by Silberman [13, 19].

The situation is less satisfactory for what concerns the attenuation of sound waves, that is described by the imaginary part of (16). The data of Silberman for $\beta = 0.53\%$ are compared with theory in Fig. 3, and a substantial discrepancy can be observed near and above the bubble resonance even for such a small void fraction.

2 Higher gas volume fraction

The literature contains several studies that purport to extend the previous theory beyond the lowest-order $O(\beta)$ terms. Unfortunately, as recently pointed out by Sangani [20], all of these results are incorrect. The basic fault can be explained by considering a simple case in which the standard "derivation" leads to incorrect results.

Suppose we try to "correct" the previous mathematical model by retaining higher-order terms as follows. In place of the momentum equation (6) we write

$$\rho(1-\beta)\frac{\partial \mathbf{u}_L}{\partial t} + \nabla P = 0 . \tag{17}$$

Here we have recognized that the density of a mixture with a gas concentration β is $\rho(1-\beta)$ rather than ρ. Furthermore, again neglecting the gas mass, the center-of-mass velocity of the mixture has just been equated to the liquid velocity $\mathbf{u}_L$. Secondly, using a well-known result [22, 23], we also introduce the bubbles' velocity $\mathbf{u}_G$ and write

$$\frac{\partial \mathbf{u}_G}{\partial t} = 3\frac{\partial \mathbf{u}_L}{\partial t} . \tag{18}$$

Retaining incompressibility of the liquid, we write the following relation expressing the fact that any volume flow is the result of the change of the bubble volume,

$$\nabla \cdot [(1-\beta)\mathbf{u}_L + \beta \mathbf{u}_G] = 4\pi a^2 n \frac{\partial R}{\partial t} . \tag{19}$$

Upon eliminating the velocity fields, one readily finds

$$\nabla \cdot [(1+3\beta)\nabla P] - \frac{3\beta}{a}\rho\frac{\partial^2 R}{\partial t^2} = 0 , \tag{20}$$

from which, upon using (8) to express the radius, one would find a dispersion relation analogous to (15) but containing higher-order terms in β. This seeming higher accuracy, however, is actually delusive for the simple reason that, as is evident from (15), κ^2 is of order β. For the correction $(1+3\beta)$ to be legitimate, therefore, also the term $\partial^2 R/\partial t^2$ should be known to better than $O(\beta^0) = O(1)$ accuracy, which implies that bubble-bubble interactions should be included in a more accurate version of the Rayleigh-Plesset equation. The use of Eq. (8) is thus seen to be the weak point of the many "derivations" of higher-order corrections found in the literature.

In a recent paper, Sangani [20] (see also [21]) has shown that the leading-order correction to Foldy's theory arises because of the fact that the bubbles are surrounded by an "effective" compressible medium, rather than by the pure liquid as assumed in writing down (8). One can thus find the lowest-order correction to Foldy's theory simply by taking the (linearized) equation of motion of a bubble in a compressible medium with speed of sound c (see e.g. [19, 12]),

$$a\frac{\partial^2 R}{\partial t^2} = \frac{1}{\rho}\left(1+\frac{a}{c}\frac{d}{dt}\right)\left(p-P-\frac{2\sigma}{R}-\frac{4\mu}{a}\frac{\partial R}{\partial t}\right), \tag{21}$$

and using the ω/κ in place of c. Proceeding in this way, in place of (15) one finds

$$\frac{\kappa^2}{\omega^2} = \frac{3\beta}{a^2}\frac{1}{\omega_0^2-(1-\kappa a)\omega^2+2ib_*\omega}, \tag{22}$$

where

$$b_* = b+\frac{1}{2}\omega\kappa a, \tag{23}$$

is the damping parameter corrected for radiation losses. If (22) is solved approximately for κ, one readily finds the leading order correction to (15) as

$$\frac{\kappa}{\omega^2} = \frac{3\beta}{a^2}\frac{1}{\omega_0^2-\omega^2+2ib\omega}\left[1-i\sqrt{3\beta}\left(\frac{\omega^2}{\omega_0^2-\omega^2+2ib\omega}\right)^{3/2}\right]. \tag{24}$$

This relation shows that the next term after the lowest order $O(\beta)$ correction is of order $\beta^{3/2}$, rather than β^2 as the previous erroneous procedure would give. The following two terms are of order $\beta^2\log\beta$ and β^2 [20].

Although not generally correct, the argument leading to (20) can be shown to give the right results at very low frequencies. This can be made apparent by considering the error terms in the previous derivation [24]. By accounting for added mass effects, the liquid momentum equation may be written

$$(1-\beta)\left(\rho\frac{\partial \mathbf{u}_L}{\partial t}+\nabla P\right) = -\frac{1}{2}\rho\beta C\left(\frac{\partial \mathbf{u}_L}{\partial t}-\frac{\partial \mathbf{u}_G}{\partial t}\right), \tag{25}$$

where a priori C is a function of β and frequency related to the usual added mass coefficient. It can be shown that, for $\omega \to 0$, $C = 1 + O(\beta)$ [24]. Similarly, the momentum equation for the bubbles is

$$\beta\nabla P = \frac{1}{2}\rho\beta C\left(\frac{\partial \mathbf{u}_L}{\partial t}-\frac{\partial \mathbf{u}_G}{\partial t}\right). \tag{26}$$

The sum of these two equations leads to (17) while, upon eliminating the pressure, we find

$$\frac{\partial \mathbf{u}_G}{\partial t} = \frac{2(1-\beta)+C}{C}\frac{\partial \mathbf{u}_L}{\partial t}. \tag{27}$$

Neglecting all terms of order β, this relation coincides with (18).

The error in the Rayleigh-Plesset equation (8) arises because of the neglect of the pressure radiated by the other bubbles and therefore it must be of the order of $\omega^2 o(\beta^0) = \omega^2 o(1)$. In the limit $\omega \to 0$ this error becomes therefore negligible and (8) can be used to express $\partial^2 R/\partial t^2$ in (20). In this limit we have from (8)

$$\frac{\partial^2 R}{\partial t^2} \to \frac{\omega^2 a(P - P_\infty)}{3p_e - 2\sigma/a}, \tag{28}$$

which can be used in (20) to find (with $C = 1$)

$$\nabla \cdot [(1 + 3\beta)\nabla P] + \frac{3\beta\rho\omega^2}{3p_e - 2\sigma/a}(P - P_\infty) = 0\,. \tag{29}$$

Correct to $O(\beta)$ and neglecting surface tension, the corresponding dispersion relation can be written

$$\frac{\omega^2}{\kappa^2} = \frac{p_e}{\rho\beta}\frac{1 + 2\beta}{1 - \beta}, \tag{30}$$

which is a result originally due to Crespo [1]. It is often stated that this expression is applicable to sound propagation in bubbly liquids at high frequencies because in this regime the bubbles cannot follow the liquid and therefore "slip" and added mass effects become important. The previous derivation shows this conclusion to be erroneous. The relative motion of the phases is always present for the simple reason that it is necessary to satisfy momentum balance, as is clear from Eq. (26). It is only at low frequency that this effect dominates over others, such as the compressibility of the mixture, so that (30) is correct.

3 Concluding remarks on the theory

In spite of its refinement, even Sangani's study fails to account for the discrepancy between theory and experiment pointed out at the end of section 1 and, in fact, gives corrections in the wrong direction.

To explain this failure of the available theoretical approaches, one might argue that things go as if the "effective" volume fraction should be estimated not on the basis of the bubble radius, but on the basis of an apparent radius given by the square root of the scattering cross section. At resonance, the single-bubble scattering cross section can be 3 or 4 orders of magnitude greater than the geometric cross section of the bubble, which would imply an apparent volume fraction up to 100 times larger than that given by (7). If this point of view is correct, then no small-β theory – whatever its order – can be hoped to agree with the available data. Direct numerical simulations seem to be the only possibility. The recent paper by Sangani and Sureshkumar [21] is a start in this direction, and this careful study can be consulted to gain a full appreciation of the difficulties involved. Another possibility must be mentioned. In the neighborhood of resonance, attenuation is very large and it could also be that Silberman's data are in error. A new and *very* careful experiment availing itself of the enormous progress in instrumentation over the last 30 years would be most welcome to help shed light on this matter.

The difficulties encountered by Sangani [20] in his theory are quite interesting and worth pointing out. In the first place, when bubble-bubble interactions are taken into account, the influence of neighboring bubbles has the effect of modifying the resonance frequency of a given bubble by an amount depending on the pair separation. Since, in a random mixture, all possible pair separations are to be expected, this effect could lead to resonance over a broad range of frequencies. Actually, this is not so due to a subtle cancellation. Only a rather weak resonance remains in correspondence of pairs of nearly touching bubbles. Secondly, if the bubbles are allowed to deform, resonances of the surface modes appear that could also lead to similar effects. As before, however, these resonant effects are weak albeit very complex.

Allowing for bubble deformations is not superfluous as, while $\omega a/c$ may be small, κa is not necessarily so in view of the drastic reduction of the speed of sound in the bubbly mixture. Because of this same circumstance, the assumption of separation of scales inherent in many averaging approaches may not be satsified except at the lowest concentrations.

As a final point we recall that, as mentioned before, the leading-order correction to Foldy's theory is due to the effect of the compressibility of the mixture. Beacuse of this fact, the use of "cell" models, or of homogeneization techniques, in which the spatial arrangement of the bubbles is assumed to be fixed, may be questionable.

4 Some applications

In spite of the limitations pointed out in the previous sections, Foldy's theory works very well at frequencies below bubble resonance and we now describe a few recent applications of it in this region.

4.1 Oceanic ambient noise

Data unequivocally show that, in the ocean, there is a substantial amount of wind-dependent noise over a very broad frequency range that extends from many hundreds of kHz down to a few tens of Hz (see e.g. [25, 26, 27]). The wind dependence rules out biological or man-made sources and strongly suggests that the origin of this ambient sound lies in physical processes at the ocean's surface. While several possibilities for the high-frequency sources readily come to mind (oscillations of single bubbles, impact of spray, rain, hale, and snow), the low-frequency range up to a few kHz is rather puzzling. Due to a curious coincidence, Dr. Carey [14] and I [9], quite independently, in the same session of a meeting of the Acoustical Society of America, suggested that the low-frequency sound could be due to collective oscillations of clouds of bubbles entrained by breaking waves.

The first step in evaluating the merits of this proposal had to be an experimental verification of the fact that bubble assemblies are indeed capable of generating low-frequency sound. This was indeed shown to be true in an experiment conducted by Dr. Yoon and Dr. Crum at the National Center for Physical Acoustics of the University of Mississippi [15]. Furthermore, in parallel theoretical work, it was shown that Foldy's theory enabled one to predict the oscillation modes of the bubble cloud with remarkable accuracy [15, 16]. More recently, it has also been shown that absolute sound levels can also be predicted by adding a suitable excitation mechanism to the theory [17].

On the basis of these concepts, very recently Dr. Oğuz has developed a model of oceanic ambient noise that fits spectral shape, levels, and wind dependence remarkably well [18].

4.2 Ocean surface scattering

It has been known for quite some time that the ocean's surface backscatters acoustic rays incident with small grazing angles (i.e., nearly horizontally) with an unexpected strength (see e.g. [7]). Theoretical estimates on the basis of wave-induced surface roughness fall far short of the data, and other explanations have been sought on the basis of bubble scattering [8, 6]. In a recent study [10] we have calculated the scattering of a hemispherical bubble cloud at the ocean surface by using Foldy's theory and we have found results quite compatible with experiment. This conclusion has been strengthened in another paper in which the sensitivity of the predictions to the cloud shape was examined [11]. A typical result from this paper is shown in Fig. 4.

4.3 Production of low-frequency sound

As any other other object whose acoustic properties differ from those of the surrounding fluid, bubble clouds possess a series of normal modes, and from the nonlinear interaction of two such modes the difference frequency can be generated. In a recent study we have proposed that, in order to enhance this low-frequency generation, it is possible to "tune" the difference frequency so that it corresponds to a resonance frequency of the bubble cloud [2]. In this work we have used a simplified version of Foldy's model and have demonstrated the effect for the simple case of a planar layer of bubbly liquid. This mechanism may furnish a practical method for low-frequency generation in the ocean.

5 Acknowledgment

This study has been supported by the Ocean Acoustics Program of the Office of Naval Research, and by grants CTS-8918144 of the National Science Foundation and DE-FG02-89ER14043 of the Department of Energy, Office of Basic Energy Sciences.

References

1. Crespo, A.: 1969 'Sound and shock waves in liquids containing bubbles', *Phys. Fluids* **12**, pp. 2274-2282.
2. Druzhinin, O.A., Ostrovsky, L.A., and Prosperetti, A.: 1994 'Low-frequency acoustic wave generation in a resonant bubble layer', submitted to *J. Acoust. Soc. Am.*
3. Foldy, L.L.: 1945, 'The multiple scattering of waves, I. General theory and of isotropic scattering by randomly distributed scatterers', *Phys. Rev.* **67**, pp. 107-119.
4. Carstensen, E.L. and Foldy, L.L.: 1947, 'Propagation of sound through a liquid containing bubbles', *J. Acoust. Soc. Am.* **19**, pp. 481-501.
5. Caflisch, R.E., Miksis, M.J., Papanicolaou, G.C., and Ting, L.: 1985 'Effective equations for wave propagation in bubbly liquids', *J. Fluid Mech.* **153**, pp. 259-273.
6. McDaniel, S.: 1987 'Vertical spatial coherence of backscatter from bubbles', *IEEE J. Ocean Eng.* **12**, pp. 349-356.
7. McDaniel, S. and Gorman, A.D.: 1982 'Acoustic and radar sea surface backscatter', *J. Geophys. Res.* **87**, pp. 4127-4136.
8. Crowther, P.A.: 1980 'Acoustical scattering from near surface bubble layers', in *Cavitation and Inhomogeneities in Underwater Acoustics*, W. Lauterborn ed., Springer, pp. 194-204.
9. Prosperetti, A.: 1985 'Bubble-related ambient noise in the ocean', *J. Acoust. Soc. Am. Suppl. 1* **78**, p. S2.
10. Prosperetti, A., Lu, N.Q., and Kim, H.S.: 1993 'Active and passive acoustic behavior of bubble clouds at the ocean's surface', *J. Acoust. Soc. Am.* **93**, pp. 3117-3127.
11. Sarkar, K. and Prosperetti, A.: 1993 'Backscattering of underwater noise by bubble clouds', *J. Acoust. Soc. Am.* **93**, pp. 3128-3138.
12. Prosperetti, A.: 1991 'The thermal behaviour of oscillating gas bubbles', *J. Fluid Mech.* **222** pp. 587-616.
13. Silberman, E.: 1957 'Sound velocity and attenuation in bubbly mixtures measured in standing wave tubes', *J. Acoust. Soc. Am.* **29**, pp. 925-933.
14. Carey, W.M. and Bradley, M.P.: 1985 'Low-frequency ocean surface noise sources', *J. Acoust. Soc. Am. Suppl. 1* **78**, pp. S1-S2.
15. Yoon, S.W., Crum, L.A., Prosperetti, A. and Lu, N.Q.: 1991 'An investigation of the collective oscillations of a bubble cloud', *J. Acoust. Soc. Am.* **89**, pp. 700-706.
16. Lu, N.Q., Prosperetti, A. and Yoon, S.W.: 1990 'Underwater noise emissions from bubble clouds', *IEEE J. Ocean Eng.* **15**, pp. 275-281.
17. Nicholas, M., Roy, R.A., Crum, L.A., Oğuz, H.N., and Prosperetti, A.: 1994 'Sound emissions by a laboratory bubble cloud', *J. Acoust. Soc. Am.* **95**, pp. 3171-3182.
18. Oğuz, H.N.: 1994 'A theoretical study of low-frequency oceanic ambient noise', *J. Acoust. Soc. Am.* **95**, pp. 1895-1912.
19. Commander, K.W. and Prosperetti, A.: 1989 'Linear pressure waves in bubbly liquids: Comparison between theory and experiments', *J. Acoust. Soc. Am.* **85**, pp. 732-746.
20. Sangani, A.S.: 1991 'A pairwise interaction theory for determining the linear acoustic properties of dilute bubbly liquids', *J. Fluid Mech.* **232**, pp. 221-284.
21. Sangani, A.S. and Sureshkumar, R.: 1993 'Linear acoustic properties of bubbly liquids near the natural frequency of the bubbles using numerical simulations', *J. Fluid Mech.* **252**, pp. 239-264.
22. Landau, L. and Lifshitz, E.M.: 1959 *Fluid Mechanics*, Pergamon.
23. van Wijngaarden, L.: 1976 'Hydrodynamic interaction between gas bubbles in liquid', *J. Fluid Mech.* **77**, p. 27-44.
24. Zhang, D.Z. and Prosperetti, A.: 1994 'Ensemble phase-averaged equations for bubbly flows', *Phys. Fludis*, in press.
25. Wenz, G.M.: 1962 'Acoustic ambient noise in the ocean: Spectra and sources', *J. Acoust. Soc. Am.* **34**, 1936-1956.
26. Perrone, A.J.: 1969 'Deep-ocean ambient noise spectra in the northeast Atlantic', *J. Acoust. Soc. Am.* **46**, pp. 762-770.
27. Wille, P.C. and Geyer, D.: 1984 'Measurements on the origin of the wind-dependent ambient noise variability in shallow water', *J. Acoust. Soc. Am.* **75**, pp. 173-185.

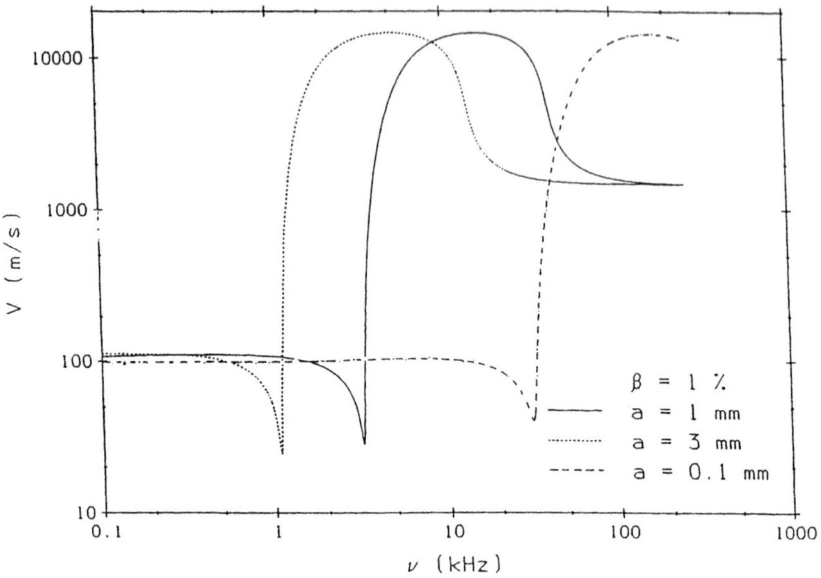
10000
1000
100
10
V (m/s)
0.1
1
10
100
1000
ν (kHz)
β = 1 %
a = 1 mm
a = 3 mm
a = 0.1 mm

Fig. 1

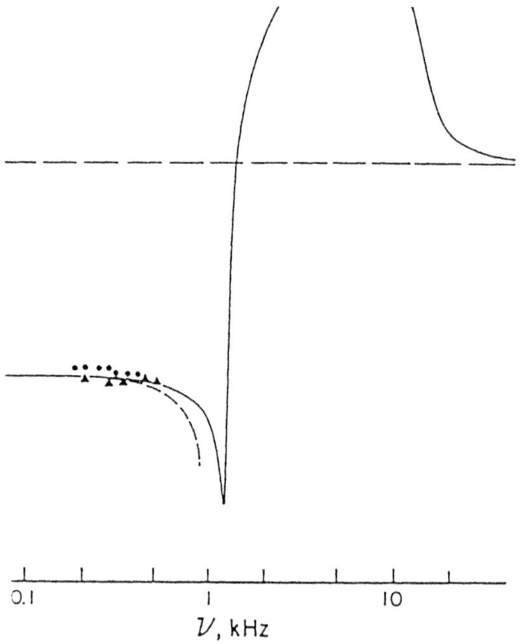
0.1
1
10
ν, kHz

Fig. 2

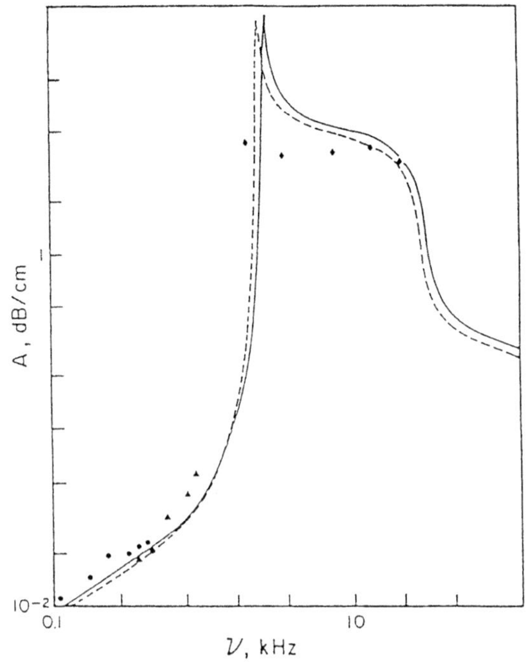
A, dB/cm
10^{-2}
0.1
10
ν, kHz

Fig. 3

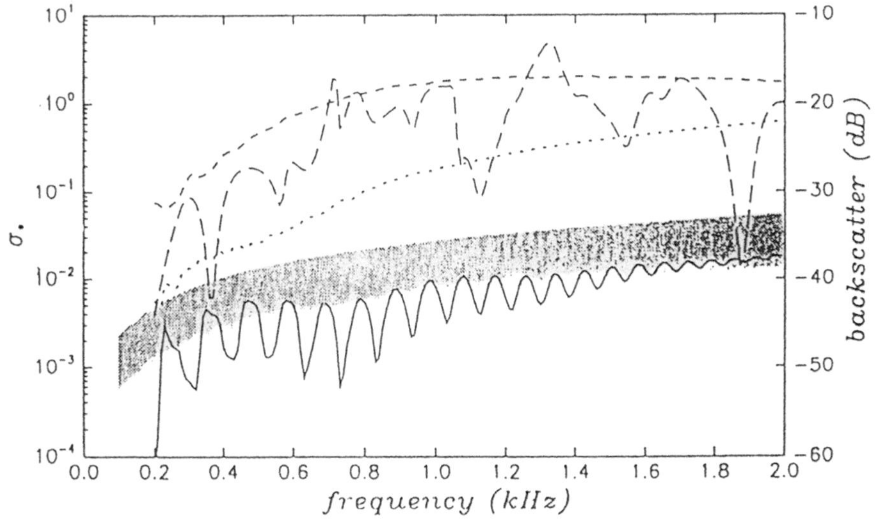
10^{1}
10^{0}
10^{-1}
10^{-2}
10^{-3}
10^{-4}
σ_s
−10
−20
−30
−40
−50
−60
backscatter (dB)
0.0
0.2
0.4
0.6
0.8
1.0
1.2
1.4
1.6
1.8
2.0
frequency (kHz)

Fig. 4

Surface Oscillations of Mutually-Interacting Gas Bubbles in a Sound Field

Hiroyuki TAKAHIRA and Teruaki AKAMATSU
Department of Mechanical Engineering, Kyoto University,
Kyoto 606-01, Japan

1. Introduction

Bubbles in a sound field emit the acoustic cavitation noise which shows chaotic behavior with the increase of the intensity of the sound [1]. To date, in order to clarify the fundamental nature of the cavitation noise, many numerical simulations were performed for the oscillations of bubbles. However, numerical models used before were virtually restricted to the single bubble case [2-4]. Since thousands of tiny bubbles are oscillating and moving in the actual cavitation state, it seems to be important to cast light on the effect of the interaction of the bubbles on their oscillatory motions.

There are generally two theoretical approaches to the dynamics of a cluster of bubbles. One is to analyze the microscopic motions of individual bubbles. The other deals with the macroscopic motions of the bubble cluster by using the averaged equations of conservation of momentum and mass of liquid and gas [5,6]. In our previous work, the former approach was chosen and the governing equations of a cluster of bubbles in a slightly viscous liquid were derived by taking account of the three-dimensional translational motion and deformation induced by the bubble-bubble interactions [7]. In the present paper, the governing equations are modified so as to consider the thermal effect of the internal gas. We particularly cast light on the interaction effects on the surface oscillations of each bubble under not so large amplitude of the sound field.

2. Equations of motion for a cluster of bubbles

Let us suppose that there are N bubbles in a liquid. The bubbles are assumed to be initially in spherical form and at rest. The oscillations of the bubbles are induced by the sinusoidal pressure change around the bubbles. The coordinate system is shown in Fig.1. We define a fixed Cartesian coordinate system having its origin at point o and also define a polar and a Cartesian coordinate system fixed at the center o_I of the observed bubble I (I=1, 2, $\cdots$, N) and moving with the velocity $\vec{v}_I$. Directions of the axes of the moving coordinate system coincide with those of the fixed coordinate system. L_{IJ} is the distance between the centers of bubbles I and J.

Our analysis is performed under the following assumptions.

(i) Wavelength of the sound field is much longer than the characteristic length of the bubble cluster. The sound field imposed on the bubbles is therefore regarded as uniform in space.

(ii) Compressibility of the liquid is disregarded and the flow field in the liquid is irrotational.

S. Morioka and L. van Wijngaarden (eds.),
IUTAM Symposium on Waves in Liquid/Gas and Liquid/Vapour Two-Phase Systems, 67-76.

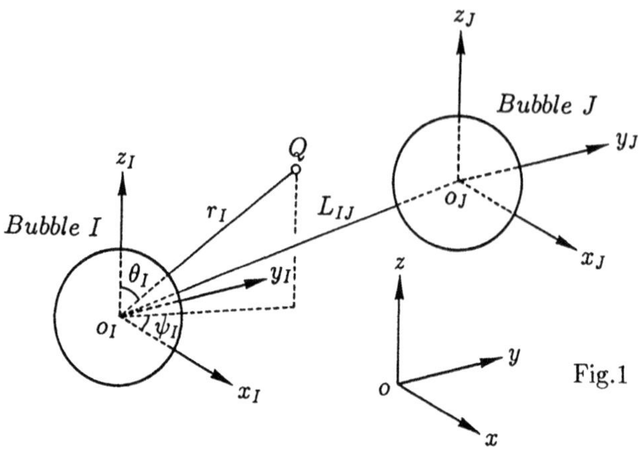

Fig.1 Coordinate system.

(iii) The interior of each bubble consists of a noncondensable gas. The gas obeys the perfect gas law.

(iv) The pressure of the internal gas is uniform. The diffusion of the gas through the bubble surface and the viscosity of the gas are disregarded.

(v) Effects of the condensation and evaporation are disregarded.

(vi) Thermal boundary layers are disregarded in the liquid.

(vii) Gravity is disregarded.

Under assumption (i), we may write the pressure variation imposed on the bubbles as

$$p_s = p_0 - p_d(t), \quad p_d(t) = p_a \sin \Omega_d t, \tag{1}$$

where p_0 is the pressure in the undisturbed liquid, t the time, p_a the pressure amplitude and Ω_d the angular frequency of the sound. Under assumption (ii), a velocity potential φ_l can be defined, which satisfies the Laplace equation on the exterior of the bubbles. The basic equations for the gas inside bubble I are given as follows:

$$\frac{\partial \rho_g}{\partial t} + [(\vec{u}_g - \vec{v}_I)\cdot\nabla_I]\rho_g + \rho_g \nabla_I \cdot \vec{u}_g = 0, \tag{2}$$

$$\rho_g C_{pg}\left\{\frac{\partial T_g}{\partial t} + [(\vec{u}_g - \vec{v}_I)\cdot\nabla_I]T_g\right\} - \dot{p}_g = \nabla_I \cdot (\lambda_g \nabla_I T_g), \tag{3}$$

where ρ is the density, $\vec{u}$ the flow velocity, $\vec{v}_I$ the velocity of the center of bubble I, ∇_I the gradient defined by the coordinate system I, C_p the specific heat at constant pressure, T the temperature, p the pressure, λ the thermal conductivity. The subscript I and g denote bubble I and the gas, respectively. The dot is the differentiation with respect to time. Equations (2) and (3) are combined into

$$\dot{p}_g/\gamma p_g + \nabla_I \cdot \{\vec{u}_g - [(\gamma-1)/\gamma p_g]\lambda_g \nabla_I T_g\} = 0, \tag{4}$$

where γ is the ratio of specific heats [8].

Disregarding the liquid viscosity, the following boundary conditions must be satisfied on the surface of bubble I:

$$\frac{\partial S_I}{\partial t} - \vec{v}_I \cdot \nabla_I S_I + \nabla_I \varphi_l \cdot \nabla_I S_I = 0, \tag{5}$$

$$\frac{\partial \varphi_l}{\partial t} - \vec{v}_I \cdot \nabla_I \varphi_l + \frac{1}{2}|\nabla_I \varphi_l|^2 = \frac{p_0 - p_d(t) - (p_g - 2\sigma H_I)}{\rho_l}, \tag{6}$$

$$\vec{u}_g = \nabla_I \varphi_l, \tag{7}$$

$$T_g = T_0, \tag{8}$$

where σ is the surface tension, H_I the mean curvature of the bubble surface and T_0 the constant temperature of the liquid. The subscript l denotes the liquid.

In order to get dynamical motions of bubbles, we may write the bubble surface and velocity potential by using the series expansion of spherical harmonics:

$$S_I = r_I - R_{I0}(t) - \sum_{n=2}^{\infty} \sum_{m=-n}^{n} R_{Inm}(t) Y_{nm}(\theta_I, \psi_I) = 0, \tag{9}$$

$$\varphi_l = \sum_{I=1}^{N} \left[\frac{q_{I0}(t)}{r_I} + \sum_{n=1}^{\infty} \sum_{m=-n}^{n} \frac{q_{Inm}(t)}{r_I^{n+1}} Y_{nm}(\theta_I, \psi_I) \right], \tag{10}$$

where

$$Y_{nm}(\theta_I, \psi_I) = P_n^{|m|}(\cos\theta_I) e^{im\psi_I}. \tag{11}$$

In Eq.(9), R_{I0} is the spherical-mode radius of bubble I and $R_{Inm} (n \geqq 2)$ the non-spherical perturbation of the surface (I, m, n : integer). The proper choice of the origin does not necessitate the R_{I1m} component [9]. The q_{I0} and q_{Inm} are the time-dependent unknown functions, $Y_{nm}(\theta_I, \psi_I)$ is the spherical harmonics defined by the moving coordinate system I, $P_n^{|m|}$ the associated Legendre polynomial of order n and degree m, and i the imaginary unit.

Viscous effects of the liquid are evaluated by using Lagrangian procedure with viscous dissipation. In order to get the viscous dissipation, the velocity gradient at the bubble wall is estimated by the solutions of the potential field. The spherical-mode radius, R_{I0}, and the displacement of the center of the bubble I, b_{Im}, and the nonspherical perturbation of the bubble surface, R_{Inm}, and their time derivatives are treated as generalized coordinates of the Lagrangian procedure. Since viscous boundary layers are neglected, the velocity gradient at the bubble wall will be overestimated. Therefore this approach tends to overestimate the viscous effects [10,11].

Equations of motion for a bubble cluster can be derived by using Eqs.(5) and (6) and the orthogonality of the spherical harmonics. Details of the formulation of the equations are referred to Reference [7]. The resulting equations of motion for a bubble cluster are

$$R_{I0}\ddot{R}_{I0} + \frac{3}{2}\dot{R}_{I0}^2 = \frac{1}{\rho_l}\left[-p_0 + p_d(t) + p_g - \frac{2\sigma}{R_{I0}}\right] - \frac{4\nu \dot{R}_{I0}}{R_{I0}} + \sum_{\substack{J=1\\J\neq I}}^{N} \frac{d}{dt}\left(\frac{q_{J0}}{L_{IJ}}\right) + G_{I0}, \tag{12}$$

$$R_{I0}\dot{v}_{Im} + 3\dot{R}_{I0}v_{Im} + \frac{18\nu}{R_{I0}} v_{Imr} = \sum_{\substack{J=1\\J\neq I}}^{N} \frac{3(1-|m|)!}{(1+|m|)!}\frac{d}{dt}\left[\frac{R_{I0}q_{J0}}{L_{IJ}^2}\overline{Y}_{1m}(\theta_{IJ}, \psi_{IJ})\right]$$

$$+G_{I1m}, \qquad -1 \leqq m \leqq 1, \tag{13}$$

$$R_{I0}\ddot{R}_{Inm} + \left[3\dot{R}_{I0} + \frac{2\nu(n+2)(2n+1)}{R_{I0}}\right]\dot{R}_{Inm}$$

$$+(n-1)R_{Inm}\left[-\ddot{R}_{I0} + \frac{(n+1)(n+2)\sigma}{R_{I0}^2\rho_l} + \frac{2\nu(n+2)\dot{R}_{I0}}{R_{I0}^2}\right]$$

$$= \sum_{\substack{J=1\\J\neq I}}^{N} \left[\frac{(2n+1)(n-|m|)!}{(n+|m|)!}\left\{\frac{d}{dt}\left[\frac{R_{I0}^n q_{J0}}{L_{IJ}^{n+1}}\overline{Y}_{nm}(\theta_{IJ}, \psi_{IJ})\right]\right.\right.$$

$$+\frac{2\nu n(n+2)R_{I0}^{n-2}q_{J0}}{L_{IJ}^{n+1}}\overline{Y}_{nm}(\theta_{IJ},\psi_{IJ})\Big\}\Big] + G_{Inm}, \qquad n \geqq 2, \quad -n \leqq m \leqq n, \tag{14}$$

where

$$q_{J0} = -R_{J0}^2\dot{R}_{J0},$$

$$v_{Imr} = v_{Im} - \frac{(1-|m|)!}{(1+|m|)!}\sum_{\substack{J=1\\J\neq I}}^{N}\frac{q_{J0}}{L_{IJ}^2}\overline{Y}_{1m}(\theta_{IJ},\psi_{IJ}),$$

$$\overline{Y}_{nm}(\theta_{IJ},\psi_{IJ}) = P_n^{|m|}(\cos\theta_{IJ})e^{-im\psi_{IJ}},$$

$(L_{IJ},\ \theta_{IJ}\ ,\ \psi_{IJ})$ are the components of $\overrightarrow{o_I o_J}$ for the polar coordinate I, ν the kinematic viscosity of the liquid, and $v_{Imr}(m = -1, 0, 1)$ the component of the relative translational velocity, in which the flow velocities induced by the other bubbles are estimated at the center of the sphere. Equations (12), (13) and (14) describe the radial motion, the translational motion and the deformation, respectively. In the equations, G_{I0} and G_{Inm} $(n \geqq 1)$ are the nonlinear terms for the effects of the translational motion on each mode. Their complete expressions are not here because of their length, but G_{I0} and G_{I1m} are available in Reference [7]. For $G_{Inm}(n \geqq 2)$, we consider higher order terms than the previous work. We disregard the terms which are smaller than or equal to the square of the quantity for the deformation. These equations are reduced to the equations of motion for a single bubble when $L_{IJ} \to \infty$ [12].

The drag coefficient for the translational motion of Eq.(13) is given by using the Cartesian coordinate as

$$C_{Dk} = \frac{48}{\mathcal{R}_{ek}}, \qquad k = x, y, z,$$

where

$$\mathcal{R}_{ek} = \frac{2R_{I0}|v_{Ikr}|}{\nu}, \qquad k = x, y, z, \qquad\qquad v_{Ixr} = v_{Ix} - \sum_{\substack{J=1\\J\neq I}}^{N}\frac{q_{J0}}{L_{IJ}^2}\sin\theta_{IJ}\cos\psi_{IJ},$$

$$v_{Iyr} = v_{Iy} - \sum_{\substack{J=1\\J\neq I}}^{N}\frac{q_{J0}}{L_{IJ}^2}\sin\theta_{IJ}\sin\psi_{IJ}, \qquad\qquad v_{Izr} = v_{Iz} - \sum_{\substack{J=1\\J\neq I}}^{N}\frac{q_{J0}}{L_{IJ}^2}\cos\theta_{IJ}.$$

Equations (12)–(14) can be solved by determining the internal pressure p_g in Eq.(12). The p_g is obtained from Eqs.(3) and (4). To get internal pressure of bubble I, we also express the velocity potential and temperature of the gas by using series expansion of spherical harmonics as

$$\varphi_g = \varphi_{g0}(t, r_I) + \sum_{n=1}^{\infty}\sum_{m=-n}^{n}\varphi_{gnm}(t, r_I)Y_{nm}(\theta_I, \psi_I), \tag{15}$$

$$T_g = T_{g0}(t, r_I) + \sum_{n=1}^{\infty}\sum_{m=-n}^{n}T_{gnm}(t, r_I)Y_{nm}(\theta_I, \psi_I). \tag{16}$$

Substituting Eqs.(15) and (16) into Eq.(4) and assuming that the λ_g is constant, we obtain

$$\frac{\partial\varphi_{g0}}{\partial r_I} = \frac{1}{\gamma p_g}\Big[(\gamma-1)\lambda_g\Big(\frac{\partial T_{g0}}{\partial r_I}\Big) - \frac{1}{3}r_I\dot{p}_g\Big], \tag{17}$$

$$\varphi_{gnm} - \frac{(\gamma-1)}{\gamma p_g}\lambda_g T_{gnm} = r_I^n g_{nm}(t), \quad n \geqq 1. \tag{18}$$

Using Eq.(7), $\dot{p}_g$ and g_{nm} are obtained from

$$\dot{p}_g = \frac{3}{R_{I0}}\left[(\gamma-1)\lambda_g\left(\frac{\partial T_{g0}}{\partial r_I}\right)_{r_I=R_{I0}} - \gamma p_g \dot{R}_{I0}\right], \tag{19}$$

$$g_{1m} = v_{Im} - \frac{(\gamma-1)\lambda_g}{\gamma p_g}\left(\frac{\partial T_{g1m}}{\partial r_I}\right)_{r_I=R_{I0}}, \tag{20}$$

$$g_{nm} = \frac{1}{nR_{I0}^{n-1}}\left[\dot{R}_{Inm} - R_{Inm}\left(\frac{\partial^2 \varphi_{g0}}{\partial r_I^2}\right)_{r_I=R_{I0}} - \frac{(\gamma-1)\lambda_g}{\gamma p_g}\left(\frac{\partial T_{gnm}}{\partial r_I}\right)_{r_I=R_{I0}}\right],$$
$$n \geqq 2. \tag{21}$$

Substituting Eqs.(15), (16) and (18) into Eq.(3) and using the orthogonality of the spherical harmonics, the governing equations for the gas temperature are

$$\frac{\partial T_{g0}}{\partial t} + \left(\frac{\partial \varphi_{g0}}{\partial r_I}\right)\left(\frac{\partial T_{g0}}{\partial r_I}\right) = \frac{(\gamma-1)}{\gamma}\frac{\dot{p}_g}{p_g}T_{g0} + \frac{(\gamma-1)\lambda_g}{\gamma}\frac{T_{g0}}{p_g}\left[\frac{1}{r_I^2}\frac{\partial}{\partial r_I}\left(r_I^2\frac{\partial T_{g0}}{\partial r_I}\right)\right], \tag{22}$$

$$\frac{\partial T_{g1m}}{\partial t} + \left[\frac{(\gamma-1)\lambda_g}{\gamma p_g}\left(\frac{\partial T_{g1m}}{\partial r_I}\right) + g_{1m} - v_{Im}\right]\left(\frac{\partial T_{g0}}{\partial r_I}\right) + \left(\frac{\partial \varphi_{g0}}{\partial r_I}\right)\left(\frac{\partial T_{g1m}}{\partial r_I}\right)$$
$$= \frac{(\gamma-1)}{\gamma}\frac{\dot{p}_g}{p_g}T_{g1m} + \frac{(\gamma-1)\lambda_g}{\gamma p_g}\left\{T_{g1m}\left[\frac{1}{r_I^2}\frac{\partial}{\partial r_I}\left(r_I^2\frac{\partial T_{g0}}{\partial r_I}\right)\right]\right.$$
$$\left. + T_{g0}\left[\frac{1}{r_I^2}\frac{\partial}{\partial r_I}\left(r_I^2\frac{\partial T_{g1m}}{\partial r_I}\right) - \frac{2}{r_I^2}T_{g1m}\right]\right\}, \tag{23}$$

$$\frac{\partial T_{gnm}}{\partial t} + \left[\frac{(\gamma-1)\lambda_g}{\gamma p_g}\left(\frac{\partial T_{gnm}}{\partial r_I}\right) + nr_I^{n-1}g_{nm}\right]\left(\frac{\partial T_{g0}}{\partial r_I}\right) + \left(\frac{\partial \varphi_{g0}}{\partial r_I}\right)\left(\frac{\partial T_{gnm}}{\partial r_I}\right)$$
$$= \frac{(\gamma-1)}{\gamma}\frac{\dot{p}_g}{p_g}T_{gnm} + \frac{(\gamma-1)\lambda_g}{\gamma p_g}\left\{T_{gnm}\left[\frac{1}{r_I^2}\frac{\partial}{\partial r_I}\left(r_I^2\frac{\partial T_{g0}}{\partial r_I}\right)\right]\right.$$
$$\left. + T_{g0}\left[\frac{1}{r_I^2}\frac{\partial}{\partial r_I}\left(r_I^2\frac{\partial T_{gnm}}{\partial r_I}\right) - \frac{n(n+1)}{r_I^2}T_{gnm}\right]\right\}, \quad n \geqq 2. \tag{24}$$

Since we neglect the the thermal boundary layers in the liquid, the boundary condition for the gas temperature at the bubble wall is simplified as

$$T_{g0}|_{r_I=R_{I0}} = T_0, \tag{25}$$

$$T_{g1m}|_{r_I=R_{I0}} = 0, \tag{26}$$

$$T_{gnm}|_{r_I=R_{I0}} = -R_{Inm}\left(\frac{\partial T_{g0}}{\partial r_I}\right)_{r_I=R_{I0}}, \quad n \geqq 2. \tag{27}$$

Under the same level of approximations as that of Eqs.(12)–(14), the nonlinear terms including T_{g1m}, which denote the effects of the translational motion on each mode like G_{I0} and G_{Inm} ($n \geqq 1$), should be considered in Eqs.(20)–(24), (26) and (27). These terms are however zero because T_{g1m} is always zero under the initial condition of T_{g1m}=0 as found from Eqs.(23) and (26). Consequently, p_g is determined by solving the only radial component of the internal gas. That is, it is not necessary to solve Eq.(24) in determining the bubble motions as long as $(R_{Inm}/R_{I0})^2$ can be neglected.

3. Instability of the bubble surface

Assuming that N bubbles oscillate with infinitesimal amplitude and disregarding their translational motions, we will investigate the linear stability of the bubble surface. For simplicity, we assume that all the initial radii of the bubbles are the same. When the bubbles oscillate with the same frequency as that of the sound, we write

$$R_{I0} = R_{00}(1 + \delta_{I0} \sin \Omega_d t), \quad \delta_{I0} \ll 1, \tag{28}$$

where R_{00} is the initial equilibrium radius of the bubbles. By linearizing Eq.(14) with respect to δ_{I0}, we obtain

$$\ddot{\delta}_{Inm} + 2\beta_n \dot{\delta}_{Inm} + (\Omega_n^2 + \alpha_n \sin \Omega_d t)\delta_{Inm}$$
$$= \sum_{\substack{J=1 \\ J \neq I}}^{N} \frac{(2n+1)(n-|m|)! \overline{Y}_{nm}(\theta_{IJ}, \psi_{IJ}) \varepsilon_{IJ0}^{n+1} \delta_{J0} \Omega_d^2 \sin \Omega_d t}{(n+|m|)!}, \tag{29}$$

where

$$\alpha_n = \left[(n+1/2)\Omega_d^2 - 3\Omega_n^2\right] \delta_{I0}, \quad \beta_n = \nu(n+2)(2n+1)/R_{00}^2,$$

$$\delta_{Inm} = R_{I0}^{3/2} R_{00}^{-5/2} R_{Inm}, \quad \varepsilon_{IJ0} = R_{00}/L_{IJ0}, \quad \Omega_n = \sqrt{(n-1)(n+1)(n+2)\frac{\sigma}{\rho R_{00}^3}},$$

L_{IJ0} is the initial distance between the centers of bubbles I and J, and Ω_n the natural frequencies for the surface oscillation of order n (n=2, 3 $\cdots$). In Eq.(29), we neglect the viscous terms with $\nu \delta_{I0} \dot{\delta}_{Inm}$, $\nu \delta_{I0} \delta_{Inm} \Omega_d$ and $\nu \delta_{J0} \varepsilon_{IJ0}^{n+1} \Omega_d$. Equation (29) is Mathieu's equation with the forced oscillation term due to bubble-bubble interactions. As well known, the parametric excitation takes place in Mathieu's equation when

$$\Omega_d = \frac{2\Omega_n}{k}, \quad k = 1, 2, \cdots. \tag{30}$$

Under the linear approximation, the thermal effect on the radial motion of an isolated bubble is evaluated by using the effective polytropic index and viscosity, which are the function of a frequency, as shown by Prosperetti [13] and Prosperetti et al.[8]. We also apply this idea to the interacting bubbles. Effective polytropic index, κ, is given by

$$\kappa(\Omega_d) = \frac{1}{3} \mathrm{Re}\Phi, \tag{31}$$

where

$$\Phi = 3\gamma / \{1 - 3(\gamma - 1)i\chi[(i/\chi)^{1/2} \coth (i/\chi)^{1/2} - 1]\}, \tag{32}$$

$$\chi = \frac{\lambda_g}{\rho_g(p_{g0}, T_0) C_{pg}} \frac{1}{\Omega_d R_{00}}, \tag{33}$$

p_{g0} is the initial pressure of the noncondensable gas. The polytropic index is evaluated by using the deriving frequency. On the other hand, the effective viscosity, μ_e, is obtained from

$$\mu_e = \mu + \mu_{th}(\Omega_0), \tag{34}$$

where

$$\mu_{th}(\Omega_0) = (p_{g0}/4\Omega_0) \mathrm{Im}\Phi(\Omega_0), \tag{35}$$

$$\Omega_0^2 = \frac{1}{\rho_l R_{00}^2(1+\zeta)}\left[3\kappa p_{g0} - \frac{2\sigma}{R_{00}}\right], \tag{36}$$

Ω_0 is the lowest natural frequency of the bubble system and ζ the constant determined by the configuration of bubbles [7]. When ζ=0, Ω_0 is reduced to the natural frequency of an isolated bubble. In general, there are several natural frequencies for multi-bubbles. When the bubbles with same radii are initially at rest, all bubbles oscillate with the nearly equal phase. Therefore, the lowest natural frequency in the system, which occurs when all bubbles oscillate with the same phase, is used in order to evaluate the effective viscosity. It should be noted that Ω_0 for multi-bubbles is lower than that for an isolated bubble since ζ is positive for multi-bubbles.

Adopting the polytropic relation for the internal gas, considering only the first unstable range for k=1 in Eq.(30), and putting $\Omega_d = \Omega_0 = 2\Omega_n$, the bubble radius with which the parametric excitation for the surface oscillation of order n occurs is

$$R_{pn} = \frac{2\sigma\left[2(1+\zeta)(n-1)(n+1)(n+2) - (3\kappa - 1)\right]}{3\kappa p_0}. \tag{37}$$

4. Numerical results

Although we use the complex number in order to formulate the governing equations of bubbles, from now on we use the following real number in order to express the bubble shape:

$$R_{Inm}^{(c)} = R_{Inm} + R_{In(-m)}, \quad R_{Inm}^{(s)} = i(R_{Inm} - R_{In(-m)}), \quad n \geqq 2, \quad 1 \leqq m \leqq n. \tag{38}$$

$R_{Inm}^{(c)}$ and $R_{Inm}^{(s)}$ denote the deviation from the axisymmetry. In the present study, computations are performed for three typical configurations of bubbles in Fig.2. One is for two bubbles on z axis. Another is for four bubbles located at the vertices of a square. The other is for four bubbles located at the vertices of a regular triangle and the center of the triangle. Every bubbles in each configuration are on $x-z$ plane and have an equal initial radius of R_{00}. The initial distance between the centers of bubbles 1 and 2 is $10R_{00}$. The R_{00} is determined by Eq.(37) for n=2 or 3. The driving frequency is therefore equal to the lowest natural frequency of the bubble system determined by Eq.(36). The ζ in Eq.(36) for (a)–(c) is ε (=R_{00}/L_{120}), $(2+\sqrt{2}/2)\varepsilon$ and $(1+\sqrt{10})\varepsilon/\sqrt{3}$, respectively. $R_{Inm}^{(c)}$ when m is odd and $R_{Inm}^{(s)}$ are always zero because of the arrangement of the bubbles. Nondimensional pressure amplitude p_a/p_0 is taken to be 0.1. Physical properties are taken to be p_0=101.3 kPa, ρ_l=998.2 kg/m^3, γ=1.4, σ=7.275×10^{-2} N/m, ν=1.004×10^{-6} m/s^2 and λ_g = 2.562×10^{-2} W/mK.

Figure 3 shows time histories of R_{10}, b_{1z}, R_{120} and R_{130} components for two bubbles in Fig.2(a) where b_{1z} is the displacement of the center of bubble 1 in the

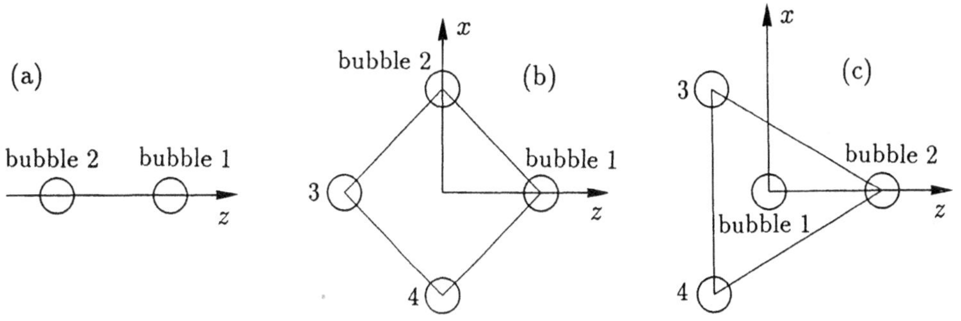

Fig.2 Configuration of bubbles.

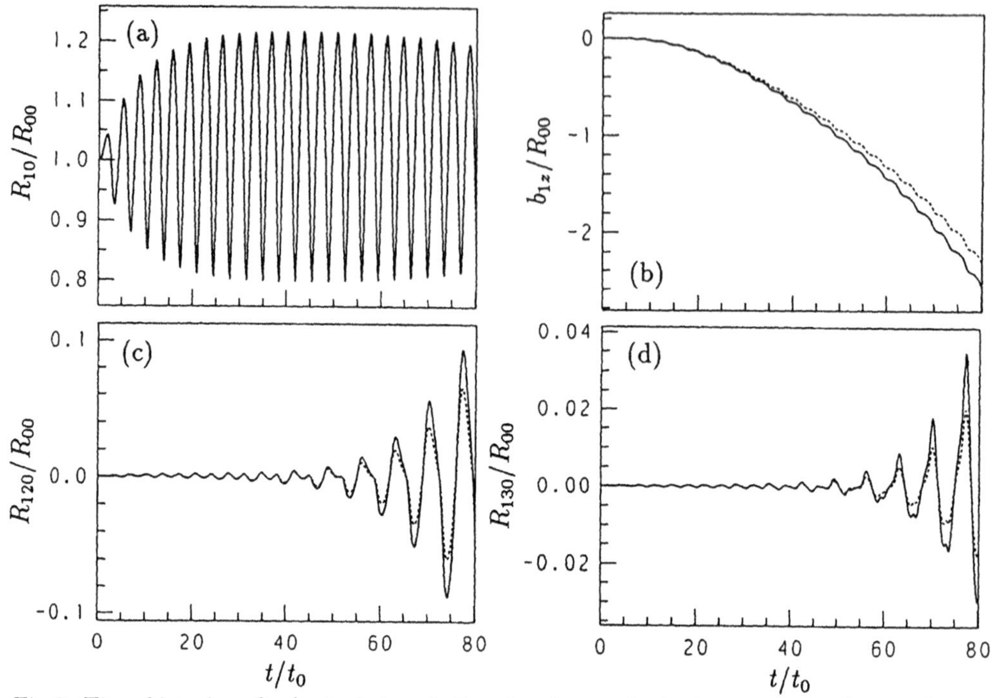

Fig.3 Time histories of spherical, translational and nonspherical components for two bubbles shown in Fig.2(a). The solid line in each figure denotes the present result in which the internal pressure is evaluated directly. The dotted line is the polytropic result. The driving frequency is 272.8 kHz.

z direction. The initial radius is taken to be 10.6 μm ($=R_{p2}$) with which surface oscillation corresponding to n=2 will be excited. The components corresponding to $m \neq 0$ are zero for two bubbles. The solid line denotes the present result in which the internal pressure is evaluated directly. On the other hand, the dotted line is the polytropic result when κ and μ_e are 1.09 and 7.10μ, respectively. t is normalized by $t_0 = R_{00}\sqrt{\rho_l/p_0}$, and the ordinate is normalized by R_{00}. Since both lines in (a) are almost overlapping, it is difficult to distinguish with each other. The polytropic result is however slightly lower than the present result. Therefore the polytropic model slightly underestimates the b_{1z}, R_{120} and R_{130}. It is found that the parametric excitation occur not only in R_{120} as predicted by the linear theory, but also in R_{130}.

Figure 4 shows the results for four bubble oscillations in Fig.2(b). R_{00} is taken to be 12.3 μm ($=R_{p2}$). The lowest natural frequency of the bubble system decreases with the increase of the number of bubbles, and therefore R_{p2} for Fig.2(b) is larger than two bubbles. Because of the symmetric arrangement, only the results for bubble 1 on z axis is shown in Fig.4. It is found that the agreement between the present results and polytropic ones (κ=1.10, μ_e=8.63μ) is rather good. The parametric excitation is clearly found in $R_{122}^{(c)}$, but not in R_{120}. This difference is due to the effect of the translational motion. That is, the G_{I20} in Eq.(14) is not zero when n=2 and m=0, while the G_{I22} is zero when n=2 and m=2 because the bubble moves in the z direction. This suggests that when the interaction is strong, the instability of the bubble surface is not predicted by the linear theory and the inclusion of the translational motion is important. Another interesting feature of this system is that the amplitude of the radial component decreases after about t/t_0=40. This is because

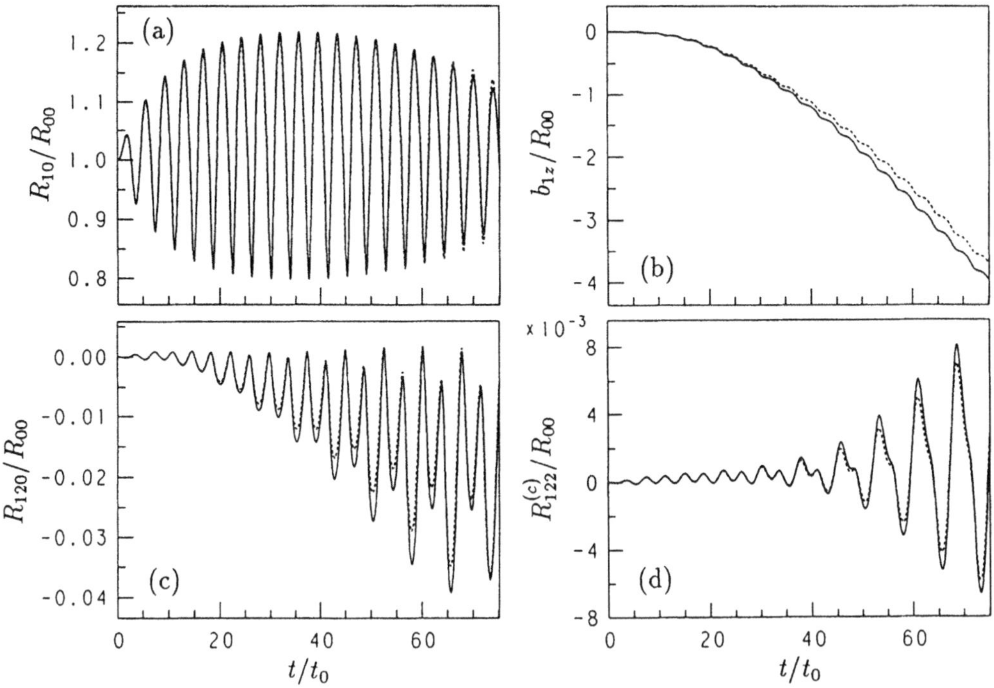

Fig.4 Time histories of spherical, translational and nonspherical components for four bubbles shown in Fig.2(b). The solid line in each figure denotes the present result in which the internal pressure is evaluated directly. The dotted line is the polytropic result. The driving frequency is 217.7 kHz.

that the natural frequency of the system decreases as the distance between the bubbles becomes shorter with time. Thus the resonance frequency of the system departs from the driving frequency which initially coincides with the resonance frequency.

Figure 5 shows that the time histories of R_{I0} and R_{I30} (I=1, 2) for four bubbles in Fig.2(c). In this configuration, three bubbles at the vertices take on a same behavior. R_{00} is taken to be 38.3 μm. This radius is the one with which the surface oscillation corresponding to n=3 will be exited. κ and μ_e are 1.21 and 21.3μ, respectively. Figure (a) and (b) are the results for the central bubble (= bubble 1), while (c) and (d) are those for the bubble on z axis (= bubble 2). It should be noted that the vertical scale in (a) is different from that in (c). Since the center of bubble 1 does not move due to the symmetric arrangement of the bubbles, the parametric excitation is found in R_{130} as predicted by the linear theory. On the other hand, R_{230} for bubble 2 grows with the same period as the spherical mode. This difference is also due to the effect of the translational motion. It is found that the agreement between the present results and the polytropic ones for bubble 2 is better than that for bubble 1 because the amplitude of radial motion of bubble 2 is smaller than bubble 1. The reason that the amplitude of bubble 1 is larger than that of bubble 2 is related to the difference of the phase between bubble 1 and 2 [11]. Since the period of bubble 1 is slightly longer than bubble 2, the final stage of the collapse of bubble 1 is performed after the other bubbles rebound. In such situation, the high pressure field is generated around bubble 1 by the other bubbles. Therefore, bubble 1 collapses more violently and its amplitude becomes larger.

5. Conclusions

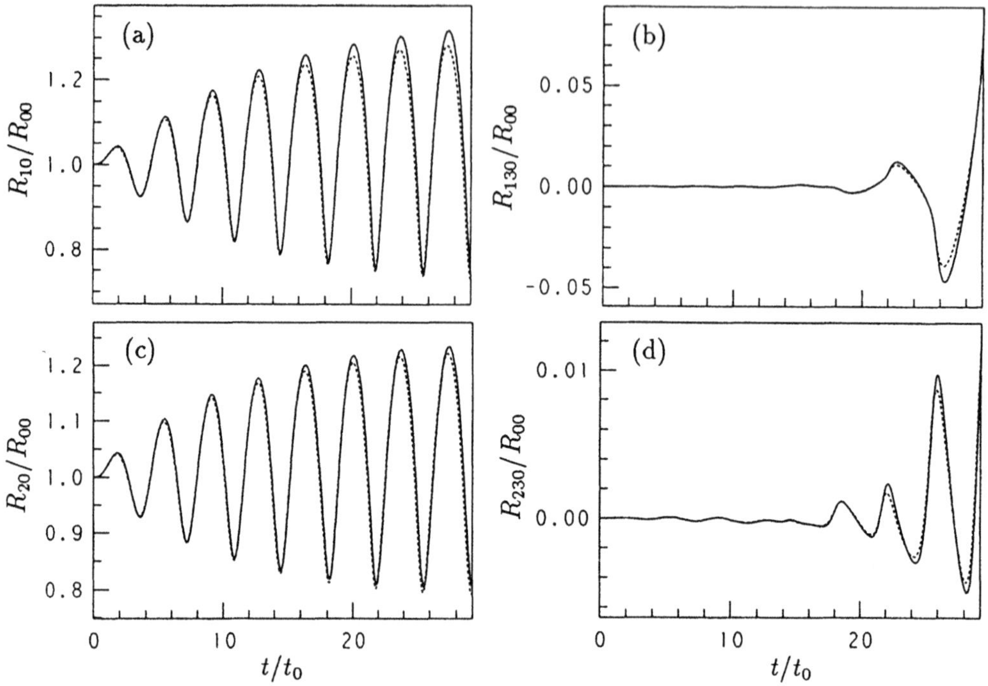

Fig.5 Time histories of spherical and nonspherical components for four bubbles shown in Fig.2(c). (a) and (b) are for bubble 1. (c) and (d) are for bubble 2. The solid line in each figure denotes the present result in which the internal pressure is evaluated directly. Dotted line is the polytropic result. The driving frequency is 72.53 kHz.

Dynamical equations of motion of a bubble cluster are derived by taking account of the temperature gradient of the internal gas. The equations deal with the three-dimensional translational motion and deformation which are induced by the bubble-bubble interactions. Results are summarized as follows:

(1) Surface oscillations depend on the configuration of bubbles. The translational motion affects the surface oscillations as well as the radial oscillations.

(2) Effective polytropic index and viscosity are useful for multi-bubbles system as long as the radial oscillations are small.

References

[1] Lauterborn, W., *Frontiers in Physical Acoustics*, (1986), 124, North-Holland.
[2] Lauterborn, W., *J. Acoust. Soc. Am.*, **59**-2(1976), 283.
[3] Parlitz, U., et al., *J. Acoust. Soc. Am.*, **88**-2(1990), 1061.
[4] Matsumoto, Y. and Watanabe, M., *JSME Int. J.*, Series II, **32**-2(1989), 157.
[5] Omta, R., *J. Acoust. Soc. Am.*, **82**-3(1987), 1018.
[6] d'Agostino, L. and Brennen, C. E., *J. Fluid Mech.*, **199**(1989), 155.
[7] Takahira, H., et al., *JSME Int. J.*, Series B, **37**-2(1994), 297.
[8] Prosperetti, A., et al., *J. Acoust. Soc. Am.*, **83**-2(1988), 502.
[9] Hermans, W. A., *Ph.D Thesis* (Technical University of Eindhoven), (1973).
[10] Prosperetti, A. and Seminara, G., *Phys. Fluids*, **21**-9(1978), 1465.
[11] Takahira, H., et al., *Proc. 3rd. Int. Conf. on Cavitation, I. Mech. E.*, Cambridge, (1992), 9.
[12] Plesset, M. S. and Prosperetti, A., *Ann. Rev. Fluid Mech.*, **9**(1977), 145.
[13] Prosperetti, A., *J. Acoust. Soc. Am.*, **61**-1(1977), 17.

ON WAVES OF THE SELF-INDUCED ACOUSTIC TRANSPARENCY IN MIXTURES OF LIQUID AND VAPOR BUBBLES

Nail A. Gumerov

Institute of Energy Resources Transportation
pr. Oktyabrya 144/3, Ufa, 450055, RUSSIA

Abstract - A theory predicting spatial and temporal changing of the average parameters of mixtures consisting of liquid and vapor bubbles under sound waves is developed. Equations describing a self-action of sound waves in such mixtures are derived. An integrability of these essentially non-linear equations is shown both for mono- and polydisperse mixtures. The waves of the self-induced transparency/non-transparency are simulated numerically. Some physical effects are found and discussed.

1. INTRODUCTION

Linear acoustic waves in bubbly liquids have been studied by numerous authors. These studies show that even small amounts of bubbles heavily influence the dispersion and dissipation of sound.

In the linear or first-order approximation bubble sizes oscillate in acoustic fields near their constant initial values. But even if the amplitude of sound is small, the non-linear second-order effects manifest themselves at large times equal to many periods of oscillations. These effects can lead to the acoustic levitation, coagulation, fragmentation and changing of the average bubble sizes due to the rectified heat and mass transfer. The changes of the main parameters of the mixture as average bubble sizes, bubble size distribution function and volume content of the bubbles, effect the amplitude of waves. We can call these effects a 'self-action' of sound and 'self-induced' parameters of the medium.

The present work focusses only on the sound self-action caused by the second-order effects of the non-linear heat and mass transfer. Moreover, the first-order effects connected with the instability of the vapor bubbles due to the surface tension effects and a non-equilibrium state of the undisturbed system are ignored.

It is remarkable that essentially non-linear system of equations describing the sound self-action is integrable in this case and a wave character of the propagation of self-induced parameters of the medium takes place.

S. Morioka and L. van Wijngaarden (eds.),
IUTAM Symposium on Waves in Liquid/Gas and Liquid/Vapour Two-Phase Systems, 77-86.

2. SINGLE BUBBLES IN ACOUSTIC FIELDS

The behaviour of a spherical vapor bubble in the limitless incompressible liquid can be described by the following equations (for example, see Nigmatulin, 1987)

$$\left.\begin{aligned}
&p_g(t) = R_g T_g(r,t)\rho_g^0(r,t), \quad \rho_l^0 = const \\
&\rho_l^0 c_l d_l T_l = \lambda_l \Delta T_l, \quad \rho_g^0 c_g d_g T_g = \lambda_g \Delta T_g - d_t p_g, \quad T_g\big|_{r=a} = T_l\big|_{r=a} = T_a, \quad T_l\big|_{r=\infty} = T_* \\
&w_l = w_{la} a^2 r^{-2}, \quad w_g = k_\gamma p_g^{-1} \lambda_g \partial_r T_g - \tfrac{1}{3} r(\gamma p_g)^{-1} d_t p_g, \quad k_\gamma = 1-\gamma^{-1} \\
&\rho_l^0[d_t w_{la} + (2d_t a - \tfrac{1}{2} w_{la}) w_{la}] = p_g - p_{l\infty} - 2\sigma a^{-1}, \quad w_{ga} = -k_\gamma p_g^{-1} q_{ga} - \tfrac{1}{3} a(\gamma p_g)^{-1} d_t p_g \\
&\rho_l^0(w_{la} - d_t a) = \rho_g^0(w_{ga} - d_t a) = -\xi, \quad q_{ga} - q_{la} = \xi l, \quad q_{ga} = -\lambda_g \partial_r T_g\big|_{r=a}, \quad q_{la} = -\lambda_l \partial_r T_g\big|_{r=a} \\
&\xi = \beta[p_S(T_a) - p_g](2\pi R_g T_a)^{-1/2}, \quad p_S(T) = p_* \exp[k_S^{-1}(1 - T_*/T)], \quad p_* = p_S(T_*), \quad k_S = R_g T_* l^{-1} \\
&d_g = \partial_t + w_g \partial_r, \quad d_l = \partial_t + w_l \partial_r, \quad \Delta = r^{-2}\partial_r(r^2 \partial_r), \quad d_t = d/dt, \quad \partial_t = \partial/\partial t, \quad \partial_r = \partial/\partial r
\end{aligned}\right\} \qquad (2.1)$$

Here r is the radial coordinate; a is the bubble radius; p, T, ρ^0, w, q and ξ are the pressure, temperature, density, radial velocity, heat flux on the interface and the rate of condensation per unit of the interface; γ, R_g, c and λ are the adiabatic exponent of the gas, gas constant, specific heat (at constant pressure) and the thermal conductivity; l and β are the latent heat of evaporation and the accommodation (condensation) coefficient; and σ is the coefficient of surface tension. The symbols k_γ and k_S denote the dimensionless angle coefficients to the adiabatic and saturation curves at $T = T_*$. The subscripts g and l refer to parameters of gas and liquid, while the subscripts a and S denote the values of parameters on the interface and on the saturation curve.

To use this closed system of basic equations we should assume that effects connected with the viscosity of liquid are negligible and the lengths of acoustic waves in gas much greater than the bubble size.

Equations (2.1) show that the equilibrium state of vapor bubble is not stable. The characteristic times of the instability development can be estimated as (Akulichev et al., 1986; Gumerov, 1991)

$$t_{instab} = k_S^{-2} k_\gamma k_\lambda^{-1} \kappa_g^{-1} p_* a_0 (\Delta p + 2\sigma a_0^{-1})^{-1}, \quad k_\lambda = \lambda_l/\lambda_g, \quad \kappa_g = \lambda_g/(\rho_{gS}^0 c_g), \quad \Delta p = p_{l\infty 0} - p_*$$

where the subscript 0 denotes the undisturbed state or quantities averaged over the period of oscillations everywhere below.

Assume that the characteristic times of the processes considered in the present work are much smaller than t_{instab}, bubbles are sufficiently large and the pressure at the infinity oscillates near the equilibrium value: $p_{l\infty} = p_*[1 + \varepsilon \operatorname{Re}\{\exp(-i\omega t)\}], \quad \varepsilon << 1$. This case

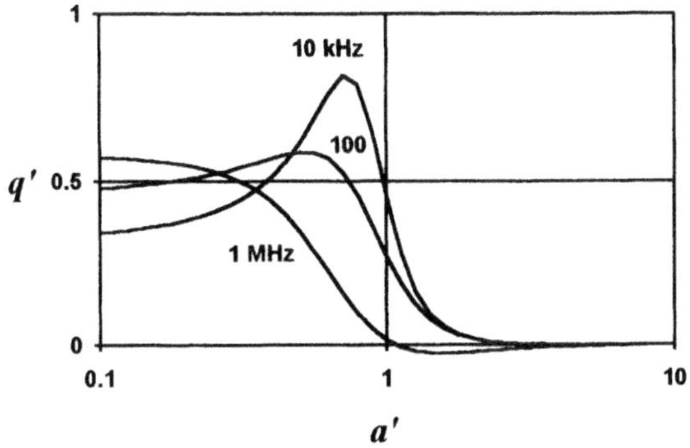

Fig.1. Dimensionless dependencies of the bubble growth rate $q'=(\rho_*/p_*)^{1/2}q$ on the average bubble size $a'=a_0/a_m$ at different frequencies of sound (water, $p_*=1$ bar, $\beta=0.04$).

was analyzed in the previous work (Gumerov, 1991), where the equation describing the dynamics of the average bubble radius was found using the multiscale technique:

$$\partial_{t_2}a_0 = q(a_0), \quad t_2 = \varepsilon^2 t \tag{2.2}$$

Here the function q depends on the frequency ω and the thermodynamic properties of the vapor and liquid. At sufficiently high frequencies or bubble sizes ($\omega a_0^2 >> \kappa_g$), when a thickness of the thermal boundary layers in gas and liquid is much less than the bubble size, the asymptotic expressions for $q(a_0)$ can be written in the form

$$\left.\begin{array}{c}
q(a_0) = \dfrac{\kappa_g k_S^2 k_\lambda (a_{st}-a_0)}{12a_m^2(\gamma-1)k_\gamma^2|D|^2}, \quad a_{st} = \dfrac{3k_\gamma(\gamma-1)}{k_S k_\lambda} a_\kappa H(\omega_\beta^{1/2}) \\
D = 1-\left(\dfrac{a_0}{a_m}\right)^2 - 3(1+i)(\gamma-1)\dfrac{a_0}{a_\kappa} I(\omega_\beta^{1/2}), \quad a_m = \dfrac{1}{\omega}\left(\dfrac{3\gamma p_*}{\rho_l^0}\right)^{1/2}, \quad a_\kappa = a_m^2\left(\dfrac{2\omega}{\kappa_g}\right)^{1/2} \\
H(\varsigma) = \dfrac{b_2\varsigma^2+2b_1\varsigma+2b_0}{K^2\varsigma^2+2K\varsigma+2}, \quad I(\varsigma) = \dfrac{(1-i)c_1\varsigma+2c_0}{(1-i)K\varsigma+2}, \quad \omega_\beta = \dfrac{4\pi k_S \kappa_g \omega}{\beta^2 l} \\
c_0 = 1+k_\gamma^S(K-2), \quad c_1 = K-k_\gamma^S, \quad K = k_\gamma^S(1+k_\lambda k_\kappa^{-1/2}), \quad k_\gamma^S = k_S/k_\gamma, \quad k_\kappa = \lambda_l/(\rho_l^0 c_l \kappa_g) \\
b_0 = k_\gamma^S[1-k_\gamma^S(1-k_c)]K + (1-k_\gamma^S)^2 - (k_\gamma^S)^2 k_c, \quad k_c = c_l/c_g \\
b_1 = k_\gamma^S K^2 + [(1-k_\gamma^S)^2 - (k_\gamma^S)^2(2-k_c)]K + (k_\gamma^S)^2(1-k_c), \quad b_2 = c_1 K
\end{array}\right\} \tag{2.3}$$

Here a_m is the Minnaert resonance radius and a_{st} is the stationary average radius.

In fig.1 the function $q(a_0)$ is shown. In fig.2 frequency dependencies of the stationary radius (2.3) are illustrated. This figure shows that the value of the stationary radius heavily depends on the value of the accommodation coefficient.

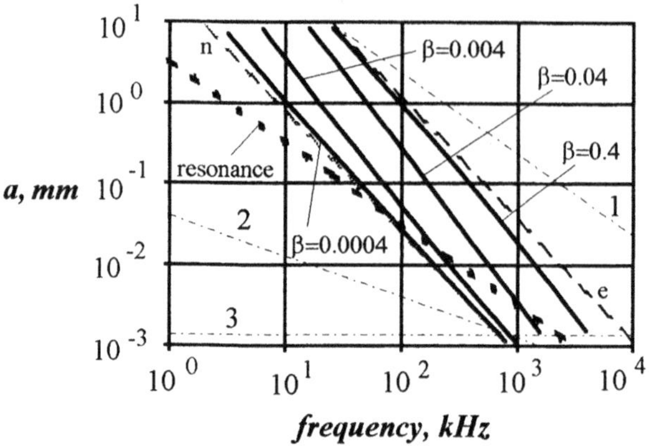

Fig.2. The stationary radius of a water vapor bubble ($p_* = 1$ bar) at various values of the accommodation coefficient (solid lines). Dashed lines marked 'e' and 'n' correspond to limiting cases of equilibrium phase transitions and $\beta \to 0$. The thick dotted line shows Minnaert resonance radius. The thin dash-dotted lines indicate the theory limits: in the area above line 1 the condition of spatial homogeneity of the pressure in the bubble is not satisfied; in the area below line 3 some additional account of surface tension effects is needed; in the area below line 2 a thickness of the thermal boundary layer in the bubble is comparable with bubble size and formulae (2.3) are not justified.

This effect can be used for non-direct measurements of the accommodation coefficient.

3. EQUATIONS OF MOTION OF BUBBLY LIQUID

Let us consider motions of a polydisperse liquid-bubble mixture. We accept the following basic assumptions: 1) the volume fraction of bubbles α_g is small ($\alpha_g << 1$); 2) the mass of bubbles are negligible if compared with the liquid mass in the mixture; 3) the carrier phase is slightly compressible and the acoustical relationship between the liquid pressure and true density holds for a macroscopic motion; 4) the one-velocity scheme of mixture motion can be accepted; 5) bubbles are spheres; 6) the initial parameters of the mixture are spatially homogeneous; 7) initially in each elementary volume of the mixture, bubbles are distributed by their radii $a = \zeta$ on a finite segment $\Delta = [a_-, a_+]$ with the number density $N_*(\zeta)$ (the quantity $dn_*(\zeta) = N_*(\zeta)d\zeta$ can be treated as a number of bubbles having radii from ζ to $\zeta + d\zeta$ per unit of mixture volume). Note that the descriptions of polydisperse mixtures can be found for example in the works of Ishii & Matsuhisa (1983), and Gumerov (1992). Thus, equations of one-dimensional motion of the mixture can be written in the following form

$$\left.\begin{array}{c} d_t\rho + \rho\partial_x v = 0, \quad \rho d_t v + \partial_x p = 0, \quad d_t N + N\partial_x v = 0, \quad d_t = \partial_t + v\partial_x \\ \mathcal{L}(d_t, a)a = p_* - p, \quad p - p_* = C_l^2(\rho_l^0 - \rho_{l*}^0), \quad \rho = (1-\alpha_g)\rho_l^0 \\ \alpha_g(x,t) = \int\limits_\Delta \tfrac{4}{3}\pi a^3(x,t;\zeta)N(x,t;\zeta)d\zeta, \quad a(x,0;\zeta) = \zeta, \quad N(x,0;\zeta) = N_*(\zeta) \end{array}\right\} \tag{3.1}$$

Here ρ and v are the density and velocity of the mixture, N is the current number density of the bubble size distribution, C_l is the speed of sound in pure liquid and $\mathcal{L}$ denotes a complex integro-differential operator connecting the bubble radius a and the pressure in the mixture p. To find the expression for this operator one should solve the problem (2.1) for a single bubble at $p_{l\infty} = p$.

For further investigations it is convenient to rewrite equations (3.1) in the following dimensionless form ($\alpha_g << 1$).

$$\left.\begin{aligned} &\partial^2_{x'}p' - \partial^2_{t'}\rho' = 0, \quad \rho' = 1 - \langle (a')^3 \rangle + b^2 p', \quad \mathcal{L}'(\partial_{t'}, a')a' = -p' \\ &a' = \frac{a}{a_*}, \quad \zeta = \frac{\zeta}{a_*} \quad p' = \frac{p}{p_*} - 1, \quad \rho' = \frac{\rho - \rho_*}{\rho_* \alpha_{g*}}, \quad t' = \frac{t p_*^{1/2}}{a_* \rho_*^{1/2}}, \quad x' = \frac{x \alpha_{g*}^{1/2}}{a_*} \\ &\omega' = \frac{\omega a_* \rho_*^{1/2}}{p_*^{1/2}}, \quad b^2 = \frac{p_*}{\rho_* \alpha_{g*} C_l^2}, \quad \langle G \rangle = \int_{\Delta'} N' G d\zeta, \quad N'(\zeta) = \frac{4\pi a_*^4 N_*(\zeta)}{3\alpha_{g*}} \end{aligned}\right\} \quad (3.2)$$

Here a_* is some representative bubble radius, ρ_* and α_{g*} are the initial density of the mixture and the volume fraction of bubbles.

4. EQUATIONS OF SELF-ACTION OF ACOUSTIC WAVES

At small amplitudes of perturbations ε solutions of equations (3.2) have the form

$$p' = \varepsilon p_1 + \varepsilon^2 p_2 + \ldots, \quad \rho' = \rho_0'(1 + \varepsilon \rho_1 + \varepsilon^2 \rho_2 + \ldots), \quad a' = a_0'(1 + \varepsilon a_1 + \varepsilon^2 a_2 + \ldots) \qquad (4.1)$$

To use the multiscale technique we introduce the set of times $\{t'_m\}, t'_m = \varepsilon^m t', m = 0, 1, \ldots$, and consider functions of t' as the functions of this set. Thus, we have the following expansion of the temporal derivative: $\partial_{t'} = \partial_{t'_0} + \varepsilon \partial_{t'_1} + \varepsilon^2 \partial_{t'_2} + \ldots$

Using the properties of the operator $\mathcal{L}$ (see (2.1)) we can show that $\rho_0' = \rho_0'(x', t'_2, \ldots)$ and $a_0' = a_0'(x', t'_2, \ldots)$. For the monochromatic main mode of perturbations:

$$p_1 = \mathrm{Re}\{A(x', t'_2, \ldots)e^{-i\omega' t'_0}\}, \rho_1 = \mathrm{Re}\{A_p(x', t'_2, \ldots)e^{-i\omega' t'_0}\}, a_1 = \mathrm{Re}\{A_a(x', t'_2, \ldots)e^{-i\omega' t'_0}\} \quad (4.2)$$

we can obtain the following closed system from (2.1), (3.1)

$$\partial^2_{x'}A + k^2 A = 0, \quad \partial_{t'_2} a_0' = q'(a_0')|A|^2, \quad k^2 = (\omega')^2 (b^2 + 3\langle (a_0')^2 / \mathcal{L}'(-i\omega', a_0') \rangle) \quad (4.3)$$

Here $q' = (\rho_* / p_*)^{1/2} q$ (see the function from (2.2)). In the linear theory ($a_0' = const$) the quantity k is the complex wavenumber. Because the average bubble radius in the present theory is changeable, this quantity will be the function of the coordinate and slow time. For high-frequency sound waves (the same assumptions as for equations (2.3)) we have

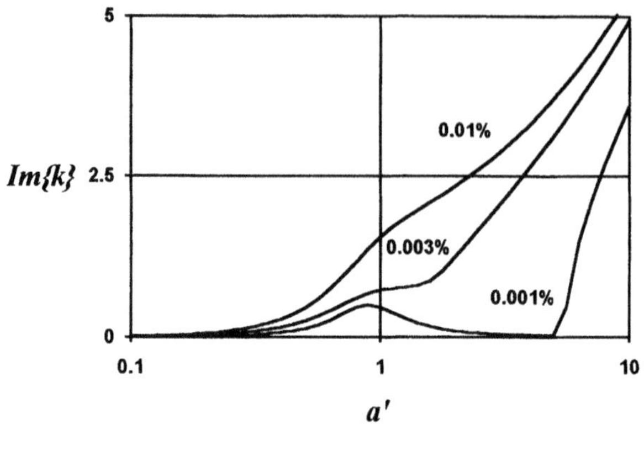

Fig.3. Dimensionless dependencies of the sound attenuation coefficient Im $\{k\}$ on the average bubble size $a' = a_0/a_m$ at different volume fractions of bubbles in the mixture (water, $p_\infty = 1$ bar, $\beta = 0.04$, sound frequency = 100 kHz).

$$k^2 = (\omega')^2 \left\{ b^2 + \left\langle \frac{(a_0')^3}{\gamma D} \left[1 + 3(1+i)(\gamma - 1) \frac{(a_m')^2}{a_0' a_\kappa'} I(\omega_\beta^{1/2}) \right] \right\rangle \right\}, \quad a_m' = \frac{a_m}{a_*}, \quad a_\kappa' = \frac{a_\kappa}{a_*} \quad (4.4)$$

In fig.3 typical dependencies $\mathrm{Im}\{k(a_0)\}$ for monodisperse mixtures are shown.

The system of equations (4.3) should be provided by initial and boundary conditions. The following conditions are most typical for semi-space

$$|A|\big|_{x'=0} = B(t'_2), \quad |A|\big|_{x'=\infty} = 0, \quad a_0'\big|_{t'_2=0} = \varsigma \quad (4.5)$$

The first of them corresponds to the case of a sound vibrator located in the origin of coordinates, whose amplitude is an arbitrary function of the slow time. Note that in spite of our consideration of small-amplitude sound waves, the system (4.3) describing self action of the waves is essentially non-linear.

5. ON INTEGRABILITY OF EQUATIONS OF SELF-ACTION

It is highly remarkable that we can get an exact analytical solution of the system (4.3) with boundary conditions (4.5). The second equation (4.3) shows that

$$F(a_0', \varsigma) = \int_{\varsigma}^{a_0'} \frac{dz}{q'(z)} = F(a_0'(x', t'_2, \varsigma), \varsigma) = f(x', t'_2) \quad (5.1)$$

From the expression (5.1) we can see that $a_0'(x', t'_2, \varsigma) = R(f(x', t'_2), \varsigma)$. Consequently, integrating this relationship over the initial radius's distribution (see the third equation (4.3)) we can obtain the function $k = k(f)$ and rewrite equations (4.3):

$$\partial^2_{x'} A + k^2(f)A = 0, \quad \partial_{t'_2} f = |A|^2 \quad (f(\infty, t'_2) = f(x', 0) = 0) \tag{5.2}$$

Searching a self-similar wave-type solution in the form

$$A = h(t'_2)P(\eta), \quad f = f(\eta), \quad \eta = x' - g(t'_2) \tag{5.3}$$

we can obtain

$$d^2_\eta P + k^2(f)P = 0, \quad d_\eta f = -|P|^2, \quad P(\infty) = f(\infty) = 0 \tag{5.4}$$

$$d_{t'_2} g = |h|^2, \quad h(t'_2)P(-g(t'_2)) = B(t'_2), \quad g(0) = -\infty \tag{5.5}$$

To solve equations (5.4) we represent $P^2 = u(f)\exp(i\varphi)$ and $k = k_R + ik_I$, where both k_R and k_I are non-negative due to waves propagate to the right. Separating real and imaginary parts of (5.4), we can get one integral of the system (5.4) and find

$$2u^3 d^2_f u + (ud_f u)^2 + 4[k^2_R(f) - k^2_I(f)]u^2 = 16\left(\int_0^f k_R(s)k_I(s)ds\right)^2, \quad u|_{f=0} = 0, \quad d_f u|_{f=0} = 2k_I(0) \tag{5.6}$$

This is closed Cauchy problem to determine the function $u(f)$. Note that the easiest way to obtain the dependence $f(x', t'_2)$ is probably to integrate the following equation directly

$$\partial_{x'} f = -u(f), \quad f|_{x'=0} = \int_0^{t'_2} B^2(\tau)d\tau \tag{5.7}$$

The found solution shows that the process of self-action of the sound in the medium has a wavy character because f is a constant on the characteristics $x' - g(t'_2) = const$.

Note that the actual number density of bubble size distribution $N'' = N''(a_0', x', t'_2)$ generally differs from the analogous function $N'(\varsigma)$. From the equation of bubble dynamics and the equation of conservation of bubble number we have

$$N''(a_0', x', t'_2) = \frac{N'}{\partial_\varsigma a_0'} = N'(\varsigma)\frac{q'(\varsigma)}{q'(a_0')} = M''(a_0', \varsigma) = M''(a_0', Z(a_0', f)) = L''(a_0', f)$$

Thus, at fixed f the actual number density depends only on a_0', and this distribution is a constant along the characteristics $x' - g(t'_2) = const$.

6. RESULTS OF NUMERICAL SIMULATIONS

To illustrate peculiarities of the sound self-action effects numerical simulations using the self-similar solution (5.6), (5.7) (B=1) were carried out for water-vapor mixtures at

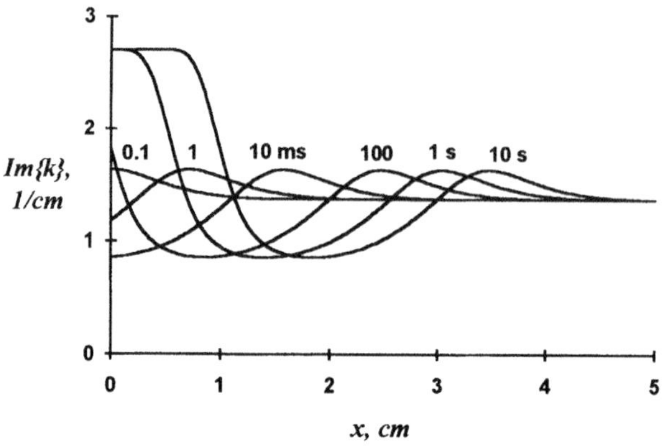

Fig.4. A propagation of the self-induced transparency wave generated by sound of frequency 300 kHz and amplitude 0.1 bar in the monodisperse water-vapor mixture at $p_* = 1$ bar, $\beta = 0.04$ (dimension variables). The initial void fraction is 0.006% and bubble radius is 7.5 μm.

$p_* = 1$ bar, $\beta = 0.04$. The initial (or characteristic) bubble radius, void fraction, bubble size distribution function and the frequency of sound were varied in the simulations.

Simulations showed various qualitative wave pictures. Even for monodisperse mixtures at fixed sound frequency qualitative spatial and temporal distributions of the self-induced parameters heavily depend on the void fraction and relative locations of the initial, resonance and stationary radii.

For example, it was discovered that spatial distributions of the imaginary part of the complex wavenumber characterizing the acoustic transparency of a medium can be non-monotonous (see fig.4). In the illustrated case, initial bubble size was less than the resonance size. The void fraction was small enough, and the function $\mathrm{Im}\{k(a_0)\}$ was as non-monotonous function shown in fig.3. Thus, initially increasing of the bubble size lead in this case to increasing of the attenuation coefficient, maximum of the attenuation at the resonance radius, decreasing after the resonance radius, and, finally, increasing to its stationary value corresponding to the stationary bubble size.

The stationary value of the attenuation coefficient can be smaller than its initial value. In this case we can say about a 'self-induced transparency'. The wave structure of this type is demonstrated in fig.5 (the initial radius is 8 μm). Also this figure illustrates a heavy influence of the initial bubble size (or void fraction) on the velocity and the structure of a self-induced transparency (or non-transparency) wave.

Numerical simulations of sound propagation in polydisperse mixtures discovered strong effects of self-organization processes -- generally, polydisperse mixtures transform to monodisperse mixtures under sound action. Note that in some variants of calculations the width of bubble size distribution spectrum initially increased and then decreased until zero.

In fig.6 the dynamics of the current bubble size spectrum near a source of sound is demonstrated. According to self-similar solutions these size distributions propagate along corresponding characteristics. In fig.7 a propagation of the spectrum width wave is shown. Thus, in the mixture with the initial parameters indicated in figures 6 and 7, the

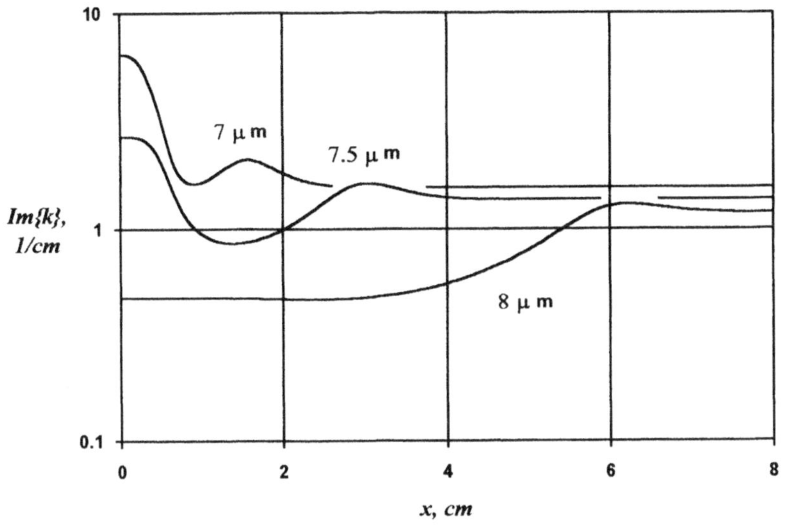

Fig.5. A comparison of three spatial distributions of the attenuation coefficient in monodisperse water-vapor mixtures ($p_* = 1$ bar, $\beta = 0.04$) induced after 1 second of sound action (the sound frequency is 300 kHz and the sound amplitude is 0.1 bar). The initial void fraction of mixtures is 0.006%, but initial bubble radii are different (indicated in figure).

monodisperse mixture with bubble radius 33 μm (the stationary radius) appears in 5 cm zone near the acoustic vibrator after 100 ms sound action (it appears from the mixture with the uniform bubble size distribution density with minimal radius 4.5 μm and maximal radius 22 μm).

7. CONCLUSIONS

The present work can be considered as the first step to understanding the complex non-linear effects of self-organization processes in bubbly liquids under acoustic fields. For liquids with vapor bubbles the surface tension effects lead to the thresholds of the acoustic cavitation (see, for example, Akulichev et al., 1986) and finite zones of the mixture stabilization near a source of sound. For liquids with gas bubbles, the same effects of sound self-action due to the effect of rectified diffusion can be observed. Also the theory can be developed if some additional effects connected with the phase slip will be taken into account.

Nevertheless, some discovered principal qualitative peculiarities of sound action on mixture parameters can take place in more complex models of the process, because effects connected with the resonance and stationary radii of the bubble in an acoustic field should manifest themselves anyway.

A general self-similar solution of equations describing a process of the wave self-action can be applied in various fields, for example in non-linear optics. An analogy with non-linear optics also allow us to expect effects of self-focusing of the acoustic waves in the case of two or three spatial dimensions.

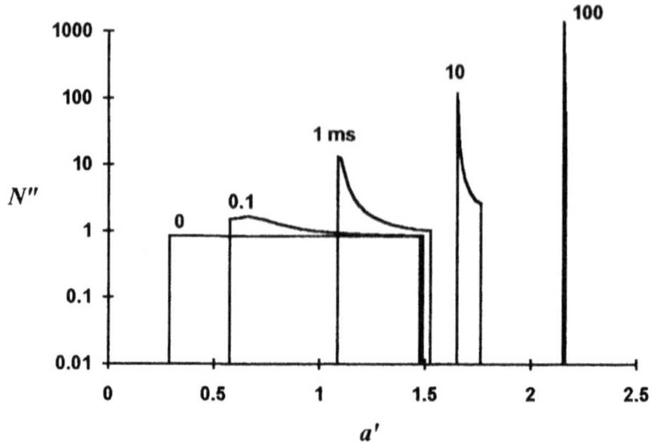

Fig.6. The dynamics of the current number density N'' near a source of sound (water, $p_* = 1$ bar, $\beta = 0.04$, the sound frequency is 300 kHz and the sound amplitude is 0.1 bar). Representative bubble size $a_* = 15\,\mu$m.

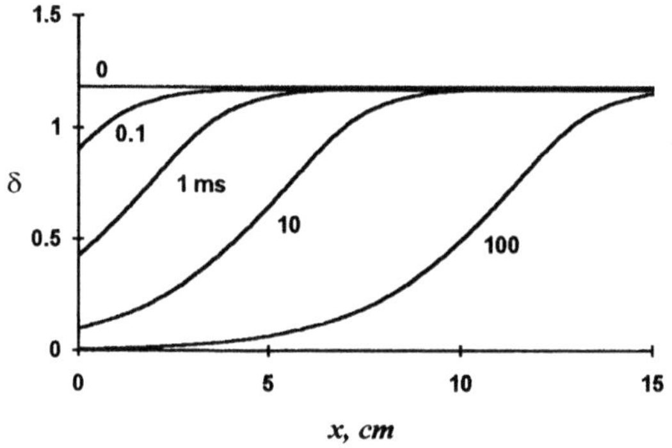

Fig.7. The spatial distributions of the width of the bubble size spectrum $\delta = (a_{max} - a_{min}) / a_*$ at moments of time indicated in the chart. Parameters of the mixture and sound the same as in fig.6. The initial void fraction $\alpha_* = 0.002\%$.

REFERENCES

Akulichev, V.A., Alekseev, V.I. & Bulanov, V.A. 1986 *Periodical Phase Transformations in Liquids.* Moscow: Nauka.

Gumerov, N.A. 1991 On weakly non-linear oscillations of a vapour bubble radius in an acoustic field. *Prikladnaya Matematika i Mekhanika,* **55**, **2**, 256-263.

Gumerov, N.A. 1992 On the propagation of finite-amplitude long waves in polydisperse mixtures of liquid and gas bubbles. *Zh. Prikladnoy Mekhaniki i Tekhnicheskoy Fiziki*, **1**, 88-95.

Ishii, R. & Matsuhisa, H. 1983 Steady reflection, absorption and transmission of small disturbances by a screen of dusty gas. *J. Fluid Mech.*, **130**, 259-277.

Nigmatulin, R.I. 1987 *Dynamics of Multiphase Media.* Moscow: Nauka.

Pressure Waves in Bubbly Liquids

Alfred E. Beylich
Stosswellenlabor, Technische Hochschule,
D–52056 Aachen, Germany

1. Introduction

Among the many varieties of phase boundary structure and shapes that may appear in gas–liquid mixtures the very unique type of a bubbly liquid, i.e. the combination of gas bubbles as the disperse phase and a liquid as the matrix, plays a special role in nature and practical applications and, consequently, has attracted researchers since many years. Even if such a bubbly liquid may exist only for volume (void) fractions of less than a few percent, the great interest into these mixtures seems to stem from their peculiar properties: This material has new properties nonexistent in any of its pure phases due to its ability to store potential energy in the gas phase of the bubbles (compressibility) and to transfer it into kinetic energy of the liquid phase (large mass). Especially dynamic phenomena, such as the ability to transport energy and information by waves and particularly their nonlinearity, due to the interaction of the spherical (three–dimensional) bubble volume changes with the translational motion of the liquid, have become a field of modern research.

The propagation of linear waves, as a limiting case, was studied by Foldy[1] using statistical averaging and a multiple scattering approach, and in the following by Carstensen and Foldy[2]. Silberman[3] and Devin[4] took into account damping effects. The speed of shock waves was derived and also experimentally studied by Campbell and Pitcher[5]. Theoretical work on the structure of shock waves was done by Crespo[6], by Nigmatulin and Shagapov[7], and by Noordzij[8]; typical for this work are two–temperature two–velocity models with non–memory type relations for the interphase momentum and energy transport.

An important theoretical study on the propagation of waves in bubbly liquids was made by van Wijngaarden[9] who derived, on the basis of a physical one–bubble model, equations of the Boussinesq and Korteweg–de Vries–type. This work found a considerable theoretical and experimental extension by the group of the Institute of Thermophysics, Novosibirsk[10–13]; they derived Burgers–Kortweg–de Vries (BKV) and Navier–Stokes–Boussinesq (NSB) type equations and found experimentally soliton structures: attempts to compare these results quantitatively with calculation were not satisfactory[14].

Considerable work with respect to the refinement of the models for bubbly liquids has been done, over many years, at Twente[15,16]; this concerns especially dissipation phenomena and bubble interaction. The very complex problem of heat transfer has been, for some time, studied by different groups[17–21]. Concerning the wave structure, endeavers to reconcile theory and experiment have been performed[13,22].

S. Morioka and L. van Wijngaarden (eds.),
IUTAM Symposium on Waves in Liquid/Gas and Liquid/Vapour Two-Phase Systems, 87-106.

The present paper will deal essentially with experiments, and only some attempts to correlate wave structures with basic sketches of models will be made. One of the simplest models consists of a set of conservation equations for the mixture and a dynamic "equation of state" for a relation between the pressures of the phases derived from a radius equation for a single (isothermal) bubble. Indeed, this model is able to describe qualitatively the fundamental structures of waves such as the evolution of positive/negative pressure steps and wave packets, respectively. For a quantitative description, this model has to be completed by several refinements, some of which are complicated. One of these most important secondary effects is bubble interaction, but also the inclusion of viscosity dissipation and heat transfer play an important role in an improved description, while liquid compressibility and the motion of the bubbles relative to the liquid are generally less important. Attempts to include these effects will be discussed.

One way to study bubbly liquids experimentally is by means of a shock tube. Therefore the tube as well as the diagnostic techniques used will be described in detail. This refers particularly to the problems of bubble production, of void fraction determination, of the measurement of an "average" pressure, and to the question of the attainability of a steady state in a two-phase shock tube.

Samples of characteristic wave patterns for step waves and wave packets will be presented. They have been studied in mixtures of non-condensable gases (He, N_2, SF_6) and (85%) glycerine with void fractions of $\varphi_0 = 0.1 - 2\%$ and bubble sizes of $R_0 = 0.115cm$ to $0.16cm$. One of the goals of these basic experiments was to test quantitatively a model including the effects of bubble interaction, heat transfer, and viscosity dissipation. For that reason, small amplitude shock waves with $\Delta p \approx 0.5bar$ were produced and averaged by superposition. Regularly shaped waves with a damped oscillation and a relaxation zone were found in cases where the distribution of bubbles was carefully adjusted, however, in order to make the numerical results fit the experiments, rather high values of the dissipation parameter have to be used.

When the pressure amplitude grows, bubble interaction becomes more and more important, even for small void fractions, and the regular oscillatory structure found for small amplitude waves is gradually replaced by a random structure of oscillation: Apparently, just as in the case to be expected at larger void fractions, new phenomena dominated by group effects replace the single bubble oscillation. We have performed some experiments in SF_6 - glycerine mixtures with larger pressure steps ($\Delta p = 1 - 10bar$) ; probing into this direction, besides the question of how to interprete single random events, we are faced with the problem of not having an adequate theoretical model.

Finally, a special case will be discussed where a precursor wave separates from the shock wave: When in a system, consisting of bubbles with a combustible mixture in a glycerine matrix, the initial amplitude of the shock wave grows beyond a certain threshold, a "detonation" wave with a very large amplitude ($\Delta p = 50 - 100bar$) will separate from the shock wave and move very stably and at a very constant speed through the mixture.

2. Definiton of the Problem

We are interested in a two-phase system consisting of gas bubbles as the disperse phase and a liquid as the matrix, and, in general, we know its static or equilibrium properties (thermal and caloric equations of state). What is of interest in a next step, are the dynamic and non-equilibrium properties that play a role in the case of wave propagation. The knowledge of the system as a continuum (as a material with averaged properties) may be our final goal, but very soon we will recognize in any kind of study that detailed investigation of the constituents of the microstructure, i.e. the single bubbles, bubble pairs, and groups of bubbles in a liquid is of fundamental importance, since results from these studies may guide us to develop a (necessarily) simplified model for the structured system as a whole.

When we think of the realization of wave phenomena in a shock tube, we have to live with many conceptual imperfections: Bubbles, created at the bottom of the vertical tube, move upwards through the liquid due to buoyancy; the barometric pressure in the liquid column will create a change in bubble size and in void fraction along the tube, and any pressure profile that is applied to the top of the two-phase column will develop modifications because of these slight changes in local state. If we investigate the evolution process of a wave phenomenon, we will have to separate these (quasi) static changes from those that we are interested in. Bubbles are created by a needle matrix at the shock tube bottom; even under ideal conditions there exists only one fixed void fraction for which we could produce an isotropic distribution (i.e. for a needle distribution with a spacing of the order of an average cell radius Λ_0). Given such a distribution initially, it would not be anymore isotropic after the passage of a shock wave.

After Stokes' relation, the bubble rising speed $u_{bb} = 2g\rho_{f0}R_0^2/(9\mu_f)$, with g as gravitational constant, ρ_{f0} as liquid density, R_0 as bubble radius, and μ_f as dynamic viscosity of the liquid, depends on its size and the liquid properties. Bubble deformation is influenced by these quantities and the surface free energy σ; this influence can be described by the Reynolds number, Re_b, the Weber number, We_b, and the Eötvös number, Eo_b, which are defined as follows

$$Re_b = \frac{2\rho_{f0}\Delta u R_0}{\mu_f}, \quad We_b = \frac{\rho_{f0}(\Delta u)^2 R_0}{2\sigma}, \quad Eo_b = \frac{2g\rho_{f0}R_0}{\sigma}.$$

with Δu as relative speed. For Re_b and Eo_b, being $O(1)$ in our case, we can expect an almost spherical bubble shape[23].

If we ask now what kind of experiment is possible and still might eventually be modelled, we may end up with the following compromise: We produce bubbles with a radius of $R_0 \approx 1mm$, a void fraction of $\varphi_0 = 0.2 - 2\%$, use a shock tube with a diameter of $D \approx 10\Lambda_0$, and a length of $L_{LP} = 3 - 4m$, a highly viscous fluid such as glycerine ($85\% C_3H_5(OH)_3 + 15\% H_2O$), a non-condensable gas like He, N_2, or SF_6. We expect a bubble rising speed of $u_{bb} \approx 5cm/s$ and assume spherically shaped bubbles.We further assume that we can describe the bubble field by one single "local" bubble radius $R(x,t)$.

2.1 Scales, dimensions, and characteristic quantities – hierarchy of description.

With the shock tube experiment in mind, we might, according to Fig.1, extract a set of length scales from our problem: On the microscale this is the bubble radius R_0 (suffix 0 refers to initial equilibrium state). On the mesoscale we may introduce the cell radius Λ_0 (or equivalently the (average) bubble distance ζ_{ij}) by the relation

$$\varphi_0 = \frac{4}{3}\pi R_0^3 n_0 = (R_0/\Lambda_0)^3, \tag{1}$$

where n_0 is the bubble number density. On the macroscale, a length L_0 will characterize the wave structure length. We can estimate L_0 by introducing the single bubble low amplitude oscillation time

$$t_0 = 2\pi R_0[\rho_{f0}/(3\gamma_0 p_0)]^{1/2}, \tag{2}$$

with γ_0 as the ratio of specific heats, and p_0 as the pressure, and an effective speed of sound of the bubbly mixture (valid for not too small φ_0 and neglecting relative speeds between bubbles and liquid)

$$c_0 = [\gamma_0 p_0/(\rho_{fo}\varphi_0)]^{1/2}. \tag{3}$$

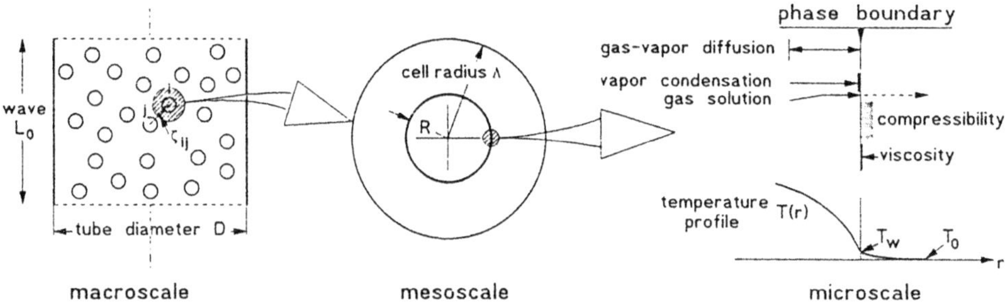

Fig.1: Length scales in a bubbly liquid and pertinent phenomena at the phase boundary.

This yields

$$L_0 = c_0 t_0 = 2\pi R_0/(2\varphi_0)^{1/2}. \tag{4}$$

We may obtain an estimate for these scales by assuming typical experimental conditions, like $R_0 = 1mm$ and $\varphi_0 = 1\%$; then, $\Lambda_0 = 4.6mm$ and $L_0 = 36.3mm$. Already here, we may question assumptions that are based on the condition $R_0 \ll \Lambda_0$ (single bubble model) or those with $\Lambda_0 \ll L_0$.

When formulating the complete set of equations for the single bubble problem[24], i.e. a gas–vapor mixture in a compressible liquid, after non–dimensionalization by the initial or equilibrium quantities, one finds, inspecting the system, a set of further typical scaling parameters, such as

$$\epsilon_D = \rho_{g0}/\rho_{f0}, \tag{5}$$

the ratio of the densities; for small ϵ_D, which will always be the case in our problem, we can introduce considerable simplifications; one of the most important is the spacial constancy of gas pressure

$$p_g = p_g(t). \tag{6}$$

Furthermore, with λ as thermal conduction coefficient and c as specific heat,

$$\epsilon_{tf} = [a_f t_0]^{1/2}/R_0, \quad \epsilon_{tg} = [a_g t_0]^{1/2}/R_0, \quad \text{with } a = \lambda/(\rho c), \tag{7}$$

determine the thermal layer in the liquid and in the gas mixture, respectively.

$$\epsilon_{df} = [D_f t_0]^{1/2}/R_0, \quad \epsilon_{dg} = [D_g t_0]^{1/2}/R_0, \tag{8}$$

with D as diffusion coefficient, scale the diffusion layer in the liquid and in the gas, respectively. The term $\epsilon_s = \sigma/(R_0 p_0)$ considers the influence of surface tension, and $\epsilon_k = [R_0/(3\gamma_0 c_f t_0)]^{\frac{1}{2}}$ describes the effect of compressibility in the liquid.

For our special problem of gas bubbles in 85% glycerine with $R_0 = 1mm$, $T_0 = 300K$, and $t_0 = 0.39ms$ we will expect a water vapor partial pressure of less than 0.35%; thus,we may neglect phase transitions at the liquid interface for several reasons: The smallness of the vapor content, the fact that $\epsilon_{dg} > \epsilon_{tg}$ which means that vapor transport to the interface is diffusion dominated (generally, a gas cushion will buildup at the interface, because of the extremely rapid process of condensation or evaporation). Since thermal diffusivity in the liquid is very small ($\epsilon_{tf} = 0.002$, about ten times smaller than ϵ_{tg} for N_2) we may, in a very good approximation, assume that the interface temperature remains constant

$$T_w = T_0. \tag{9}$$

Since gas diffusion in the liquid is also small ($\epsilon_{df} = 0.0002$), we may neglect it for the wave transition process that we are interested in. With $\epsilon_s = 0.002$ we may also neglect the influence of surface free energy.

Even after these simplifications, there remains still the difficult task to include a set of phenomena that cannot be neglected; these are: viscosity, the heat transfer problem, and bubble interaction. We shall discuss the influence of these phenomena in the following.

2.2 Viscosity - a microscale effect

For the case of a spherical single bubble, the momentum equation contains the viscosity term[25,21]

$$\frac{1}{\rho_f}\frac{4}{3}\mu_f\frac{\partial}{\partial r}\left[\frac{1}{r^2}\frac{\partial}{\partial r}(r^2 v_f)\right] \tag{10}$$

which vanishes within the incompressible liquid. However, at the phase boundary, the difference term yields a contribution

$$\frac{4}{3}\mu_f\left(\frac{\partial v_f}{\partial r} - \frac{v_f}{r}\right) \tag{11}$$

to the momentum balance. Since μ_f is a typical (continuum) volume quantity, we may question its applicability to a surface layer from which we know that it modifies typical quantities like entropy or pressure. In addition, it is an open question whether glycerine is still a Newtonian fluid for the extremely rapid changes that take place in a bubble oscillation. Nevertheless, the Hagen–Poiseuille value[26] for μ_f, which is strongly dependent on temperature and water content, seems to have the right order of magnitude, if we compare, for instance, experimental results of shock induced bubble oscillations with calculation, Fig.2. In these experiments, we have used SF_6 which changes almost isothermally due to a $\gamma_0 = 1.09$. Thus, a pressure step of $4bar$ will cause an isentropic temperture rise of only $47K$ compared to $470K$ for a noble gas. Note that a drastic change in μ_f does not result in a dramatic improvement of the damping properties in Fig.2.

When, for an order-of-magnitude estimation, we take the damping coefficient of Devin[4]

$$\delta_0 = \delta_{thermal} + \delta_{radiation} + \delta_{viscosity}$$

that consists of terms for thermal, acoustic radiation, and viscous damping, we will find for our example that the thermal term $\delta_{thermal}$ is about ten times smaller than δ_{visc}, and $\delta_{rad} \approx \delta_{visc}$, which justifies the approach used in the numerical calculation

Fig.2. It should be noted, however, that Devin's approach is by no means applicable to a realistic quantitative treatment of transient phenomena such as shock wave profiles.

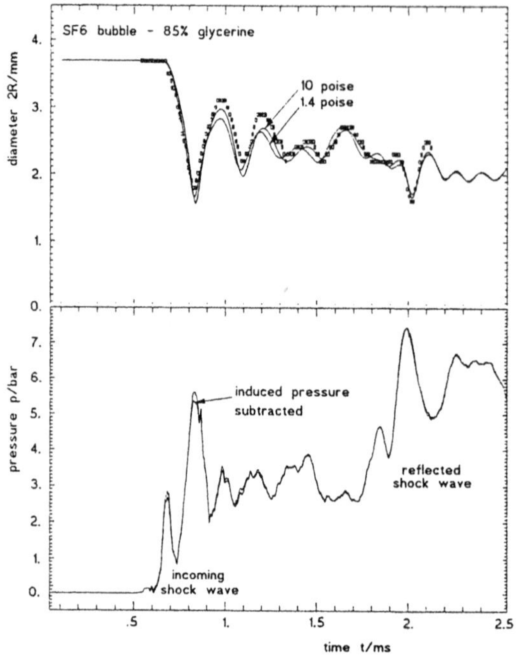

Fig.2: Single SF_6 bubble in glycerine. Symbols: Experimentally obtained radius as function of time. Lower curve: Pressure signal used in the calculation of $R(t)$, upper diagram, for a viscosity of $\mu_f = 1.4 poise$ and $10 poise$, respectively. By an iterative process, the pressure induced by the bubble at the shock tube wall (gauge) was subtracted from the signal. Equations used: See Reference (27).

Unsatisfactory results for the damping in bubbly liquids, when comparing experiments with calculations that consider only the term (11) have forced us to search for other phenomena that might cause dissipation. In a real bubbly liquid we may have a modification of the viscosity due to a random distribution of the bubbles causing a random relative motion and deviation from spherical shape, and due to long living micro bubbles; since the non–dimensional viscosity term is $\mu \sim 1/R$, even a small amount of void fraction of, say, $0.01mm$ bubbles will cause a considerable increase in μ_{total}. We have found that, after a few shots, there exists a considerable number of microbubbles visible in scattered laser light; the void fraction of these bubbles, however, seems to be within the error margin that we find when reading the averaged overall φ_0.

In a rather heuristic way we have tried to include these (by no means separated or even identified) effects in the averaged momentum equation (22) for the bubbly liquid. We introduce (with $\langle\ \rangle$ for a volume averaged quantity) into the flux term, which can be transformed as $\langle \rho u u\rangle = \langle \rho u\rangle\langle u\rangle + \langle(\rho u)\Delta u\rangle$, a Boussinesq–type ansatz

$$\langle(\rho u)\Delta u\rangle = -\epsilon_t \frac{\partial}{\partial x}\langle u\rangle, \tag{12}$$

and try to find the quantitative value of the coefficient ϵ_t by experiment. When introducing this term in this way, we notice that the (thermodynamic) force $\partial\langle u\rangle/\partial x$ is rather inefficient, since changes in $\langle u\rangle$ are of the order of φ_0, and it would be considerably more "efficient" to relate this term, just as the molecular viscosity μ, to the bubble radius change $\partial R/\partial t$. However, the argument of inefficiency also holds for a turbulent gas flow where, due to this, the turbulent viscosity may become some orders of magnitudes larger than the molecular term, but still $\partial\langle u\rangle/\partial x$ is the force used in the model of the Reynolds stress term.

2.3 The heat transfer problem - a microscale effect

From previous considerations we know that heat flux is confined to the inside of the bubble. It is, therefore, the gas pressure p_g by which the action of the heat flux enters the problem. Assuming $T_w = T_0$, simplifies the task considerably, because we can look for a single bubble solution of the heat flux problem. When we introduce a new related void fraction variable

$$\psi = \frac{\varphi}{\varphi_0} - 1, \tag{13}$$

we have two very simple limiting cases for the relation between p_g and ψ for isentropic and isothermal changes, respectively,

$$p_g/p_0 = (1+\psi)^{-\gamma_0}, \quad p_g/p_0 = (1+\psi)^{-1}. \tag{14}$$

For the real transient process, however, p_g is related to ψ by a complex memory integral. Combining the continuity and the energy equations for a spherical bubble (coordinate is r), we find, after integration,

$$\frac{d}{dt}p_g = -3\gamma_0 p_g \frac{R_t}{R} - 3(\gamma_0 - 1)\frac{q_R}{R}, \tag{15}$$

where $q_R = -\lambda_g(\partial T/\partial r)_R$ is the heat flux at $r = R^-$. We may further integrate with respect to time and obtain (for small φ_0)

$$\frac{p_g}{p_0} = \left(\frac{R}{R_0}\right)^{-3\gamma_0}(1+J) = (1+J)(1+\psi)^{-\gamma_0}, \tag{16}$$

with

$$J = 3\epsilon_{tg}^2\gamma_0 \int_0^t \frac{\tilde{R}^2}{\tilde{T}_{is}}\left(\frac{\partial \tilde{T}}{\partial \tilde{r}}\right)_{\tilde{R}} d\tilde{t}', \tag{17}$$

where non-dimensional quantities (marked by a tilde sign) have been introduced. Sophisticated techniques have been developed[19] to solve these equations. In order to have some solution for comparison with our experiments at hand, we have developed a crude but efficient solution[22] which consists of a combination of the solution for short times (thin thermal layer) and large times (bubble filled with temperature profile). It is interesting to note that, across a shock wave, the memory integral has a well defined known value: Combining (21) and (23), we obtain for $x \to \infty$

$$J_{+\infty} = (1+\psi_{+\infty})^{(\gamma_0 - 1)} - 1, \tag{18}$$

where $\psi_{+\infty}$ is known from the Hugoniot relations. For the isothermal case we have $J = 0$.

For the present case we need an estimate of the relaxation time for transient processes like shock waves. Treating the shock wave as a step type transition, we can find a solution for the average over-temperature in the bubble with a relaxation time characterized by

$$t_{thermal} = (R_0/\pi)^2/a_g. \tag{19}$$

The dilemma for the shock tube experiment, where the wave passes through a mixture with p and φ slightly changing, has been mentioned before. One might expect that only for a small $t_{thermal}$ (in the sense that quasi-steady profiles are reached) simple results and quantitative comparison with steady state solutions are possible. To avoid this problem of heat transfer completely we have used, in many experiments, the gas SF_6.

2.4 Bubble interaction - a mesoscale effect

To obtain a complete set of equations, we need an equation that relates the pressure $\langle p \rangle$ to the void fraction φ or ψ. In the limit of extremely small void fractions ($\varphi \ll 0.1\%$) this is achieved by the classical Rayleigh–Plesset equation[25]. However, with increasing φ_0 the liquid cell volume for a bubble is reduced. For $\varphi_0 = 1\%$ we have already a mean distance between bubbles of $\Lambda_0 = 4.6R_0$, and one certainly may not speak anymore of a single bubble in an infinite medium. It has been shown in studies of bubble pairs[28,29] that the (breathing) monopole fields of neighbor bubbles increase the oscillation time. This is also shown for bubble clouds[16,30,31]. If one includes the influence of nearest neighbor bubbles upon the bubble cell, one obtains correction terms of $O(\varphi_0^{1/3})$ for the breathing terms and of $O(\varphi_0)$ for bubble motion (which can consequently be neglected for all present applications). This interaction has to be distinguished from a volume averaging procedure[32], which also yields terms of $O(\varphi_0^{1/3})$, but of opposite (negative) sign. In a simplified approach we might, therefore, introduce a modified Rayleigh–Plesset equation of the type

$$\alpha_1 R R_{tt} + \frac{3}{2}\alpha_2 R_t^2 = \left(p_g - \langle p \rangle - 4\mu_f \frac{R_t}{R} \right) / \rho_f, \tag{20}$$

with $\alpha_1 = 1 + (b_0 - 2)\varphi^{1/3}$, $\alpha_2 = 1 + \frac{4}{3}(b_0 - 2)\varphi^{1/3}$, where the term (-2) is due to the averaging process and b_0 due to the monopole bubble interaction. We leave the determination of b_0 to the experiment.

3. The theoretical model - a description on the macroscale

We shall now, very briefly, describe the development of a theoretical model that might be able to describe nonlinear wave phenomena in a shock tube. We begin with the averaged[32,33] equations of total mass and momentum conservation for the mixture

$$\frac{\partial}{\partial t}[(1-\varphi)\rho_f] + \frac{\partial}{\partial x}[(1-\varphi)\rho_f \langle u_f \rangle] = 0, \tag{21}$$

$$\frac{\partial}{\partial t}\langle \rho u \rangle + \frac{\partial}{\partial x}\langle \rho u u \rangle = -\frac{\partial}{\partial x}\langle p \rangle, \tag{22}$$

with

$$\langle p \rangle = \varphi \langle p_g \rangle + (1-\varphi)\langle p_f \rangle. \tag{23}$$

Here we have neglected gravitational forces, assumed that $\varphi \ll 1$, and $\rho_g \ll \rho_f$, with small changes in ρ_f. Using in (1) the new variable

$$\psi = \frac{\varphi}{\varphi_0} - 1 = \frac{n}{n_0}\left(\frac{R}{R_0}\right)^3 - 1, \tag{24}$$

neglecting now compressibility terms for the liquid, and terms $O(\varphi_0)$, we obtain for (21,22)

$$\frac{\partial}{\partial \tilde{t}}\psi = \frac{\partial}{\partial \tilde{x}}\langle \tilde{u}_f\rangle + O(\varphi_0), \tag{25}$$

$$\frac{\partial}{\partial \tilde{x}}\langle p\rangle = 3\gamma_0\varphi_0^{1/3}\frac{\partial}{\partial \tilde{t}}\langle \tilde{u}_f\rangle + O(\varphi_0), \tag{26}$$

where we have introduced nondimensional variables, using for $\langle u\rangle$ the speed $u_0 = \varphi_0\Lambda_0/t_0$. Combining (25) and (26) yields

$$\langle \tilde{p}\rangle_{xx} = -\frac{\gamma_0}{\tilde{c}_0^2}\psi_{tt} + \gamma_0\tilde{\epsilon}_t\psi_{xxt}, \tag{27}$$

with the characteristic speed and the "bulk" viscosity term, respectively,

$$\tilde{c}_0^2 = (c_0 t_0/\Lambda_0)^2 = \frac{1}{3\varphi_0^{1/3}}, \qquad \tilde{\epsilon}_t = \epsilon_t\varphi_0/(\gamma_0 t_0 p_0).$$

This basic equation has to be completed by (20), however, in order to be able to combine both equations into one fundamental equation, we have to replace R by φ. This is possible only by means of an expansion which limits the result to small void fractions

$$\frac{\tilde{R}_t}{\tilde{R}} = \frac{\varphi_t}{3\varphi} + O(\varphi_0), \qquad \frac{\tilde{R}_{tt}}{\tilde{R}} = \frac{\varphi_{tt}}{3\varphi} - \frac{2\varphi_t}{9\varphi} + O(\varphi_0). \tag{28}$$

Limiting at this point the equations to small pressure steps by keeping in the expansion in ψ only terms up to second order, we obtain

$$\alpha_{10}\psi_{tt} - \frac{1}{3}\psi\psi_{tt} - \alpha_{20}\psi_t^2 + \tilde{\mu}_f\psi_t = \frac{1}{\gamma_0}(\tilde{p}_g - \langle\tilde{p}\rangle), \tag{29}$$

with $\tilde{\mu}_f = (4\mu_f)/(3p_0t_0)$, and combination with (16) and (27) yields

$$(\alpha_{10}\psi_{tt} - \frac{1}{3}\psi\psi_{tt} - \alpha_{20}\psi_t^2 + \tilde{\mu_{tot}}\psi_t)_{xx} = \frac{1}{\gamma_0}[\frac{(1+J)}{(1+\psi)^{\gamma_0}} - 1]_{xx} + \frac{1}{\tilde{c}_0^2}\psi_{tt}. \tag{30}$$

If we expand the first rhs–term up to quadratic terms, omitting J for the moment, we have

$$\alpha_{10}\psi_{ttxx} + \tilde{\mu}_{tot}\psi_{txx} - (\frac{1}{3}\psi\psi_{tt} - \alpha_{20}\psi_t^2) = -(\psi_{xx} - \frac{1}{\tilde{c}_0^2}\psi_{tt}) + \frac{\gamma_0+1}{2}\psi_{xx}^2 + \cdots. \tag{31}$$

Thus, for $\psi < 1$, i.e. for small pressure amplitudes, this equation is characterized by a dispersion term ψ_{ttxx}, a damping term ψ_{txx}, a wave propagating term $(\psi_{xx} - \psi_{tt}/\tilde{c}_0^2)$, and nonlinear terms. With further simplifications, (31) is reduced to a Burgers–Korteweg–de Vries type equation[13].

3.1 Steady state: Hugoniot relations

The two equilibrium states before (index 0) and after (index 1) a shock wave running steadily with the speed V are related by simple conservation equations, and none of the highly complex non–equilibrium phenomena mentioned before play a role. These relation, connecting the two states, are

$$n_0 V = n_1(V - u_1), \tag{32}$$

$$(1-\varphi_0)V = (1-\varphi_1)(V-u_1), \tag{33}$$

$$(1-\varphi_0)V^2 + \frac{p_0}{\rho_f} = (1-\varphi_1)V^2 + \frac{p_1}{\rho_f}. \tag{34}$$

We still need a thermodynamic relation. For an equilibrium (isothermal) state change and for a "frozen" (isentropic) change we have, respectively,

$$\frac{p}{p_0} = \left(\frac{R}{R_0}\right)^3, \quad \frac{p}{p_0} = \left(\frac{R}{R_0}\right)^{3\gamma_0}. \tag{35}$$

Introducing a Mach number $M = V/c_0$, we obtain for the equilibrium case

$$M_e^2 = \frac{1}{\gamma_0}(1+\Delta\tilde{p}), \tag{36}$$

with $\Delta\tilde{p} = (p_1/p_0 - 1)$. For the adiabatic case

$$M_f^2 = \frac{\Delta\tilde{p}}{\gamma_0}\frac{(1+\Delta\tilde{p})^{1/\gamma_0}}{(1+\Delta\tilde{p})^{1/\gamma_0}-1}. \tag{37}$$

The relation for small $\Delta\tilde{p}$ is of interest

$$M_e = 1 + \frac{1+\gamma_0}{4\gamma_0}\Delta\tilde{p} + O(\Delta\tilde{p}^2), \tag{38}$$

$$M_f = \frac{1}{\sqrt{\gamma_0}}(1+\frac{1}{2}\Delta\tilde{p}) + O(\Delta\tilde{p}^2). \tag{39}$$

We see that, for a small amplitude, we have a conflict when relating M_f to c_0 which is based on an isentropic process. We might expect that, during the transfer from state 1 to state 2 inside the shock wave, the bubbles initially have an isentropically changing kernel which is, by the heat transfer process, reduced leading finally to a constant temperature $T_{+\infty} \approx T_0$ for the whole mixture. When comparing experimental results for He, N_2 , and SF_6 bubbles with (36) and (37), we found[22] that for He the results follow closely M_e, for N_2 they are closer to M_f, while for SF_6 the difference between M_e and M_f is too small to be discriminated by experiment. Some results for SF_6 are shown is Fig.3.

Apparently the problem is far more complex than the Hugoniot relations are able to describe, since they tell nothing about the evolution time that it takes to reach an equilibrium state[34]. Facing the before mentioned practical problem of state change due to the barometric pressure in a vertical shock tube, it may be impossible ever to reach M_e for large $t_{thermal}$; this seems to be the case for N_2, when interpreting the results[22].

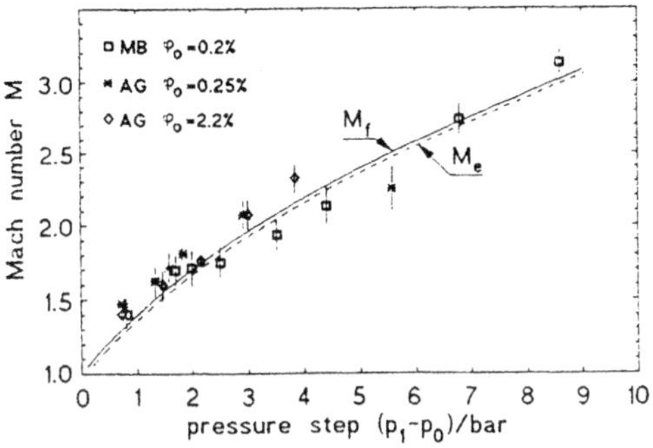

Fig.3: Mach number M as function of pressure step Δp. Symbols are from experiments in SF_6 - glycerine mixtures for different void fractions. Lines refer to (36,37) with $\gamma_0 =$ 1.09.

An investigation of the linearized forms[22] of (30) around the equilibrium points shows that the upstream point is a node, whereas the downstream point is a focus with spiraling trajectories, provided that the dissipation is not too small. Thus, for very small radii R_0 and for large heat transfer at the phase boundary we may expect a smooth transition in the shock wave profile.

4. Experiments in bubbly liquids

Thinking of performing experiments with wave phenomena being one–dimensional on the macroscale, probably a conventional (vertical) shock tube[11] is the best compromise, and we believe that the classical tube, as it is schematically shown in Fig. 3, is most versatile: It permits to study, besides the step (shock) waves, also the evolution of positive and negative wave packets, and a very well defined gas shock wave is applied to the top surface of the two–phase mixture column. There must be a transition zone for the change from the step wave to the (quasi) steady profile in the upper part of the two–phase system; we believe that this zone is shorter when the initial profile is a step than if it is some ramp of unknown properties as it is to be expected in an atmospherically driven shock tube[5,8].

The most important part of the two–phase shock tube is the bubble generator; if one is interested in producing bubbles with small dispersion in size, probably the best way is to use a needle matrix and control the gas pressure upstream of each needle individually. The spacing of the needles depends on the void fraction φ_0, if one wants to create an isotropically homogeneous distribution. The bubble size depends rather weakly on the (inner) needle diameter D_n; we obtained with $D_n = 0.2mm$ a bubble radius of $R_0 = 1.15mm$, and with $D_n = 1.4mm$ a radius of $R_0 = 1.6mm$. The adjustment that produces finally a monodisperse bubble distribution is rather difficult to obtain. After each shot, especially for large amplitude wave, a readjustment is required, tiny bubbles have to be removed, and a steady state of bubbles with the same size and rising speed has to be established again. In order to be sure that a steady state is achieved at least in the lower part of the tube, one would prefer to make the tube very long, typically several meters in length. However, we have noticed that, due to wall effects, bubbles tend to accumulate near the axis of the tube already beyond a distance of about two meters from the bottom, and the wave profiles taken by pressure gauges in the upper part of the tube are generally of a poor quality.

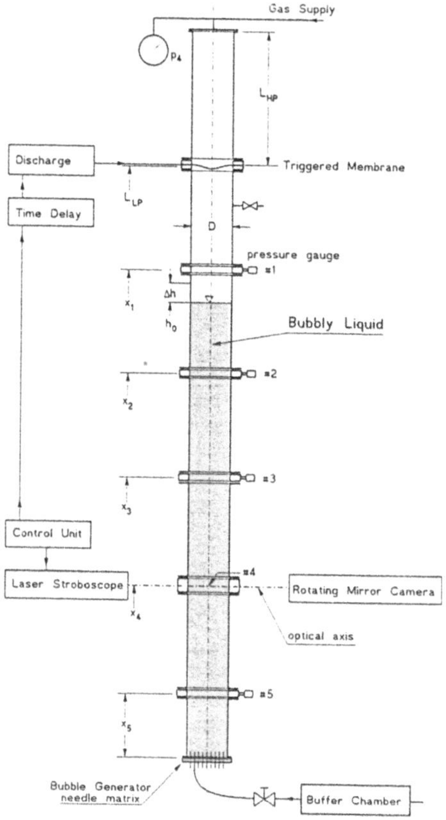

Fig.4: Left: Schematic setup of the experimental system used for the study of wave phenomena in bubbly liquids.

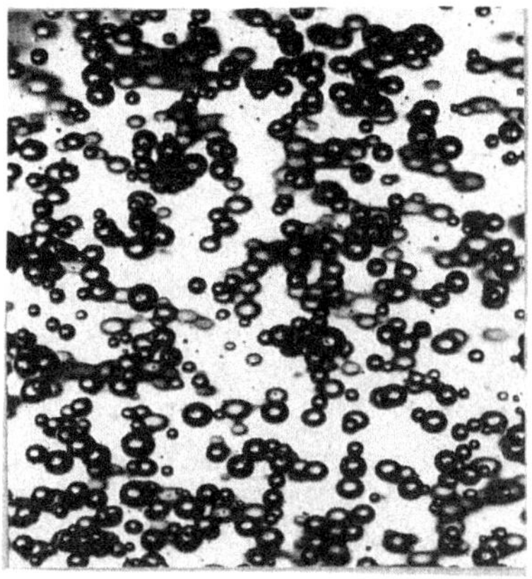

Fig.5: Top: Clustering of N_2 bubbles in glycerine at large void fraction φ_0 .

All this worsens considerably with increasing void fraction φ_0, and for $\varphi_0 > 2-4\%$ an additional problem arises, see Fig.5; in this case bubbles show the tendency to build clusters, they are certainly not an average distance Λ_0 apart, and one may expect that these clusters will behave like one large bubble. We might have to consider these effects in addition to strong bubble interaction as dominating effects for large φ_0.

Whereas we are accustomed to relate a signal from a piezo-electric gauge taken in a gas flow to a "pressure", we should, at least, question this habit when looking at a profile of some transient in a bubbly liquid. A pressure signal taken at the wall of the shock tube is just a local signal, and to obtain the pressure $\langle p \rangle$ that appears in the model equations one would have to take randomly samples on the microscale level at different locations in the "cell". This is not possible in a real experiment, but by repeating the experiment and by superimposing several shots one might imagine that this procedure is equivalent to taking samples at different points in the "cell". This problem is not so evident for small amplitudes (i.e. for $\Delta p < 0.5 bar$) and the first oscillation of a pressure wave, but it becomes a fundamental problem for the damped oscillating part of a shock wave and for large pressure amplitudes.

In the following, several sets of experimental results will be presented. We begin with some typical examples of profiles for small amplitude waves, in order to show that a BKdV-equation, as discussed in the context of (31), is able to describe these phenomena at least qualitatively. Then, an approach to describe quantitatively small amplitude waves in systems with gas bubbles of largely varying thermal diffusivity and the influence of void fraction will be mentioned, and, finally, some first attempts to understand large amplitude waves will be reported on.

4.1 Some typical wave patterns

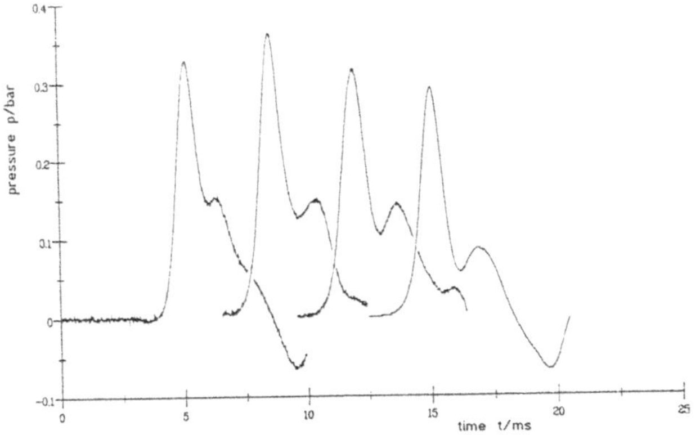

Fig.6: SF_6-glycerine. Evolution of a very short wave packet. $\varphi_0 = 0.35\%$, radius $R_0 = 1.6mm$. Spacings $x_{2,3,4,5} =$ 210, 163, 105, 460 cm.

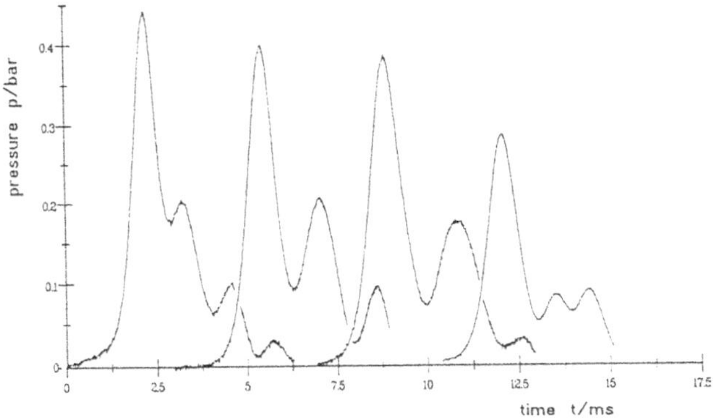

Fig.7: SF_6-glycerine. Evolution of a small amplitude wave packet. Parameters as in Fig.6.

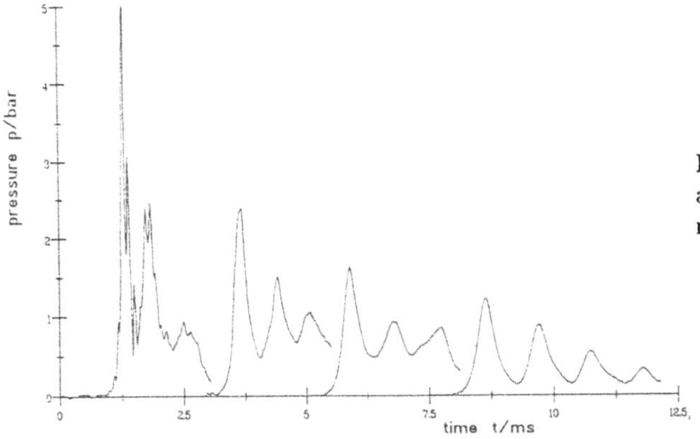

Fig.8: SF_6-glycerine. Evolution of a large amplitude wave packet. Parameters as in Fig.6.

From (31) we have seen that, for small amplitudes, an equation of the following type exists

$$\psi_{ttxx} + \tilde{\mu}\psi_{txx} - \kappa\psi_{xx}^2 + (\psi_{xx} - \psi_{tt}/\tilde{c}_0^2) = 0,$$

which contains the elements of dispersion, damping, nonlinear steepening of the wave, and a wave operator, respectively. The examination of the equilibrium points for the steady case indicated that the upstream part of the wave merges into a node and the downstream part ends in a focus with spirals. We can find these properties in every detail in the experimental results.

In Fig.6 we have the evolution of a low amplitude very short wave packet. It is produced by means of a very short high pressure tube of a few cm length. At the last pressure gauge almost one single soliton has evolved.

In Fig.7 a slightly larger wave packet evolves with about tree peaks. When we increase the amplitude, new effects gradually become important: Bubble interaction and chaotic phenomena play a role, see Fig.8. Note that the initial amplitude is about ten times that of the wave in Fig.6. The damping is very pronounced in this case.

When we produce a negative step, the prevoiusly derived equations predict an evolution into a ramp, similar to a rarefaction wave in a gas. In Fig.9 we can follow the evolution of a negative wave packet; the negatively sloped ramp flattens, and at the positive step we find an oscillation just as it is to be expected at the downstream part of a shock wave profile.

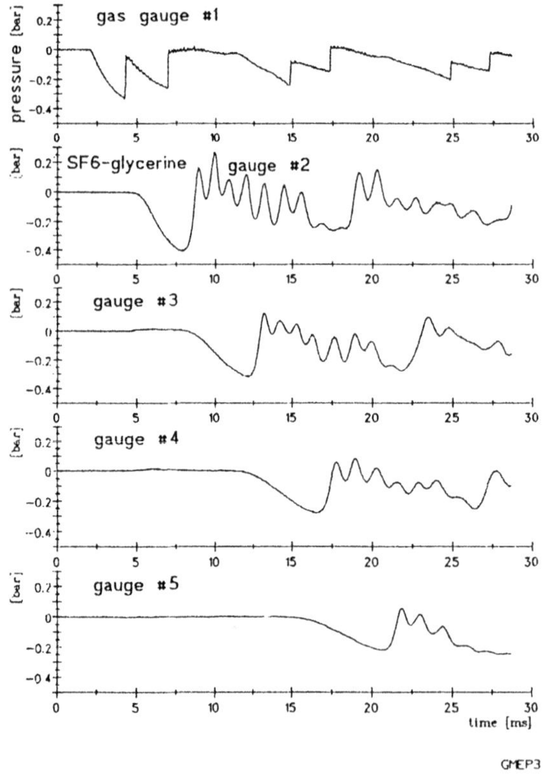

Fig.9: Evolution of a negative wave packet.

Typical shock wave profiles are reproduced in Fig.10 for a small pressure amplitude. Note the smooth transition at the foot and the damped oscillation downstream the shoulder of the wave.

4.2 Quantitative study of low amplitude waves

Wave profiles in bubbly mixtures with the gases He, N_2, and SF_6, having almost an order of magnitude difference in their thermal diffusivity a_{tg}, were investigated. Estimating the relaxation time for these gases, following (19), we find for the relaxation time $t_{thermal}(He, N_2, SF_6) \approx 1,\ 8,\ 45\ ms$ which means that we may expect a profile close to a steady state for He–bubbles but not necessarily for N_2. For SF_6 we are not so much interested in the value of $t_{thermal}$ which indicates that we should expect a frozen state, but, since the transfer is almost isothermal, we would not be able to distinguish between frozen and the equilibrium state.

We have attempted to find, from the experiment, a total viscosity value that describes the damping properly. Fair agreement was achieved for a $\mu_{total} \approx 12\mu_f$ for all data, including SF_6 bubbles. We have seen before, that similar adjustments have to be made for single SF_6–bubble shock wave experiments. In spite of the good agreement with different experiments, we are not satisfied with this result. We tried to consider other effects like micro–bubbles but their void fraction seems to be too small to justify such a large viscosity increase. It is still uncertain that micro–bubbles are responsible for the large damping.

Detailed comparison of averaged profiles and numerical results have been given elsewhere[22]. Here, only two typical profiles in SF_6–glycerine for $\varphi_0 = 0.25\%$ and 2.2%, respectively, are presented, see Fig.10.

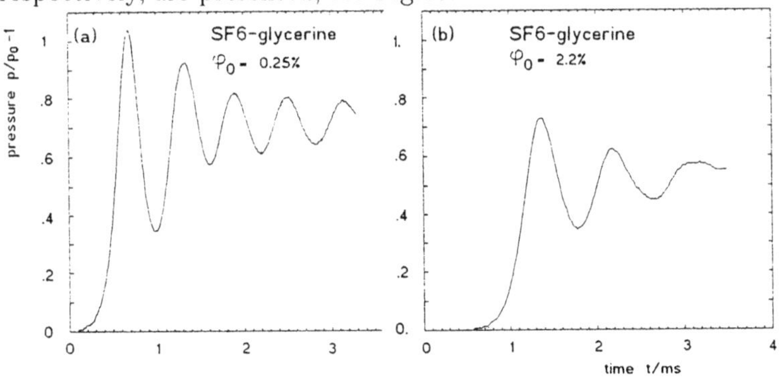

Fig.10: Shock wave profiles (20 shots superimposed) in a SF_6-glycerine mixture. $p_0 = 1.11 bar$, $T_0 = 296K$, $\rho_{f0} = 1.22 g/cm^3$, $\gamma_0 = 1.09$, $R_0 = 1.15mm$. Left: $\varphi_0 = 0.25\%$, $\Delta p = .68bar$. Right: $\varphi_0 = 2.2\%$. $\Delta p = .57bar$.

When comparing the oscillation time, one may notice an increase for the larger void fraction. A good agreement between experiment and calculation was achieved using $b_0 = 15$ in (20).

4.3 Large amplitude shock waves

When increasing the pressure step to, say, $\Delta p = 1 - 10bar$, many of the simplifying assumptions that led to a fundamental equation in ψ, (30), cannot be justfied any more. One has to start again with the conservation equations (21,22) and the bubble–dynamics equation (20), and the result will be a more complicated set of simultaneous partial differential equations. We shall not pursue this procedure here, but rather we want to present some experimental material that might be helpful when developing models for these waves. In Fig.11 the pressure amplitude Δp is increased, while φ_0 is kept constant. Note the growing complexity of the oscillations, also visible in the auto-correlation functions. Still, the fundamental oscillation time t_{10} can be extraced. It decreases with pressure amplitude and is larger than the single bubble oscillation time, see Fig.12. For $\Delta p = 0.8 bar$ kept constant, t_{10} grows with φ_0, Fig.13. The unexpected increase for $\varphi_0 > 1.5\%$ seems to be due to bubble clustering.

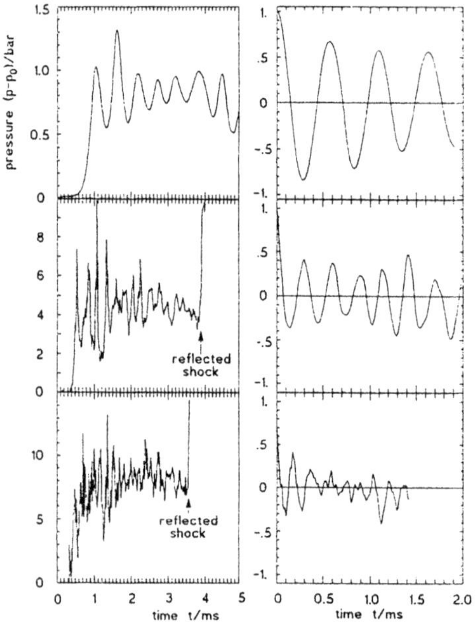

Fig.11: SF_6–glycerine, $\varphi_0 = 0.25\%$, $R_0 = 1.45mm$. Left: Pressure; right: Autocorrelation function of $p(t)$. Top to bottom: $\Delta P = 0.8,\ 4.4,\ ,8.6\ bar$.

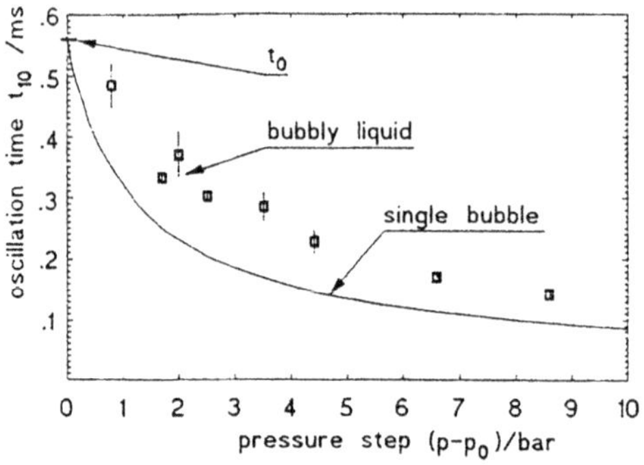

Fig.12: SF_6-glycerine, $\varphi_0 = 0.25\%$, $R_0 = 1.45mm$. Symbols: Fundamental oscillation time t_{10} obtained from auto-correlation function as function of pressure step Δp. Full line: Calculation for single bubble.

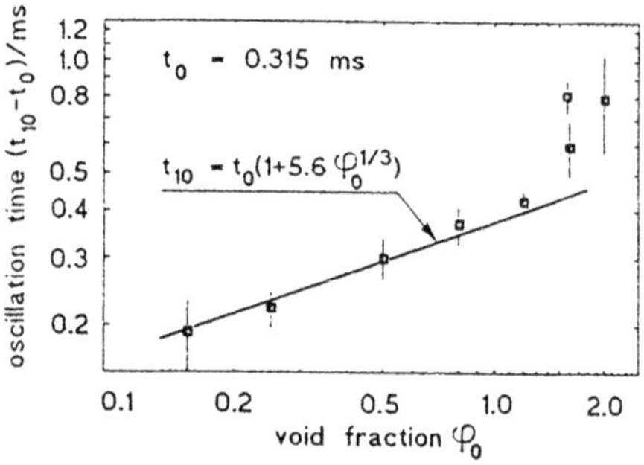

Fig.13: SF_6–glycerine, $R_0 = 1.45mm$. Variation of oscillation time t_{10} with void fraction φ_0 for a constant pressure step of $\Delta p = 0.8bar$.

We have also studied bubble group dynamics by using the laser–stroboscope and the rotating mirror camera. In Fig.14 the averaged radius $\langle R \rangle$, obtained from $R_m = (R_{max} + R_{min})/2$, and the deformation $\Delta R = (R_{max} - R_{min})$ are plotted for different pressure steps. Note the growing chaotic behavior, announced by increasing deviation from spherical shape and deformation; u_b is the bubble speed.

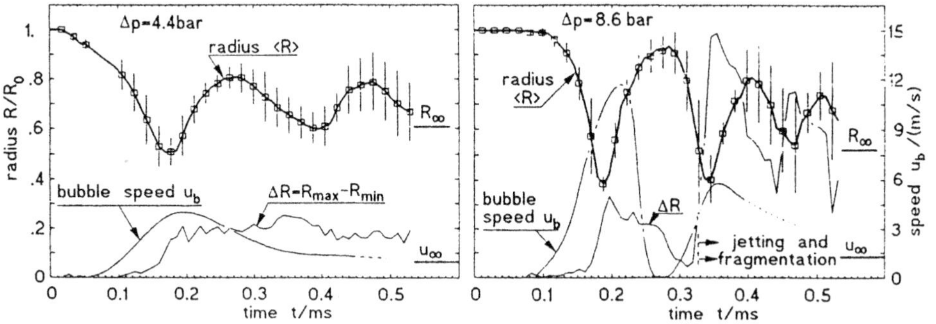

Fig.14: SF_6-glycerine, $\varphi_0 = 0.25\%$, $R_0 = 1.45mm$. Bubble group dynamics in the shock profile. Average radius, deformation and bubble speed as function of time. Pressure steps $\Delta p =$ 4.4, and $8.6bar$.

The series of Fig.15 provide a further insight into the bubble group dynamics during the shock wave transition; note the jetting, the fragmentation, and the increasing random orientation of the jet axis due to bubble interaction with increasing pressure step.

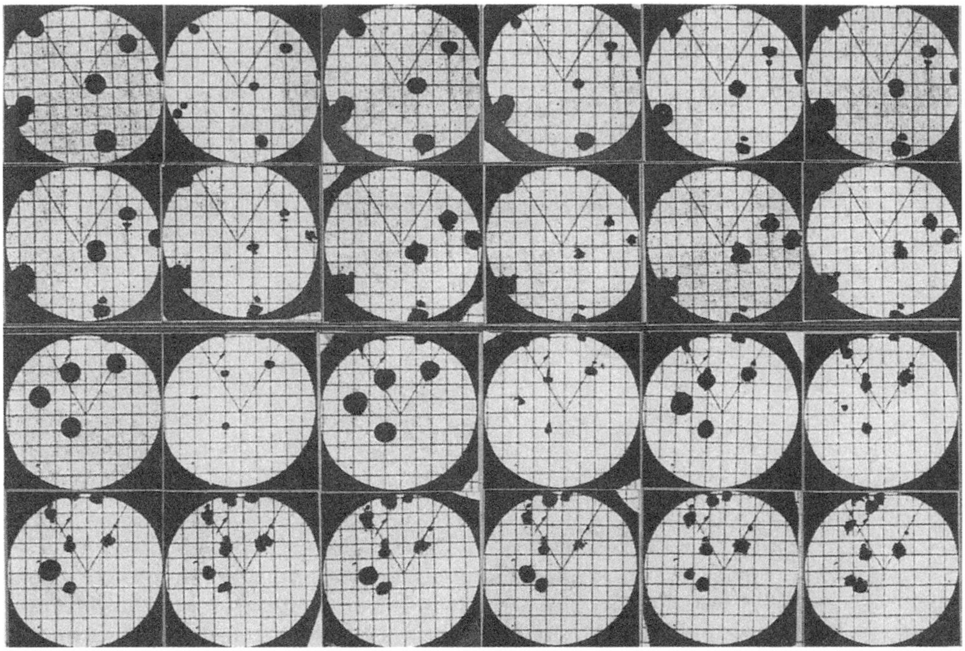

Fig.15: SF_6-glycerine, $\varphi_0 = 0.25\%$, $R_0 = 1.45mm$. Dynamics of bubble groups in the shock profile. Top: $\Delta p = 4.4bar$, Frame # (left to right): 1, 11, 21, 31, 41, 50, 54, 65, 75, 84, 93, 102. Bottom: $\Delta p = 8.6bar$, Frame #: 1, 8, 19, 26, 33, 40, 47, 54, 61, 68, 75, 95. Time interval: $8.8\mu s$. Window diameter is 21 mm.

For larger pressure steps, even for small void fractions, group effects cause a gradual replacement of the regular oscillatory structure by a random structure of oscillation; this must not be confused with the bifurcation transitions of forced large amplitude (single) bubble oscillations[27], because there is not sufficient time for these processes.

4.4 Waves in reactive bubbly liquids

It has been verified by several investigators[35–38] that combustion can be started in bubbles filled with a reactive mixture, when the amplitude of the initiating shock wave is increased beyond a certain threshold. Using a premixed system of $70\% Ar + 30\%(2H_2+O_2)$, we found that for $\Delta p > 8bar$ some kind of detonation soliton separates from the shock wave and moves at about twice its speed. From optical measurements of the luminescence[38] the combustion process seems to be of the order of $1 - 2\mu s$. Pressure spikes of more than $100bar$ can be measured. In Fig.16 the evolution of the waves for a typical system are plotted.

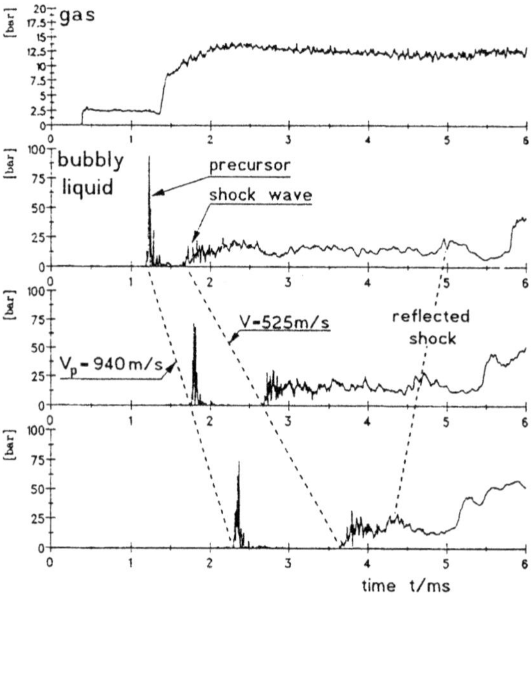

Fig.16: Evolution of shock and detonation wave in a bubbly liquid with $70\% Ar + 30\%(2H_2+O_2)$ bubbles in glycerine. $\varphi_0 = 0.4\%$, $T_0 = 293K$, $R_0 = 1.23mm$. Speed of the detonation wave $V_d = 940m/s$, shock wave speed $V_s = 525m/s$.

The behavior of the bubbles, while the detonation waves passes, can be studied in the series of Fig.17: The crossing of the detonation wave can be seen in frames #4 to #8. It is interesting to note that the individual bubbles emit spherical shock waves that seem to feed a more or less planar detonation zone and merge into it. If one follows the spherical wave in frame $\#4 - 5$ or $\#6 - 7$, its speed is $1.9mm/\mu s$ which corresponds to the sound speed in glycerine ($c_f = 1.895mm/\mu s$); here the estimated bubble speed is $50 - 100m/s$. After combustion, very typically, the bubbles are "jetting" into forward (flow) direction. Some fragmentation can be seen, but this is not much more dramatic than for a normal inert gas shock wave of $\Delta p = 5 - 10bar$.

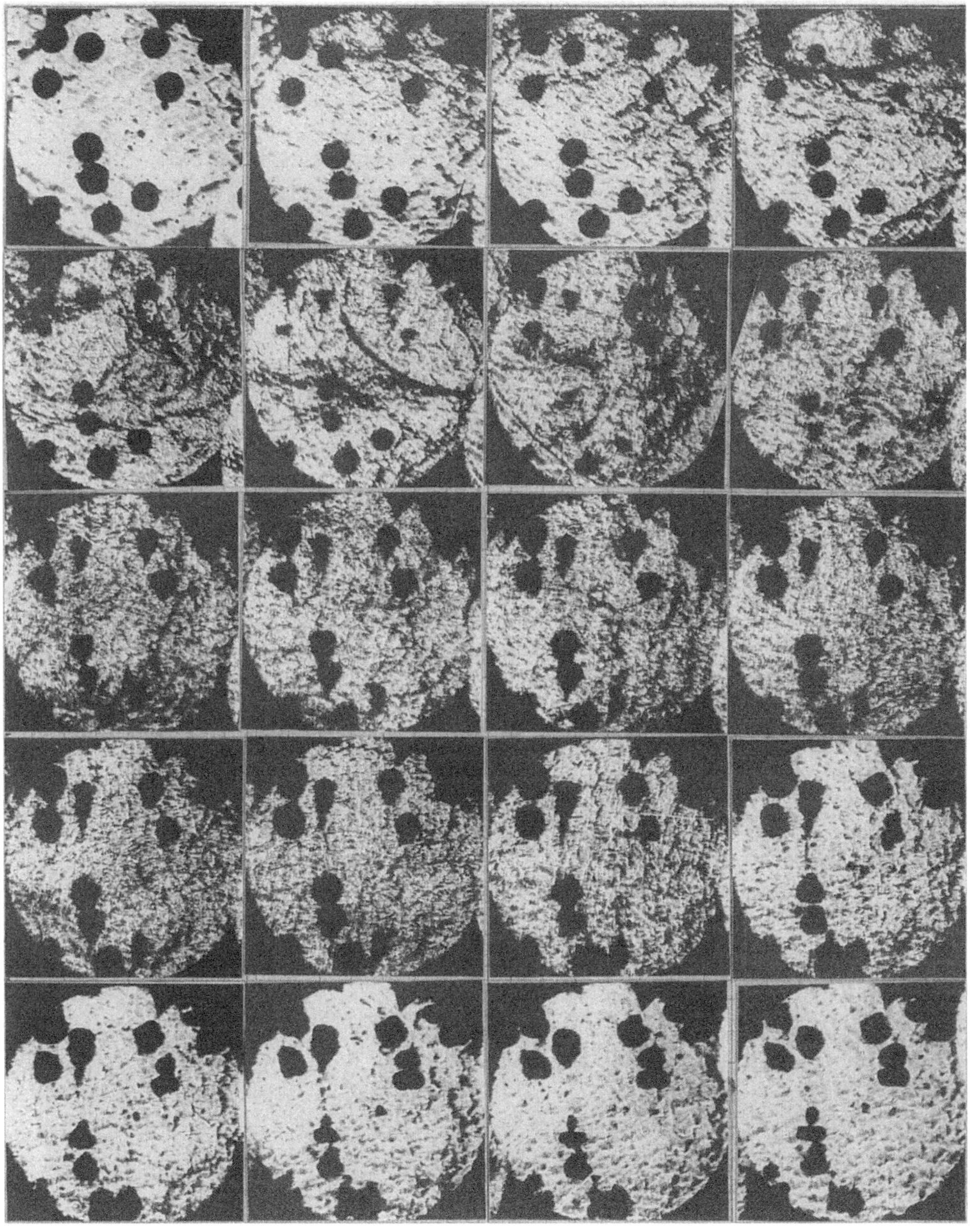

Fig.17: Behavior of a bubble group in the reacting zone of the detonation wave. Window diameter is $21mm$. Sequence of frames: Top row, left to right, top second row, etc. Time intervals: 0, 20, 25, 30, 35, 40, 45, 50, 55, 60, 65, 70, 75, 80, 100, 120, 140, 160, 180, 200 μs. Same parameters as in Fig.16

Acknowledgement

Especially Wolfgang Tepper, Ali Gülhan, and Martin Blanke have contributed to this work; this is gratefully acknowledged.

References

1 L.L.Foldy, Phys. Rev.**67**, 109(1945).
2 E.L.Carstensen and L.L.Foldy, J.Acoust.Soc.Am.**19**, 481(1947).
3 E.Silberman, J.Acoust.Soc.Am.**28**, 925(1957).
4 C.Devin, Jr., J.Acoust.Soc.Am.**31**, 1654(1959).
5 I.J.Campbell and A.S.Pitcher, Proc.R.Soc.London Ser.A**243**, 534(1958).
6 A.Crespo, Phys.Fluids**12**, 2273(1968).
7 R.I.Nigmatulin and V.Sh.Shagapov, Izv.Akad.NAUK SSSR, Mekh.Zhidk.Gaza **6**,30(1974).
8 L. Noordzij, Ph.D. dissertation, Technische Hogeschool, Twente 1973.
9 L.van Wijngaarden, J.Fluid.Mech.**33**, 465(1968).
10 S.S.Kutateladze,Editor,*Nonlinear wave processes in two-phase media.* XXth Sibirian Seminar on Thermophysics, Novosibirsk, 1976.
11 V.V.Kusnetsov, V.Ye.Nakoryakov, B.G.Pokusaev and I.R.Shreiber, J.Fluid.Mech.**85**, 85(1987).
12 V.Ye.Nakoryakov, I.R.Shreiber and V.G.Gasenko, Fluid Mech.Sov.Res. **10**,51(1981).
13 V.Ye.Nakoryakov, B.G.Pokusaev and I.R.Shreiber, *Wave propagation process in gas- and vapor-liquid media* (Institute f. Thermophysics, Novosibirsk, 1983).
14 D.S.Drumheller, M.E.Kipp and A.Bedford, J.Fluid Mech.**119**, 347(1982).
15 A.Biesheuvel, Ph.D. dissertation, Technische Hogeschool, Twente, 1984.
16 R.Omta, J.Acoust.Soc.Am.**82**, 1018(1987).
17 M.J.Miksis and L.Ting, J.Acoust.Soc.Am.**76**, 897(1984).
18 A.Prosperetti, L.A.Crum and K.W.Commander, J.Acoust.Soc.Am.**83**, 502(1989).
19 V.Kamath and A.Prosperetti, J.Acoust.Soc.Am.**85**, 1538(1989).
20 A.Prosperetti, J.Fluid Mech.**222**, 587(1991).
21 A.E.Beylich, VDI Forschungsh.**630**(1985).
22 A.E.Beylich and A.Gülhan, Phys.Fluids A**2**, 1412(1990).
23 R.Clift, J.R.Grace and M.E.Weber,*Bubbles, Drops and Particles* (Academic, 1978).
24 A.E.Beylich, in Proceed. Int. Symp. on Propeller and Cavitation, Wuxi, 1986, 274.
25 D.Y.Hsieh, J.Basic Engrg., Transact. ASME, 991(1965).
26 D'Ans–Lax, *Taschenbuch für Chemiker und Physiker* Bd.I (Springer, Berlin, 1967).
27 U.Parlitz, V.Englisch, C.Scheffczyk & W.Lauterborn, J.Acoust.Soc.Am.**88**, 1061(1990).
28 M.Morioka, J. of Nuclear Sci. and Techn.**11**, 554(1974).
29 A.Gülhan and A.Beylich, Acustica**63**, 276(1987).
30 H.Takahira, T.Akamatsu and S.Fujikawa, FED–Vol.116, Cavitation'91, ASME 1991.
31 L.D'Agostino and C.E.Brennen, J.Fluid Mech.**199**, 155(1989).
32 R.I.Nigmatulin, Fluid Mech.Sov.Res.**10**, 72(1981).
33 W.G.Gray and P.C.Y.Lee, Int.J. Multiphase Flow**3**, 333(1979).
34 Y.Matsumoto and M.Kameda, Pressure wave phenomena in bubbly liquid. Jap.–U.S. Seminar on Two–Phase Flow Dynamics, Berkeley, July 5–11, 1992.
35 V.K.Kedrinskii and C.L.Mader, in *Shock Tubes and Waves*, H. Grönig, Ed., p.371 (VCH, Weinheim,1988).
36 T.Hasegawa and T.Fujiwara, 19th Symp.on Comb. The Combustion Inst.,675(1982).
37 A.V.Pinaev and Sychev, Fizika Goreniya i Vzryva**22**, 109(1986).
38 A.E.Beylich and A.Gülhan, in *Adiabatic Waves in Liquid–Vapor Systems*, IUTAM Symp. Göttingen, p.39 (Springer, Berlin, 1990).

PRESSURE WAVES IN TWO-PHASE BUBBLE/SLUG FLOWS AND INTERPHASE PROCESS

B.G.Pokusaev*, N.A.Pribaturin, E.S.Vasserman, S.I.Lezhnin

Institute of Thermophysics, Siberian Branch of Russian Academy of Sciences, 630090, Novosibirsk, Russia

* *Moscow State Academy of Chemical Mechanical Engineering 107884, Moscow, Russia*

Abstract This paper considers the experimental results and physical effects on the pressure waves dynamics of vapour-liquid two-phase medium of bubble and slug structure. The general mechanisms of wave formation, behaviour and instability of two-phase flow under pressure waves, basic peculiarities of the interphase heat transfer is obtained. The role of destruction and collapse of bubbles and slugs, phase transition (condensation and evaporation), structural transition on wave dynamics is studied also. Experiments were carried out at the 52 and 8 mm ID tubes with Freon-11 and water as a test liquid. High-speed filming, laser optical methods, together with pressure history recording used in experiments. Theoretical models are suggested for wave evolution in media with nonuniform void fraction, which describe amplification of reflected waves and pressure pulses generation.

Introduction

Due to recently increased interest to investigations of pressure waves propagation in two phase media we present in this report our recent results alongside with the data obtained earlier. Up to the moment we have studied in details propagation of waves in gas-liquid media of bubble and slug structures [1,2,3]. Uncleared remained the problem of phase transition effect on the pressure waves evolution. Our earlier investigations [2] for the case of rather weak waves showed that "switching on" phase transition can result both in essential damping of pressure waves and their amplification. As is shown in work [2] for the case of rather weak waves ($\Delta p_0 / p_0 \leq 0.5$) effect of thermal dissipation is so strong that sharp pressure step is not forming. As the intensity of initial shock wave increases, oscillatory shock waves can form in vapour-liquid mixture as well as in bubble medium without phase transition [1]. Regularities of propagation of weak shock waves in vapour-liquid bubble medium being analysed in [1,2], based on assumption of insignificant change of void fraction after the wave front. It is not correct for the case of rather strong waves where partial or

S. Morioka and L. van Wijngaarden (eds.),
IUTAM Symposium on Waves in Liquid/Gas and Liquid/Vapour Two-Phase Systems, 107-116.

complete condensation is possible after the pressure pulse front. This leads to sharp change of acoustical properties of the medium in a wave and hence it should effect on wave propagation. To reveal the regularities and the mechanism of amplification a series of experiments in vapour-liquid flow of bubble and slug structures was conducted.

Propagation of pressure waves in two-phase medium of slug structure without phase transition taking into account had been also studied earlier [3,4]. It was shown that the waves evolution there is similar to that in a bubbly liquid. Investigations of phase transition effect on the wave dynamics were absent to the moment of beginning of this work.

Bubble structure with phase transition

Experiments were carried out on the vertical thermostated two-phase shock tube [5], where working liquid (water, Freons-11, 113) was kept on saturation point. Bubbles were generated by barbotage of vapour-liquid through capillaries mounted at the tube bottom. Shock wave initiated by breaking of membrane separating the test section from high pressure chamber. Original optical technique based on a simple shadow scheme allowed on-line monitoring of main parameters of the bubble structure: void fraction and characteristic size of the bubbles as well as their dynamics in a pressure wave.

In the figures below the experimental results on dynamics of shock waves and vapour bubbles are showed. As is seen from Fig.1 (designated by "r" here and below are the waves reflected from the tube bottom) weak pressure perturbations are damped much while propagating and their profile becomes monotone. Front of condensation looks a like -Fig.2, curve $\varphi = \varphi(t)$. The dependence of wave front width $\Delta = u\Delta p_0 / (dp/dt)_{max}$ the distance of wave propagation is shown in Fig.2. One can see an increase of Δ with the distance, which results in pressure jump decay.

Stronger waves acquire oscillatory structure at early stages of their evolution. Nonmonotone also is the condensation front here. As is seen here bubbles make mainly 1.5 cycles of oscillations before practically absolute disappearing. An increase of the shock wave amplitude up to $1.5p_0$ results in qualitative changes. Experiments have shown that shock waves of such intensity may transform into short shock pulses of 300-500 mcs duration; their amplitude significantly exceeds amplitude of incident wave (Fig.3). In opposite to the previous case width of pressure pulse decreases with the distance of pulse propagation. Vapour bubbles are collapsing in such pulse and can appear again after its propagation. Distance of the order of its length the pulse propagates without significant change of its form, that indicates of its quasistationary character. Such change in a pressure wave profile is directly connected with dynamics of bubbles collapse. It is known that collapse of solitary bubble is followed by radiation of short high amplitude pressure pulse [6]. Such reradiation should essentially effect on pressure wave evolution in vapour-liquid medium, here bubbles can interfere with each other and collective effects can appear.

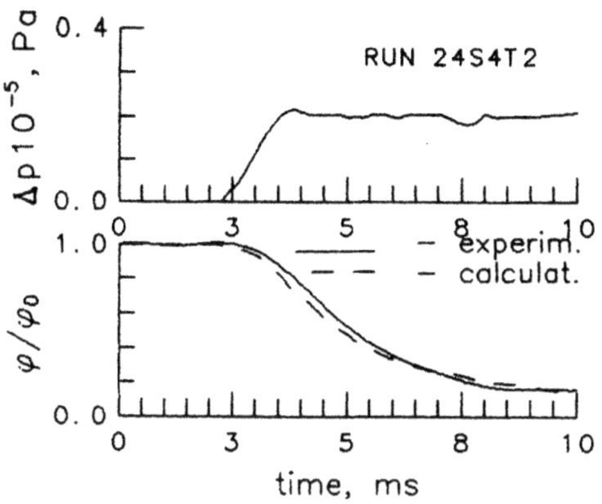

Fig.1 Profiles of pressure and void fraction for weak shock wave. Water, p_0=0.1 MPa, φ_0=0.005, R=1.7 mm

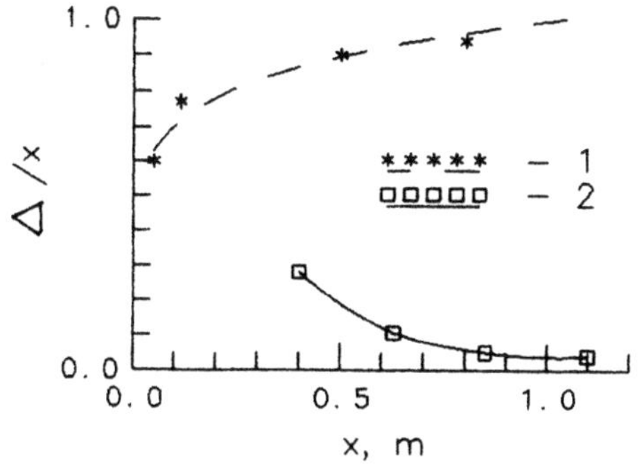

Fig.2 Width of pressure wave front vs. distance. Water, p_0=0.1 MPa, φ_0=0.01, R=1 mm, 1 - weak wave, $\Delta p_0 / p_0$=0.08-0.25, 2 - pressure pulse, $\Delta p_0 / p_0$=4.

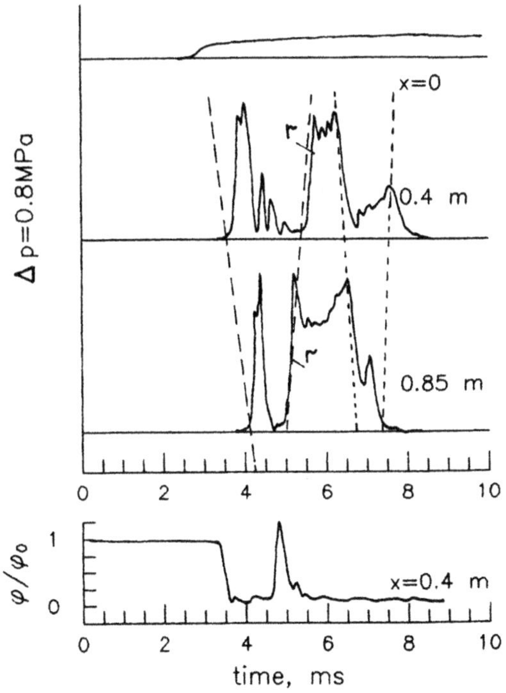

Fig.3 Evolution of pressure pulse in vapour-water bubble mixture, p_0=0.1 MPa, φ_0=0.002, R=0.6 mm

Theoretical model Most our early papers on the wave dynamics in bubbly media study the processes of perturbation propagation, with the value of the void fraction along the channel is constant. For the pressure waves of moderate intensity the approximation of the incompressible carrying liquid (for example [1, 2])is usually employed. It is quite appropriate for a higher vapour void content $\varphi > \rho_g c_g^2 / \rho_l c_l^2$. This limiting condition over φ naturally fails for the processes, accompanied by the full condensation of vapour bubbles. In more general case the problems are solved in the framework of following assumptions:

a) While the wave propagates the phase transitions take place on the bubble-liquid boundary. They are determined by the liquid thermal conductivity [1].

b) Inside the spherical bubbles, there is saturated vapour, for which the state equation of the ideal gas holds true.

c) The functional connection between perturbations of liquid pressure and density is considered to be linear $\delta\rho_l = \delta p_l / c_l^2$.

d) Oscillations of a single bubble are described by the Raleigh equation.

Taking into account the above assumptions the equations of momentum, continuity and the equation of acoustic compressibility of liquid have the form

$$\frac{\partial p}{\partial t} + \rho_l c_l^2 \frac{\partial u}{\partial t} = 4\pi n \rho_l c_l^2 Q \tag{1}$$

$$\rho_l \frac{\partial u}{\partial t} + \frac{\partial p}{\partial z} = 0 \tag{2}$$

Here $Q = (1/3)dR^3/dt$. Equations describing the behaviour of the homogeneous saturate vapour have the following form [7]

$$\frac{dQ}{dt} = R\frac{p_g - p}{\rho_\ell} + \frac{Q^2}{2R^3} \tag{3}$$

$$\frac{dp}{dt} = -\frac{3\gamma p_g}{R}\left[\frac{Q}{R^2} - \frac{q_l}{\rho_g h_{g,l}}\right] \tag{4}$$

Here q is the heat flux density into the liquid phase. Temperature distribution in the liquid is described by the thermal conductivity equation

$$\frac{\partial T}{\partial t} + \frac{Q}{r^2}\frac{\partial T}{\partial r} = a_l\left[\frac{\partial^2 T}{\partial r^2} + \frac{2}{r}\frac{\partial T}{\partial r}\right] \quad , \quad r > R \tag{5}$$

with $T(r = R) = T_s(p_g)$, $T(r \Rightarrow \infty) = T_0$.

The system of equations (1)-(5) describes the wave dynamics in the bubble vapour-liquid medium with nonuniform void fraction $\varphi(z)$ and $R = R(z)$. In the case of small amplitude wave propagation in a uniform bubbly media the set of equation (1)-(6) can be reduced to a one model evolution equation

$$\frac{\partial \tilde{p}}{\partial \tau} + \frac{\partial \tilde{p}}{\partial \zeta} + M\tilde{p}\frac{\partial \tilde{p}}{\partial \zeta} + \frac{M}{\sigma^2}\frac{\partial^3 \tilde{p}}{\partial \zeta^3} = -WM^{1/2}\int_0^\tau \frac{\partial \tilde{p}/\partial \tau'}{\sqrt{\tau - \tau'}} d\tau' \tag{6}$$

where $\tau = c_0 t / L$, $\xi = z / L$, $\tilde{p} = \Delta p / \Delta p_0$, L-is the length of initial perturbation, Δp_0 is the amplitude of initial perturbation, Similarity criteria W, M, σ define the contribution into distortion of wave profile of disperse, nonlinear effects and interphase heat transfer process[2]. This equation has been obtained earlier for bubble vapour-liquid medium [2] and with zero right-hand side for the wave propagation in the bubble structure [8].

Fig.4 demonstrate the results of calculations of a pressure wave evolution with the initial void fraction decaying toward the wave propagation according to the law $\varphi(z) = \varphi_0 \exp(-z/l)$ in the case of stepwise increase of the external pressure to $p_0 + \Delta p_0$. Characteristics of the wave propagating in nonuniform bubble medium depend essentially on the perturbation amplitude Δp_0 and on the void fraction distribution $\varphi(z)$. The front oscillatory structure (Fig.4a) is characteristic only for intensive enough shock waves ($\Delta p_0 / p_0 \geq 1$), weak waves have a monotonous structure (Fig.4b) that is due to the dissipative losses in the phase transition. The mean pressure gradient behind the wave front depends only on the vapour content $\varphi(z)$, the structure of the wave front is also determined by the "microparameters" of medium (radius and concentration of vapour bubbles) One can see from these results the effect of void fraction nonunifomity on wave amplification.

Let us consider the shock wave which reflected from the bottom of the shock tube. Fig.5 shows the amplification of such reflected shock wave. For initial amplitude of shock wave of 0.2 MPa the maximum amplitude of reflected shock wave is 2.2 MPa. The comparison of experimental and calculated by (1)-(5) results is well. The propagation of the pressure waves in the dead-end pipeline of length H is considered under the following boundary conditions $z = 0,\ p = p_0 + \Delta p_0;\quad z = H,\ u = 0$. Time interval between the pressure maximums is $2\Delta t_0$ where $\Delta t_0 = H / c_0$ is the time of the wave propagation along the channel in the medium with void fraction φ. The vibration period decreases gradually as a result of the decrease of void fraction and asymptotically approaches to 4H/c.

Slug structure

The experiments in slug regime have revealed, alongside with general regularities, also essential differences in the evolution of shock wave compared to a bubbly flow. The wave structure here also differs much from that reported for a gas-liquid slug flow [3,4]. Relatively weak waves $\Delta p_0 / p_0 < 1$ as well as for a bubbly medium acquire the oscillatory structure in the front at early stages of formation (Fig.6), further oscillations attenuate and the front flattens. The reflected waves designated by the symbol "r" have the same character and nature as in a bubbly medium. An increase in the amplitude of an incident wave up to 1.2 atm results in qualitative changes, i.e. random high-amplitude pulses appear first at the background of a step, which further aquire a rather regular comb-wise structure (Fig.7). According to the filming of the process an intensive destruction of vapour slugs starts with approximately these amplitudes of the incident wave. In this case a highly turbulized two-phase cloud with the developed interphase surface forms. This results in a sharp intensification of heat and mass transfer and acceleration of complete condensation of vapour slug. Filming together with pressure history recording of the rather strong wave ($\Delta p_0 / p_0 > 1.2$) shows that

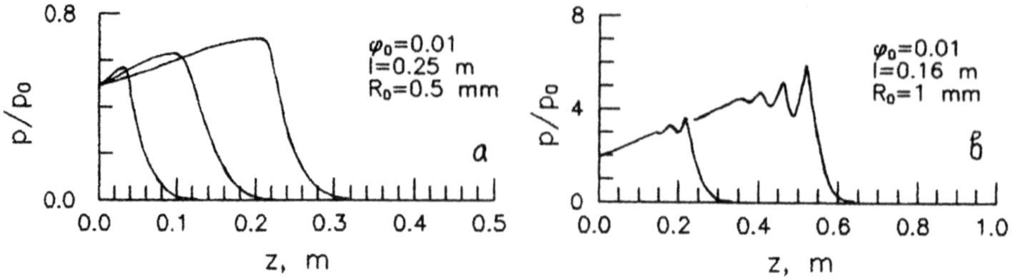

Fig. 4 Profiles of shock wave in bubble medium with nonuniform void fraction. a- $\Delta p_0 / p_0$=0.5, t=0.75, 1.5, 2.25 ms, b- $\Delta p_0 / p_0$=2, t=1, 1.5 ms

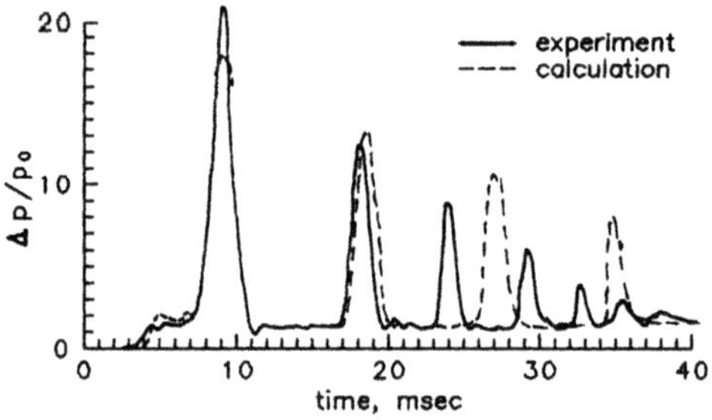

Fig.5 Amplification of reflected shock wave. Water, p_0=0.1 MPa, φ_0=0.008, R=1.5 mm

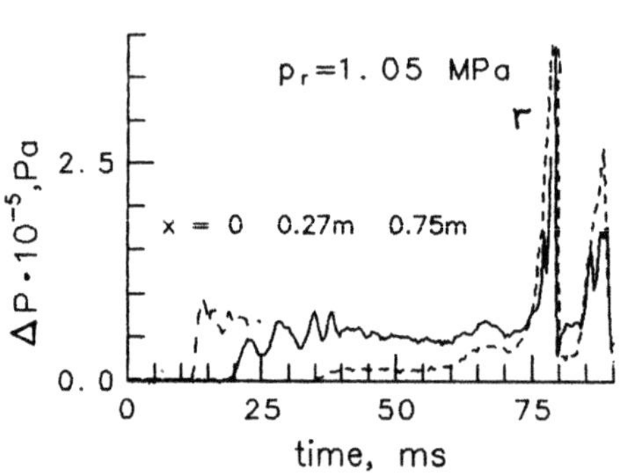

Fig.6 Evolution of weak shock wave in slug structure. Freon-11, D=8 mm, φ_0=0.23

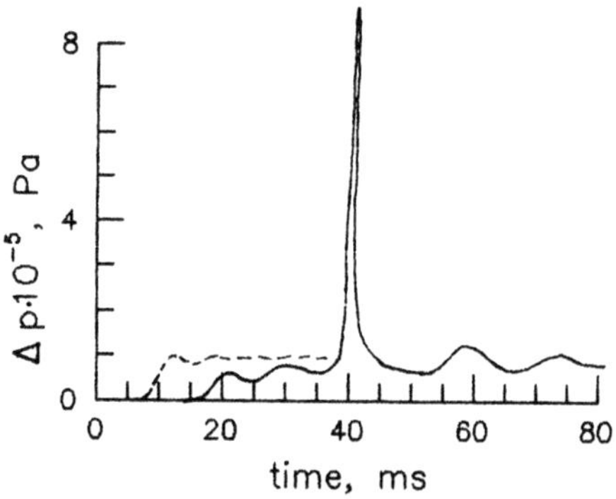

Fig.7 High amplitude pulse generated by weak shock wave. Freon-11, D=8 mm, φ_0=0.23, x = 0, 0.27 m

practically complete slug collapse accompanied by the strong pulse generation occurs already at the first compression stage. This pulse is shielded by the next slug, which is not collapsed yet, but it can propagate freely in the reverse direction since the proceeding slugs have condensed practically completely up to this moment. The threshold of the comb generation 1.1-1.2 is equal approximately to specific heats ratio, i.e. corresponds to a supersonic regime of the propagation of stationary wave with complete condensation, so as for a solitary shock pulse in a bubbly medium.

The further increase in the amplitude of initial pressure jump results in a rapid formation of a regular "comb" (Fig.8), each pulse of which corresponds to the collapse of the next slug and moves in the inverse direction (upwards) with the sound velocity in a pure liquid. In this case the first pulse corresponds to the front of complete condensation wave moving downwards. Its velocity agrees well with the expression for a stationary wave with complete condensation which in the approximation of incompressible liquid is of the form: $U^2 = \Delta p_0 / [\rho_l \varphi_0 (1 - \varphi_0)]$ A low-frequency rapidly attenuating precursor propagating with higher velocity can be singled out in front of this wave. Its velocity is close to the value for a shock wave in a gas-liquid mixture $u = c_0 [\Delta p_0 / p_0 + 1]^{1/2}$, where $c_0 = [\gamma p_0 / (\rho_l \varphi_0 (1 - \varphi_0))]^{1/2}$ is the sound velocity in homogeneous two-phase mixture.

Essential difference in the strong shock wave structure between bubbly and slug vapour-liquid flows can be qualitatively accounted for as follows: backward secondary waves radiated by collapsing bubbles are distributed uniformly and randomly in time (space) in contrast to the slug flow, where they are desecrate and definite in character. So for bubbly flow the backward waves superposition forms the shock pulse tail as a surplus pressure fluctuating near incident step value. Forward secondary waves quickly slow down in bubbly flow to the main front velocity because of void fraction increase and superpose into the solitary shock pulse.

Mathematical model and calculation The well-known [3,9] mechanical analogy ("mass + spring" chain) is used while constructing a mathematical model of the process of waves propagation in a two-phase media of slug structure. According to the model inertialess gaseous (vapour) slugs of cylinder form transfer the momentum to the inertial liquid plugs. Between the vapour slugs and the channel wall there is a thin liquid film of constant thickness. The film flow is neglected. The dynamics equation for the n-th liquid plug has a form:

$$\rho_l l (1 - \varphi) \frac{d^2 \xi_n}{dt^2} = p_n - p_{n+1} \tag{7}$$

The following assumptions are used:

1. The vapour pressure throughout the vapour slug is uniform in volume. The vapour is an ideal gas. The equation of state of n-th vapour slug is

$$p_n (l_g + \xi_n - \xi_{n-1}) = \frac{\tilde{R} T_n}{\mu} \tag{8}$$

2. The liquid in the process is not compressible.

3. There is a local thermodynamic equilibrium of vapour and liquid on the interface. The connection between a temperature of the interface and pressure is determinated by the Clapeyron-Clausius equation

$$\frac{dp_{n,s}}{dT_s} \cong \frac{h_{g,l}\rho_{g,n}}{T_s} \tag{9}$$

4. Depending on the process conditions, the vapour temperature and density are connected with pressure according to the polytropic law. At the earlier stage of the dynamic process when a heat layer does not intergrow the vapour slug adiabatic law is valid. At asymptotically high values of t the vapour within the vapour slug should be considered as saturated. A detailed analysis of the vapour bubble dynamics showed that these asymptotic models differ only in effective polytropic exponent in the case of weakly nonlinear processes. For saturated vapour the exponent $\gamma^* = (1 - p_0 / \rho_g L)^{-1}$

5. Viscous and mass forces are negligible in the process of a liquid plug motion in the channel.

6. The change of vapour mass is determined only by the heat flux q in the liquid

$$m_{g,n} = m_{g,0} + \frac{1}{h_{g,l}} \int_0^t q_{l,n} dt \tag{10}$$

The heat flux q is determinated by means of the two methods:

a) numerically from the solution of the thermal conductivity equation

$$\frac{\partial T_n}{\partial t} = a_l \frac{\partial^2 T_n}{\partial y^2} \tag{11}$$

with the boundary conditions on the vapour-liquid interface $T_n = T_{s,n}(p)$ and on the channel wall $T_n = T_0$ (isothermal wall) or $\partial T_n / \partial y = 0$ (adiabatic wall);

b) analytically from generalised Duamel`s theorem of a flux on a flat interface within a changing surface. When a surface area increases

$$q_{l,n} = -\lambda \left[\pi \frac{D^2}{2} \int_0^t \frac{\partial T_s}{\partial t'} \frac{dt'}{\left(\pi a (t - t')\right)^{1/2}} + \int_0^t \frac{\partial}{\partial t'} \left[\pi D \ell_g \Delta T_s\right] X_i (t - t') dt' \right] \tag{12}$$

when a surface square decreases

$$q_{l,n} = -\lambda \left[\pi \frac{D^2}{2} \int_0^t \frac{\partial T_s}{\partial t'} \frac{dt'}{\left(\pi a (t - t')\right)^{1/2}} + \pi D \ell_g \int_0^t \frac{\partial T_s}{\partial t'} X_i (t - t') dt' \right] \tag{13}$$

where X(t)-are the Theta functions. The different notation of expressions (12) and (13) is connected with the fact that with the surface extension the liquid with the initial temperature T is involved into the heat exchange.

It has been shown that the behaviour of pressure waves of small ($\Delta p_0 / p_0 < 1$) intensity , their structure, attenuation in a slug vapour-liquid medium are well described by a one model equation (6). In this case the similarity criteria is

$\sigma = (24M)^{1/2} L / 1$ $W = \left(a_l l_0 / \pi c_0 l^2 \varphi_0^2 M\right)^{1/2} c_s^2 \rho_l C_{p,l} T / \rho_g h_{g,l}^2$ $M = (\gamma + 1)\Delta p_0 / 2\gamma p_0$

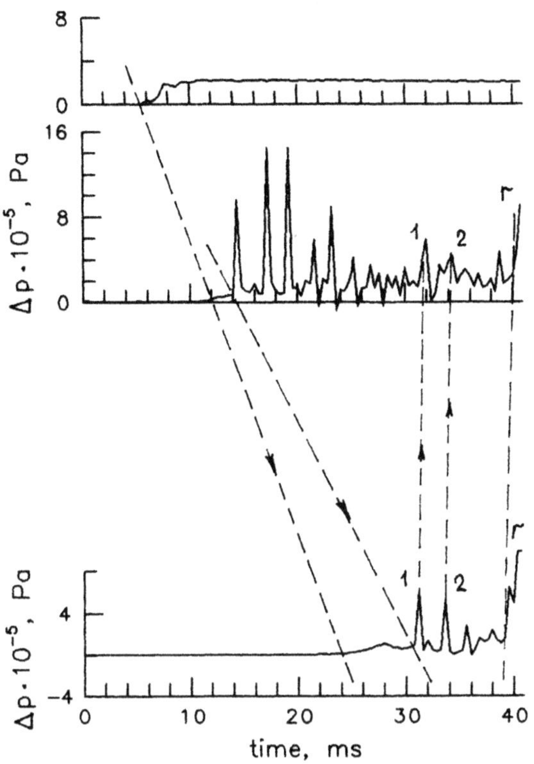

Fig.8 Evolution of strong shock wave in vapour-slug structure. Freon-11, p_0=0.1 MPa, D=8 mm, φ_0=0.23, x=0, 0.2, 0.65 m

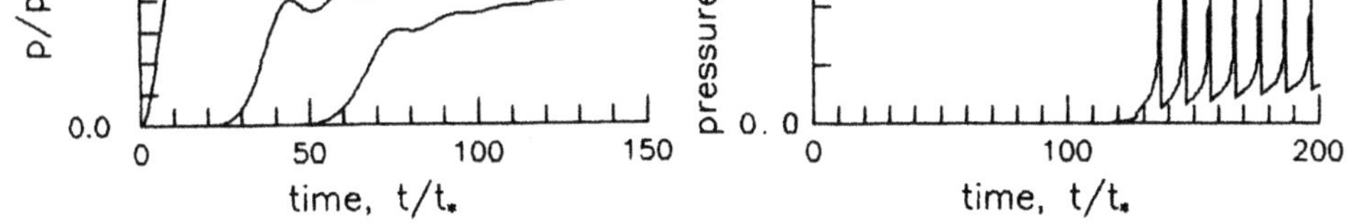

Fig.9 Calculation of pressure wave profiles for slug structure. φ_0=0.25, a- weak wave, x=0.05, 0.3, 0.5, b- strong wave, $\Delta p_0 / p_0$=1.5, x=0.5 m

Depending on values of σ, M, W the wave profile may take on following structures: attenuated solitons, oscillatory shock waves, quasilinear wave packets.

While calculating more strong wave dynamics it is necessary to model complete collapse of vapour slugs and liquid plugs interaction along with the common equations system (7) - (13). Two asymptotic models are used:

a) liquid plugs interaction is assumed to be absolutely nonelastic; compression and decompression wave propagation through liquid is neglected.

b) liquid plugs interaction is assumed to be absolutely elastic; compression wave propagation through liquid plugs is taken into account here. The calculation of the weak shock wave evolution is shown in Fig.9a. Comparison of numerical results for strong shock wave by model (a) with experimental data are shown in Fig.9b. One can see that even neglecting of the local waterhammers high pressure pulses appear here.

Acknowledgement

Part of this work was supported by the Russian foundation investigation, grant N 93-01-17599 and International Sciences Foundation, grant N RPIOOO.

References

1. Nakoryakov V.E., Pokusaev B.G., Shreiber I.R. Wave dynamics of gas-and vapour-liquid media. - New York a.o. Begell House Publishers, 1992.

2. Nakoryakov V.E., Pokusaev B.G., Pribaturin N.A., Shreiber I.R. (1988) The wave dynamics of vapour-liquid medium. Intern. J. of Multiphase Flow 1988, v.14, N6, pp.655-677.

3. Pokusaev B.G., Lezhnin S.I., Pribaturin N.A. Waves in a gas-liquid medium of slug structure. Russian J. Eng. Thermophysics, 1991, v.1, N 4.

4. Nakoryakov V.E., Pokusaev B.G., Lezhnin S.I., Pribaturin N.A., Shock waves structure and their propagation in two-phase slug flow. Proc. 16th ISSW,1987, Aachen, Germany, pp.305-309.

5. Pokusaev B.G., Vasserman E.S., Mulliajanov I.I., Pribaturin N.A. (1989) Bubbles collapse and destruction in compression waves propagating in two-phase medium. In: Nestacionarnye processy v dvuchfaznych potokah, 1989, Novosibirsk, pp.3-27 (in Russian).

6. Nakoryakov V.E., Pokusaev B.G., Pribaturin N.A., Lezhnin S.I., Vasserman E.S. Nonstationary wave processes in boiling media. Adiabatic Waves in Liquid-Vapor Systems IUTAM Symp., 1989, Gottingen, Germany pp.381-391.

7. Nigmatulin R.I.,Dynamics of multiphase media, 1987, Moskow, Nauka

8. Wingaarden L., One-dimensional flow of liquids containing small gas bubbles, Annual rev. of fluid mechanics, 1972, v.45, pp.369-396.

9. Matsui G., Sugihara M., Arimoto A., Propagation characteristics of pressure wave through gas-liquid plug systems, Bull. JSME, 1979, v.22, pp.1562-1569.

Structure of Shock Waves in a Liquid Containing Gas Bubbles

Masaharu KAMEDA† and Yoichiro MATSUMOTO
Department of Mechanical Engineering, The University of Tokyo
7-3-1 Hongo, Bunkyo-ku, Tokyo 113, Japan

Abstract. Transient shock wave phenomena in a liquid containing noncondensable gas bubbles are investigated experimentally and numerically. In the experiment, time evolution of the shock wave is measured using two vertical glass shock tubes with different diameters, 18 mm and 52 mm. In the numerical simulation, thermal processes inside each bubble are directly calculated by using full equations for mass, momentum and energy conservation, and the results are combined with the averaged conservation equations of the bubbly mixture to simulate the propagation of the shock wave. Initial spatial distribution of the bubbles is also taken into account. The numerical results reveal that the spatial distribution greatly affects the structure of the shock wave, such as the pressure peak of the shock front and the period of the relaxation oscillation. The present numerical results agree very well with the experimental results, in which the bubbles do not distribute uniformly and are relatively concentrated on the axis of the shock tube. However, the experimental data is much different from the numerical results with uniform spatial distribution of the bubbles.

Key words: Bubble, Shock Wave, Shock Tube Experiment, Numerical Analysis, Spatial Distribution of Bubbles

1 Introduction

The bubbly mixture has some unique features as far as wave phenomena are concerned. For example, even though the concentration of bubbles may be quite small, the compressibility of the mixture increases so greatly that the speed of sound decreases much below that in either the gas or the liquid alone. Also the mixture is a dispersive and dissipative medium due to the motion of the bubbles (van Wijngaarden 1968, 1972; Kuznetsov *et al.* 1978). On the other hand, the relaxation effect caused by various kinds of phenomena, such as heat transfer through the bubble wall and the translational relative motion between the bubbles and the liquid, changes the waveform as the wave propagates. Noordzij & van Wijngaarden (1974) state that the relative motion plays a major role in the relaxation process. However, as already pointed out by some investigators (Nigmatulin & Shagapov 1974; Matsumoto & Kameda 1993; Watanabe & Prosperetti 1994) heat transfer, rather than relative motion, mainly affects the relaxation process. The heat transfer is governed by the thermal behavior of the bubble interior (Prosperetti 1991). Experimentally, the propagation velocity of shock waves in bubbly mixtures was first measured by Campbell & Pitcher (1958), and the time evolution of waveform was first observed by Noordzij and van Wijngaarden (1974). Moreover, the propagation of small perturbations (Kuznetsov *et al.* 1978) and relatively strong shock waves (Tan & Bankoff 1984) were observed. Owing to these efforts, the basic features of shock waves in bubbly mixtures have been demonstrated qualitatively. However, there are few experimental results which are quantitative enough to compare with the numerical result, except for the result by Beylich & Gülhan (1990). In consequence of the comparison between their theory and experiment, Beylich & Gülhan pointed out that the influence of neighboring bubble monopole fields cannot be negligible even though the gas volume fraction is very small. Recently, Watanabe & Prosperetti (1994) calculated

† Present address: Department of Mechanical Systems Engineering, Tokyo University of A & T (Agriculture and Technology), 2–24–16, Nakamachi, Koganei, Tokyo 184, Japan.

S. Morioka and L. van Wijngaarden (eds.),
IUTAM Symposium on Waves in Liquid/Gas and Liquid/Vapour Two-Phase Systems, 117-126.

the non-steady and steady profiles of the shock waves using a mathematical model, in which the thermal exchange between the gas bubbles and the liquid is described accurately. Nevertheless, they faced a discrepancy between their numerical results and the experimental data by Beylich & Gülhan and did not identify the cause of the discrepancy.

In this paper, we deal with the propagation of shock waves in liquids containing noncondensable gas bubbles numerically and experimentally. Governing equations for the bubbly mixture are formulated with emphasis on the radial and translational relative motion of the bubbles. The conservation equations for mass, momentum and energy inside the bubbles are solved directly in order to precisely estimate the effects of internal phenomena on bubble motion. The effect of neighboring bubble monopole fields introduced by Beylich & Gülhan is not included in this mathematical formulation. A numerical method, where individual bubbles are tracked to estimate the effects of volumetric changes and translational relative motion on wave phenomena, is developed. The propagation of the shock waves is calculated not only in the one-dimensional (1-D) situation, but also in the two-dimensional (2-D) one. In these calculations, the non-uniform spatial distribution of the bubbles is considered. Two vertical shock tubes with different inner diameter are used in the experiment. The time evolution of the waveform is measured by pressure transducers mounted at several points along the tube. Finally, the experimental results are compared with the numerical ones.

2 Experiments

2.1 Experimental apparatus

Figure 1 shows the schematic diagram of the experimental apparatus. Two vertical glass shock tubes with inner diameters D of 18 and 52 mm are used for the present experiment. The total length is 5.4 m. Each tube is separated into two parts: a 1.6 m long high pressure gas chamber and a 3.8 m long low (atmospheric) pressure gas section. The liquid is introduced in the low-pressure section up to a certain level, about 2 m. A shock wave is generated by rupturing a thin plastic membrane (material: Lumirror) separating the high pressure chamber from the low pressure gas section. Five piezoelectric pressure transducers (Kistler 603B) are flush mounted on the wall of the tube to measure the shock waves. The signals from the transducers are recorded in a 1 MHz digital transient memory through charge amplifiers. A number of glass needles with 0.1 mm inner diameter and 80 mm in length are mounted at the bottom of the tube to serve as bubble generators. The gas volume fraction is almost proportional to the number of glass needles. To make the variance in size distribution of the generated bubbles small, the pressure difference between the liquid at the bottom of the shock tube and the gas chamber is carefully maintained constant. A photograph is taken by a 35 mm framing camera every time just before rupturing the membrane. The initial bubble radius and local gas volume fraction at the photographic observation section are measured from the photograph. The initial bubble radius and local gas volume fraction at other points are obtained by accounting for the hydrostatic effect.

In the present experiments, the liquid used is silicone oil (Shin-etsu Chemical Co. KF96-50: density $\rho_l = 960$ kg/m^3; surface tension $\sigma = 20.8 \times 10^{-3}$ N/m; kinematic viscosity $\mu_l/\rho_l = 50\times10^{-6}$ m^2/sec at 298.1 K). The gas inside the bubbles is nitrogen. The vapor pressure of silicone oil is extremely low so that the bubble consists of only noncondensable gas without vapor. The liquid viscosity is high enough that the bubbles remain spherical during the interaction with shock. The initial mean gas volume fraction $\alpha_{L0,mean}$ is approximately 0.2%, the pressure step ΔP is 100 kPa, and the mean bubble radius $R_{L0,mean}$ is about 0.6

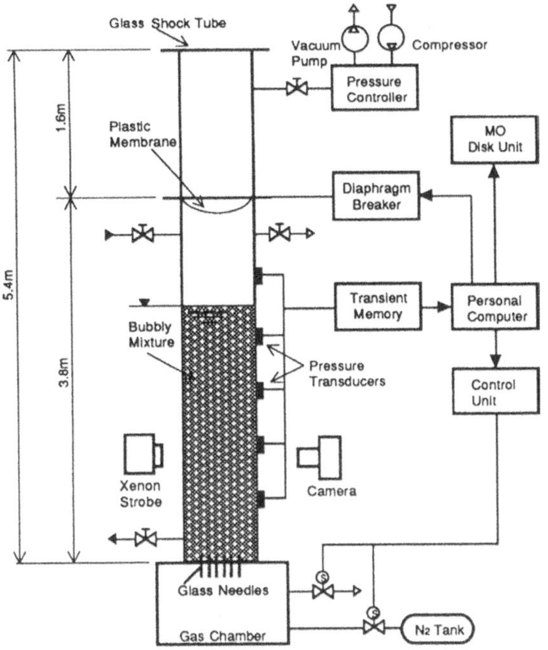

Fig. 1. Schematic diagram of the experimental apparatus

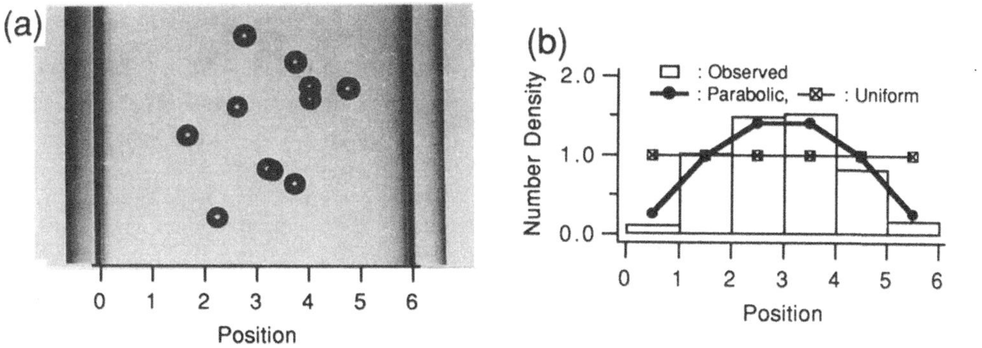

Fig. 2. Spatial distribution of bubbles in the radial direction of the 18 mm-shock tube.

mm in all the experiments. The standard deviation of $R_{L0,mean}$ is less than 5%.

2.2 SPATIAL DISTRIBUTION OF THE BUBBLES IN THE SHOCK TUBE

Figure 2 shows a typical spatial distribution of the bubbles in the radial direction of the shock tube. An example of the photographed bubbles is shown in Fig. 2(a). In Fig. 2(b), an experimental histogram is displayed with two lines which represent two ideal cases; the parabolic and uniform distributions in the radial direction. To obtain the experimental data, the pictures are divided into six parts, then the number of bubbles in each part is counted up. The two lines are obtained by the same means as the experimental data. The histgram

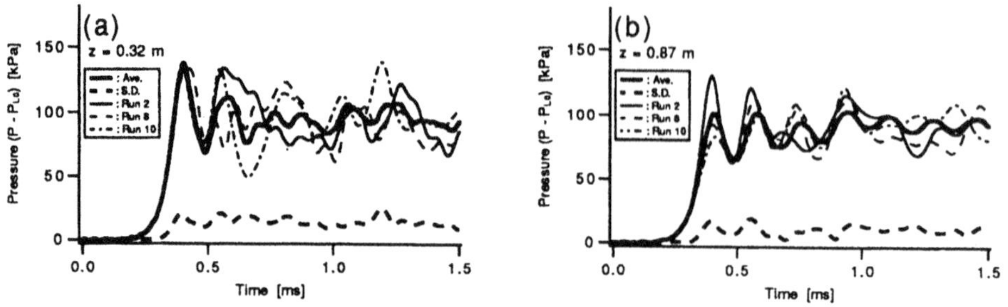

Fig. 3. Average and overlay of pressure profiles with standard deviation. $R_{L0} = 0.661$ mm, $P_{L0} = 105.2$ kPa (left figure); $R_{L0} = 0.650$ mm, $P_{L0} = 110.4$ kPa (right one). $D = 18$ mm, $\alpha_{L0,mean} = 0.15\%$, $T_l = 291.1$ K, $\Delta P = 101.2$ kPa.

and the two lines are normalized by the values of the uniform distribution. The experimental spatial distribution is very close to the parabolic one in the radial direction of the shock tube rather than the uniform one, as observed from the picture shown in Fig. ??(a). There is the same tendency in all experimental conditions. The spatial distribution of the bubbles is not uniform in the axial direction of the shock tube either. This non-uniformity has a strong effect on the shock wave.

2.3 Development of the waveform

Figure 3 shows the pressure profiles as functions of time obtained by averaging 10 shots with an overlay of a few individual profiles. In this case, the inner diameter of the shock tube is 18 mm, the mean gas volume fraction $\alpha_{L0,mean}$ is 0.15%, the pressure step ΔP is 101.2 kPa, and the liquid temperature T_l is 291.1 K. The standard deviation of the averaged profiles (S.D.) is also shown in Fig. 3. The left and right figures represent the profiles obtained from the pressure transducer mounted at about 0.32 m and 0.87 m below the free surface, respectively. The initial bubble radius R_{L0} and pressure P_{L0} of the left figure are 0.661 mm and 105.2 kPa, respectively. Those of the right one are 0.650 mm and 110.4 kPa.

As the shock wave propagates from the free surface to the bottom of the shock tube, the first maximum of the average pressure profile decreases and the amplitude of the relaxation oscillation behind the shock front becomes smaller due to the thermal dissipation of the bubbles. These phenomena have already been found in previous investigations, e.g. Noordzij & van Wijngaarden (1974). As seen in Fig. 3, the individual profiles are very different from each other except at the shock front. This difference mainly comes from the non-uniform distribution of the bubbles in the axial direction. The standard deviation of the pressure profile is about 10% of the average value. It increases as the pressure step ΔP becomes large or the inner diameter of the shock tube D becomes small. The smaller the diameter of the tube, the smaller the total number of bubbles in the tube. This smaller number causes the standard deviation to increase, that is, the individual pressure profiles are more different from each other.

3 Numerical calculation

The bubbly mixture is defined as a liquid with many dispersed small bubbles. The gas volume fraction α is defined as the ratio of the volume of an unit cell ΔV to the sum of the volumes of all the bubbles v included in it. It is written as

$$\alpha = \sum_{k=1}^{K} v_k/\Delta V, \tag{1}$$

where K denotes the total number of bubbles in the unit cell. Here we assume that all the bubbles retain their spherical shape, so that the volume of the k-th bubble is written as $v_k = \frac{4}{3}\pi R_k^3$ for bubble radius R_k. In terms of the gas volume fraction α, density ρ and velocity $\boldsymbol{U}$, the mixture density ρ_m, and mixture momentum $(\rho \boldsymbol{U})_m$ are written as

$$\rho_m \simeq (1-\alpha)\rho_l, \tag{2}$$

$$(\rho \boldsymbol{U})_m \simeq (1-\alpha)\rho_l \boldsymbol{U}_l. \tag{3}$$

where the subscript l denotes the liquid phase. The density and momentum of the gas phase are neglected because they are very small against those of the liquid phase.

The bubbly mixture can be treated as a continuum fluid because the size of the bubble is assumed to be small compared to the characteristic length in the bubbly mixture. Using the mixture density and the mixture momentum mentioned above, the equations for the conservation of mass and momentum of the bubbly mixture are given as

$$\frac{\partial}{\partial t}[(1-\alpha)\rho_l] + \nabla \cdot [(1-\alpha)\rho_l \boldsymbol{U}_l] = 0, \tag{4}$$

$$\frac{\partial}{\partial t}[(1-\alpha)\rho_l \boldsymbol{U}_l] + \nabla \cdot [(1-\alpha)\rho_l \boldsymbol{U}_l \boldsymbol{U}_l] = -\nabla P + (1-\alpha)\rho_l \boldsymbol{g}, \tag{5}$$

where t, P and $\boldsymbol{g}$ denote time, averaged mixture pressure, and gravitational acceleration, respectively. In Eq. (5), the mixture viscosity is ignored because it has little influence on the wave phenomena.

To approximate the real experimental conditions as closely as possible, bubbles should be dealt with individually. Though the effect of relative motion between bubbles and liquid play a minor role on the wave phenomena in the bubbly mixture, it has some influence on the wave propagation process. Therefore, the effect of relative motion is included in this mathematical formulation (Matsumoto & Kameda 1993). On the other hand, the equation of motion of a single bubble in an infinite incompressible liquid, called the Rayleigh-Plesset equation (Plesset & Prosperetti 1977), is used as the equation for the radial motion of a bubble in the bubbly mixture. It is written as

$$R\ddot{R} + \frac{3}{2}\dot{R}^2 = \frac{1}{\rho_l}\left\{p_{gw} - \frac{2\sigma}{R} - 4\mu_l\frac{\dot{R}}{R} + \frac{\rho_l|\boldsymbol{U}_g - \boldsymbol{U}_l|^2}{4} - P\right\}, \tag{6}$$

where the gas phase denoted by the subscript g, the gas pressure at the bubble wall is denoted by p_{gw}, and the surface tension by σ. The gas pressure p_g depends on the thermal behavior of the bubble interior, that is, the effect of the internal processes appears on the wave propagation phenomena in the bubbly mixture through p_{gw}. p_g is estimated by solving the conservation equations for mass, momentum and energy of the bubble interior. On solving these equations, the following assumptions are employed: (1) The gas inside the bubble, consisting of noncondensable gas without vapor, obey the perfect gas law; (2) The temperature at the bubble wall is equal to the constant liquid temperature T_l; (3) The amount of gas inside the bubble remains constant; (4) The viscosity of the gas is ignored.

The finite difference method is used to solve the set of governing equations numerically. A staggered grid is used to discretize Eqs. (4) and (5), and a semi-implicit method, which is similar to the HSMAC method (Hirt *et al.* 1975), is applied for numerical integration. The staggered mesh is also used to discretize the conservation equations inside the bubbles and these equations are numerically integrated by the CIP(Cubic Interpolated pseudo-Particle) method (Yabe & Takei 1988; Matsumoto & Takemura 1994). The details of the numerical procedure are discussed in a recent paper (Matsumoto & Kameda 1993).

4 Results and discussions

4.1 Effect of the spatial distribution of bubbles

Propagation of shock waves in a bubbly mixture is calculated in two-dimensional Cartesian coordinate system as a kind of piston problem, where the pressure at the free surface is raised from the initial value P_{L0} to the high value P_{H0} at $t = 0$ and is then maintained constant. In this section, we try to give an answer to the question of how the spatial inhomogeneity of the bubbles like the experiment seen in Fig. 2 affects the wave phenomena in the bubbly mixture. The effect of body force (gravity) is ignored for the moment.

In this calculation, the pressure at one end of the tube is raised from the initial value P_{L0} to the high value P_{H0} at time zero and is then maintained constant. Initially, all the bubbles have the same size R_{L0}. Time t, transverse coordinate y and streamwise coordinate z are nondimensionalized as follows,

$$t^* = t/\tau_0, \qquad y^* = y/L, \qquad z^* = z/L.$$

$$\tau = R_{L0}\sqrt{\frac{\rho_l}{P_{L0}}}, \qquad C_0 = \sqrt{\frac{P_{L0}}{\alpha_{L0,mean}(1-\alpha_{L0,mean})\rho_l}},$$

$$L = C_0 \cdot \tau = \frac{R_{L0}}{\sqrt{\alpha_{L0,mean}(1-\alpha_{L0,mean})}}, \tag{7}$$

where an asterisk indicates the nondimensional variable, and $\alpha_{L0,mean}$ denotes the initial gas volume fraction assumed spatially uniform. τ is the characteristic time which has the same order of the period of free oscillation of a bubble. C_0 is the speed of sound in a bubbly mixture in isothermal equilibrium conditions. Thus L has the meaning of the characteristic wave length in the bubbly mixture.

An example of the results of the 2-D calculation is illustrated in Fig. 4. In this figure, the initial distribution of gas volume fraction, the instantaneous pressure contours at $t^* = 2$, 4, 6, 8 with the velocity vectors, and the time history of the pressure at $z^* = 10$, are shown for two cases, where the nondimensional width y^*_{max} are 4.0 and 2.0, respectively. The initial pressure P_{L0}, high pressure P_{H0}, mean gas volume fraction $\alpha_{L0,mean}$, liquid temperature T_l, and bubble radius R_{L0} are 100 kPa, 200 kPa, 0.2%, 298.1 K and 0.50 mm, respectively. The characteristic time τ, the speed of sound C_0 and the characteristic wave length L are 49.0 μs, 228 m/s, and 11.2 mm, respectively.

In this case, the bubbles are distributed uniformly only in the region from $0.25y^*_{max}$ to $0.75y^*_{max}$, so that the gas volume fraction in this "central" region is equal to 0.4%.

When the high pressure reaches the free surface, the front of the generated shock wave curves toward the high gas volume fraction region due to the difference of propagation velocity. As a result, a pressure "focusing" occurs. The amplitude of the pressure oscillation

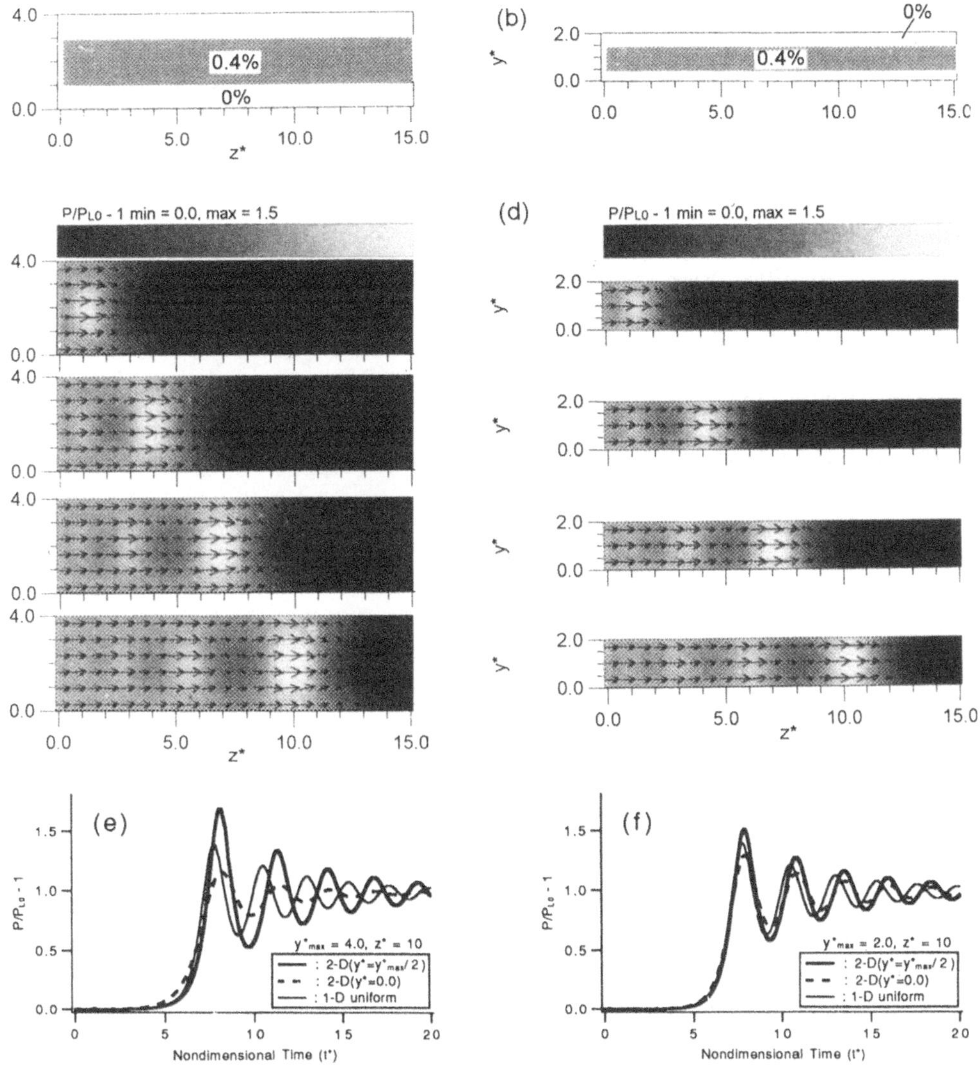

Fig. 4. Two-dimensional structure of shock waves. (a), (b) initial gas volume fraction; (c), (d) instantaneous velocity vectors with pressure contours at $t^* = 2, 4, 6, 8$; (e), (f) time history of pressure measuring at $z^* = 10$. $\alpha_{L0,mean} = 0.2\%$, $P_{L0} = 100$ kPa, $P_{H0} = 200$ kPa, $R_{L0} = 0.50$ mm. (a), (c), (e) $y^*_{max} = 4.0$, (b), (d), (f) $y^*_{max} = 2.0$.

behind the shock front in the high gas volume fraction region is larger than the amplitude in the low gas volume fraction region because of this focusing.

In the figures of the time history of the pressure (Fig. 4 (e), (f)), the results of the 2-D calculation are compared with those of the 1-D calculation with an uniform bubble distribution. First maximum of the pressure profile at the center ($y^* = 2.0$ for $y^*_{max} = 4.0$ and $y^* = 1.0$ for $y^*_{max} = 2.0$) is larger than the maximum at $y^* = 0$. Furthermore, the period of pressure oscillation behind the shock front according to the 2-D calculation is larger than that given by the 1-D uniform calculation. In the case of $y^*_{max} = 2.0$, on the other hand, the

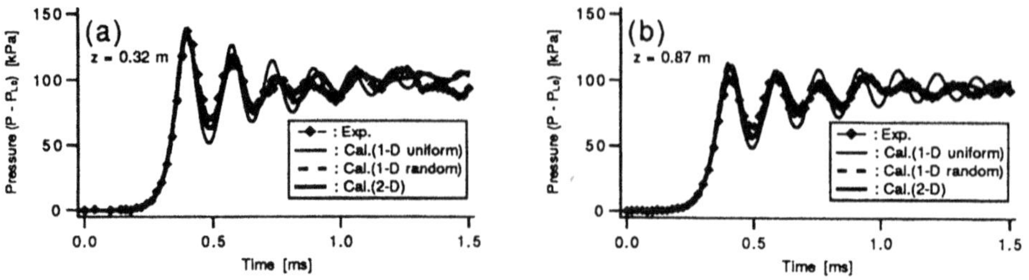

Fig. 5. Comparison of the numerical results with the experimental data shown in Fig. 3. Symbols represent the experimental data. Solid, broken and thin lines show the numerical results of the 2-D, 1-D random, and 1-D uniform cases, respectively.

discrepancy between the 2-D and 1-D uniform calculations is smaller than that of $y^*_{max} = 4.0$. Although the spatial distribution is similar, the two-dimensional structure of the shock wave mentioned above is found more clearly in the case where $y^*_{max} = 4.0$ than in the case of $y^*_{max} = 2.0$.

As seen in the figures of the velocity vectors (Fig. 4 (c), (d)), the magnitude of the transverse velocity obviously depends on the nondimensional width y^*_{max}. The magnitude of the transverse velocity U_{ly} at the shock front is of the same order as the streamwise velocity U_{lz} in the case $y^*_{max} = 4.0$.

Fluid motion in the transverse direction is restricted because the transverse velocities at $y^* = 0$ and y^*_{max} are forced to be zero. As y^*_{max} becomes smaller, therefore, the structure of the shock waves in the 2-D calculation, irrespective of the spatial distribution of the bubbles in the transverse direction, is similar to the 1-D uniform case. On the other hand, as y^*_{max} becomes larger, the wall effects become so small that the degree of freedom of the fluid motion in the transverse direction appears. As a result, the structure of the shock wave can undergo a large transformation.

4.2 Comparison of the numerical results with the experimental data

In this section, we compare the experimental data introduced in Section 2.3 with the numerical results. Only the averaged pressure profiles from the experiments are compared to the numerical results. The numerical calculation is carried out in cylindrical polar coordinates taking the gravitational force into account. To approximate the experimental conditions as closely as possible, the total number of bubbles, height of the bubbly mixture, and inner diameter of the shock tube are set to be equal to the experimental values. The pressure at the free surface is given as a function of time as measured in the experiment. According to the observations shown in Fig. 2, the initial arrangement of the bubbles is given randomly in the axial direction, and parabolically in the radial direction. Of course, hydrostatic pressure effects are taken into account to set the initial numerical conditions.

In Fig. 5, the numerical result of the 2-D calculation is compared to the experimental result as shown in Fig. 3. In this case, the characteristic wave length L defined by Eq. (7) is about 15 mm, so that the nondimensional tube diameter D/L is equal to 1.2, which is relatively small. This numerical result is also the averaged profile obtained by superposing 10 profiles with different arrangement of the bubbles.

In addition to the experimental and the 2-D results, two numerical results of the 1-D calculations are superposed in Fig. 5. One is the result with uniform distribution of the bubbles in the axial direction (after here this is called the "1-D uniform"), the other is the

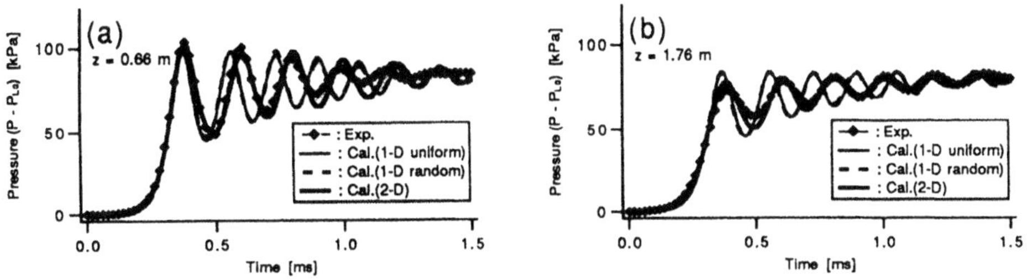

Fig. 6. Comparison the numerical results with the experimental data. $R_{L0} = 0.587$ mm, $P_{L0} = 105.2$ kPa (left figure); $R_{L0} = 0.569$ mm, $P_{L0} = 117.3$ kPa (right one). $D = 52$ mm, $\alpha_{L0,mean} = 0.19\%$, $T_l = 298.9$ K, $\Delta P = 90.3$ kPa.

results with the same random distribution of bubbles in the axial direction as for the 2-D calculation (called the "1-D random").

Good agreement is seen between the experimental result and the 2-D numerical results. There is also a good agreement between the 1-D random and 2-D numerical results. However, the numerical results of the 1-D uniform case is different from the others.

The waveform behind the shock front of the experimental, 2-D numerical, and 1-D random numerical results deform more strongly than the 1-D uniform case. On the other hand, the discrepancy in the period of the pressure oscillation due to the spatial distribution in the transverse (radial) direction is very small for such smaller shock tube. This discrepancy is more remarkable for the larger shock tube.

Figure 6 shows the pressure profiles for the case where the inner diameter of the tube D is 52 mm. In this case, the mean gas volume fraction $\alpha_{L0,mean}$ is 0.19%, the pressure step ΔP is 90.3 kPa, and the liquid temperature T_l is 298.9 K. The left and right figures represent the profiles obtained from the pressure transducer mounted at about 0.66 m and 1.76 m below the free surface, respectively. The initial bubble radius R_{L0} and pressure P_{L0} of the left figure are 0.587 mm and 107.0 kPa, respectively. Those of the right one are 0.569 mm and 117.3 kPa. The characteristic wave length L is 13 mm, so that the nondimensional diameter D/L is equal to 4.0, which is relatively large. As in Fig. 5, four results, the experimental, 2-D, 1-D random, and 1-D uniform numerical, are shown.

Again, there is a good agreement between the experimental data and the 2-D numerical results. The experimental data are different not only from the numerical results of the 1-D uniform case but also from those of the 1-D random case. As mentioned in Section 4.1, the wave structure behind the shock front is affected by the width in the transverse (radial) direction. The effect of the spatial distribution in the radial direction is no longer negligible for such larger shock tube.

As contrasted with Fig. 5, the results of the 1-D random case is very close to those of the 1-D uniform case. The number density of bubbles in the case of Fig. 6 is about 5000 per 1 m in length of the shock tube, on the other hand, that in Fig. 5 is only about 450. Notice that the discrepancy between the 1-D uniform and 1-D random results comes from only the randomness. Therefore, it may be concluded that the deformation of the pressure profiles behind the shock front is caused by the smaller total number of bubbles in the shock tube. This fact reveals that the spatial distribution of the bubbles in the streamwise (axial) direction has a significant influence on the wave propagation process.

5 Conclusions

Transient shock wave propagation phenomena in a liquid containing noncondensable gas bubbles is investigated numerically and experimentally

Governing equations for the bubbly mixture are formulated with emphasis on the radial and translational motion of the bubbles. The thermal behavior of bubble interior is carefully simulated and the effect of the spatial distribution of the bubbles is taken into account in the present numerical calculation. The numerical results are compared with the experimental data obtained from two shock tubes with different inner diameters.

A good quantitative agreement between the experimental data and the numerical results of the calculation, in which the effect of the spatial distribution of bubbles is taken into account, is found. On the other hand, the numerical results with an uniform bubble spatial distribution are much different from the experimental data. The results of the present investigation suggest that the spatial distribution of the bubbles has a considerable influence on the wave phenomena.

Acknowledgements

The authors wish to express their appreciation to Y. Haneda, Y. Kurashiki, and H. Kurita for their help with the experiment. This study was supported partially by the Grant-in-Aid for Scientific Research on Priority Areas (No. 04222103) of the Japanese Ministry of Education, Science and Culture. One of the authors (MK) was also supported by JSPS Fellowships for Japanese Junior Scientists (No. 05003524).

References

Beylich, A.E. & Gülhan, A. 1990 On the structure of nonlinear waves in liquids with gas bubbles. *Phys. Fluids* A **2**, 1412–1428.

Campbell, I.J. & Pitcher, A.S. 1958 Shock waves in a liquid containing gas bubbles. *Proc. R. Soc. Lond.* Ser. A, **243**, 534–545.

Hirt, C.W., Nichols, B.D. & Romero, N.C. 1975 SOLA — A numerical solution algorithm for transient fluid flows. *Los Alamos Sci. Lab. Rep.* LA-5852.

Kuznetsov, V.V., Nakoryakov, V.E., Pokusaev, B.G. & Shreiber, I.R. 1978 Propagation of perturbations in a gas-liquid mixture. *J. Fluid Mech.* **85**, 85–96.

Matsumoto, Y. & Kameda, M. 1993 Pressure wave phenomena in bubbly liquid. *Nucl. Engng. & Des.* **141**, 123–134.

Matsumoto, Y. & Takemura, F. 1994 Influence of internal phenomena on gas bubble motion. (Effect of thermal diffusion, phase change on the gas-liquid interface and mass diffusion between vapor and non-condensable gas in the collapsing phse) *JSME Int. J.* Ser. B **37**, 288–296.

Nigmatulin, R.I. & Shagapov, V.Sh. 1974 Structure of shock waves in a liquid containing gas bubbles. *Fluid Dyn.* **9**, 890-899.

Noordzij, L. & van Wijngaarden, L. 1974 Relaxation effects, caused by relative motion, on shock waves in gas-bubble/liquid mixtures. *J. Fluid Mech.* **66**, 115–143.

Plesset, M.S. & Prosperetti, A. 1977 Bubble dynamics and cavitation. *Annu. Rev. Fluid Mech.* **9**, 145–185.

Prosperetti, A. 1991 The thermal behaviour of oscillating gas bubbles. *J. Fluid Mech.* **222**, 587–616.

Tan, M.J. & Bankoff, S.G. 1984 Strong shock wave propagating through a bubbly mixture. *Exp. Fluids* **2**, 159–165.

van Wijngaarden, L. 1968 On the equations of motion for mixtures of liquid and gas bubbles. *J. Fluid Mech.* **33**, 465–474.

van Wijngaarden, L. 1972 One-dimensional flow of liquids containing small gas bubbles. *Annu. Rev. Fluid Mech.* **4**, 369–396.

Watanabe, M. & Prosperetti, A. 1994 Shock waves in dilute bubbly liquids. *J. Fluid Mech.* in press.

Yabe, T. & Takei, E. 1988 A new higher-order Godunov method for general hyperbolic equations. *J. Phys. Soc. Japan* **57**, 2598–2601.

HIGH FREQUENCY NONLINEAR WAVES IN BUBBLY LIQUID

D.G. Crighton & P. Sionoid

Department of Applied Mathematics
&
Theoretical Physics,
University of Cambridge
Cambridge CB3 9EW, UK

1. Background

Consider a monodisperse bubbly liquid, with volume concentration of bubbles α_0, bubble resonance frequency ω_0, and mean density ρ_0, and let c_0, c_∞ stand for the sound speeds at zero and infinite frequency (equilibrium and frozen sound speeds), respectively. Also let Γ stand for an effective adiabatic exponent in a local approximation $p/p_0 = (\rho/\rho_0)^\Gamma$ for the equation of state in pure water ($\Gamma \approx 7$). Then (cf Crighton 1991 for a simple derivation) the equation for weakly nonlinear one-dimensional wave propagation through the mixture, at frequencies well above the resonance, is

$$\frac{1}{c_\infty^2}\frac{\partial^2 p}{\partial t^2} - \frac{\partial^2 p}{\partial x^2} + \frac{\omega_0^2}{c_0^2}p = \frac{(\Gamma + 1 - 2\alpha_0)}{2(1-\alpha_0)}\frac{1}{\rho_0 c_\infty^4}\frac{\partial^2 p^2}{\partial t^2}\,, \tag{1.1}$$

where p is the pressure fluctuation. In terms of the velocity fluctuation $u = p/\rho_0 c_\infty$, the reduction of this compound wave equation to an equation for a predominantly right-propagating wave is

$$\frac{\partial}{\partial x}\left(\frac{\partial u}{\partial t} + c_\infty\frac{\partial u}{\partial x} + \frac{(\Gamma + 1 - 2\alpha_0)}{2(1-\alpha_0)}u\frac{\partial u}{\partial x}\right) - \frac{\omega_0^2 c_\infty}{2c_0^2}u = 0\,. \tag{1.2}$$

If we take the units of velocity and time (or length) to be those introduced naturally by the boundary or initial data, then with a suitably stretched length (or time) scale this equation can be put in the standard form

$$\frac{\partial}{\partial x}\left(\frac{\partial u}{\partial t} + u\frac{\partial u}{\partial x}\right) - \lambda u = 0 \tag{1.3}$$

for some $\lambda > 0$. The equation (1.3) will be referred to as the Nonlinear Klein-Gordon equation, NLKG.

S. Morioka and L. van Wijngaarden (eds.),
IUTAM Symposium on Waves in Liquid/Gas and Liquid/Vapour Two-Phase Systems, 127-136.

A fair amount of work has already been done on solutions of this equation, though not nearly enough to be commensurate with the assertion (cf Hunter 1990) that this equation is a canonical one for weakly dispersive weakly nonlinear waves in the high frequency limit, just as KdV is a canonical one for the low frequency limit. Several other applications should be noted, in addition to the present one concerning multiphase liquids. It arises in the case of waves on shallow layers of rotating liquid at frequencies $\omega/\Omega \gg 1$, well above the rotation rate Ω; in nonlinear internal waves at frequencies $\omega/N \gg 1$, well above the Brunt-Väisälä frequency N (Ostrovsky 1978, Grimshaw 1985); and in the case of acoustic propagation in turbulent media (Benilov & Pelinovsky 1988). In the last case the general one-dimensional equation is, in the coherent field approximation,

$$u_t + uu_x = \int_0^\infty W(\tau)u_{xx}(x+2\tau,t)d\tau\,, \tag{1.4}$$

where $W(\tau)$ is the autocorrelation function for fluctuations in the sound speed. At frequencies $\omega/\omega_T \gg 1$ well above the characteristic turbulence frequency ω_t, the integral term can be simplified to λu, where $\lambda = -W''(0)/8 > 0$.

The purpose of this study is to present exact analytical solutions to NLKG for periodic waves, to show how they reduce to the well known sawtooth wave profiles of the familiar Riemann simple wave equation as $\lambda \to 0$, and to show how at large times the new solutions approach a periodic travelling wave solution of *limiting* amplitude possessing sharp corners. First we review the known facts about travelling waves for NLKG.

2. Travelling Waves for NLKG

In (1.3) take $u(x,t) = u(\theta), \theta = x - ct$. An integration gives, for $v = u - c$,

$$(vv')^2 - \tfrac{2}{3}\lambda v^3 - \lambda c v^2 = d\,, \tag{2.1}$$

and further integration can be completed in elliptic integrals of the first kind. From that representation, or from phase plane analysis of (2.1), one finds (cf Ostrovsky 1978, Leonard 1988, Sionoid 1994) that for $c > 0$ there is a family of continuous periodic solutions of different wave length and amplitude, of which the *limiting* wave, of maximum amplitude for given $c > 0$, has a corner at the crest, at which $[u'] = 2(\lambda c)^{1/2}$. For $c < 0$ there are no periodic solutions.

For $c > 0$ the general periodic solution can be expressed via

$$\lambda^{1/2}(\theta - \theta_0) = 6\zeta\left\{\mathcal{P}^{-1}(\hat{v})\right\} - \frac{\lambda c}{2}\mathcal{P}^{-1}(\hat{v}) \tag{2.2}$$

where $\mathcal{P}$ is the Weierstrass $\mathcal{P}$ function and ζ the Riemann zeta function, related by $\mathcal{P}(z) = -\zeta'(z)$. The argument here is $\hat{v} = \frac{1}{6}(u - \frac{1}{2}c)$. In the special case when the discriminant of the parameters of $\mathcal{P}$ vanishes we have (cf Ostrovsky 1978) a solution in the form of a parabolic arc,

$$u(\theta) = \frac{\lambda}{6}\theta^2 - \frac{c}{2}, \qquad -3\left(\frac{c}{\lambda}\right)^{1/2} \leq \theta < +3\left(\frac{c}{\lambda}\right)^{1/2}. \tag{2.3}$$

These facts can all be determined from reference to the works by Ostrovsky (1978), Hunter (1990), Grimshaw (1985), Leonard (1988), and Sionoid (1994).

Figure 1 shows travelling periodic wave solutions of NLKG with $\lambda = 1$, and for $c = 1$. Included among them is the limiting wave of maximum amplitude.

3. Time-Dependent Periodic Solutions

Numerical integration of NLKG for given periodic $u(x,0)$ shows that for large λ, although shock-like features may appear in the waveform, they rapidly disappear, and as $t \to +\infty$ the waveform corresponds to one of the smooth periodic travelling waves of less than maximal amplitude. See, for example, Figure 2, from Sionoid (1994), where $u(x,0) = (2\pi)^{-1}\sin 2\pi x$ and $\lambda = 2$. Such cases are not usually relevant to the bubbly liquid, however, where the relevant values of λ cannot (formally at least) be large. For small λ the typical evolution involves the formation of stable shocks and their monotonic decrease in strength as time increases, the shocks being separated by regions of apparently parabolic shape. See Figure 3, for example, in which the initial condition is as in Figure 2, but now $\lambda = 1/2$.

We propose now a "weak shock theory" for the stages and parameter regions exemplified by Figure 3. (Just what these stages and parameter regions might be is not known. Hunter (1990) has obtained sufficient conditions on λ and $u(x,0)$ to guarantee shock formation in finite time, but these conditions are known to be far from necessary: see Sionoid (1994).) For weak shock theory we need a description, via exact analytical solutions, of the continuous portions of the wave, and conditions determining the propagation of discontinuities.

The numerical results suggest that the large-t form of the continuous portions is parabolic. Accordingly we seek a solution in the form

$$u(x,t) = f(t)[x - c(t)]^2 + b(t)$$

and find that this satisfies NLKG (1.3) if

$$f(t) = \lambda/6, \dot{c} = -2b\,.$$

Thus, for some $c(t)$ to be determined by considerations of the shock conditions, each arc between shocks is taken to be given by

$$u(x,t) = \frac{\lambda}{6}[x - c(t)]^2 - \tfrac{1}{2}\dot{c}(t)\,. \tag{3.1}$$

For the shock conditions, assume a discontinuity at $x = X(t)$. Then from (1.3) we have

$$[u_t] + [uu_x] = 0\,,$$

or

$$\dot{u}_2 - \dot{u}_1 - \dot{X}(t)[u] + [uu_x] = 0 \tag{3.2}$$

with u_1, u_2 the values on either side of $X(t)$. A second condition arises from writing (1.3) in integrated form

$$u_t + uu_x = Lu\,, \tag{3.3}$$

$Lu = \lambda \int_0^x u(y,t)dy - \frac{\lambda}{p}\int_0^p \int_0^y u(z,t)dzdy$, where p is the period. Integrating (3.3) over a period containing $x = X(t)$, we have

$$-\dot{X}(t)[u] + \tfrac{1}{2}[u^2] = 0\,, \quad \text{or} \quad \dot{X}(t) = \tfrac{1}{2}(u_1 + u_2), \tag{3.4}$$

just the standard shock propagation speed of weak shock theory for the Riemann simple wave equation ($\lambda = 0$). Eliminating $\dot{X}(t)$ between (3.2) and (3.4) we have

$$\dot{u}_2 - \dot{u}_1 + \tfrac{1}{2}(u_2 - u_1)(u_{1x} + u_{2x}) = 0\,. \tag{3.5}$$

Take (3.1) to hold for

$$-p/2 < x - c(t) < +p/2\,,$$

this interval containing a shock at $x = X(t)$. The same solution is periodically continued elsewhere. Then we find

$$u_2 - u_1 = \frac{\lambda p}{3}(X - c - p/2)\,, \tag{3.6}$$

$$\dot{X} = \frac{\lambda}{6}\left((X - c - p/2)^2 + p^2/4\right) - \frac{\dot{c}}{2}\,, \tag{3.7}$$

on use of periodicity requirements. In terms of

$$Y(t) = X(t) - c(t) - p/2\,,$$

(3.2), (3.4), (3.6) and (3.7) give

$$\dot{Y} = -\frac{\lambda}{3}Y^2\,,$$

$$Y = \frac{3}{\lambda}\cdot\frac{1}{t + t_0}\,, \tag{3.8}$$

with t_0 a constant of integration. Then

$$u_2 - u_1 = \frac{p}{t + t_0} \tag{3.9}$$

and

$$c(t) = Ut - \frac{3}{\lambda} \cdot \frac{1}{t + t_0}, U = \frac{\lambda p^2}{36}. \tag{3.10}$$

Thus the shock strength decreases as t^{-1}, and ultimately the shocks drift through the waveform at the constant velocity U. Figure 4 (from Sionoid 1994) shows that the weak shock theory provided here agrees very well with numerical calculation for a sinusoidal input and $\lambda = 1/2$.

Now let $\lambda \to 0$. We find from (3.1),

$$u(x,t) \sim \frac{x}{(t + t_0)}, \tag{3.11}$$

$$\dot{X} \to 0, \tag{3.12}$$

$$u_2 - u_1 \sim \frac{p}{t + t_0}, \tag{3.13}$$

which are the results of weak shock theory for the Riemann equation $u_t + uu_x = 0$, describing periodic sawtooth ramps separated by fixed shocks decaying in strength as t^{-1}. Thus the description in terms of periodic parabolic arcs (3.1) with unsteadily propagating shock discontinuities located by (3.2) and (3.4) is seen to provide the natural extension of weak shock theory to the nonlinear dispersive wave equation NLKG.

4. Conclusions

This note has provided an analogue for the NLKG of weak shock theory for the Riemann simple wave equation, for periodic waves. It is found that shocks, once fully developed in the waveform, decay just as do those for the Riemann equation ($\lambda = 0$ in NLKG), but drift through the waveform at an effectively constant velocity. Separating the shocks are parabolic arcs given by an exact unsteady solution of the full NLKG.

It would be interesting to find experimental confirmation of the features of NLKG predicted here. There is certainly available evidence concerning the predictions of linear theory for high frequency waves in bubbly liquid (Nakoryakov et al 1993, pp 88-90), but the parabolic waveforms have not yet been confirmed. They are qualitatively different from both the sawtooth signals well known in the case of a monophase fluid and from the solitons which develop in the low frequency limit in bubbly liquid.

References

Benilov, E.S. & Pelinovsky, E.N. 1988 Dispersionless wave propagation in nonlinear fluctuating media, *Soviet Physics JETP* **67**(1), 98–103.

Crighton, D.G. 1991 Nonlinear acoustics of bubbly liquids, in *Nonlinear Waves in Real Fluids, CISM Courses and Lectures No. 315*, ed. A. Kluwick, Springer Verlag pp. 45–67.

Grimshaw, R. 1985 Evolution equations for weakly nonlinear, long internal waves in a rotating fluid, *Stud. Appl. Math.* **73**, 1–33.

Hunter, J.K. 1990 Numerical solution of some nonlinear dispersive wave equations, *Lect. Appl. Math.* **26**, 301–316.

Leonard, S.R. 1988 *The propagation of nonlinear waves in a bubbly liquid*, Ph.D. thesis, University of Leeds.

Nakoryakov, V.E., Pokusaev, B.G. & Shreiber, I.R. 1993 *Wave Propagation in Gas-Liquid Media*, 2nd Ed., CRC Press.

Ostrovsky, L.A. 1978 Nonlinear internal waves on a rotating ocean, *Oceanology* **18**(2), 120-125.

Sionoid, P. 1994 *Nonlinear wave equations with diffusion, diffraction and dispersion*. Ph.D. dissertation, University of Cambridge.

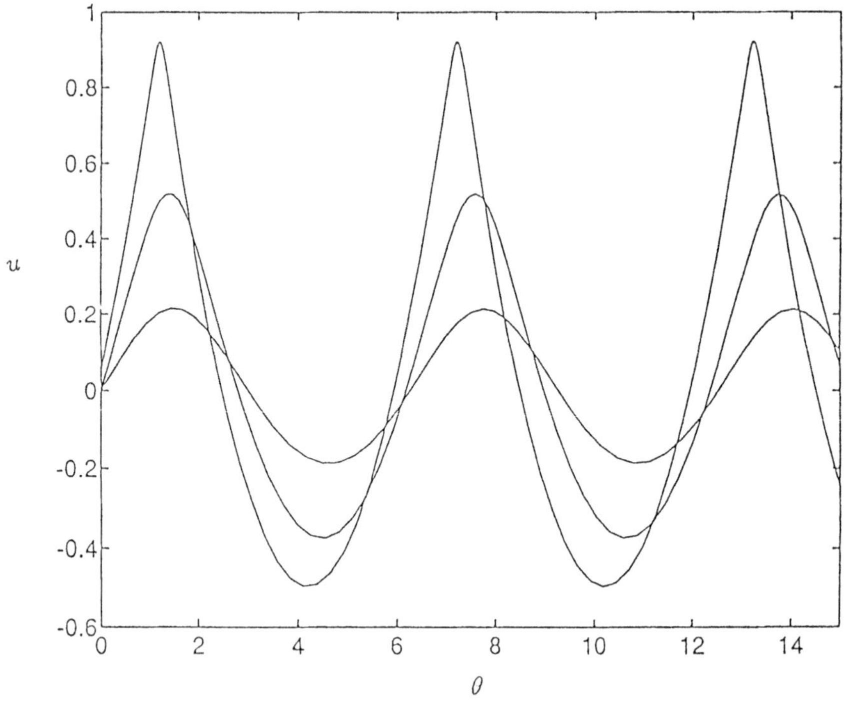

Figure 1: Travelling wave solutions of NLKG for $c = 1$ including wave of maximum amplitude.

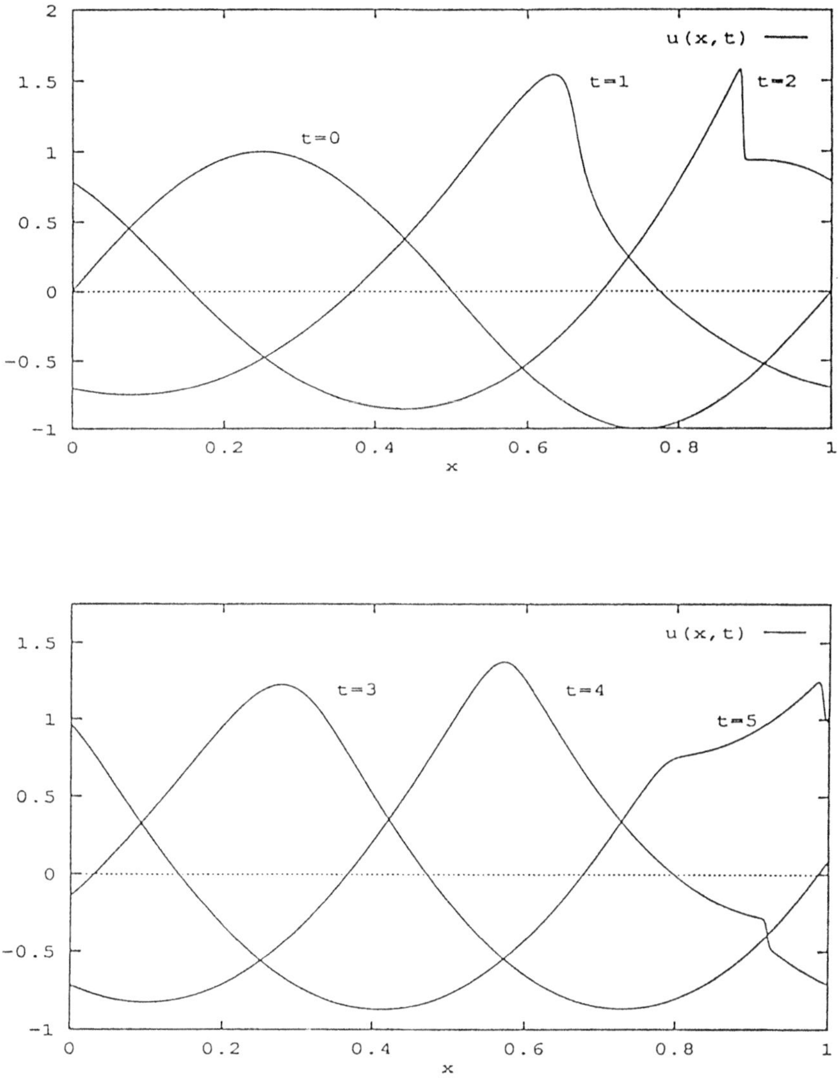

Figure 2a: Numerical solution of NLKG: $u(x,0) = (2\pi)^{-1} \sin 2\pi x$, $\lambda = 2$.

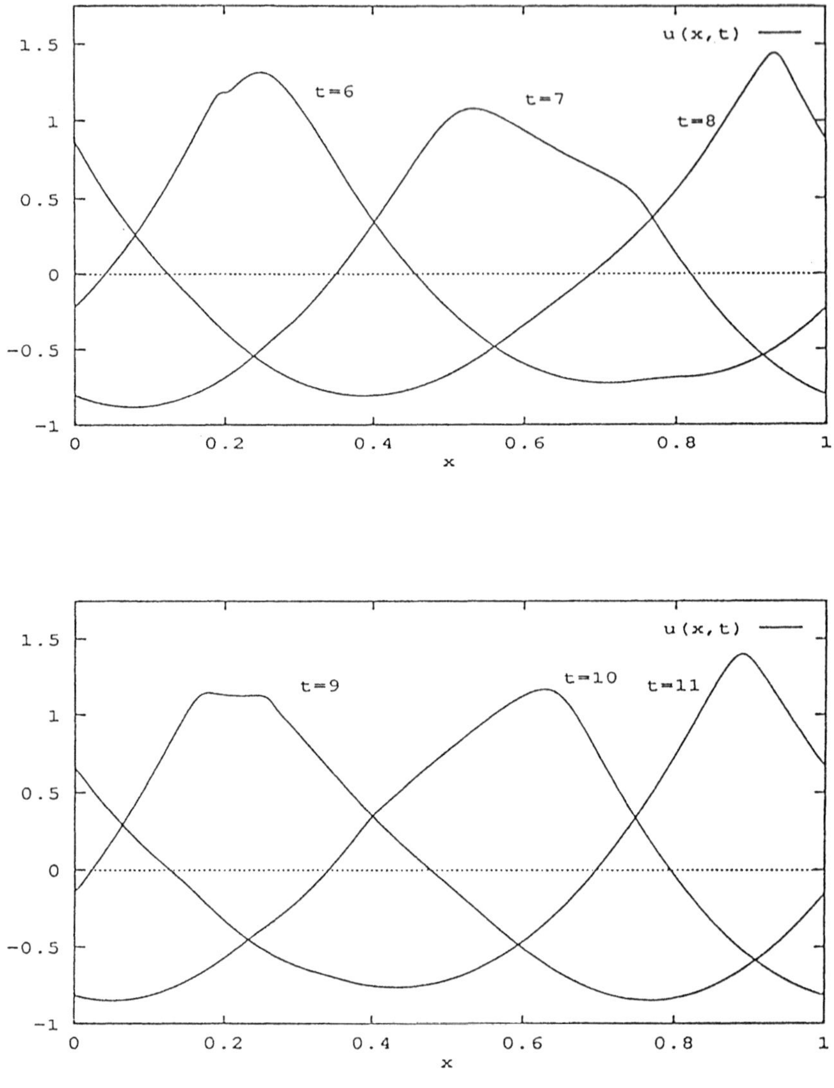

Figure 2b: Numerical solution of NLKG: $u(x,0) = (2\pi)^{-1}\sin 2\pi x$, $\lambda = 2$.

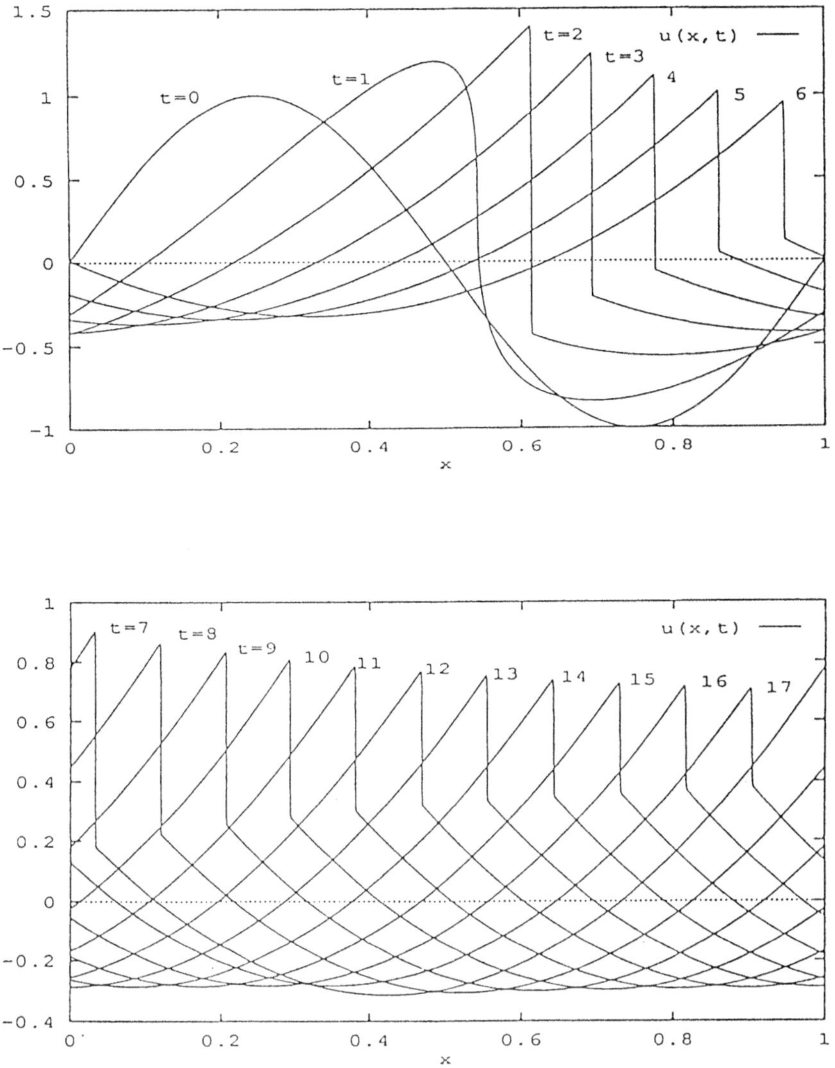

Figure 3: Numerical solution of NLKG: $u(x,0) = (2\pi)^{-1}\sin 2\pi x$, $\lambda = 0.5$.

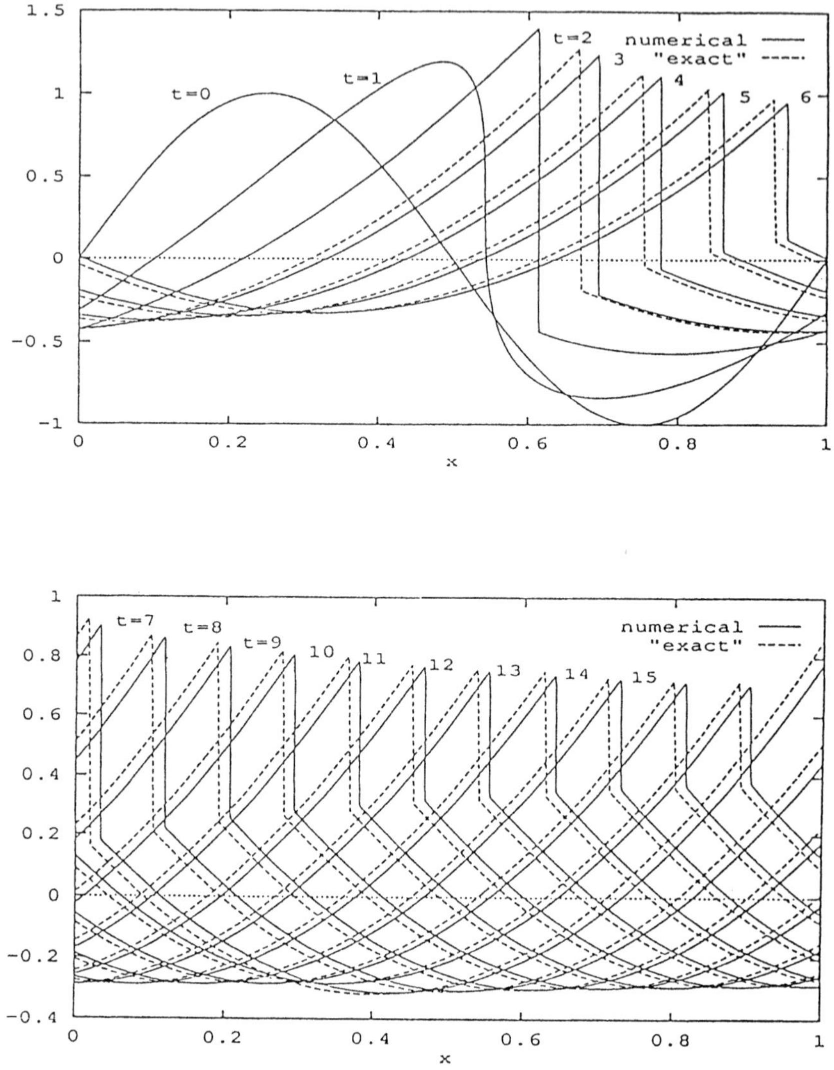

Figure 4: Comparison of numerical with exact solution calculated in §3 (offset for clarity); $\lambda = 0.5$.

Numerical simulations and kinetic theory for simple shear motion of bubbly suspensions

A. S. Sangani[1], S.-Y. Kang[1], D. L. Koch[2] and H.-K. Tsao[2]

1. Department of Chemical Engineering, Syracuse University, Syracuse, NY 13244.
2. School of Chemical Engineering, Cornell University, Ithaca, NY 14853.

Abstract

Simple shear flows of dilute suspensions of spherical bubbles at large Reynolds numbers are studied using numerical simulations and kinetic theory. It is shown that the mean-square bubble velocity is very sensitive to the volume fraction and Reynolds number of the bubbles as well as on initial conditions. The balance of energy contained in bubble velocity fluctuations plays an important role in the rheology of the dispersed phase, which is generally non-Newtonian.

1 Introduction

Numerical simulations and theoretical analyses of detailed hydrodynamic interparticle interactions have provided a sound basis for understanding the rheological behavior of suspensions in which the particle Reynolds number is much less than one. However, our understanding of suspensions in which the inertia of the continuous phase is important is much more rudimentary and is based primarily on studies based on the potential flow approximation. A series of recent papers (Bieshuvel and Gorissen 1990; Sangani and Didwania 1993a, which will be refered to as I; Zhang and Prosperetti 1994; and Bulthuis, Prosperetti, and Sangani 1994) provide a rigorous derivation of the equations of motion for a suspension of bubbles with potential flow interactions. An important property arising in these equations is the dispersed-phase pressure tensor, which depends on hydrodynamic and collisional interactions between the bubbles and also on their velocity distribution.

A uniform suspension of bubbles rising due to their buoyancy is subject to instabilities that lead to volume fraction variations. One mechanism that has long been studied is the instability to void fraction waves resulting from the bubble's added mass and the dependence of the drag force on volume fraction (Bieshuvel 1994). Sangani and Didwania (1993b; henceforth refered to as II) and Smereka (1993) have recently observed a second instability mechanism in dynamic simulations of potential flow. The bubbles rising due to buoyancy are attracted into a horizontal orientation and form clusters in these simulations. In either case, the mechanism that may tend to stabilize the homogeneous suspension is the positive bubble pressure created by fluctuations in the bubble velocities.

The purpose of the present work is to determine the distribution of bubble velocities in a suspension of bubbles subject to simple shear flow with potential flow

S. Morioka and L. van Wijngaarden (eds.),
IUTAM Symposium on Waves in Liquid/Gas and Liquid/Vapour Two-Phase Systems, 137-150.

interactions among the bubbles. This will be accomplished using a numerical simulation method similar to that developed in II and the results of the simulation will be interpreted using kinetic theories. It will be seen that the bubble collisions induced by shearing motion can provide a quite effective mechanism for enhancing the bubble velocity fluctuations.

The application of results based on the potential flow interactions to physical systems must be approached with caution. The potential flow approximation is rigorously valid for bubbles with slip boundary conditions in the dual limits of high Reynolds number and low Weber number (so that bubble deformation can be neglected). The best system for approaching these limits is bubbles with diameters of about 1 mm in water for which the Reynolds number is about 130 and the Weber number about 0.5. Even in this case the limits are only approximately satisfied and the value of the potential flow approximation may depend upon the nature of the interactions being described. For example, potential flow would be expected to be more accurate for collision between bubbles moving with substantially different speeds in different directions than when the bubbles rise at nearly the same velocity. In more general circumstances, such as solid-liquid or liquid-liquid suspensions and suspensions of bubbles with larger diameters (4 mm) which are typical of most of the experimental literature, boundary layer separation plays an important role in the dynamics.

Other methods currently under development (Ladd 1990; Ladd 1994; Zaleski and Zanetti 1994) show promise for the possibility of simulating suspensions including bubble deformation and continuous phase vorticity. However, we believe that the relatively extensive simulation results and the mechanistic understanding that can be developed for bubbles with potential flow interactions will provide a useful reference for understanding these more complex suspensions.

In section 2 we briefly describe the numerical simulation procedure and in section 3 we present the results of simulations and an approximate kinetic theory of dilute bubbly liquids. We find that the steady state velocity variance depends on the volume fraction ϕ of bubbles and the Reynolds number Re_s based on shear rate in a rather complicated manner. At large Re_s and small ϕ, we find multiple steady states: if the initial variance is relatively large, then the final state variance is very large, of $O((Re_s/\phi)^2)$. We refer to this as an ignited state. On the other hand, if the initial variance is small, then the steady state, which we refer to as a quenched state, has a vanishingly small variance.

The multiple steady states are shown to arise due to nonlinear dependence of the dispersed-phase shear viscosity on the velocity variance of bubbles or, equivalently, the temperature of the dispersed-phase. We also find that the multiple steady states are observed are only when $Re_s > 88$ and ϕ is sufficiently small ($\phi < \phi_c(Re_s)$). For smaller Re_s, the final state is quenched regardless of the initial conditions, and for $Re_s > 88$ and $\phi > \phi_c(Re_s)$, the final steady state is always ignited. We also find that the dispersed-phase rheology exhibits normal stress differences. Finally, in section 4 we assess approximate conditions under which the shear-induced variance may be significant in stabilizing flows of bubbly liquids through pipes.

The results presented here are preliminary and limited to dilute bubbly liquids. We plan to report more complete investigation including the simulations and theory for non-dilute bubbly liquids and the question of stability of bubbly liquids under simple shear and gravity in a future publication.

2 The simulation method

The simulation method is described in detail in II. Here, we briefly summarize the method indicating some modifications we have made in the present study. We consider motion of N spherical bubbles of radius a placed within a unit cell of periodic array. The velocity $\mathbf{w}^\alpha$ of a representative bubble α is written as a sum of the ensemble-averaged mixture velocity $< \mathbf{u} >$ and a relative velocity $\mathbf{V}^\alpha$:

$$\mathbf{w}^\alpha(t) =< \mathbf{u} > (\mathbf{x}^\alpha, t) + \mathbf{V}^\alpha(t), \tag{1}$$

where $\mathbf{x}^\alpha$ is the position of the center of the bubble at time t. Similarly, the velocity of the fluid is written as a sum of $< \mathbf{u} >$ and $\mathbf{u}'$ where $\mathbf{u}'$ is assumed to depend only on the position and the relative velocities of the bubbles. Thus, by definition $< \mathbf{u}' >= \mathbf{0}$. In simulations we enforce this condition by requiring that the average of $\mathbf{u}'$ over the unit cell vanishes at any given instant. We are interested in a large Reynolds number situation where the hydrodynamic interactions are dominated by a potential flow, and therefore we write $\mathbf{u}' = \nabla\varphi$ and solve Laplace equation for φ subject to the boundary condition $\nabla\varphi \cdot \mathbf{n} = \mathbf{V}^\alpha \cdot \mathbf{n}$ on the surface of the bubbles. Here, $\mathbf{n}$ is the unit outward normal vector on the surface of the bubble α.

As shown in Sangani, Zhang, and Prosperetti (1991) and in II, the velocity potential can be determined to a very good accuracy with a point-dipole approximation

$$\varphi(\mathbf{x}) = \mathbf{G} \cdot \mathbf{x} - \sum_{\alpha=1}^{N} \mathbf{D}^\alpha \cdot \nabla S_1(\mathbf{x} - \mathbf{x}^\alpha), \tag{2}$$

where S_1 is a Green's function for Laplace equation in a periodic domain (Hasimoto 1959, Sangani, Zhang, and Prosperetti 1991) and $\mathbf{D}^\alpha$ is the dipole strength. The condition that the average of $\mathbf{u}'$ over the unit cell must vanish is satisfied by taking (cf. II)

$$\mathbf{G} = \frac{4\pi}{\tau} \sum_{\alpha=1}^{N} \mathbf{D}^\alpha, \tag{3}$$

where τ is the volume of the unit cell. To calculate the trajectories of the bubbles we apply the force balance on each bubble (assumed to be massless). The impulse defined by

$$\mathbf{I}^\alpha = -\rho \int_{S^\alpha} \varphi \mathbf{n} dS \tag{4}$$

plays a role analogous to the momentum of a particle in Newtonian mechanics. We write the force balance as

$$\frac{d\mathbf{I}^\alpha}{dt} = \mathbf{F}^\alpha = \mathbf{F}_g + \mathbf{F}_v^\alpha + \mathbf{F}_{<\mathbf{u}>}^\alpha + \mathbf{F}_p^\alpha + \mathbf{F}_c^\alpha, \tag{5}$$

where $\mathbf{F}_g = -4\pi a^3/3\mathbf{g}$ is a force due to buoyancy, $\mathbf{F}_v$ is a viscous force, $\mathbf{F}_{<\mathbf{u}>}$ is a force due to temporal and spatial variations in the ensemble-averaged velocity $< \mathbf{u} >$ of the mixture, $\mathbf{F}_c^\alpha$ is a force on the bubble during its collision with other bubbles, and $\mathbf{F}_p^\alpha$ is a force due to potential flow interactions. As shown in I, the last quantity is evaluated from

$$\mathbf{F}_p^\alpha = -4\pi\rho \sum_{\gamma=1}^{N} \mathbf{D}^\alpha\mathbf{D}^\gamma : \nabla\nabla\nabla S_1(\mathbf{x}^\alpha - \mathbf{x}^\gamma), \tag{6}$$

where the singular part of S_1 must be removed from S_1 for $\gamma = \alpha$ before evaluating the third derivative of S_1. The viscous force is evaluated using a method described in II. The main modifications in the present study are concerned with the evaluation of $\mathbf{F}_{<\mathbf{u}>}$ and $\mathbf{F}_c$. In II we considered a special case of constant average velocity whereas in the present study we are interested in the case of simple shear. For this purpose we use a slightly modified version of the expression proposed by Auton, Hunt, and Prud'homme (1988)

$$F^{\alpha}_{<\mathbf{u}>,i} = m\frac{D < u_i >}{Dt}(\mathbf{x}^{\alpha}) - I^{\alpha}_j \frac{\partial < u_j >}{\partial x_i}(\mathbf{x}^{\alpha}), \tag{7}$$

where $m = 4\pi\rho a^3/3$ is the mass of fluid having the same volume as the bubble and $D/Dt = \partial/\partial t + < \mathbf{u} > \cdot\nabla$ is a time derivative following the average motion of the mixture. The above expression can be shown to give the correct force on a bubble for the cases of (i) a small amplitude oscillatory flow examined in Sangani, Zhang, and Prosperetti (1991) and (ii) pure extensional flow, at least to the point-dipole approximation. No restrictions on the magnitude of ϕ are needed for the validity of (7) in the above two cases. Finally, (7) also gives the correct lift force on a bubble in dilute dispersions in the presence of simple shear, for which the the magnitude ω of vorticity is small compared to V/a where V is a characteristic magnitude of bubbles' relative velocity. Thus, we expect it to provide a very good approximation for the ignited states. Note that for the case of an isolated bubble with velocity V^{α}_j, the expression (7) reduces to the one given by Auton, Hunt, and Prud'homme (1988) upon substituting $I^{\alpha}_j = mV^{\alpha}_j/2$.

In simulations of non-deformable bubbles with potential flow, it is commonly observed that the bubbles will come into contact while still moving at a finite relative velocity. In a physical suspension such a collision could lead to a bubble bounce or to coalescence depending on the Weber number based on the relative velocity. For the sake of simplicity in the simulation, it is convenient to assume that the bubbles always bounce and this can be achieved in practice without violating the free slip boundary condition on the bubble surface through the addition of salt to the suspending water (Lessard and Zieminski 1971; Tsao and Koch 1994). Thus, we include a collisional force in the simulations to achieve an energy and momentum conserving bubble bounce. The collisional force $\mathbf{F}_c$ was evaluated in II by assuming the collision process to be instantaneous. This introduces some difficulties in numerical implementation and to overcome these we have experimented in the present simulations with using a soft core repulsive potential to model the collision process. Specifically, the collision force was taken to be

$$\mathbf{F}^{\alpha}_c = -\nabla_{\mathbf{x}^{\alpha}}\varphi_c \tag{8}$$

with

$$\varphi_c = \sum_{\alpha=1}^{N}\sum_{\gamma=1}^{N}\Gamma_{\alpha\gamma}[2a - |\mathbf{x}^{\alpha} - \mathbf{x}^{\gamma}|]^3. \tag{9}$$

Here, $\Gamma_{\alpha\gamma} = 0$ if bubbles α and γ are not overlapping. Otherwise, $\Gamma_{\alpha\gamma} = \Gamma_c\Delta^{\alpha\gamma}_{ni}$ where Γ_c is a constant and $\Delta^{\alpha\gamma}_{ni}$ is the component of the relative velocity $\mathbf{V}^{\alpha} - \mathbf{V}^{\gamma}$ along the line joining the position of the bubbles at the onset of the overlap.

The numerical algorithm consisted of determining the force on each bubble given the position and impulse of all the bubbles in the suspension and integrating $\dot{\mathbf{I}}^{\alpha} = \mathbf{F}^{\alpha}$ and $\dot{\mathbf{x}}^{\alpha} = \mathbf{v}^{\alpha}$ using a fourth-order Runge-Kutta scheme. The time step for integration was chosen to scale with the root-mean-squared velocity of the bubbles.

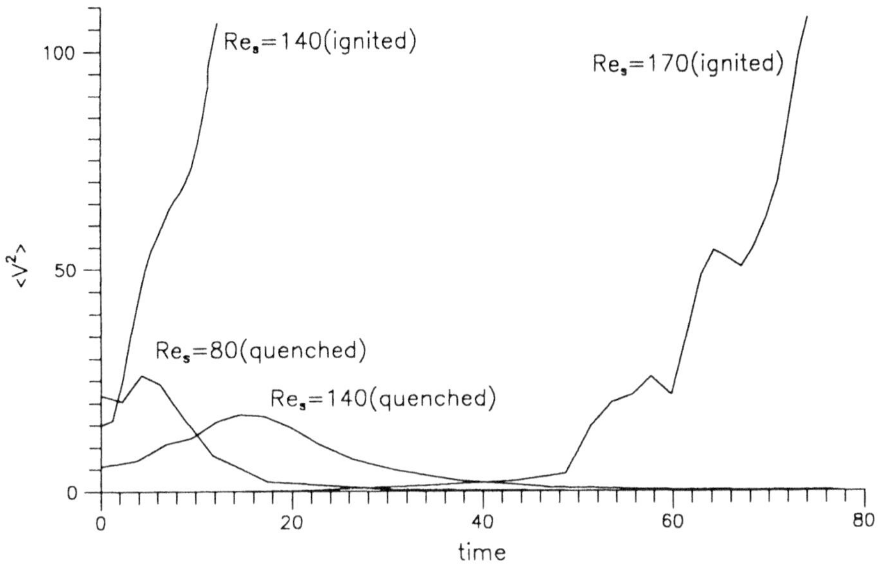

Figure 1: Velocity variance as a function of time, initial conditions, and Re_s for $\phi = 0.005$ and $N = 32$.

3 Simulation results and kinetic theory for simple shear motion

In this paper we shall consider simple shear flows of dilute bubbly liquids with $\mathbf{g} = \mathbf{0}$ and $< u_i >= \gamma x_2 \delta_{i1}$. The distances are rendered non-dimensional with a, the velocity with γa, and the time with $1/\gamma$. Typical simulation results are illustrated in Figure 1 which shows velocity variance $< V^2 >$ as a function of time for $\phi = 0.005$ at three different Reynolds numbers $Re_s = \gamma a^2 \rho/\mu$. These results are summarized as follows: (i) the final state for Re_s less than about 90 is quenched (very low variance) regardless of the initial conditions; (ii) the multiple steady states exist for $90 < Re_s < Re_c(\phi)$ (e.g. $Re_s = 140$ in Figure 1) – the final state is quenched or ignited depending on the initial conditions; (iii) the final state for $Re > Re_c(\phi) > 90$ (e.g. $Re_s = 170$ in Figure 1) is ignited regardless of the initial conditions. Here, $Re_c(\phi)$ is given approximately by $Re_c^3\phi = 2 \times 10^4$.

Before presenting an approximate kinetic theory and a more detailed comparison between the theory and numerical simulations, it will be helpful to have a qualitative understanding of the phenomenon. The steady state variance is determined by the balance between the energy input in shearing the dispersion and the viscous energy dissipation. The former is approximately equal to $\mu_s^*\gamma^2$ while the latter to $12\pi\mu n a < V^{*2} >$. Here, μ_s^* is the (dimensional) dispersed-phase shear viscosity, n is the number density of bubbles, and $< V^{*2} >$ is the dimensional velocity variance. In the ignited state, the collision time $\tau_c = a/(\phi < V^{*2} >^{1/2})$ is much smaller than the viscous relaxation time $\tau_v = \rho a^2/(18\mu)$, and the leading order velocity distribution as $Re_s \rightarrow \infty$ is an isotropic Maxwellian. Thus we can estimate μ_s^* from the kinetic theory of gases by taking the mass of bubbles to be their virtual mass $m/2$ and the mean free path to be a/ϕ to yield $\mu_s^* \sim \rho a < V^{*2} >^{1/2}$. The energy balance then shows that

$< V^2 > \sim (Re_s/\phi)^2$. In the quenched state, $< V^2 >$ is very small, and, consequently, $\tau_v \ll \tau_c$. Thus, the majority of the bubbles move with the velocity of the fluid. The initial conditions influence the final steady state for intermediate values of Re_s by setting up the initial value of the collision time. At smaller Re_s, the viscous relaxation time is small enough to dissipate the fluctuations leading always to the quenched state while at large enough Re_s the fluctuations induced by the imposed shear are sufficient to eventually make $\tau_c \ll \tau_v$ corresponding to the ignited state.

3.1 Kinetic theory

For a spatially homogeneous dispersion of bubbles, the velocity distribution function $f(\mathbf{V})$ satisfies

$$\frac{\partial f}{\partial t} + \nabla_{\mathbf{V}} \cdot (\dot{\mathbf{V}} f) = 0. \tag{10}$$

Since ϕ is very small, the bubbles only undergo occasional interactions. It is well known that potential flow bubbles often undergo actual collisions and Tsao and Koch (1994) showed that for low Weber number bubbles with short-range repulsive forces these collisions are nearly elastic. In this simple kinetic theory we will neglect the hydrodynamic interactions between bubbles and treat the collisions as perfectly elastic.

Substituting in (5) $\mathbf{F}_g = \mathbf{F}_p = \mathbf{0}$, $\mathbf{F}_v = -12\pi\mu a\mathbf{V}$, $\mathbf{I} = m\mathbf{V}/2$, $\mathbf{F}_{<\mathbf{u}>,i} = -(1/2)mV_1\delta_{i2}$ (lift force), expressing the contribution due to the collisional force in the notation of Chapman and Cowling (1970), and non-dimensionalizing we obtain

$$\frac{\partial f}{\partial t} - \frac{\partial}{\partial V_k}[(\delta_{k2}V_1 + St^{-1}V_k)f] = \frac{\partial_e f}{\partial t}, \tag{11}$$

where the Stokes number, $St = Re_s/18 = \gamma\tau_v = \rho\gamma a^2/(18\mu)$, is a non-dimensional viscous relaxation time and $\partial_e f/\partial t$ is the rate of change in f at a fixed point due to collisional encounters. As shown in Chapter 3 of Chapman and Cowling, this latter quantity is expressed in terms of an integral

$$\frac{\partial_e f}{\partial t} = \int\int d\mathbf{w}_1 d\mathbf{k} k_1 [f(\mathbf{w}')f(\mathbf{w}'_1) - f(\mathbf{w})f(\mathbf{w}_1)]. \tag{12}$$

Here, $\mathbf{w} =< \mathbf{u} > +\mathbf{V}$ is the velocity of a representative bubble in collision with another bubble with velocity $\mathbf{w}_1$, the velocities of the bubbles at the end of the collision encounter being $\mathbf{w}'$ and $\mathbf{w}'_1$, and $\mathbf{k}$ and k_1 are parameters that depend on the relative velocities and orientation of the bubbles at the onset of the collision.

For small ϕ we expect the dispersed-phase pressure to be dominated by its kinetic part and therefore $P_{ij} \simeq P^k_{ij} = n < I_i V_j > \simeq (1/2)\rho\phi < V_i V_j >$ (cf. I). We shall non-dimensionalise pressure by $\rho\gamma^2 a^2/2$. To determine the pressure and velocity variance, we multiply (11) by $\phi V_i V_j$ and integrate over the velocity space to obtain

$$\frac{\partial P_{ij}}{\partial t} + [P_{j1}\delta_{i2} + P_{1i}\delta_{j2} + 2St^{-1}P_{ij}] = \Delta_e(P_{ij}), \tag{13}$$

where, in obtaining the terms inside the square brackets on the left-hand side, use has been of integration by parts, and

$$\Delta_e(P_{ij}) \equiv \phi \int d\mathbf{V} V_i V_j \frac{\partial_e f}{\partial t}. \tag{14}$$

To make further progress we need to determine f and $\triangle_e(P_{ij})$. Exact solution for f is rather difficult to obtain and hence we have developed approximate methods. First we develop an approximation in which the shear-induced collisions are neglected in evaluating $\partial_e f/\partial t$, i.e. the actual velocities $\mathbf{w}$, $\mathbf{w}_1$, etc. in (12) are replaced by the relative velocities $\mathbf{V}$, $\mathbf{V}_1$, etc.. The resulting theory will explain the origin of multiple steady states but not the absence of quenched state at high enough Re_s. The theory will be subsequently modified to include the shear-induced collisions which play an important role in the behavior of quenched states.

We have developed two approximate theories for evaluating $\triangle_e(P_{ij})$. Both give identical results. One is based on the method due to Grad (1949) in which f is expanded in a series of Hermite polynomials

$$f(\mathbf{V}) = \left[1 + \frac{1}{2}(a_{ij} - \frac{1}{3}\delta_{ij}a_{kk})\frac{\partial^2}{\partial V_i V_j} + \ldots\right] f_M, \tag{15}$$

where f_M corresponds to an isotropic Maxwellian distribution. The series is truncated keeping only the first two terms and (12) and (14) are evaluated in terms of a_{ij}. Substituting for $\triangle_e(P_{ij})$ in (13) and solving the resulting equations then yields a_{ij} and P_{ij}. In the other method, which we describe in more detail here, we model the collision process as similar to that between Maxwell molecules. Thus, we assume that the force F between two bubbles is repulsive and along the line joining the center of the two bubbles separated by a distance r with $F = \kappa r^{-5}$ where κ is a constant of proportionality. For this special case it turns out that the collision term (12) assumes a particularly simple form and one finds

$$\triangle_e(P_{ij}) = \lambda^* \phi(p\delta_{ij} - P_{ij}), \tag{16}$$

where $p = 1/3P_{kk}$ and λ is a constant related to κ. Now in the limit $St \to \infty$, we expect the variance to become very large and the collision term on the right-hand side of (11) to dominate leading to an isotropic Maxwellian velocity distribution. The dispersed-phase viscosity will then be expected to approach the viscosity of dilute gas consisting of hard-sphere molecules. Matching with the known expression for that viscosity requires

$$\lambda^* = \lambda T^{1/2} \quad \text{with} \quad \lambda = \frac{24}{5\sqrt{\pi}}, \tag{17}$$

$T \equiv < V^2 > /3$ being the dispersed-phase temperature. This choice of λ^* is equivalent to choosing the force law constant κ of Maxwell molecules to be proportional to T.

Since $T^{1/2}$ in due course will be shown to be $O(1/\phi)$, it should be noted that $\lambda^*\phi = O(1)$ and the right-hand side of (13) is the same order of magnitude as the other terms in that equation in the limit $\phi \to 0$. Thus, we need to keep the collision term in our analysis even for very dilute suspensions. Now substituting (16) and (17) into (13), using $p = \phi < V_k V_k > /3 = \phi T$, and solving for the steady state conditions, we obtain

$$P_{11} = P_{33} = P_{22} + \frac{P_{12}}{St^{-1} + \phi\lambda T^{1/2}} = \frac{\lambda\phi^2 T^{3/2}}{St^{-1} + \phi\lambda T^{1/2}}, \tag{18}$$

$$P_{12} = -\frac{P_{11}}{2(St^{-1} + \phi\lambda T^{1/2})} = -\frac{\lambda\phi^2 T^{3/2}}{2(St^{-1} + \lambda\phi T^{1/2})^2}, \tag{19}$$

and $P_{23} = P_{31} = 0$. To complete the solution we need to determine T. This is accomplished by taking the trace of (13) to yield at steady state the energy equation for the dispersed phase

$$P_{12} + 3\phi St^{-1} T = 0 \tag{20}$$

according to which the work done in shearing the suspension is dissipated by the viscous drag. Substituting for P_{12} from (19) into (20) yields a cubic equation for $T^{1/2}$ whose three roots are given by

$$T_1 = 0, \quad T_{2,3}^{1/2} = \frac{5\sqrt{\pi}\, St}{288\ \phi}\left[1 - \frac{12}{St^2} \pm \sqrt{1 - \frac{24}{St^2}}\right], \tag{21}$$

where we have used the numerical estimate of λ from (17).

It is easy to show that T_2 corresponds to an unstable state so that the quenched and ignited states we found in numerical simulations (cf. Figure 1) correspond respectively to T_1 and T_3. Moreover we see that the ignited state exists only when $Re_s = 18St$ exceeds a critical value given by $Re_{cr} = 18\sqrt{24} = 88.18\ldots$. This explains why the final state is the quenched state regardless of the initial variance for $Re_s = 80$. For $Re_s = 140$, the variance corresponding to the unstable state 2 is approximately 4.3 according to (21). Thus, if initial variance is smaller than this value, the final state must be the quenched state, and a higher initial variance should lead to the ignited state. Simulations for $Re_s = 140$ qualitatively agree with this prediction although we find that even a slightly higher initial variance of 6 leads to a quenched state. In fact, the variance reaches a maximum of about 15 before eventually decreasing to a vanishingly small value corresponding to the quenched state. This quantitative discrepancy may arise due to a number of reasons including (i) the neglect of shear-induced collisions in the theory, (ii) finite number of bubbles ($N = 32$) used in the numerical simulations, (iii) the use of soft core repulsive potential in simulations, and (iv) the neglect of hydrodynamic interactions in the theory. Finally, we see that the theory we have presented fails completely in the case of $Re_s = 170$ in predicting the existence of only one stable state.

The theory we have presented so far is adequate for determining the steady state variance in the ignited state for which the root-mean-squared velocity is much greater than γa and for giving the criterion for extinction of the ignited state, i.e, $Re_s < Re_{cr}$. However, the preceeding theory gives poor estimates for T_1 and T_2 because it neglects shear-induced collisions which are important for these two states.

To improve the theory, we now consider the limit in which the root-mean-square velocity is much smaller than γa, a situation applicable to the quenched state. Since the collisions are infrequent in this state, $\tau_v \ll \tau_c$ and the majority of bubbles travel with the velocity of the fluid. Therefore, in this limit $\Delta_e(P_{ij})$ (cf. (14)) can be determined from simple geometric considerations using $\mathbf{w} =< \mathbf{u} >$ to yield

$$\Delta_e(P_{11}) = 2\Delta_e(P_{22}) = 8\Delta_e(P_{33}) = \frac{512}{315\pi}\phi^2, \qquad \Delta_e(P_{12}) = -\frac{8}{35}\phi^2 \tag{22}$$

and $\Delta_e(P_{13}) = \Delta_e(P_{23}) = 0$. Substituting for $\Delta_e(P_{ij})$ from (22) into (13) and solving for steady state conditions yield

$$P_{11} = 4P_{33} = \frac{256}{315\pi} St\phi^2, \quad P_{22} = \frac{64}{315\pi} St^3\phi^2\left[1 + \frac{4\pi}{St} + \frac{9}{8St^2}\right], \tag{23}$$

$$P_{12} = -\frac{64}{315} St^2\phi^2\left[1 + \frac{9\pi}{16St}\right]. \tag{24}$$

Thus, the quenched state variance is dominated by the value of $< V_2^2 >$ and equals roughly $(64\pi/315)St^3\phi$ in the limit of small ϕ for Stokes numbers of magnitude 5 or greater. At $\phi = 0.005$ and $Re_s = 18St = 140$, the conditions for the simulations shown in Figure 1, this gives an approximate variance of 1.5 whereas the simulation gave a vanishingly small number. This occurs because the bubbles arrange themselves eventually in positions which avoid collisions making their velocities the same as the local fluid velocity and it is therefore an artifact of the simulation with periodic boundary conditions. It is possible to avoid this problem in simulations that neglect hydrodynamic interactions by using the Direct-simulation Monte Carlo method (Kumaran, Tsao, and Koch 1993).

We have shown that the ignited state exists for all $Re_s > Re_{cr}$. Now we shall determine in what part of this regime, one has multiple steady states and in what portion only the ignited state exists. If the shear-induced variance $Re_s^3\phi$ is greater than the variance of the unstable state 2, i.e. $O(Re_s^{-3}\phi^{-1})$ (cf. (21)), then the imposed shear will create enough velocity fluctuations to take the suspension past the unstable state 2 even when the intial variance is zero. Consequently, only the ignited state will exist when $Re_s^3\phi$ exceeds a certain $O(1)$ number. To estimate this number we constructed an *ad-hoc* approximation for $\triangle_e(P_{ij})$ by superimposing its values in the two limits as given by (16) and (22). Solving the resulting equations for P_{ij} at steady state yielded a quartic equation in $T^{1/2}$. One root of this equation is always negative and the other three correspond qualitatively to the three solutions (quenched, unstable, and ignited) given by (21). However, when ϕ is increased from zero at a fixed value of $St = Re_s/18$ that is greater than $\sqrt{24}$, we now find that the quenched and unstable state variances approach each other. The variances of these two states become equal at

$$\phi_c = \left(\frac{7875\pi^2}{2304}\right)^{1/3} St^{-3}, \quad \text{or} \quad St^3\phi_c = 3.231\ldots \tag{25}$$

For $\phi > \phi_c$ the two roots become complex so that the only physically meaningful solution to the equations corresponds to the ignited state.

The above criterion can also be used for estimating Re_s beyond which the only steady state is the ignited state for given ϕ. At $\phi = 0.005$, this yields the transition Re_s of 155.6. This is in agreement with the simulations shown in Figure 1 for which the multiple states are observed at $Re_s = 140$ but only the ignited state at $Re_s = 170$.

3.2 Comparison with simulations

We now compare the theory and simulations in more detail. Figure 2 shows the regimes of quenched, ignited, and multiple (quenched plus ignited) states. For small ϕ we expect only the quenched state when Re_s is less than 88 and the multiple (quenched plus ignited) states for $Re_s > 88$ and $\phi Re_s^3 > 3.231 \times 18^3 = 18843$. For each value of ϕ, we carry out simulations at different values of Re_s with an initial variance of zero and determine the critical value of Re_s for which the final state is ignited. The pluses are the results obtained by simulations with $N = 32$ and full hydrodynamic interactions together with the soft core repulsive potential for overlapping bubbles as described in section 2. For the purpose of comparison with the theory which neglected the hydrodynamic interactions altogether and modeled the collision process as that corresponding to hard spheres, we also carried out another set of simplified simulations in which these conditions were satisfied exactly. These simulations with $N = 100$ are shown by circles. The latter results were also confirmed to be free from finite N effects by another method (Direct-simulation Monte Carlo, Kumaran, Tsao, and Koch 1993). We believe that the better agreement obtained with the full hydrodynamic interaction calculations represented by the pluses is fortuitous.

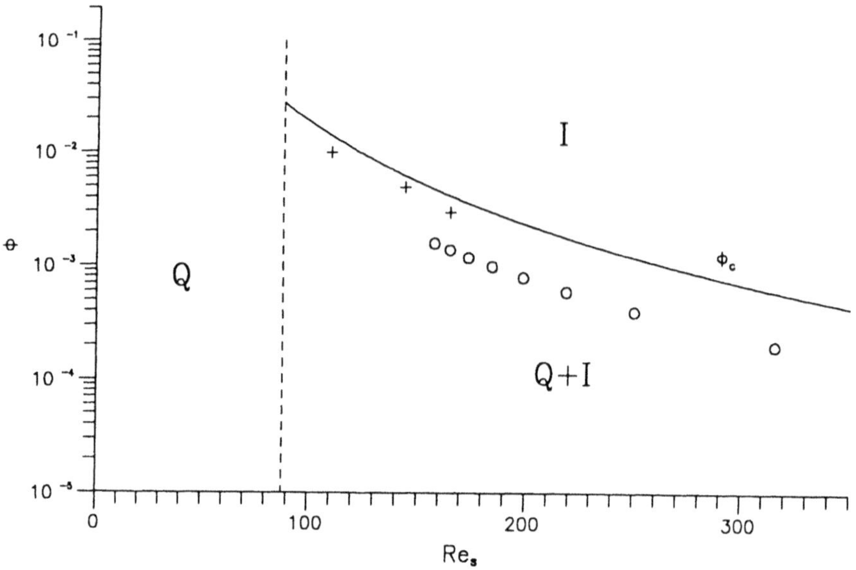

Figure 2: The multiple states-ignited state transition. The solid curve is the theoretical prediction and the circles and pluses are, respectively, the results of simulations with and without hydrodynamic interactions.

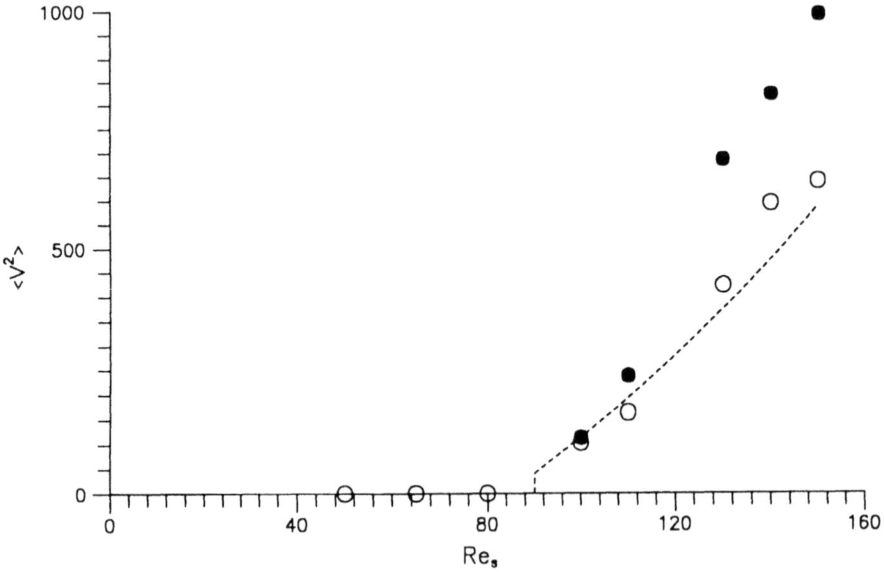

Figure 3: Velocity variance as a function of Re_s for $\phi = 0.03$. The dashed line is theory, the filled circles are the results of simulations with the soft core repulsive potential, and the open circles are the results for the hard core potential.

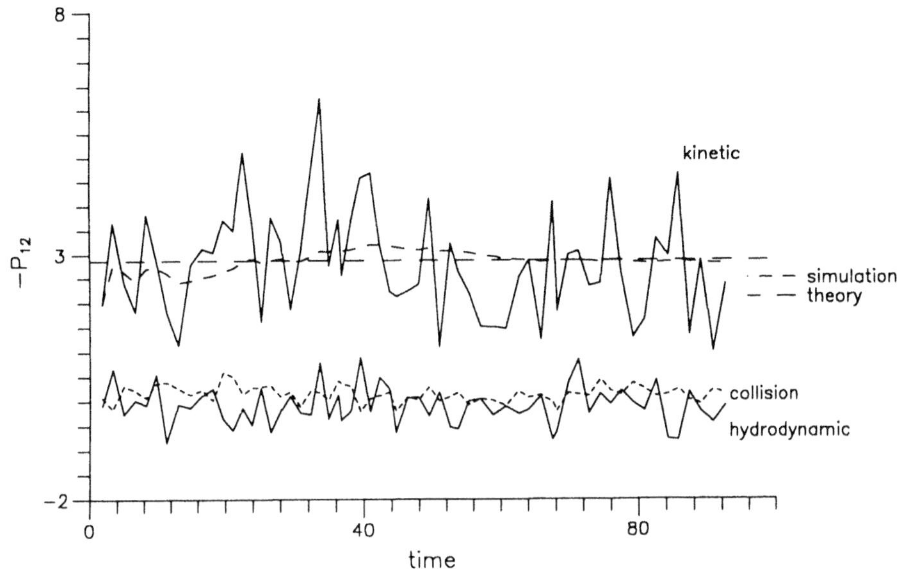

Figure 4: The kinetic, collisional, and hydrodynamic contributions to the dispersed-phase shear viscosity $\mu_s \equiv -P_{12}$ for $Re_s = 150$ and $\phi = 0.03$.

Figure 3 shows a comparison between the theory (cf. (21)) and simulations at $\phi = 0.03$. For this ϕ, the critical Re_s for multiple steady states to exist is about 86 which approximately coincides with the extinction of the ignited state branch so that we observe only the ignited state at larger Re_s. The simulation results indicated by filled circles were obtained with full hydrodynamic interactions and soft core potential. We find that there is considerable discrepancy between the theory and simulations. This results from the use of soft core potential in simulations which allow bubbles to overlap considerably resulting in a decrease in the apparent volume fraction of bubbles. Since the variance varies roughly as $1/\phi^2$, the use of soft core potential results in a higher variance. To correct for this effect we carried out simulations with no hydrodynamic interactions with both soft as well as hard core potentials. The open circles in Figure 3 represent the results obtained by multiplying the results of full hydrodynamic interactions with the correction ratio accounting for the use of soft core potential. We see that with this correction, the theory and simulations are in very good agreement with each other.

Figure 4 shows the kinetic, collisional, and hydrodynamic (or potential) contributions (cf. I and Bulthuis, Prosperetti, and Sangani 1994 for definitions) to the dispersed-phase shear viscosity (non-dimensionalised by $1/2\rho\gamma a^2$) $\mu_s = -P_{12}$. As expected, the collision and hydrodynamic parts are seen to make negligible contributions to the overall value of shear viscosity at $\phi = 0.03$. The average value of the kinetic part is seen to be in a very good agreement with the value predicted by theory (cf. (19)) provided that we use the value of average variance computed in simulations to substitute for T instead of using (21). This distinction is necessary to make since T for the soft core potential is different from the theoretical estimate.

Figure 5 shows a comparison between the theory and simulations for P_{22}/P_{11}. The computed values include the collision and hydrodynamic parts also. Once again we see a reasonable agreement between the two. More importantly, it must be noted

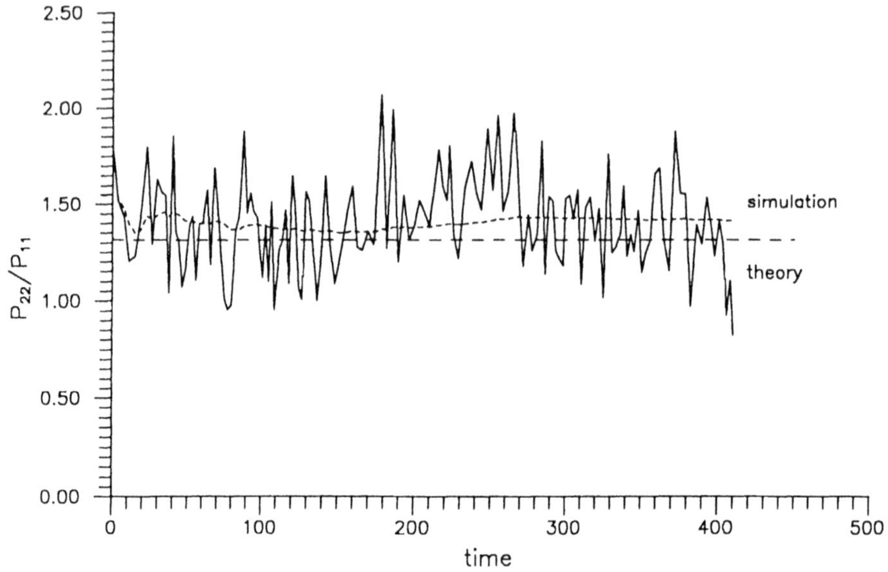

Figure 5: P_{22}/P_{11} as a function of time for $\phi = 0.03$ and $Re_s = 150$.

that the dispersed-phase exhibits considerable normal stress differences.

Finally, we note that a discrepancy that still remains between the theory and simulations is the rather high value of the maximum variance seen in Figure 1 for the quenched state simulation with $Re_s = 140$. Even after accounting for the shear-induced variance, our approximate theory estimates the variance of the unstable state 2 to be approximately 5 which is much lower than the maximum value of about 15 obtained in the simulation. Thus, it appears that our theory underestimates the magnitude of the variance in the unstable state.

4 Concluding Remarks

In this paper we have addressed the problem of the dispersed-phase rheology in suspensions of spherical bubbles at relatively large Reynolds numbers. We found that the rheology is quite complex even when the microscale physics governing the bubble motion is considerably simplified in terms of lift and viscous forces. The key to understanding the results of simulations has been to appreciate the dependence of the dispersed-phase viscosity on its temperature. This dependence is nonlinear and gives rise to multiple steady states. The calculations presented here also show a need to include the energy balance and temperature in the averaged description of flows of suspensions.

Our simulations and theory have thus far been restricted to small ϕ where the shear-induced fluctuations are the largest. We have extended the theory to higher volume fractions using the theory of dense gases and Grad's moment expansion approximation (cf. (15)). The predictions of this theory are being currently tested for nondilute suspensions.

One motivation of this study was to investigate the possibility that the presence of shear may stabilize bubbly liquids. We can now make a rough estimate of when stabilization is possible. At present we are studying the flow of bubbly suspension under simple shear in the presence of gravity. In that case the dispersed-phase pressure depends on both the mean relative velocity of the bubbles induced by buoyancy and lift forces and mean shear rate. The mean relative motion gives a negative contribution to the pressure via hydrodynamic contribution as shown in I while the shear gives a positive contribution via the kinetic and collision contributions. Exact criterion for the stability has not been determined yet but preliminary calculations (simulations) already show that the suspension is stable at least when the Reynolds number based on shear is in the ignited state regime. Thus, let us take $Re_s = \rho\gamma a^2/\mu = 100$ for the purpose of estimating when shear may play an important stabilizing role. For 1 mm radius bubbles in water, this requires $\gamma \simeq 100$ s^{-1}. For air-water system with $\gamma a = 10$ cm/s, the small We approximation is justified since $We \simeq 0.14$. To provide γ of 100 s^{-1} in a flow of bubbly suspension through a pipe of radius R (in cm) we need the mean flow rate to be $100R$ (in cm/s). Since we are dealing with large Reynolds numbers, one question that immediately comes to mind is if the turbulence would set up much before the high shear required for the ignited state. Taking the mean flow rate to equal γR, the Reynolds number Re_p based on the pipe radius will be $100(R/a)^2$. This will be indeed quite large for many practical applications. However, it should also be noted that the turbulence will be considerably delayed due to high Reynolds stress created by the velocity fluctuations induced by the presence of shear. A parameter that is expected to be more important in determining the onset of turbulence is $Re_p^* = Re_p\mu/\mu^*$, the Reynolds number based on the effective viscosity μ^* of the mixuter. From the definition of the mixture stress given in Biesheuvel and Wijngaarden (1984) and I, we find that the effective viscosity of the mixture in the ignited state is close to the dispersed-phase viscosity, or, in the present example, $\mu^* \simeq 100\mu$. Thus, in fact, it is possible to have a significant range of R/a and Re_p values over which the mean flow will be steady and one-dimensional. We hope to carry out a more detailed analysis based on averaged equations to test this speculation in our future work.

This work was supported by grants from the National Science Foundation (CBT-9307723) to ASS and the Office of Naval Research (N00014-91-J- 1790) to DLK. The computations were performed on the Cornell National Supercomputer Facility.

5 References

Auton T. R., Hunt J. C. R. & Prud'homme M. 1988 *J. Fluid Mech.* **197**, 241.
Biesheuvel A. 1994 a paper in this proceedings.
Biesheuvel A. & Gorissen W. C. M. 1990 *Int. J. Multiphase Flow*, **16**, 211.
Biesheuvel A. & Wijngaarden L. van 1984 *J. Fluid Mech.* **148**, 301.
Bulthuis H. F., Prosperetti A. & Sangani A. S. 1994 *J. Fluid Mech.* (submitted).
Chapman S. & Cowling T. G. 1970 *Mathematical Theory of Nonuniform Gases*, Cambridge University Press, Cambridge.
Grad H. 1949 *Comm. Pure Appl. Math.* **2**, 331.
Hasimoto H. 1959 *J. Fluid Mech.* **5**, 317.
Kumaran V., Tsao H.-K. & Koch D. L. 1993 *Int. J. Multiphase Flow* **19**, 665.
Ladd A. J. C. 1990 *Phys. Fluids A*, **2**, 1921.
Ladd A. J. C. 1994 *J. Fluid Mech.* (submitted).
Lessard R. R. & Zieminski S. A. 1971 *Ind. Engng Chem. Fund.* **10**, 260.
Sangani A. S. & Didwania A. K. 1993a *J. Fluid Mech.* **248**, 27.
Sangani A. S. & Didwania A. K. 1993b *J. Fluid Mech.* **250**, 307.

Sangani A. S., Zhang D. Z. & Prosperetti A. 1991 *Phys. Fluids A* **3**, 2955.
Smereka P. 1993 *J. Fluid Mech.* **254**, 79.
Tsao H.-K. & Koch D. L. 1994 *Phys. Fluids* (to appear)
Zaleski S. & Zanetti G. 1994 a paper in this proceedings.
Zhang D. Z. & Prosperetti A. 1994 *J. Fluid Mech.* **267**, 185.

Bouncing and coalescence of two bubbles in pure water.

P.C. Duineveld
Department of Applied Physics, University of Twente,
PO.Box 217, 7500 AE Enschede, The Netherlands.

Abstract

We have investigated encounters between two bubbles rising at high Reynolds number in pure water. The purity of the water was verified by measuring the rise velocity of one bubble. If the equivalent radius is larger than 0.91 mm the trajectorie becomes unstable and the bubble performs a zig-zag motion, most likely caused by viscous effects. Above a critical Weber, based on the approach velocity bubbles were found to bounce in pure water. If their size and initial separation do not exceed a critical value bubbles approach again and coalesce after bouncing, which is in agreement with potential flow calculations. Above a mimimum equivalent radius of $\approx$ 0.86 mm, which is near the onset of path instability, the behaviour of the bubble pair becomes totally different. The bubbles then perform a symmetrical zig-zag motion with respect to each other, which can not be described by potential flow theory.

1 Introduction

In the past decades bubbly flows have been studied by various researchers. These flows can be divided in a number of different regimes. The first regime is one in which the gas fraction of the bubbles is very small, the relative distance between bubbles is large and therefore they do not interact. This regime holds for gas fractions up to $\approx 1\%$. A second regime is obtained for gas fractions from 1% to ≈ 10 %, where interactions between two bubbles are dominant. The speed of rise of bubbles significantly deviates from the single bubble result due to the mutual interactions, (see e.g. Van Wijngaarden 1993). For larger gas fractions multiple bubble interactions occur, which can considerably complicate the behaviour of bubbly flows. In this contribution we consider interactions between two bubbles with equivalent radii in the range 0.5-1.0 mm.

Recently several authors (Van Wijngaarden 1993, Kumaran & Koch 1993, Sangani & Didwania 1993 and Smereka 1993) have noticed a discrepancy be-

S. Morioka and L. van Wijngaarden (eds.),
IUTAM Symposium on Waves in Liquid/Gas and Liquid/Vapour Two-Phase Systems, 151-160.

tween their calculations and experiments of e.g. Lammers (1994) concerning bubbly flows in water. In the calculations bubbles were found to rapidly aggregate themselves in horizontal planes, however this does not occur in experiments of bubbly flows at low gas fractions, where bubbles were found to be homogeneously dispersed. All these authors assumed the liquid to be pure, i.e. no influence of surfactants, in that case a zero tangential shear stress holds on the bubble surface. Bubbles were all taken to be spherical and bounce elastically at encounter. The liquid motion was described by potential flow theory.

In this contribution we try to understand the deviation between theory and experiment. Therefore we have tested the assumptions made in the theories. First the purity of the liquid was tested by measuring the rise velocity of one bubble. We also discuss the stability of the bubble path of one bubble. Next the behaviour of two, equally sized bubbles rising under buoyancy was studied experimentally. Here we are specially interested what will happen at encounter; will bubbles coalesce or bounce, what will happen with them after bouncing and how does this depend on bubble size and initial distance. We also discuss the influence of bubble encounters on bubbly flows.

This paper is an extension and refinement of Duineveld (1994b). Here only main features will be discussed. Details can be found in Duineveld (1994a).

2 The rise velocity of a bubble in pure water

It is well known that water is very sensitive to surface active impurities. If a bubble is rising through contaminated water its rise velocity is lower than that in pure water. This is caused by a tangential stress due to a concentration gradient of surfactants on the bubble surface. This stress has to be compensated by a viscous shear stress which increases the drag on the bubble. Therefore if the experimentally determined rise velocity of a bubble is lower than that given by the zero shear stress calculation of Moore (1965), one can conclude that the water contains impurities. Hence the rise velocity of a bubble provides a way to test the purity of the water.

One may note why so much effort is put in obtaining pure water while another liquid, which is less sensitive to impurities, may also be used. Besides the fact that water is by far the most common liquid on our planet it has the advantage of an extremely low Morton number. This Morton number depends on the liquid properties only and is defined as:

$$M = \frac{g\mu^4}{\rho\sigma^3}, \tag{1}$$

where g is the accelaration of gravity, μ the dynamic viscosity, ρ the density and σ the surface tension. For water at 20 °C M has a value of $2.4 \cdot 10^{-11}$. In a liquid with a low Morton number bubbles can be produced which rise at a large

Reynolds number while their deformation is reasonably small. These bubbles are essential to test the zero stress theory of Moore (1965).

Our experiments were performed in a cubic tank with glass walls of 50 cm side length and thickness of 18 mm. The tank was filled with "hyper clean" water, which was produced with a Millipore purification system. The quality after purification was: specific resistance 18.2 MΩ·cm and the number of organic particles was less than 10 ppb. The tank was rinsed and refilled every day. Bubbles were produced with a specially designed system with which they could be made very accurately. Bubble sizes were obtained by measuring the length of the air plug in a feed capillary. In this manner the equivalent bubble radius can be determined with an accuracy between 0.5% and 0.3 % for an equivalent radius of 0.5 mm and 1.0 mm respectively.

The bubble motion was recorded with a NAC high speed video recorder with a frame rate of 500 frames/sec. The rise velocity and the shape of the bubbles were determined using image software routines. The rise velocity was determined at a distance of 10 to 12 cm from the release point, and proved to be perfectly stationary in this range.

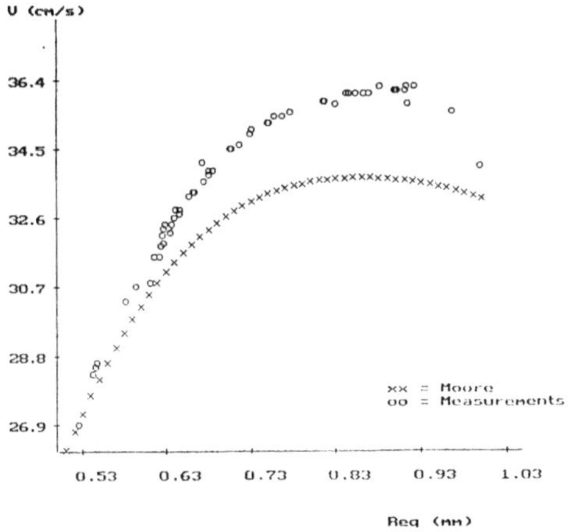

Figure 1: Vertical rise velocity of a bubble.

In figure 1 the experimentally determined vertical rise velocities are given as a function of the bubble size. The temperature of the water was 19.6±0.2 °C during all experiments. The experimental results are compared with the theoretical rise velocity calculated by Moore (1965). Up till an equivalent radius of 0.63 mm there is a good agreement between theory and experiment, indicating that our water can be considered as pure. For larger sizes the experimental rise velocity is slowly becoming larger than the theoretically determined one, but the shape of the two curves are more or less similar, with both a maximum at R=0.9 mm, where R denotes the equivalent bubble radius.

The theoretical calculations are based on an oblate ellipsoidal bubble shape. Above $R = 0.63$ mm we have found that the experimental bubble shape deviates from this oblate ellipsoidal shape and the difference between theory and experiment is caused by a theoretical overestimation of the deformation, which causes a larger drag force and hence a lower theoretical rise velocity as follows from figure 1.

Usually experiments of bubbly flows are not performed in this "hyper clean" water but in tap water. We have measured the rise velocity of bubbles in fresh tap water and found that above an equivalent bubble radius of 0.7 mm there is no noticeable difference between the rise velocity in fresh tap water and pure water. In experiments in bubbly flows bubbles have a minimum equivalent radius of $\approx$ 1.4 mm (e.g. Lammers 1994) and therefore the behaviour of these bubbles can be treated with pure liquid theory.

3 Path instability of a bubble

The sudden decrease of the experimental vertical rise velocity is due to a path instability of the bubble. At $R = 0.91$ mm we first noticed this instability. At the recorded distance the bubble path was still rectilinear, but higher up in the tank the bubble starts to zig-zag. The critical Weber number at path instability was found to be 3.3, where the Weber number is defined as:

$$We = \frac{2\rho U^2 R}{\sigma}. \tag{2}$$

Here U is the velocity of rise. The bubble size and Weber number are in good agreement with the experimentally determined results of Hartunian & Sears (1957). Note that the path instability is different from a shape instability of a bubble, which occurs at higher Weber number (Duineveld 1994a).

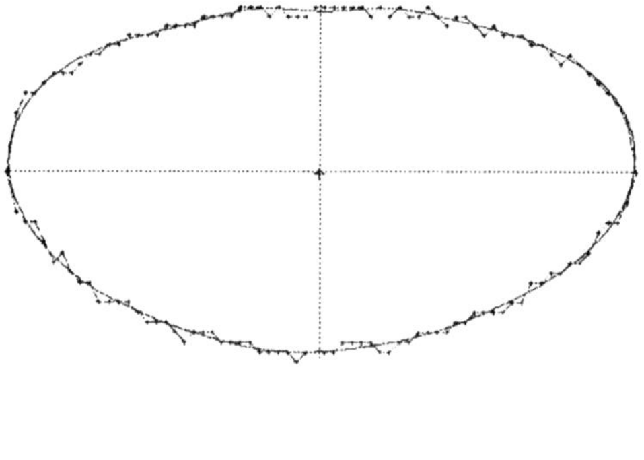

Figure 2: Contour of a bubble with $R = 0.91$ mm.

At the moment the cause of this path instability is not completely clear. Meiron (1989) showed that the interaction of hydrodynamic pressure and surface tension can not be the cause of path instability. Amongst others Ryskin & Leal (1984) suggested that it was caused by vortex shedding. Fore this to happen

R (mm)	V_0 (cm/s)	U_t (cm/s)	A (mm)	f (Hz)
1.15 (0.05)	29.2 (0.3)	31.6 (0.5)	3.5 (0.3)	6.4 (0.1)
1.34 (0.05)	29.4 (0.3)	29.5 (0.5)	3.3 (0.3)	7.0 (0.1)

Table 1: Vertical rise velocity, radius and frequency of zig-zagging bubbles

a standing eddy must exist behind the bubble. The eddy is not created by boundary layer separation, as is the case for a solid sphere, but by accumulation of vorticity at the rear of the bubble, brought to this area by convection. At the bubble surface vorticity is created with magnitude :

$$\omega_s = \frac{2u_\theta}{R_c}, \tag{3}$$

where ω_s is the azimuthal component of vorticity at the interface, u_θ the tangential velocity at the surface and R_c the local radius of curvature (Leal, 1989). With increasing bubble size bubble deformation grows and so R_c decreases at the bubble midplane, hence increasing the vorticity production. On the other hand from figure 1 the rise velocity is nearly maximum at path instability, therefore convection of vorticity is not increasing. As a consequence vorticity can accumulate at the rear of the bubble and cause an eddy. An indication for this eddy may be found from figure 2. This is the projection of a bubble with $R = 0.91$ mm, near path instability. Clearly visible is a larger radius of curvature at the front than at the rear of the bubble. The normal stress condition on the bubble surface,

$$p_g = p_l + 2\sigma/R_c, \tag{4}$$

where p_g and p_l are the gas and liquid pressure respectively, can indicate a wake with the subsequent low pressure at the rear of the bubble.

Still it is yet not clear why a bubble performs a path instability and a solid sphere not, while for both a standing eddy appears and both have similar effective mass. Ryskin & Leal (1984) suggested this difference due to be by different moment of inertia.

In our experiments the motion of zig-zagging bubbles was studied for two different bubble radii. The bubbles were produced by blowing air through a stainless steel hollow needle in a plexiglass tank. The results are shown in table 1, with in between brackets the errors. In this table V_0 is the vertical rise velocity of the bubble, A is the radius of the zig-zag motion and f is the frequency of this zig-zag.The measured frequency of the zig-zag is in agreement with what was found by Saffman (1956), however the radius of the zig-zag is larger in our experiments by a factor two. This is probably caused by impurities in the water used by Saffman. The vertical rise velocity is lower than predicted by the theory because the bubble also has a horizontal velocity. Therefore we have measured for $R = 1.15$ mm the path velocity of the bubble. This path velocity is nearly constant and 33.5 cm/s, again larger than the theoretical value because of the overestimation of the deformation in the theory. In the experiments it is also observed that the velocity of the bubbles is parallel to the axis of symmetry, as was already argumented by Saffman.

4 Encounter between two bubbles

Kok (1993a) calculated the trajectories of two, equally sized spherical bubbles with viscous potential flow theory. He found that spherical bubbles attract each other if the angle between their line of centre and the vertical is above a critical value, otherwise they repel. A hydrodynamic moment is working on the bubble pair as well, forcing them in a cross stream orientation. Kok confirmed his calculations with an experiment with two bubbles with R=0.5 mm (Kok 1993b). These bubbles rise at a large Reynolds number but deformation is still reasonably small.

In this contribution we report on encounters of bubbles with an equivalent radius ranging from 0.5-1.0 mm. These bubbles are not spherical but have approximately an oblate ellipsoidal shape. The trajectories appear to be qualitatively similar as those for spherical bubbles. Here we do not further discuss trajectories, but concentrate on the behaviour of bubble pairs at encounter.

At encounter bubbles can both coalesce and bounce, depending on the approach velocity of the bubble pair. When the bubbles become close, the pressure in the film between the bubbles increases, resulting in a deformation of the bubbles. At the same time this increasing pressure causes film drainage and a repelling force on the bubbles. Because of inertia the bubbles continue to approach each other, which increases the radius of the film and decreases the film thickness and the velocity of the bubbles. This thinning continues until a film thickness of order 100 A^0 is reached. At this separation an attracting molecular force, the Van Der Waals force, becomes important, resulting in coalescence of the two bubbles. So bubbles can coalesce in a pure liquid if the film reaches a critical thickness before the motion of the bubble has stopped. Bubbles can bounce if the motion of the bubble has arrested before the film is thinned to a critical thickness. Inertia plays a dominant role in this process. Increasing bubble inertia induces an increasing approach velocity, favoring coalescence, on the other hand the repelling force due to the increased pressure is increasing more significantly, which can cause the motion to be stopped before coalescence. A bouncing critirium was proposed by Chesters & Hofman (1982), which is that the Weber number, based on the approach velocity V,

$$W = \frac{\rho V^2 R_{eq}}{\sigma}, \tag{5}$$

exceeds a critical value.

R mm	Init sep.	V_a m/s	U_a m/s	W	bouncing
0.63	$8.3\,R$	0.13 ± 0.01	0.26 ± 0.02	0.15 ± 0.02	no
0.71	$7\,R$	0.13 ± 0.01	0.27 ± 0.02	0.16 ± 0.02	no
0.71	$8 - 10.6\,R$	0.14 ± 0.01	0.30 ± 0.02	0.19 ± 0.03	yes
0.78	$5.7\,R$	0.13 ± 0.01	0.25 ± 0.02	0.18 ± 0.03	no
0.78	$7.7\,R$	0.14 ± 0.01	0.29 ± 0.02	0.21 ± 0.03	yes
0.85	$4.6\,R$	0.11 ± 0.01	0.25 ± 0.02	0.14 ± 0.02	no
0.85	$6.1\,R$	0.13 ± 0.01	0.29 ± 0.02	0.20 ± 0.03	yes
0.91	$4.0\,R$	0.08 ± 0.01	0.17 ± 0.04	0.08 ± 0.02	no
0.91	$5.0\,R$	0.12 ± 0.01	0.26 ± 0.03	0.18 ± 0.03	yes
0.97	$4.4\,R$	0.11 ± 0.01	0.21 ± 0.04	0.16 ± 0.03	yes

Table 2: Critical Weber number for bouncing-coalescence

Kok (1993a) and Van Wijngaarden (1993) found that before encounter a bubble pair accelerates considerably. Therefore the approach velocity in expression (5) has to be determined just before encounter. Encounters were recorded with the NAC high speed video recorder, this time with a frame speed of 1000 frames/sec. With the image software routines it was possible to measure the approach velocity with an accuracy of $\approx 10\%$.

In our experiment we have indeed found coalescence of bubbles only if the Weber number above was below a critical value. In table 2 the critical Weber number (W), the pertinent approach velocity (V_a) and rise velocity (U_a) at encounter are given for the pertinent bubble size and initial distance.

The critical Weber number was found to be $W_{cr} = 0.18 \pm 0.03$. In Duineveld (1994a) numerical calculations of this critical Weber number are discussed, which show reasonably good agreement with the experiments.

It follows from table 2 that for all cases considered the approach velocity at encounter is $\approx 0.5\,U_a$. This is only half of the approach velocity following from the potential flow calculation of Van Wijngaarden (1993). This experimental result seems to be in agreement with the result of Kok (1993b), who finds that the experimental time scale was a factor 2 larger than in theory. The cause of this discrepancy is not yet clear. The experimental relation $V_a \approx 0.5\,U_a$ can help us predicting the minimum bubble size where bouncing occurs for other low Morton liquids. From the data of Moore (1965) we find for instance that methyl alcohol will give a comparable bubble size (≈ 0.6 mm).

After bouncing the bubbles repel each other, but due to the attractive hydrodynamic force they encounter again and coalesce, which is due to the reduced approach velocity. This scenario is in agreement with the potential flow theories mentioned before but holds only if the bubble size and initial distance are below critical values. If the bubble size and initial distance exceed a critical value the behaviour of the bubble pair can become totally different. Then bubbles bounce and separate considerably. This is shown in figure 3 for a bubble pair with $R = 0.86$ mm. The bubbles first approach each other in a cross-stream orientation (figure 3a, 3b), bounce and encounter again (fig 3c), the approach velocity is again too large and the bubbles separate and approach again (fig 3d, 3e), once more the approach velocity is too large and the bubbles bounce, but after this bounce the bubbles separate considerably and do not coalesce (fig 3f). Larger bubbles may separate already at one bounce.

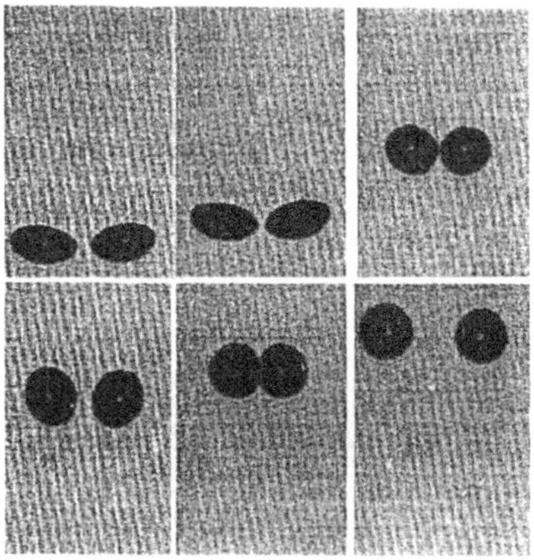

Figure 3: Bouncing-separation for two bubbles with eq. radius of 0.86 mm.

The minimum bubble size for bouncing-separation is $R \approx 0.86$ mm. For this size the initial distance is nearly at the maximum separation where bubbles still attract each other (8.9 R). However the initial distance between the bubbles, where they bounce-separate, decreases quickly with increasing bubble size, e.g. for $R = 1.4$ mm the initial distance is only $3.3R$. This minimum bubble size is very close to the minimum size where path instability for one bubble occurs, $R = 0.91$ mm in pure water. Path instability was already discussed to be most likely caused by an instability of the wake behind the bubble, by wich vortices are shed off the bubble, and induce the zig-zag motion. Here we suggest that the bounce of bubbles triggers the instability of the wake and the shedding of vortices. This bounce causes a large distortion and can probably therefore cause a path instability at already smaller bubble size than for one bubble. Apparently the two bubbles perform a symmetrical zig-zag motion with respect to each other. This was verified by recording the motion of the bubbles after the separation. There it was found that two bubbles indeed perform a symmetrical zig-zag motion with respect to each other. The bubbles first separate as is shown in fig 3f until a maximum separation of ≈ 8.3 mm, which is about two times the diameter of the zig-zag path of one bubble though somewhat smaller because of the attracting hydrodynamic force between the bubbles. After this maximum separation the bubbles approach and encounter again. The time between the two encounters was found to be ≈ 150 msec, which is about the time for a completete zig-zag of one bubble ($f_z \approx 6.5$ Hz).

With a view to these observations we suggest that the separation of the bubbles after bouncing is caused by the shedding of vortices. This is a very important physical effect, because it dominates the relative motion. After the first bounce the shedding feeds extra energy in the repelling relative motion. Therefore the maximum separation and the time before the bubbles approach again is much larger than calculated with potential flow theory (Van Wijngaarden 1993; Kumuran & Koch 1993). They found the time between two encounters to be of order R/U, which is orders of magnitude smaller than $1/f_z$. This can have important consequences for the stability of bubbly flows.

In the potential flow calculations of Van Wijngaarden (1993), Kumaran & Koch (1993), Sangani & Didwania (1993) and Smereka (1993) it was found that from an initially randomly distributed set of bubbles they orient themselves rapidly in horizontal planes transverse to gravity. In the experiments in bubble columns of Lammers (1994), where bubble sizes could be controlled carefully and the spread in size was small, no clustering was observed at low gas fractions, in contrast with the above mentioned theoretical results.

In bubbly flows it is very hard to produce small bubbles and a sufficient gas flow (e.g. Lammers 1994). In these flows bubbles have a minimum radius, $R \approx 1.4$ mm. These bubbles bounce-separate at already very low initial distance and therefore will not rapidly orient themselves into horizontal planes, as was found from the calculations based on potential flow. If it is possible to produce bubbles with very small radii in pure water then these will coalesce rapidly and finally one ends up with bubbles with a minimum radius of ≈ 1 mm in water. These bubbly flows will be much more stable because of the bouncing-separation mechanism. Therefore a minimum stable bubble size in bubbly flows in water will probably be of this order.

Our results show that for the modelling of bubble trajectories in bubbly flow it will be esential to incorporate vorticity production at the boundary of the bubble. However this will be a formidable task; even the behaviour of one zigzagging bubble is not yet fully understood !

Acknowledgements

I would like to thank prof. L. van Wijngaarden for many helpful discussions and a continuous interest in this work.

References

[1] Chesters, A.K. & Hofman,G. 1982 Bubble coalescence in pure liquids *Appl.Sci.Res.* **38**, 353-361.

[2] DUINEVELD, P.C. 1994a *Bouncing and coalescence of two bubbles in water* PhD thesis, University of Twente.

[3] DUINEVELD, P.C. 1994b *Bouncing and coalescence phenomena of two bubbles in water* Blake, J.R. (ed), Bubble dynamics and interface phenomena, IUTAM symposium, Birmingham 1993, Kluwer Academic publishers.

[4] HARTUNIAN, R.A. & SEARS, W.R. 1957 On the instability of small gas bubbles moving uniformly in various liquids. *J.Fluid Mech.* **3**, 27-47.

[5] KOK, J.B.W. 1993a Dynamics of a pair of gas bubbles moving through liquid, I Theory. *Eur.J.Mech.,B/Fluids* **4**, 515-540.

[6] KOK, J.B.W. 1993b Dynamics of a pair of gas bubbles moving through liquid II Experiment. *Eur.J.Mech., B/Fluids* **4**, 541-560.

[7] KUMARAN, V. & KOCH, D.L. 1993 The rate of coalescence in a suspension of high Reynolds number and low Weber number case *Phys. Fluids A* **5**, 1135-1140.

[8] LAMMERS, J.H. 1994 *The stability of bubbly flows* PhD thesis, University of Twente.

[9] LEAL, L.G. 1989 Velocity transport and wake structure for bluff bodies at finite Reynolds number. *Phys Fluids A* **1**, 124-131.

[10] MEIRON, D.I. 1989 On the stability of gas bubbles rising in an inviscid liquid *J.Fluid Mech.* **198**, 101-114.

[11] MOORE, D.W. 1965 The velocity of rise of distorted gas bubbles in a liquid of small viscosity *J.Fluid Mech.* **23**, 749-766.

[12] RYSKIN, G. & LEAL, L.G. 1984 Numerical solution of free-boundary problems in fluid mechanics. Part 2 Bouncy motion of gas bubble through a quiscent liquid. *J.Fluid Mech.* **148**, 19-35.

[13] SAFFMAN, P.G. 1956 On the rise of small air bubbles in water. *J.Fluid Mech.* **1**, 249-275.

[14] SANGANI, A.S. & DIDWANIA, A.K. 1993 Dynamic simulations of flows of bubbly liquids at large Reynolds numbers *J.Fluid Mech.* **250**, 307-337.

[15] WIJNGAARDEN, L. VAN 1993 The mean rise velocity of pair wise interacting bubbles in liquid *J.Fluid Mech.* **251**, 55-78.

Interaction of an underwater shock wave with a gas bubble

Kiyoshi Yamada and Kazuyoshi Takayama
Shock Wave Research Center, Institute of Fluid Science,
Tohoku University,
2-1-1, Katahira, Aoba-ku, Sendai 980, Japan

Abstract
Interactions of a shock wave with a gas bubble were investigated experimentally for four combinations of liquid/gas which are water/air, water/helium, silicone-oil/air, and silicone-oil/mixture of air and helium. In the experiments a gas bubble was exposed to a spherical shock wave generated by detonation of a micro-explosive, and the shock-bubble interactions were visualized using double exposure holographic interferometry and shadowgraphy. The gas bubble contracts on being hit by the shock, and a liquid jet forms. Eventually, rebound shock waves occur. It was found that there are at least two types of rebound shocks. The generation of these for each type is explained by sequential flow visualizations.

1. Introduction
The interaction of a shock wave with a gas bubble in liquid is one of the important basic research topics of shock dynamics as well as of bubble dynamics. This problem is also related closely to underwater shock focusing applied to extracorporeal shock wave lithotripsy (ESWL) (Chaussy et al. 1982). It is well recognized today that the interaction of shock waves with cavitation bubbles induced by shock focusing is somehow responsible for tissue damage occurring in ESWL. Initiation of detonation in emulsion explosives is also sustained by successive interaction of shock waves with micro-bubbles (Bourne and Field 1991).

In conjunction with these application projects, a flow visualization study was conducted for quantitative observation of the interaction between a spherical shock wave and a gas bubble by means of double exposure holographic interferometry. The entire set of processes of the collapse of a gas bubble from the shock impingement to the generation of rebound shocks was quantitatively visualized. Four combinations of liquid/gas which are water/air, water/helium, silicone-oil/air, and silicone-oil/mixture of air-helium were examined.

2. Experiments
The test liquids were water and dimethylsilicone-oil (Shin-Etsu Chemical Industry Co., KF96-1.0), and the test gases were air, helium, and mixture of air-helium in which helium concentration was 40 to 60 % volume. The major physical properties of these liquids and gases are tabulated in Table 1 (Shin-Etsu Chemical Industry Co. 1989; Weissler 1949).

S. Morioka and L. van Wijngaarden (eds.),
IUTAM Symposium on Waves in Liquid/Gas and Liquid/Vapour Two-Phase Systems, 161-170.

Table 1: Major physical properties of test liquids and test gases at 293 K. ν is kinematic viscosity.

Test fluid	Sound speed a_0 (m/s)	Density ρ_0 (kg/m^3)	Remark
Water	1482.9	998.20	
Dimethylsilicone-oil	936.5	821	$\nu = 1.0 \pm 0.1 \times 10^{-6}$ m^2/s (298 K)
Air	343.6	1.205	
Helium	1001	0.1663	
Mixture of air-helium	4.3–5.0 $\times 10^2$	0.58–0.79	Helium 40 to 60 % volume

Figure 1a shows an experimental setup for trapping gas bubbles having predetermined volume in a capillary tube. When the test gas was helium or mixture of air-helium, helium in a cylinder (1) was released in a liquid tank (2) and was collected in a PMMA pot (3), to which the capillary tube (4) and a flexible tube (5) were connected. When the test gas was air, apparatus (1) and (3) were not needed. The test liquid was filled in the capillary tube whose inside diameter was uniform throughout, and the length of the gas bubble in the capillary tube was measured so that the volume of the gas bubble was precisely determined. The gas bubbles were sucked separately and trapped in the capillary tube by a roller pump (6).

The setup of releasing the gas bubble in liquid and exposing it to a shock wave is illustrated in Fig. 1b. The capillary tube was set at the bottom of a test tank. A micro-explosive that was a 10 mg silver azide (AgN_3) pellet (manufactured by Chugoku Kayaku Co.) was stuck on a thin cotton thread and immersed in the test liquid. A small volume of test liquid was supplied from the beaker (7) to the capillary tube (4) again by means of the roller pump (6), and a gas bubble was readily produced in the test tank (8). The gas bubble released from the capillary tube moved upward. A He-Ne laser beam monitored the gas bubble. The change in the He-Ne laser beam when the gas bubble traversed it triggered an oscillation of a ruby laser to photograph and also an ignition of the micro-explosive via a delay circuit (Fig. 1c). As soon as the gas bubble arrived at a specified standoff distance from the explosion center, the micro-explosive was ignited and subsequently a spherical shock wave was loaded on the gas bubble. The specified distance was 20 mm for the silicone-oil/air combination and 10 mm for all the other cases. The micro-explosive was detonated by radiating a pulsed Nd: YAG laser beam (Laser Photonics, MYL-100, 20 mJ/pulse and 8 nsec pulse duration). The silver azide pellets were 10.5 ± 0.5 mg in weight, and the contribution of the scatter of the weight of the explosives to resulting shock overpressure was of the order of magnitude of the cube root of the weight. Consequently, the present explosives could produce spherical shock waves with a high degree of repeatability. The ignition delay time was estimated to be of the order of magnitude of the Q-switched ruby laser pulse (Takayama et al. 1983).

The arrangement of the double exposure holographic interferometric flow visualization (e.g. Takayama 1988) for the silicone-oil/air combination is shown in Fig. 1c. The light source was a ruby laser (Apollo Lasers, Model 22HD, approximately 25 nsec pulse duration and approximately 1 J/pulse). In addition to holographic interferometric observations, high-speed photography, using shadowgraphy and an

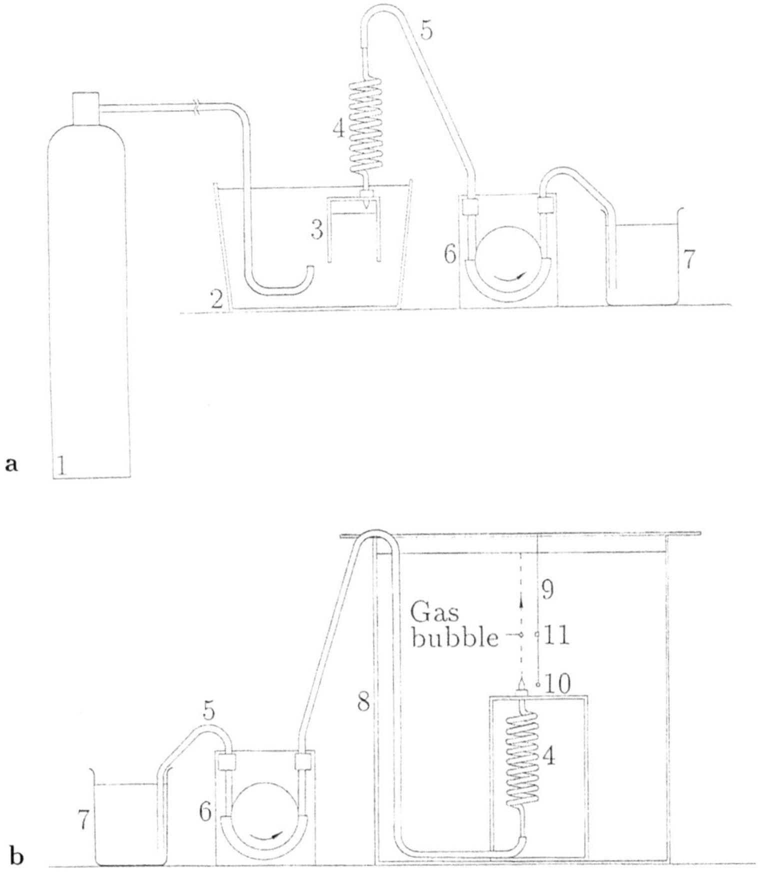

Fig. 1a, b: a Trapping gas bubbles in a capillary tube and **b** releasing the gas bubble in liquid: 1 helium cylinder; 2 see-through tank; 3 PMMA pot; 4 spiral capillary glass tube; 5 flexible tube; 6 roller pump; 7 beaker; 8 test tank; 9 cotton thread; 10 sinker; 11 micro-explosive

image converter camera (John-Hadland Photonics, ImaCon 790), was employed for all the other cases.

A series of shock overpressure measurements for the explosion of a 10 mg silver azide pellet was carried out by varying the distance of a pressure transducer from the explosion center. The pressure transducer was a PVDF pressure gauge which was manufactured and calibrated in home. Equation (1) is an empirical formula for the overpressure-distance relation.

$$\frac{\Delta P_{\mathrm{r}}}{\Delta P_{\mathrm{r1}}} = \left(\frac{x}{x_1}\right)^{\alpha} \tag{1}$$

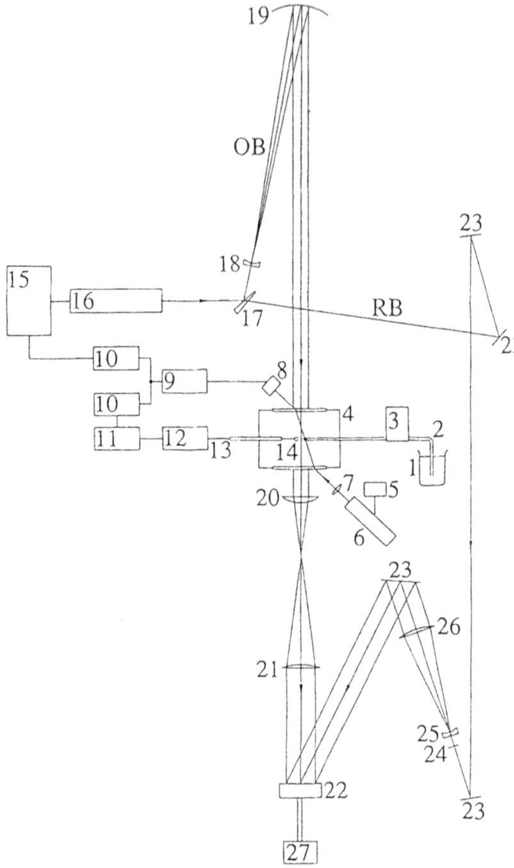

Fig. 1c: Experimental setup and optical arrangement of a double exposure holographic interferometry: OB – object beam; RB – reference beam; 1 beaker; 2 flexible tube; 3 roller pump; 4 test tank; 5, 6 He-Ne gas laser; 7, 20, 21, 26 convex lens; 8 photodiode; 9 high gain amplifier; 10 retarder; 11, 12 pulsed Nd: YAG laser; 13 optical fiber cord; 14 micro-explosive; 15, 16 holographic ruby laser; 17 half mirror; 18, 25 concave lens; 19 paraboloidal mirror; 22 film holder; 23 plane mirror; 24 filter; 27 air pump

where $\Delta P_{\rm r}$ is the peak overpressure measured by the pressure gauge, x is the distance from the explosion center, α is a self-similarity exponent, and $\Delta P_{\rm r1}$ is the peak overpressure at the distance from the explosion center x_1. For water, $\alpha = -1.05$ and for $x_1 = 10\,\text{mm}$, $\Delta P_{\rm r1} = 137.3\,\text{MPa}$. The overpressure behind the reflected shock wave was measured so that the overpressure of the incident shock wave was estimated by equations of Rankine-Hugoniot relations and equation of state of water. Therefore the overpressure of the incident shock wave was estimated to be 47.4 % of that of the reflected shock wave, i.e. 65.1 MPa where the distance from the explosion center is 10 mm.

Table 2: Typical dimensions of the gas bubble immediately before the interaction with a shock wave. a, b, f, and d are major diameter, minor diameter, flattening factor, and equivalent diameter of sphere, respectively.

Combination of liquid/gas	a (mm)	b (mm)	$f = (a-b)/a$	$d = \sqrt[3]{a^2 b}$ (mm)
Water/air	3.0	2.3	0.23	2.8
Water/helium	3.9	1.5	0.61	2.8
Silicone-oil/air	2.0	1.0	0.50	1.6
Silicone-oil/mixture of air-helium	1.2	0.82	0.29	1.0

3. Results and discussions

The shape of the gas bubble before the interaction with the shock wave was not spherical. It deformed roughly into an oblate spheroidal shape as a result of balancing of the surface tension of the liquid/gas interface and the buoyancy of the gas bubble. The axis of the minor diameter of the oblate spheroidal gas bubble was nearly parallel to the direction of the buoyancy. Typical dimensions of the gas bubbles are tabulated in Table 2.

Figures 2a and 2b are shadowgraphs of shock-bubble interaction for the water/helium combination, in which the incident shock propagated from left to right. A dark belt in frames 1 and 2 in Fig. 2a shows an incident spherical shock. The incident shock interacted with the helium bubble, and the wave reflected from the interface was an expansion wave (frame 2 in Fig. 2a), since the acoustic impedance of the liquid is in most of cases larger than that of the gas. The pressure behind the expansion wave is lower than that ahead of it. However, the pressure was found to increase along the periphery of the bubble where the shock first hit (e.g. Takayama 1989). Due to this increase of pressure the bubble contracted. A liquid jet was formed simultaneously, because liquid particles near the point where the shock first hit were accelerated and liquid was entrained into the bubble (frames 3–6 in Fig. 2a). The liquid jet penetrated the interface (frame 6 in Fig. 2a). From these sequential shadowgraphs in Fig. 2a the average speed of the liquid jet was estimated to be $275 \pm 10\,\mathrm{m/s}$ so that the stagnation pressure was approximately 400 MPa. Spherical rebound shock waves were generated, and the bubble motion changed from contraction to expansion almost immediately (frame 7 in Fig. 2a). In frames 1–3 in Fig. 2b the group of rebound waves displays a structure containing major waves A and B. In frames 1 and 2 in Fig. 2b, for first wave A, the wave was stronger at the downstream side and weaker as it approached the explosion center. The first wave could not be distinguished at the upstream side. The strength of the second wave B was more uniform compared to the first wave.

Figures 3a and 3b show shock-bubble interactions for the water/air combination. The air bubble deformed most remarkably at the part of the bubble where the shock first hit. This initiated a liquid jet. The formation of the liquid jet occurred due to the effect of the curvature of the interface. The process of the collapse of the bubble and the appearance of rebound shocks were similar to that for the water/helium combination. The average speed of the liquid jet in Fig. 3a was estimated to be $2.6 \times 10^2\,\mathrm{m/s}$. In frame 8 in Fig. 3a and frames 3–5 in Fig. 3b, two major rebound waves can be seen. In frame 8 in Fig. 3a and frame 3 in Fig. 3b, the shape

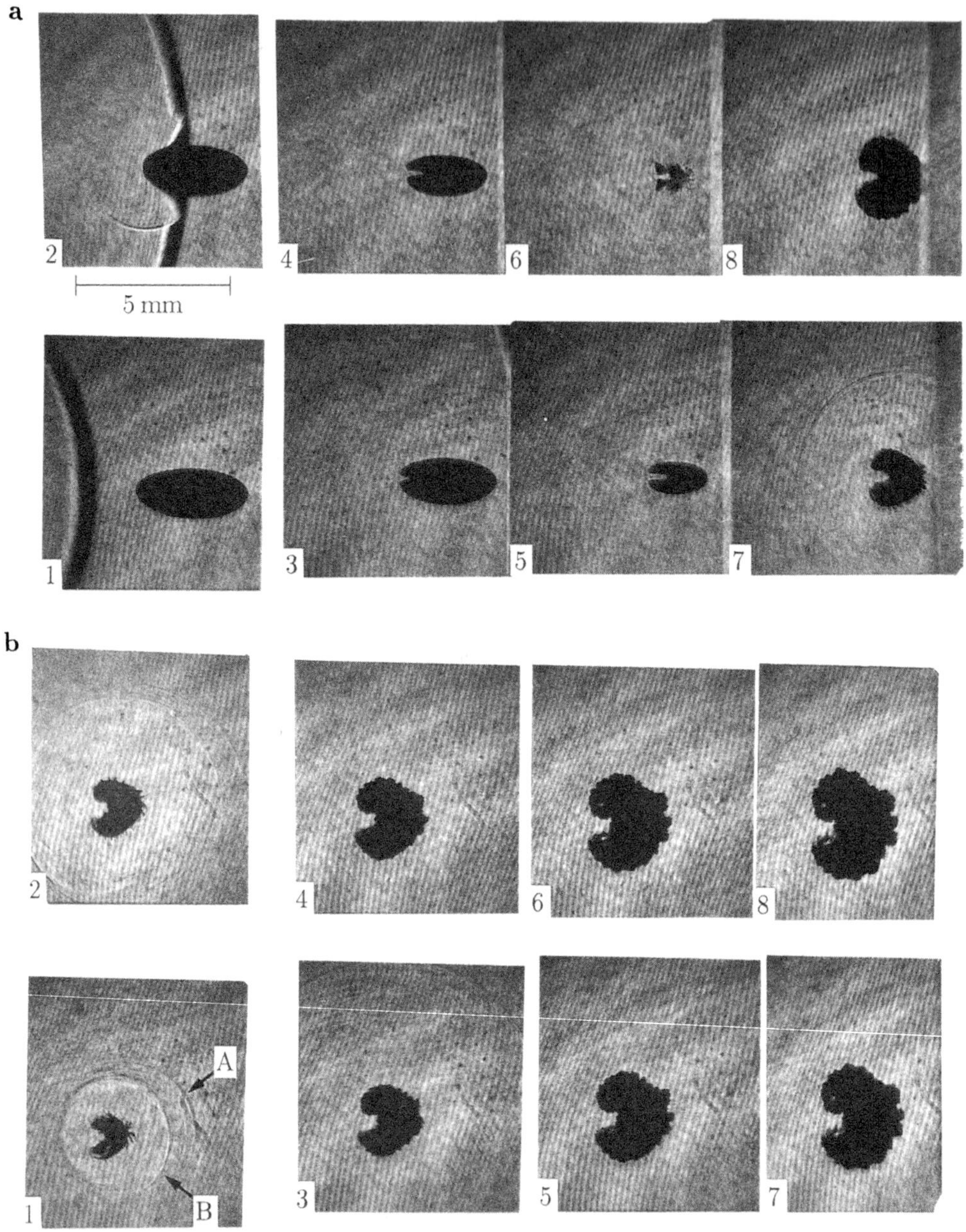

Fig. 2a, b: Shock-bubble interactions for water/helium combination. A spherical shock propagated from left to right. **a** The interframe time is 2.0 μs and the exposure time for each frame is 0.4 μs. **b** Later stage than **a**. The spherical shock already passed. The interframe time is 1.0 μs and the exposure time for each frame is 0.2 μs.

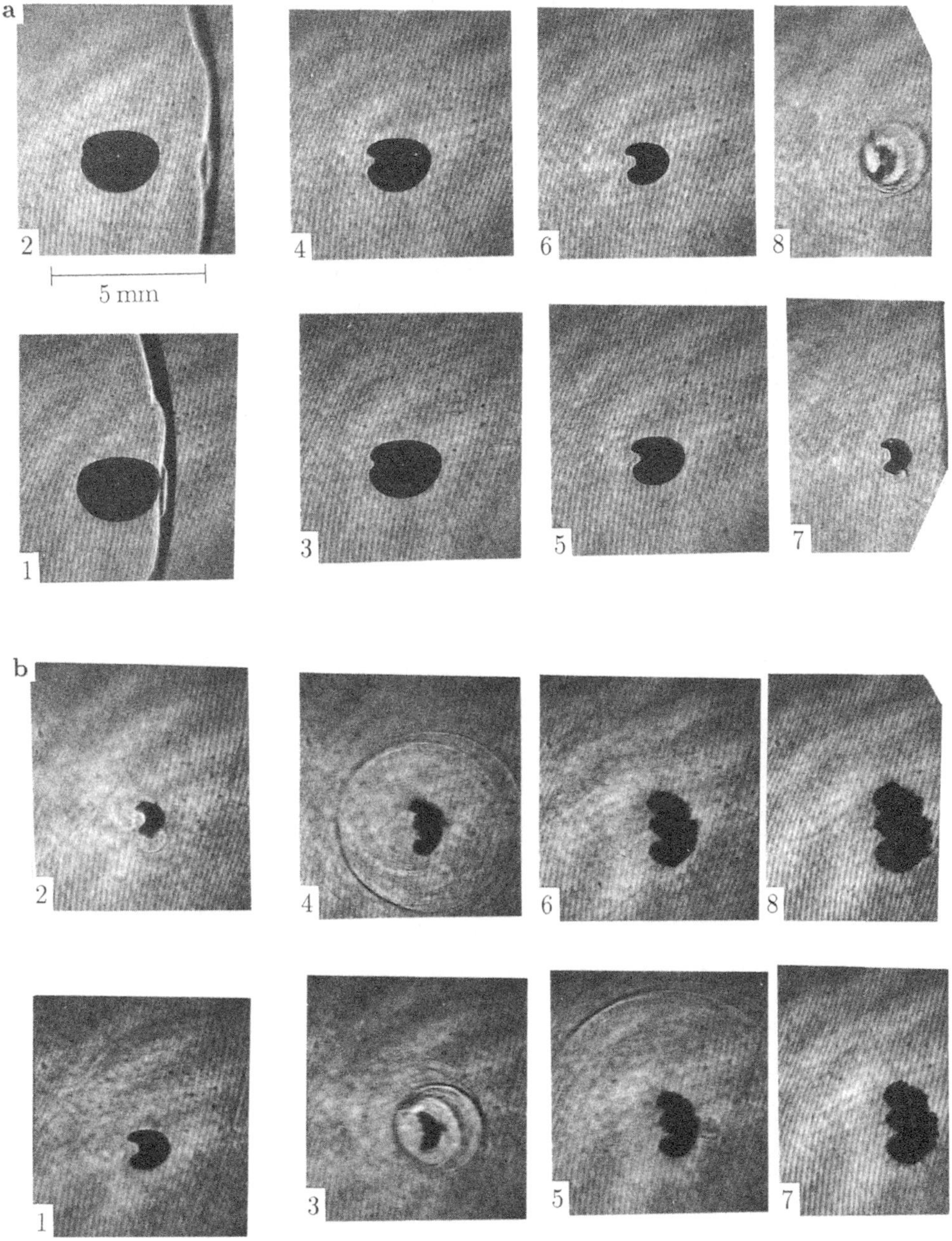

Fig. 3a, b: Shock-bubble interactions for water/air combination. A spherical shock propagated from left to right. The interframe time is $1.0\,\mu$s and the exposure time for each frame is $0.2\,\mu$s. **b** Later stage than **a**.

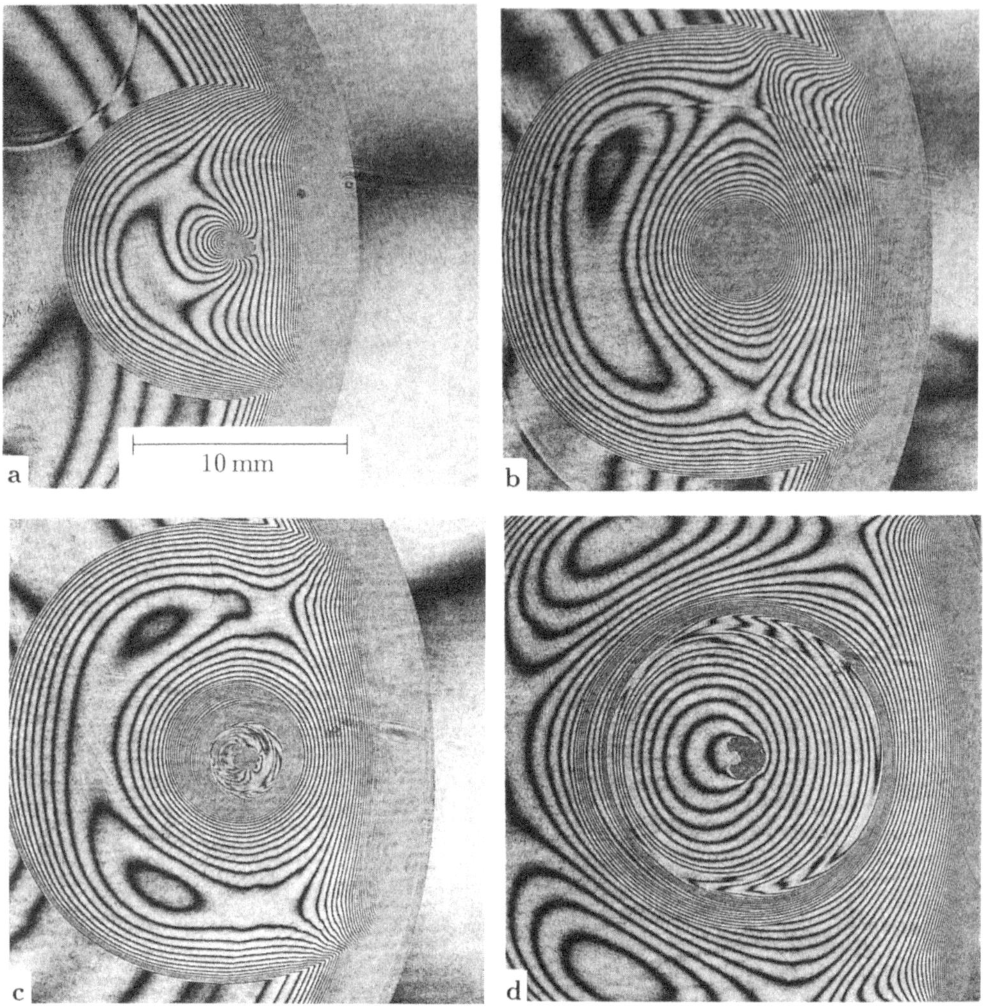

Fig. 4a–d: Shock-bubble interactions for silicone-oil/air combination. A spherical shock propagated from left to right.

of the wave front of the first wave appears to be nearly circular, while that of the second wave is elliptic with the major axis roughly perpendicular to the direction of shock propagation.

Figures 4a–4d show sequential interferograms for the silicone-oil/air combination. It is noted that fringes do not correspond to isopycnics but their distribution indicates the trend of density variation. The local area at which fringes are concentrated simply indicate that the density variation is large. In Fig. 4a concentric fringes gradually increasing to the shock front indicate a very steep pressure rise. An expansion wave reflected from the bubble is revealed by the direction of the fringe shift

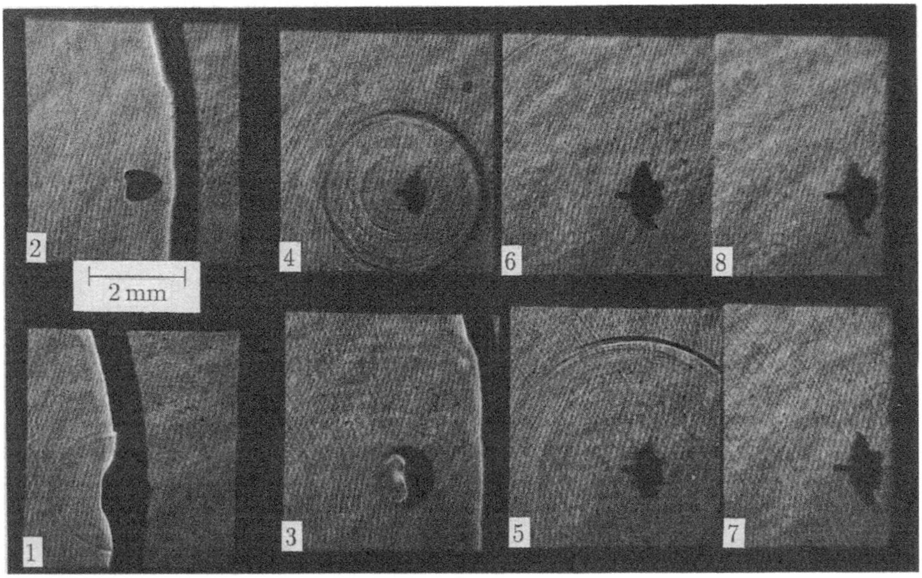

Fig. 5: A shock-bubble interaction for silicone-oil/mixture of air-helium combination. A spherical shock propagated from left to right. The interframe time is 1.0 μs and the exposure time for each frame is 0.2 μs.

opposite to that of the shock wave. The fringes were concentrated at the side of the bubble where the shock first hit. This implies that the pressure was enhanced there.

In Fig. 4b the fringe concentration at the bubble surface is a maximum, which at the final stage of the pressure increase, will be similar to the accumulation of energy at a point, that is, a point explosion. At the time instant when the pressure was a maximum, the volume of the bubble was a minimum.

In Fig. 4c spherical compression waves are seen. Later, the compression waves became rebound shocks (Fig. 4d). When the bubble expanded, its shape appeared to be jagged. At the upstream side the first rebound shock and the second rebound shock were observed, and they coalesced at the downstream side.

Figure 5 shows a shock-bubble interaction for the silicone-oil/mixture of air-helium combination. Several rebound shocks are observed as concentric circles (frame 4–6 in Fig. 5). It was similar to water/air and water/helium combinations in that for first wave of the rebound shocks, the wave was stronger at the downstream side and became weaker as it approached the explosion center. The expanding bubble moved away from the explosion center and consequently appeared to take a shape of a broken umbrella. It is now under investigation why this unusual bubble shape appears.

For the water/helium combination one of the reasons for the generation of the rebound shock is the high pressure at the neighborhood of the point where the liquid jet penetrated the interface (Fig. 2a). For the water/air and the water/helium combinations the center of the first wave was almost on the downstream interface and the center of the second wave was in the bubble (frames 1–2 in Fig. 2b). Therefore, the source pressure of the first wave was high along the downstream periphery of the bubble. Since the first wave had to go around the bubble, it would then be weak on the upstream side. The second wave was generated by the expansion of the bubble.

For the silicone-oil/water combination the center of the first wave was at the hollow of the bubble, and the center of the second wave was in the bubble (Fig. 4c and 4d). Therefore, it would be that the source pressure of the first wave was the intense pressure created at the point where shock first hit the bubble. For the second wave the mechanism is similar to that discussed earlier.

4. Conclusion

The interaction of a shock wave with a gas bubble was investigated experimentally using double exposure holographic interferometry and shadowgraphy. It was found that rebound shock waves are generated at least by two different mechanisms or by combination of these mechanisms depending upon the liquid/gas combination, overpressure of the incident shock, and gas bubble diameter. One rebound shock was generated by high pressure created at the neighborhood of the point where shock first hit the gas bubble or at the neighborhood of the point where liquid jet penetrated the downstream interface. Another rebound shock was generated by rapid expansion of the gas bubble.

References

Bourne NK, Field JE (1991) Bubble collapse and the initiation of explosion. Proc R Soc Lond A 435:423

Chaussy Ch, Schmiedt E, Jocham D, Walther V, Brendel W, Forssmann B, Hepp W (1982) Extracorporeal shock wave lithotripsy. Ernst Kieser GmbH, Augsburg

Shin-Etsu Chemical Industry Co. (1989) Explanatory note for KF 96. Japan (Japanese)

Takayama K, Esashi H, Sanada N (1983) Propagation and focusing of spherical shock waves produced by underwater microexplosions. In: Archer RD, Milton BE (eds) Shock tubes and waves, proceedings of the 14th international symposium on shock tubes and shock waves, University of Sydney, New South Wales, Australia, August 1983. Sydney Shock Tube Symposium Publishers, Australia, pp 553–562

Takayama K, Abe A, Sanada N, Sugiyama H (1988) Holographic interferometric observation of underwater shock wave with air bubbles. In: Wood PA (ed) Proceedings of the 9th international symposium on jet cutting technology, Sendai, Japan. BHRA, The Fluid Engineering Centre, Cranfield, Bedford, UK, pp 641–650

Takayama K, Abe A, Tanaka K (1989) Interaction of underwater shock wave with air bubbles. In: Meier GEA, Thompson PA (eds) Adiabatic waves in liquid-vapor systems, IUTAM Symposium Göttingen, Germany, 1989. Springer-Verlag, Berlin Heidelberg, pp 91–100

Weissler A (1949) Ultrasonic investigation of molecular properties of liquids. III. Linear polymethylsiloxanes. J Ame Chem Soc 71:93

WAVE PHENOMENA DURING DROPLET IMPACT

Martin Rein*)
Max-Planck-Institut für Strömungsforschung
Bunsenstraße 10, 37073 Göttingen, Germany

Summary. Wave phenomena that govern different processes during droplet impact are introduced and discussed. These include surface waves, in particular capillary waves, that are important for the spreading motion of droplets impinging onto solid walls. The rim of the advancing thin liquid film is subject to an instability that causes the formation of well-known lobes ("ink blob"). Surface waves on a liquid surface are observed to enhance floating and bouncing of droplets impinging on such surfaces. Kinematic effects play an important role for splashing. In the case of high-speed impacts the compressibility of the liquid is no longer negligible. On impact shock waves are formed that increase the pressure. Expansion waves later reduce the pressure. Under certain circumstances tension waves occur that cause the formation of cavitation zones. Finally, the collision of meteoroids with planetary surfaces can be considered as a droplet impact. During planetary impacts conditions may be such that evaporation waves are formed. These issues are all discussed.

1. Introduction

When a droplet impinges on a liquid or solid surface a wide variety of physical processes may occur. What actually happens depends on the detailed circumstances and conditions under which the collision takes place. A discussion of the phenomena of droplet impact was given in a previous review (Rein, 1993). Particular emphasis was laid on distinguishing between different impact scenarios and the conditions under which a certain phenomenon appears. Important parameters for impact processes are the kinetic and surface energy of the approaching drop. The ratio of these energies is proportional to the hydrodynamic Weber number that is often stated in works on droplet impact. This number alone, however, is not sufficient to classify different classes of droplet impact. In some cases the compressibility of the liquid is also of importance. Further, the unsteadiness inherent to the process of droplet impact may lead to the occurrence of kinematic effects. Hence, it is not surprising that many different phenomena may be observed during droplet impact. Some of these are directly connected with waves. In other cases this is less obvious. The aim of this presentation is to give a survey of wave phenomena that are important during different cases of droplet impact.

In the beginning let us briefly review the different cases that may be observed during droplet impact. A droplet that falls on a solid wall may bounce off the surface, spread

*) *present address: Institut für Angewandte Mechanik und Strömungsphysik, Universität Göttingen, Bunsenstraße 10, 37073 Göttingen, Germany*

S. Morioka and L. van Wijngaarden (eds.),
IUTAM Symposium on Waves in Liquid/Gas and Liquid/Vapour Two-Phase Systems, 171-190.

without disintegrating, or splash. In some cases droplets stick to the surface. This is illustrated in Fig. 1. After a first contact is formed between the liquid and the solid the droplet usually starts spreading out. A thin liquid film is formed that is called a lamella. The lamella increases until it reaches its maximum extension. It then starts to shrink and finally assumes an equilibrium shape. In a few cases the droplet recovers its spherical shape and bounces off the wall. Under favorable conditions, typically when the Weber number exceeds a critical value, the rim of the lamella detaches from the surface and splashing is initiated. The expanding lamella forms a supercritical flow with respect to surface waves. Capillary waves are of some importance for the development of the lamella. On its rim characteristic disturbances of a regular form are present. These appear to be due to an instability that is present right after spreading commencement. The lobes on the rim of the lamella may be predecessors of splash droplets in the splashing case. In the case of droplets impacting periodically on a fixed position splashing has recently been explained in terms of kinematic effects. These issues will be discussed.

When droplets fall into a liquid pool they may smoothly coalesce with the liquid or splash. Further, they may bounce off the liquid surface or float on it. These phenomena are shown in Fig. 2. Coalescing drops usually result in vortex rings that propagate into the pool liquid. Splashing drops produce large craters. At the circumference of the crater a liquid sheet is thrown out of the crater. The sheet becomes unstable and droplets are shed from its rim. It is therefore also called a crown. When the crater eventually disappears a liquid jet rises out of its center. From the tip of the jet droplets may be shed that have a similar size as the impacting droplet. Splashing occurs typically at large Weber numbers while coalescence is usually present at small Weber numbers. In a transitional region a droplet impact may cause the entrainment of a gas bubble into the pool liquid. Several wave processes that are important in this case will be discussed. Finally, floating and bouncing of droplets can be enhanced by the presence of capillary waves on the impacted liquid surface.

2. Compressible impact

2.1 Shock waves

When the impact velocity is high compressibility effects can no longer be neglected. These are reviewed in Lesser and Field (1983). The Mach number formed with the impact velocity and the sound speed in the liquid of the drop becomes important for the impact process. On impact shock waves are formed in both the droplet and the target. The pressure increase across the shock waves can be very large and a high pressure region is formed inside the drop.

The initial stage of a compressible impact is illustrated in Fig. 3 for the case of a droplet impacting on a rigid surface. Right after the drop hits the wall a contact zone is formed between the liquid and the wall which increases with time. Its circumference that is also called a contact edge, moves radially out. As long as the shock is not too strong its front can be constructed based on Huygens' principle. When the undisturbed free surface of the droplet first touches the rigid surface wavelets are emitted. The envelope of these wavelets is the shock front (Fig. 4a). Based on these ideas Lesser (1981) calculated the shape of the shock front. Using an geometrical optics approach he also determined the pressure behind it.

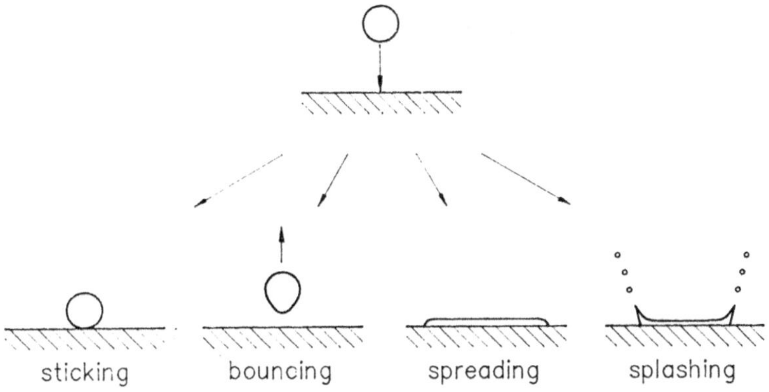

Fig. 1: Survey of phenomena that may occur on impact of a droplet onto a solid surface.

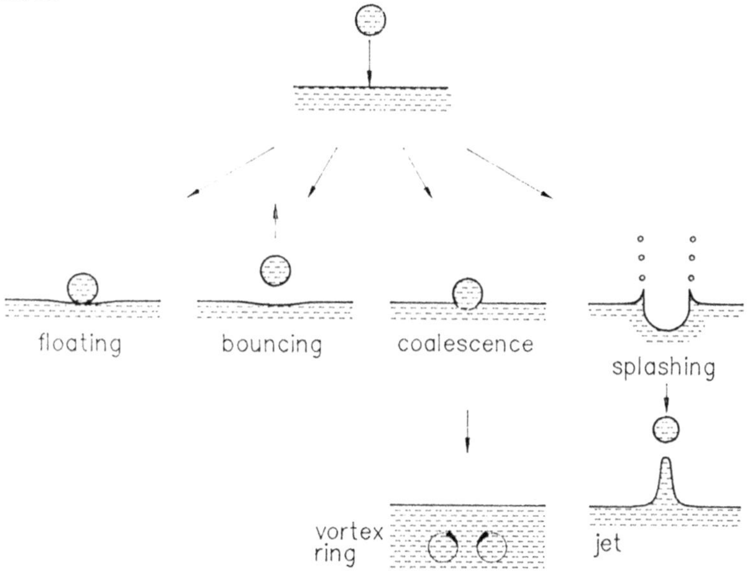

Fig. 2: Survey of phenomena that may occur on impact of a droplet into a deep liquid.

In the reference frame of the unshocked drop liquid the contact edge moves up the surface of the drop. Its velocity is then $u_e = u_i / \sin\beta$ where u_i is the impact velocity and the angle β between the free surface of the drop and the wall increases with time. This is shown in Fig. 4b. When u_e becomes smaller than the sound speed c, i.e., when $\beta \geq \arcsin(u_i/c)$, the shock moves ahead of the contact edge. Then the high pressure region is no longer enclosed by the wall and the shock. Expansion waves propagate into the droplet and fluid is accellerated radially outward. This is the beginning of a jetting motion.

The picture put forward so far is no longer adequate when the shocks are strong, i.e., when the shock speed is much larger than the sound speed of the liquid. In this case

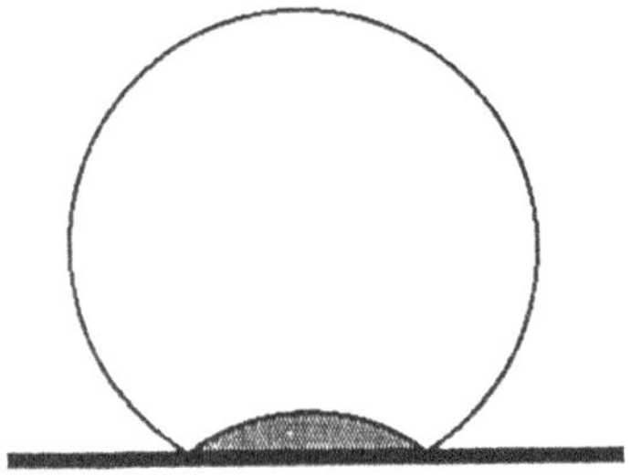

Fig. 3: High-speed droplet impact on a solid surface: sketch of shock wave propagating into the droplet.

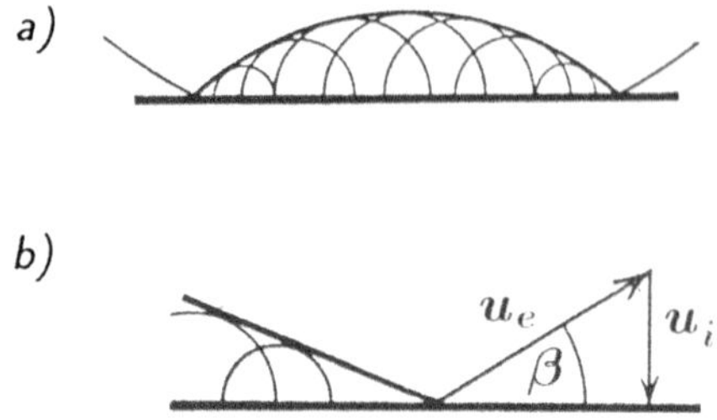

Fig. 4: High-speed droplet impact on a solid surface:
(a) Front of shock wave after Huygens' principle.
(b) Section showing shock wave at contact edge.

it is suitable to refer locally to the reference frame of the contact edge. In this frame the drop liquid approaches the contact edge tangentially to the drop's surface with a velocity u_e (see again Fig. 4b). The shock is oblique and deflects the flow so that it is parallel to the wall after the shock. This is like a wedge flow in gasdynamics and it is clear that the shock will move ahead of the contact edge when the deflection angle, i.e. the contact angle β, becomes too large for an attached shock. This nonlinear approach was used by Heymann (1969) and Lesser (1981) to calculate the edge pressure in the case of drop impact on solid walls. We have applied this approach to the case of liquid-liquid impacts. This is subject of a separate contribution to these proceedings (Rein, 1994). In all cases the peak pressure that is reached just before shock detachment is much greater than the one obtained from a one-dimensional approximation considering the impact of planar liquid surfaces.

2.2 Expansion and tension waves

After the shock has detached from the contact edge the high pressures are relaxed by expansion waves. A contact edge with a detached shock, an expansion wave and the initial jet is sketched in Fig. 5. Just before shock detachment the highest pressure within the liquid is present right at the contact edge. This maximum pressure thus determines the amplitude of the expansion wave. The expansion wave then propagates into the droplet where the pressure is lower than at the contact edge and the wave may eventually turn into a tension wave. This is enhanced by the axial symmetry of the geometry which leads to an increase in the amplitude of the expansion wave when it moves into the droplet. The appearance of tensions in the liquid can cause a rupture of the liquid. This was actually observed in the experiments by Brunton and Camus (1970). Typically, a number of cavitation bubbles is produced. A subsequent collapse of these bubbles may cause severe erosion to the impacted surface.

There is another possible mechanism leading to cavitation inside the drop that was also observed by Brunton and Camus (1970). After detachment the shock propagates through the droplet. When it is reflected at its apex, i.e., at an interface of low impedance, tensions are created. The curvature of the drop's surface further leads

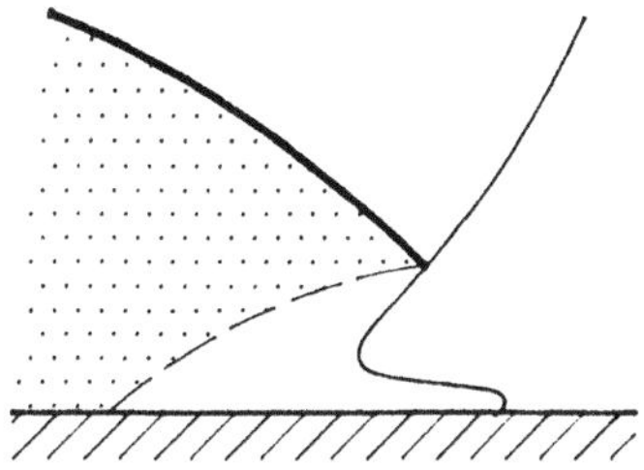

Fig. 5: Detached shock wave, expansion wave and jetting commencement: section showing contact edge.

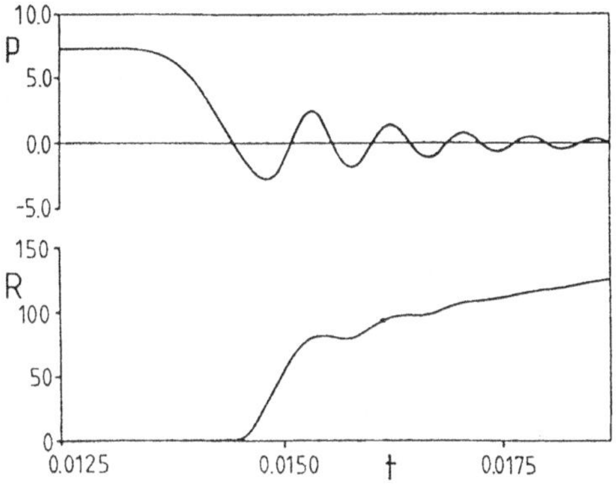

Fig. 6: Cavitation caused by the reflection of a strong expansion wave at a solid wall: pressure and radius versus time at the wall (from Rein and Meier, 1988).

to a focussing of the tension waves. The cavitation zone caused by these waves lies somewhat below the apex of the drop. Both mechanisms for the creation of tensions were already proposed by Engel (1955)

The formation of cavitation bubble fields was succesfully described using a microbubble model. Cavitation nuclei are approximated by tiny gas bubbles. The bubbly mixture is then treated as a homogeneous fluid according to the van Wijngaarden-Kogarko-Jordansky approach (cf. van Wijngaarden, 1968). The dynamics of the gas bubbles is taken into account via a relation between the radius of a single bubble and the pressure far away from the bubble as usual. A tension wave propagating through the mixture causes a violent expansion of the bubbles. This leads to a quick decrease of the tensions. Kedrinskii (1976, 1994) modelled the reflection of shock waves at the sea surface and Rein and Meier (1990a/b) considered the reflection of strong expansion waves at the closed end of a pipe. The latter authors found that tensions created at the reflection of the expansion wave are quickly annihilated and the pressure begins to oscillate about $p \approx 0$. The results of a sample calculation are shown in Fig. 6. The pressure and radius scaled by the ambient pressure $p_0 = 1.01 \cdot 10^5 Pa$ and the corresponding equilibrium radius $R_0 = 10^{-6} m$, are plotted versus a nondimensional time. The initial void fraction is $\alpha = 10^{-9}$. The oscillations are reminiscent of the oscillatory pressure increase behind shock waves in bubbly liquids (van Wijngaarden, 1972). However, there is an important difference since in the case of cavitation the

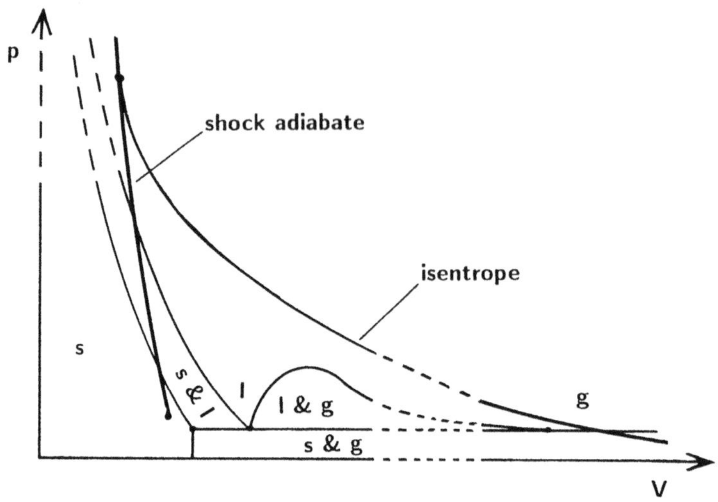

Fig. 7: Stages of meteoroid impact in a pressure versus volume diagram: shock compression, liquefaction, evaporation and desublimation.

bubbles are expanded to an unstable equilibrium size while they oscillate about their stable equilibrium size in the case of shock propagation in bubbly liquids. Oscillations in the pressure behind tension waves were also observed in experiments by Bode *et al.* (1990). The frequency and amplitude of the oscillations depend on the nuclei's number density. Qualitatively the experimental results agree well with the results of the theoretical simulation.

2.3 Evaporation waves

In the case of extremely strong impacts the entropy increase across the shock wave which propagates into the droplet becomes appreciable. Events where this plays a role are planetary impacts. When a meteoroid collides with a planet or moon pressures can be so high that the materials behave hydrodynamically and the process can be considered as a liquid-liquid impact. The initial stage of hypervelocity liquid-liquid impacts is considered at a different place in these proceedings (Rein, 1994). It is similar to the case of compressible drop impact on solid surfaces in that a shock is formed in the drop that first remains attached to the contact edge and later becomes detached. Thus, first a high pressure zone is build up by shock compression. Then, the high pressures are adiabatically relaxed. This may lead to the formation of evaporation waves. This is most easily seen by considering the process in a pressure versus volume diagram (Fig. 7). Initially, the material of the projectile (meteoroid) is in the solid state. It is then compressed to very high pressures up to its liquid state via a shock adiabate. When the shock detaches from the contact edge the pressure is reduced by expansion waves which is a process that follows an isentrope in the pressure versus volume diagram. If the conditions are such that this isentrope enters the gas-liquid two-phase region the projectile will partially evaporate. It may even happen that the isentrope passes above the critical point into the gaseous region. At very large expansions a partial condensation of the vapor into the solid state may occur.

In order that the collision of a meteoroid with the earth results in the formation of vapor the collison speed should exceed a value of about $15 km/s$ (Melosh, 1989). When the size of the meteoroid is large enough a huge vapor plume is formed that expands into the atmosphere where it will desublimate. This may cause disturbances of the climate. It has been suggested that the extinction of dinosaurs and other animals some 65 million years ago was caused by such an event (Alvarez *et al.*, 1980). This hypothesis is based on a characteristic anomaly in the concentration of iridium at a geological layer that corresponds with the Cretaceous-Tertiary boundary. Iridium is found in meteoroids but is not very abundant on the surface of the earth. A vapor plume that is formed after a meteoroid impact and shoots up into the stratosphere, is thought to be the reason for the global distribution of the iridium. (In the original paper by Alvarez *et al.* (1980) evaporation and desublimation are not considered but it is assumed that dust from pulverized rock is distributed in the atmosphere after a meteoroid impact.) Recently, a crater that was caused by a giant impact at the appropiate time has been detected in the Gulf of Mexico region (Levi, 1992). Finally, however, it should be mentioned that there are also other hypotheses for the distinction of dinosaurs still under discussion. For example, in one it is assumed that disturbances in climate were caused by a period of increased vulcanological activity.

3. Droplets and liquid films

3.1 Spreading of liquid lamella

The collision of a droplet with a solid wall resembles a stagnation point flow of a liquid jet with a time dependent radius and, hence, flux. Right after impact the flow direction is changed to be parallel to the wall and a thin liquid film is formed. In the case of normal impacts the film is circular. This film is also called a lamella. The radially expanding flow of the liquid film is called spreading of the droplet. This type of spreading needs to be distinguished from the spreading of droplets that are carefully placed onto a solid surface so that the kinetic energy is negligible. In the latter case molecular forces determine the process of spreading.

The motion of the liquid lamella is time dependent. First it starts spreading out. It then reaches a maximum size, begins to contract and eventually reaches an equilibrium size. In some cases further expansion and contraction phases are observed. Loehr (1990) took photographs of different stages of the motion of the lamella. Three examples are shown in Fig. 8.

Initially, the flow in the lamella is supercritical with respect to surface waves (Fig. 8a). This is well-known from daily experience when one considers the impinging flow of a water jet onto a kitchen sink. There, behind small disturbances, structures can be seen that resemble Mach cones. It appears not to be clear, however, what kind of waves are responsible for the formation of these "Mach cones." A first guess might be capillary waves but in the case of radially expanding liquid sheets with free surfaces on both sides, wave patterns due to symmetric capillary waves that are caused by small disturbances look different (Taylor, 1959). Rather, wave patterns due to antisymmetric capillary waves are similar to what can be seen in lamella flows or in a kitchen sink. The two cases may not be comparable though, since in the present case viscous forces will no longer be negligible.

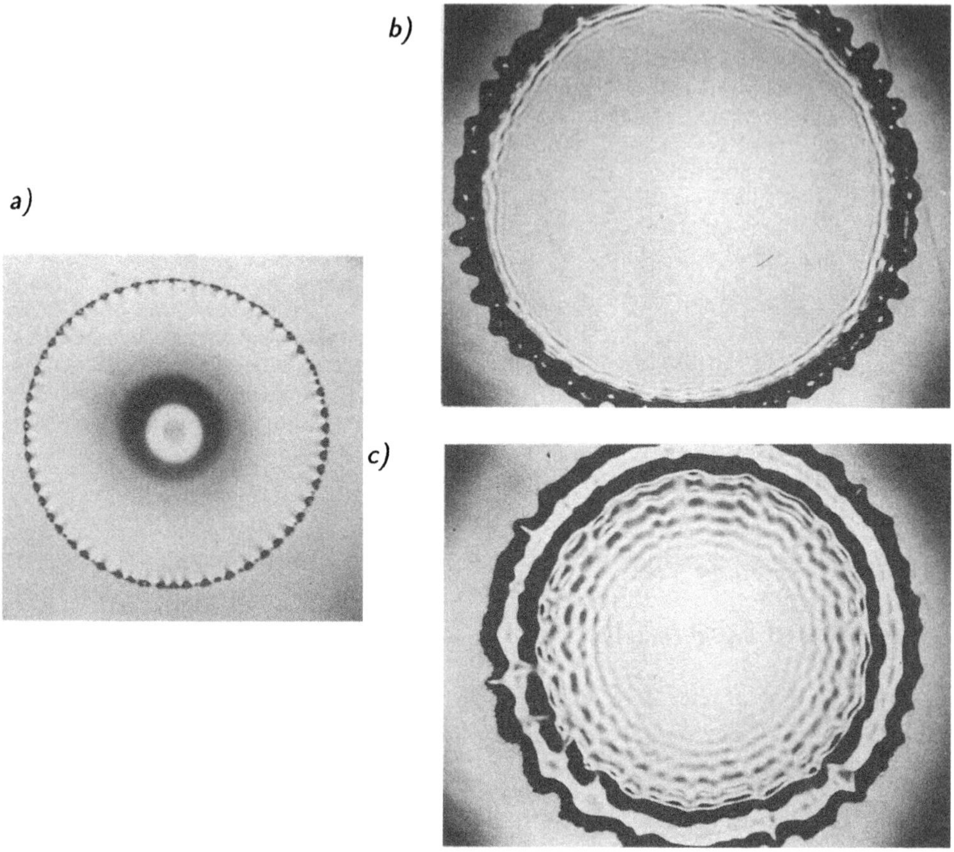

Fig. 8: Droplet impact onto a solid surface: Spreading and contraction of lamella. (a) supercritical flow in radial direction, (b) maximum size of lamella, (c) capillary waves initiating the contraction phase (from Loehr, 1990).

When the liquid of the droplet is used up the supply of liquid becomes exhausted and the expansion stops. In the supercritical flow this information can be thought of being conveyed along characteristic lines from the center to the rim of the lamella. The lamella at its maximum extension is shown in Fig. 8b. The lamella stays at its maximum size for a while and its rim thickens. Then, capillary waves begin to move radially towards the center of the lamella (Fig. 8c). These waves initiate the contraction phase. This process can be compared to a choked Laval nozzle flow when the plenum pressure is decreased. The supersonic region behind the throat will disappear without shocks propagating upstream and annihilating the supersonic region. Eventually, the whole flow is subsonic and pressure waves can propagate upstream.

3.2 Perturbations on the rim of the lamella

Characteristic perturbations are present on the rim of the expanding lamella. These can well be seen in Fig. 8. The first investigation into these patterns was performed

Fig. 9: Perturbations on the rim of the spreading lamella of a mercury droplet as observed by Worthington (1877).

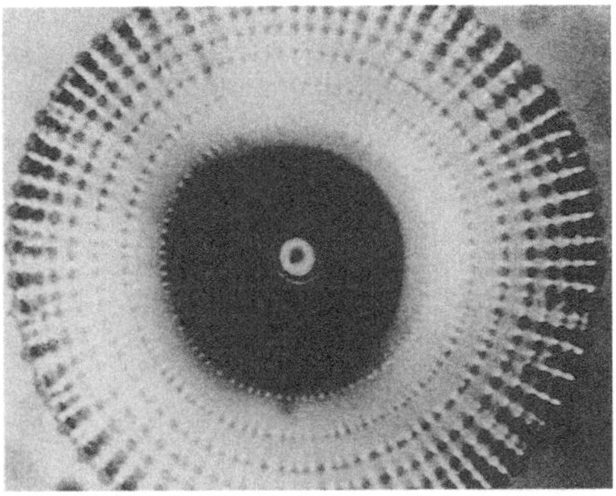

Fig. 10: Perturbations on rim of spreading lamella of a water droplet. Multiexposure photography showing constancy in the number of disturbances (from Loehr, 1990).

by Worthington (1876/77). He observed the spreading lamella of droplets of different liquids by illuminating the process with electric sparks. The results were represented in the form of drawings, one of them is reproduced in Fig. 9. Loehr (1990) who was interested in the spreading behaviour of the lamella also looked at the perturbations on its rim. He reports the following findings. First, perturbations are present on the rim as soon as the lamella can be seen to come out from beneath the droplet. Second, the number of lobes is constant with time. This is apparent in Fig. 10 where the expanding lamella was photographed using multiexposure. The technique of multiexposure photography yields pictures similar to Worthington's figures (cf. Fig. 9). Therefore the ray-like structures on the lamella reported by Worthington are probably due to his observation technique (Loehr and Lasek, 1990). Finally, Loehr observed a linear dependence on the Reynolds number of the number of perturbations. He formed the Reynolds number with the impact velocity, the radius of the drop and the viscosity of the liquid. However, in his experiments the viscosity was changed by a factor of two only. Thus, the correlation may not hold on a larger scale and the viscosity may be of minor importance.

Although the phenomenon of the disturbed lamella rim is well-known (just think of an ink droplet falling onto a sheet of paper) there is little work on it. In particular, a satisfactory explanation of the instability of the rim is missing. A mechanism proposed by Allen (1975) appears to point in the right direction. Allen considers the rim to be an accelerated interface between two fluids of different densities that is subject to the Rayleigh-Taylor instability. In the present case surface tension acts on the interface. The surface tension selects a mode of maximum instability. This mode is said to be amplified and thus determines the number of lobes on the rim. Using several approximations and a rough estimate for the acceleration caused by the retardation of the spreading motion of the lamella Allen (1975) arrives at a wave number that he says is in the right order of magnitude. However, the acceleration will change during the process of spreading

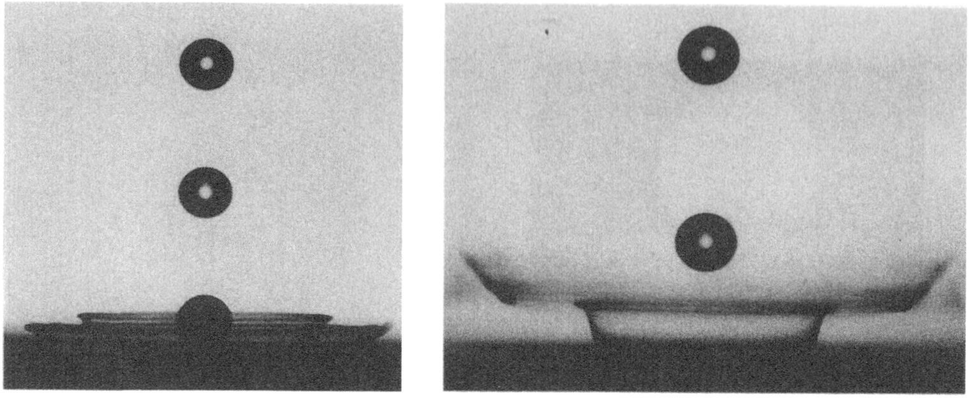

Fig. 11: Spreading and splashing of a droplet stream on a thin liquid film: stroboscopically illuminated photographs. (a) spreading, (b) splashing (from Weiss, 1993).

and it is not clear why this approach should yield a wave number that is constant in time. We will therefore discuss a slightly different mechanism. Kivity and Hanin (1981) investigated the stability of the two-dimensional contact surface in a Riemann problem. They found that the interface is unstable and that the growth rates of perturbations are constant at large times. As in the case of the Rayleigh-Taylor instability surface tension can be expected to stabilize perturbations of large wave numbers on the contact surface. Further, again a most unstable wave number might be found. In a detailed calculation of this stability problem the whole flow field - from the leading edge of the perturbed expansion wave up to the perturbed shock wave - needs to be solved for. This has not yet been done. In the case of spreading droplets the sudden start of the radially expanding lamella right after impact can be though of being analogous to the start of the flow in a Riemann problem. The rim of the spreading lamella can then be compared to a contact surface with forces due to surface tension acting on it. Our picture yields both, the quick selection of perturbation of a certain wave number on the rim of the lamella and the constancy in time of the dominant wave number.

3.3 Splashing of droplets on thin liquid films

The conditions under which droplets spread or splash were investigated by Weiss (1993) for the case of droplet streams impinging normally onto solid surfaces. The continuous supply of liquid causes the formation of a thin liquid film on the surface. Therefore the interaction of droplets with thin liquid films is actually considered. During the spreading motion of the droplets the rim of the advancing lamella may separate from the film. Two different cases can be distinguished. First, the rim becomes only slightly elevated and the lamella does not disintegrate. Eventually the elevated rim coalesces with the liquid film. This is called spreading. Second, the separated part of the lamella disintegrates into many splash droplets that are ejected away from the wall. This is called splashing. Examples of the two cases are shown in Fig. 11. The photographs were illuminated stroboscopically, individual splash droplets can therefore not be seen. Weiss (1993) who used different liquids and varied both the droplet diameter and impact velocity over a wide range, found that the transition from spreading to splashing depends on a single dimensionless number only. The frequency and velocity of the impacting droplets, and

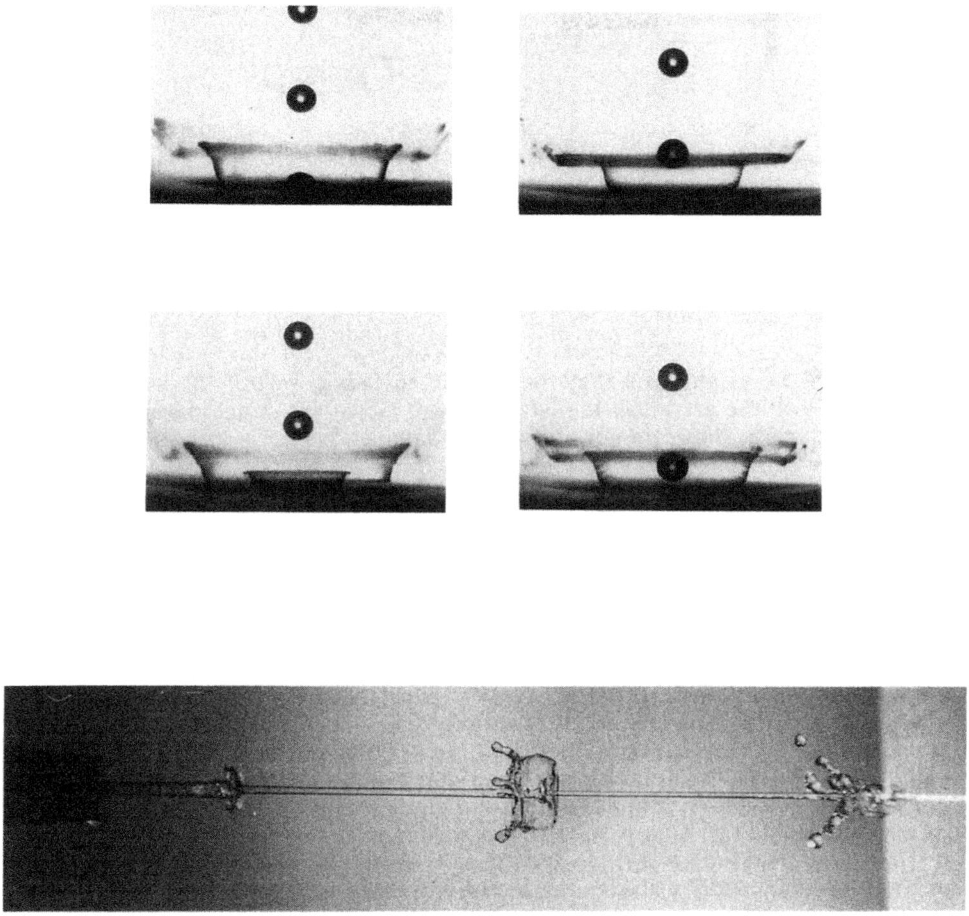

Fig. 12: Comparison of splashing droplet stream (top, from Weiss, 1993) and forced disintegration of a liquid jet (bottom, from Grabitz et al., 1993).

the viscosity and surface tension of the liquid enter this number but the radius of the droplets does not (see also Yarin and Weiss, 1994).

The mechanism leading to splashing is based on the unsteadiness of the process. Every time when a droplet impinges onto the liquid film a new supply of liquid is provided. Initially, the spreading velocity is high and the expanding liquid catches up with the one of preceding droplet impacts. Under suitable conditions this results in splashing. A related phenomenon is present in the case of a forced breakup of liquid jets. When the initial velocity of a liquid jet is periodically modulated the faster liquid can be thought of passing the slower one. In reality this causes the formation of droplets and a disintegration of the jet (Grabitz and Meier, 1983). In Fig. 12 the two phenomena are compared. In its upper part different phases of a splashing droplet stream are shown. The disintegration of a liquid jet that is caused by a strong velocity modulation is presented in the lower part of Fig. 12. The jet emanates from a nozzle on the left

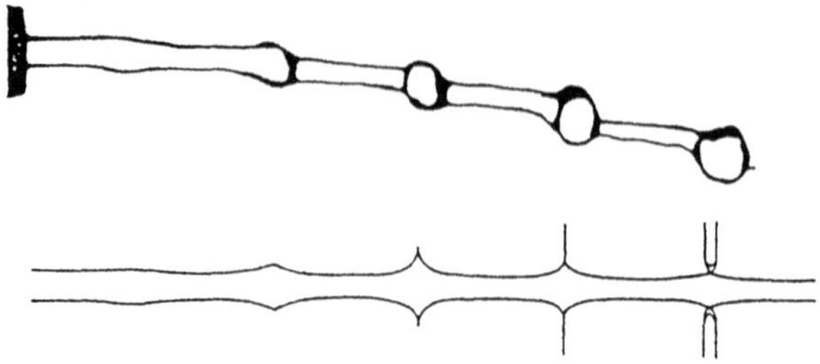

Fig. 13: Forced disintegration of liquid jet: comparison of experiment and theory (from Grabitz et al., 1993).

side. It should be emphasized that the radially spreading flow of the lamella should be compared with the axial flow the jet. Although crown like structures are eventually formed in both cases these are not directly comparable. Similar processes that are also caused by kinematic effects can be found in separated Laval nozzle flows (Meier *et al.*, 1991) and in sprays (Meier, 1994).

The processes of both splashing droplet streams and disintegrating jets can be described by quasi one-dimensional models. The disintegrating of a liquid jet caused by velocity modulations was modelled by Grabitz and Meier (1983) (see also Meier *et al.*, 1992) using a Lagrangian approach. In the momentum equation the pressure term can be neglected. In this manner a continuity equation for the velocity is obtained. Liquid 'particles' that emerge from the nozzle with a certain velocity move downstream with this velocity. It is therefore appropiate to introduce Lagrange-coordinates. A solution of the velocity and the cross-sectional area of the jet as a function of time and space is then readily obtained. The solution of the cross-sectional area becomes triply valued and contains poles. At these positions droplets are formed. A comparison of the theoretical solution with a liquid jet that is subject to initial velocity modulations is shown in Fig. 13. Here, the amplitude of the perturbations is small (its exact value is not known) and the frequency of the disturbances scaled with the nozzle radius R_a and the flow velocity U is given by $\omega = 0.6\, U/R_a$.

Yarin and Weiss (1994) considered the flow in the expanding liquid lamella produced after droplet impacts using a different approach. They found that under certain conditions a kinematic discontinuity is formed. This discontinuity has some interesting properties that are discussed in the original paper. It corresponds with the separation and disintegration of the rim of the spreading lamella, i.e., with splashing. The discontinuity is directly related to the triply valued solution of the cross-sectional area in the case of the perturbed jet. The theory reproduces the experimental results very well. Whether a kinematic discontinuity is formed or not depends on one dimensionless number only. This number is the same that was found to describe the transition from spreading to splashing in the experiments.

The related problem of splashes from liquid jets falling onto a shallow layer of water was considered by Peregrine (1981). He presented a straightforward approach towards its solution by considering both the impacting jet and the separating splashes as symmetrical impacts of jets. Further, he addressed questions concerning the influence of viscosity and the formation of vorticity at the separation point.

4. Droplets impinging onto wavy liquid surfaces

4.1 Floating droplets and anti-bubbles

Under certain circumstances droplets may float on liquid surfaces for a few seconds before they coalesce with the liquid. This phenomenon that is known since long was probably first mentioned in the scientific literature by Reynolds (1881). He stressed that the cleanliness of the surface is of great importance for obtaining floating droplets. This is to be expected since any small dust particle will puncture the air layer between the droplet and the supporting surface. An attempt at a physical explanation for the suspension of the droplets on liquid surfaces was not given and it appears that the support mechanism enabling drops to float is still not exactly known. A major point, however, will be the drainage of entrapped air out of the thin gap between the two liquids.

More recently, interesting experiments on floating droplets were reported in the *Amateur Scientist's* corner of *Scientific American* (Walker, 1978). A method is described by which droplets can be made to last for a much longer time than usually. First, the suspension time of water droplets on a water surface will be increased when a detergent, i.e., a surfactant, is added to the liquids. A detergent decreases the surface tension but the addition of the surfactant increases Boussinesq's coefficient of surface viscosity (Levich, 1962). Because the gas adheres to the liquid surfaces the drainage of the air out of the gap between the liquids will be retarded. Second, the suspension time can be increased to several minutes when, in addition to adding a detergent, the surface of the supporting liquid is vibrating. Best results are obtained for standing surface waves, i.e., for certain frequencies, and for certain amplitudes. However, when we checked on this experiment we successfully increased the lifetime of floating droplets even with very unregular surface waves that were produced by holding some vibrating apparatus against a beaker filled with a mixture of water and detergent. Walker (1978) suggests two possible mechanisms for increasing the suspension time by surface waves. First, he observed that small drops slightly rebound during every oscillation cycle. On the succeeding impact an air layer is trapped anew. Larger drops sitting on a wavy surface oscillate themselves. Following Walker on average the resulting thickness of the narrowest gap through which the air must escape from beneath the drop is smaller than in the steady case. Both mechanisms cause the time for draining the air out of the gap to increase. It is not clear, however, whether the second mechanism may also be applicable in the case of unregular surface waves.

Closely related with floating droplets are anti-bubbles. Anti-bubbles are submerged droplets that are surrounded by a thin film of gas. This is shown in Fig. 14. They are also called inverse bubbles or inverted soap bubbles since their appearance is like the counterpart to soap bubbles. Anti-bubbles are most easily obtained by slowly pouring liquid from one beaker into another one that is filled with the same liquid. In the case of water it is important, however, that a surfactant is added to the liquids. Like in the case of floating droplets detergent and soap are recommended in the literature. When droplets fall onto a liquid surface from a small height they might float. But it may also happen that on impact they penetrate the surface deeply enough so that the receiving liquid closes above them. When the air film does not break an anti-bubble is formed. Otherwise the droplet coalesces with the receiving liquid. Spectacular photographs of the air film between the submerging droplet and the supporting liquid can be found in Sigler and Mesler (1990). Since these authors worked with pure water the air film eventually broke and no anti-bubble was formed. The air film surrounding anti-bubbles

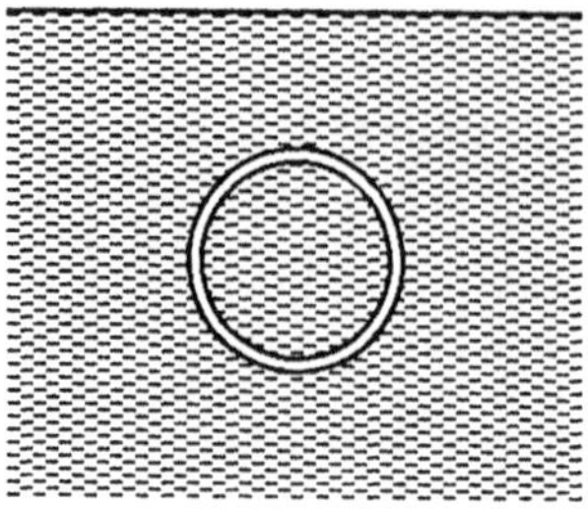

Fig. 14: Anti-bubble.

is very thin. Therefore their buoyancy is small and anti-bubbles move slowly through the bulk liquid. Further, the thinness of the air film may give rise to colorfull interferences as with soap bubbles. Anti-bubbles are not well-known. The first comment on anti-bubbles that came to our attention is by Hughes and Hughes (1932). A more detailed describtion was then given by Skogan (1956). Baird (1960) who looked at the drainage of the gas film separating the liquids, provides a brief overview of earlier work. Those who like to play around with anti-bubbles themselves should have a look at Stong (1974).

4.2 Bouncing droplets

Droplets impinging onto liquid surfaces may be reflected. Often the reflected droplets are smaller than the incident ones. Then a contact was formed between the liquids and partial coalescence took place. When the reflected droplets have the same size as the incident ones it is likely that no contact is made between the liquids. As in the case of floating an air film cushions the incident droplet. Again, surfactants may be of great importance. Rayleigh (1882) who was interested in the scattering behaviour of liquid jets due to rebounding droplets, observed that the addition of a small amount of soap or milk to the liquid causes colliding droplets to coalesce and thus reduces scattering. This appears to be contradictory to what was said about floating droplets and anti-bubbles but the exact conditions under which Rayleigh did the experiments are not known.

In experiments bouncing droplets were usually obtained with droplet streams. Single droplets were observed to bounce off a liquid surface in a few cases only (Rodriguez and Mesler, 1985). In the case of droplet streams it is obvious that oblique impacts are considered. Further, the impacted surface is disturbed by preceding droplets so that the interaction of droplets with wavy surfaces is actually considered. This was pointed out by Ching *et al.* (1984) who never noticed single droplets to be reflected under the conditions of their experiments. A wavy surface was also present in the reflection experiments of Jayaratne and Mason (1964). This is evident on their photographs although they state that the surface regained its equilibrium position at the point of impact before the arrival of the next droplet. Ching *et al.* (1984) observed that reflected droplets normally leave the liquid surface with different velocities. However, under certain conditions all droplets are reflected in the same manner. This is the case when the impact frequency coincides with the one of the waves. They further considered an energy balance and suggested that single droplets loose too much energy in forming a crater while in the case of droplet streams conditions close to resonance may be established and the droplets retain enough kinetic energy to be reflected. However, the detailed mechanism of bouncing remains unclear and the importance of the air film between the liquids is not considered at all.

Jayaratne and Mason (1964) and, later, Zhbankova and Kolpakov (1990) found that in the case of low impact velocities droplet streams rebound at small and large impact angles. In an intermediate range of impact angles the droplets coalesce with the receiving liquid. An explanation was not given but there may be a connection between this phenomenon and the observation of Ching *et al.* (1984) that resonance and, thus, reflection is obtained for certain normal impact velocities. Finally, it should be mentioned that bouncing is also observed with droplet streams impinging obliquely on solid walls when the temperature of the wall is above the Leidenfrost temperature (Anders *et al.*, 1993). In this case the mechanism preventing wetting is obvious.

5. Bubble entrainment and formation of thin jet

Besides floating and bouncing two other phenomena that may be observed when droplets fall into a deep liquid pool are coalescence and splashing. In the former case the free surface of the pool liquid is only little disturbed and a vortex ring that propagates downward is formed. In the latter case a crater is produced. At its rim a crown like liquid sheet rises above the original level of the pool liquid and, later, when the crater recedes, a liquid jet emerges out of its center. Which one of the two cases will actually appear depends on the particular conditions under which a collision process takes place. An important parameter will be the viscosity of the liquid even though the source of vorticity has not yet been explained satisfactorily (Rein, 1993). Only few quantitative data are available about the influence of viscosity. Other important parameters are taken into account in the Weber and Froude number, $We = \rho u_i^2 D/\sigma$ and $Fr = u_i^2/gD$, respectively. Here, ρ and σ are the density and surface tension of the liquid, D is the diameter of the droplet, u_i its impact velocity and g the acceleration of gravity. In the case of water there exist regions in a Weber versus Froude number diagram where the impact of a droplet results in a splash and a vortex ring, respectively. Pumphrey and Elmore (1990) found that these regions are separated by another region where a bubble is entrained into the pool liquid. These regions are shown in Fig. 15. The experimental data for the lower and upper boundary of the region with bubble entrainment are well approximated by $We_l = 41.2\ Fr^{0.179}$ and $We_u = 48.2\ Fr^{0.247}$ (Prosperetti and Oğuz, 1993). These relations were used in plotting Fig. 15. The entrainment of a bubble is accompanied by the emission of sound that is caused by oscillations of the bubble (Pumphrey and Walton, 1988). The underwater noise of rain could be explained in terms of this source of sound. This is discussed in detail in the review by Prosperetti and Oğuz (1993) and will not be repeated here.

Oğuz and Prosperetti (1990) suggested possible mechanisms for the lower and upper boundaries of the region of bubble entrainment. In the context of wave phenomena the argument put forward for the lower boundary is of interest. The main idea is that a capillary wave is formed on impact at the bottom of the expanding crater. Initially it produces a downward motion of the liquid. Depending on whether the crater begins to shrink before or after the phase of the capillary wave is reversed, a bubble will be entrained or not. The characteristic time scale for the crater to reach its maximum size is approximated by an empirical relation and the time scale of the capillary wave is its period. In the dispersion relation of capillary waves a characteristic length scale for spreading of the drop liquid over the crater surface is used as the wave length. A

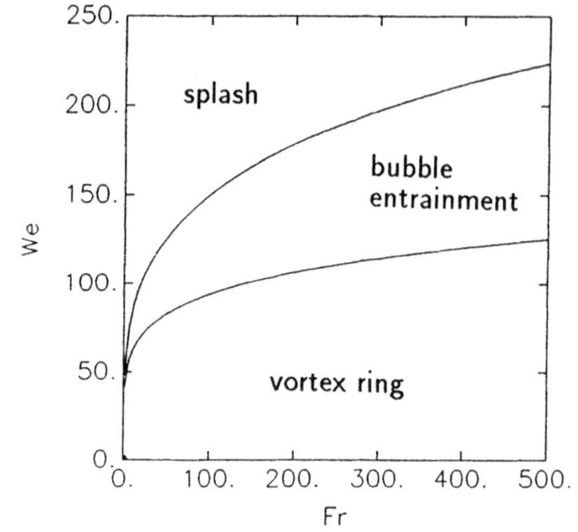

Fig. 15: Droplet impact into a deep liquid: regions of vortex ring formation, bubble entrainment and splashing in a Weber versus Froude number diagram. The asterisk shows the conditions under which the photographs of Fig. 16 were taken.

comparison of these time scales leads to $We_l \sim Fr^{1/5}$ which is close to the scaling found in the experiments.

Hallett and Christensen (1984) studied the maximum height to which the liquid jet rises when water droplets splash on a deep layer of water. They observed that a high narrow jet is formed when the impact velocity lies within a certain range. At the same time they noticed a characteristic sound the source of which they did not explain. At impact velocities just above the range where this happens the jet is wide and rises to a small height only. This height then increases only gradually with the impact velocity. At impact velocities below the range producing narrow jets a jet was not observed. A comparison of the data of Hallett and Christensen (1984) with Fig. 15 shows that the range of impact velocities that yield narrow high jets, coincides with the regime of bubble entrainment. Hence, simultanously with the entrainment of a bubble, a thin liquid jet may be ejected out of the crater formed on droplet impact. This connection between the two phenomena was first pointed out by Rein (1993).

In order to clearly show the connection between the formation of a bubble and a thin high rising liquid jet we have taken high-speed photographs of this event. A sequence of photographs is shown in Fig. 16. In this sequence a water droplet of radius $R = 1.15\ mm$ impinges upon a water surface with a velocity $u = 2.19\ m/s$. The Weber and Froude number are $We = 155$ and $Fr = 213$. In the Weber versus Froude number diagramm of Fig. 15 these conditions are marked by an asterisk. The sequence shown is well within the regime of bubble entrainment. The droplet is colored with $KMnO_4$ that leaves the surface tension virtually unchanged. The time increment between successive photographs is $\Delta t \approx 2\ ms$ (the white patches at the upper boundary of some pictures are time marks $10\ ms$ apart). The horizontal black line in the middle of the photographs is due to the meniscus.

Directly after impact a hemispherical crater is formed in the target liquid. At its floor liquid of the droplet can be seen. The size of the crater increases and its shape changes. Eventually it assumes a conical shape with a small cylindrical stem at its bottom. Note

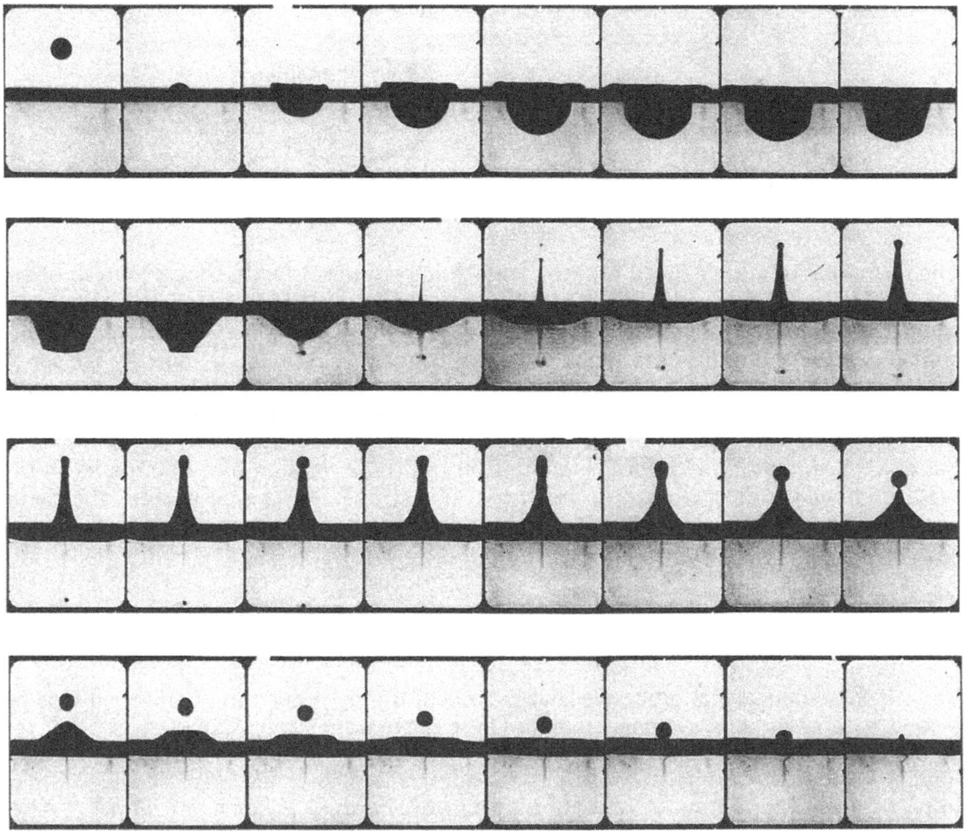

Fig. 16: High-speed photographs of droplet impact into a deep liquid: entrainment of a bubble and formation of a thin liquid jet ($We = 155$, $Fr = 213$, $\Delta t \approx 2ms$).

that almost no wave swell is formed around the periphery of the crater. When the crater relaxes a part of the cylindrical stem is sealed off and a bubble is entrained. Only a very small time later a thin liquid jet is ejected out of the crater. Tiny droplets are shed from the tip of this jet. The velocity of these droplets is approximately twice the impact velocity. They thus rise to a great height. It is this height that was measured by Hallett and Christensen (1984) as the maximum jet height. The main body of the jet quickly reaches a maximum height and starts to thicken. At the same time the bubble moves downward and eventually leaves the section shown on the photographs. On the axis of symmetry there is thus a flow of liquid up into the jet and also a small flow velocity downwards. This may be due to the unsteadiness of the event, but it also reminds one of a Munroe jet, as in the case of shaped explosive charges. – Some colored liquid of the droplet can be seen to extend down to the bubble while other droplet liquid is distributed around the receding crater. The liquid jet becomes unstable and when its height decreases a big droplet separates from its tip. This drop oscillates strongly in its higher order modes. When it touches the liquid surface it quietly disappears. On the photographs following the last of this sequence the liquid surface is flat.

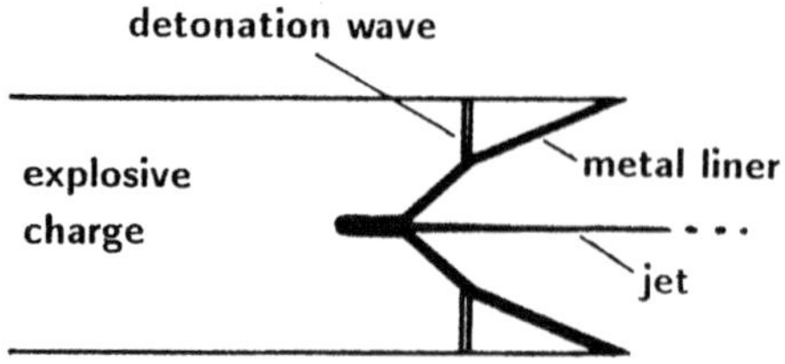

Fig. 17: Chaped explosive charge.

The formation of a high-speed Munroe jet is due to compressibility effects (Walsh *et al.*, 1953). These may also play a role during the process of bubble formation. When the liquid closes above the bubble its velocity is quite large and a high-speed liquid-liquid impact is present. In this case shock waves propagate into the liquid. Initially the shock is attached to the contact line between the impacting liquids. Later, when the angle between the colliding liquid surfaces exceeds a certain threshold a liquid jet is ejected. These processes are subject of our contribution on hypervelocity liquid-liquid impact in these proceedings (Rein, 1994). The principles of the formation of a munroe jet in the case of chaped explosive charges are shown in Fig. 17. At the point were the metal liners collide a thin jet of high velocity is produced in one direction and a thick slug of low velocity in the other.

6. Conclusions

Various fluid mechanical processes that can occur during droplet impact were discussed from the viewpoint of wave phenomena. These range from strong shock waves produced by hypervelocity impacts to capillary waves and interfacial processes in the case of small impact velocities. Straightforward explanations of processes like compressible droplet impact and the formation of cavitation zones inside a drop are at hand. In other cases like the instability of the rim of spreading lamellas, and floating and bouncing of droplets on wavy surfaces, the importance of waves is obvious but the exact mechanisms are not understood. We have suggested possible approaches towards the explanation of such processes. It is hoped that this survey provides some insight into the fascinating and various phenomena of droplet impact.

7. References

Allen, R. F. (1975) The Role of Surface Tension in Splashing, *J. Colloid Interface Sci.*, Vol. 51 (2), 350-351

Alvarez, L. W., Alvarez, W., Asaro, F. and Michel, H. V. (1980) Extraterrestrial cause for Cretaceous-Tertiary extinction, *Science*, **208**, 1095-1108

Anders, K., Roth, N. and Frohn, A. (1993) The velocity change of ethanol droplets during collision with a wall analyzed by image processing, *Exp. in Fluids*, **15**, 91-96

Baird, M. H. I. (1960) The stability of inverse bubbles, *Trans. Faraday Soc.* **56**, 213-219

Bode, J., Meier, G. E. A. and Rein, M. (1990) Cavitation behind tension waves, in: Meier, G. E. A. and Thompson, P. A. (eds.), *Adiabatic Waves in Liquid-Vapor Systems*, IUTAM Symp. Göttingen/Germany 1989, Berlin, Heidelberg, 341-350

Brunton, J. H. and Camus, J. J. (1970) The Flow of a Liquid Drop During Impact, in: A. A. Fyall and R. B. King (eds.), *Proceedings of the 3rd International Conference on Rain*

Erosion and Associated Phenomena, Meersburg, West Germany, RAE, Farnborough, U.K., 327-352

Ching, B., Golay, M. W. and Johnson, T. J. (1984) Droplet Impacts Upon Liquid Surfaces, *Science*, **226**, 535-537

Engel, O. G. (1955) Waterdrop Collisions With Solid Surfaces, *J. Research NBS*, **54**, 281-298

Grabitz, A., Grabitz, G. and Meier, G. E. A. (1993) Water Crowns in Pulsed Jets, in: Visualization Soc. of Japan (ed.), *Atlas of Visualisation - Progress in Visualisation*, **1**, Pergamon Press, Oxford, 7-18

Grabitz, G. and Meier, G. E. A. (1983) Über Laufzeitinstabilitäten in Flüssigkeitsstrahlen, *ZAMM* **63**, T255-T257

Hallett, J. and Christensen, L. (1984) Splash and penetration of drops in water, *J. Rech. Atmosph.* **18**, 225-242

Heymann, F. J. (1969) High-Speed Impact between a Liquid Drop and a Solid Surface, *J. Appl. Phys.*, **40**, 5113-5122

Hughes, W. and Hughes, A. R. (1932) Liquid drops on the same surface, *Nature*, **129**, 59

Jayaratne, O. W. and Mason, B. J. (1964) The coalescence and bouncing of water drops at an air/water interface, *Proc. R. Soc. Lond. A*, **280**, 545-565

Kedrinskii, V. K. (1976) Negative pressure profile in cavitation zone at underwater explosion near free surface, *Acta Astronautica*, **3**, 623-632

Kedrinskii, V. K. (1994) Behaviour of Liquid at Dynamic Loading, in: Morioka, S. and van Wijngaarden, L. (eds.), *Waves in liquid/gas and liquid/vapor two-phase system*, IUTAM Symp. Kyoto/Japan 1994, Kluwer Academic Publishers

Kivity, Y. and Hanin, M. (1981) Stability of interface and shock wave driven by initial pressure discontinuity, *Phys. Fluids*, **24**, 1010-1016

Lesser, M. B. (1981) Analytic solutions of liquid-drop impact problems, *Proc. R. Soc. Lond. A*, **377**, 289-308

Lesser, M. B. and Field, J. E. (1983) The impact of compressible liquids, *Ann. Rev. Fluid Mech.*, **15**, 97-122

Levi, B. G. (1992) Twelve-year trail of clues leads to impact crater from the K-T boundary *Phys. Today*, **55**, 17-19

Levich, V. G. (1962) *Physicochemical Hydrodynamics*, Prentice-Hall, Englwood Cliffs, NJ

Loehr, K. F. (1990) Etalement et éclatement de gouttes, *Diplom De Doctorat de l'Université Pierre et Marie Curie*, Paris 6

Loehr, K. F. and Lasek, A. (1990) Splashing of drops, *Arch. Mech.*, **42**, 507-513

Meier, G. E. A. (1994) On Kinematic Waves in Two-Phase Systems, in: Morioka, S. and van Wijngaarden, L. (eds.), *Waves in liquid/gas and liquid/vapor two-phase system*, IUTAM Symp. Kyoto/Japan 1994, Kluwer Academic Publishers

Meier, G. E. A., Klöpper, A. and Grabitz, G. (1992) The influence of kinematic waves on jet break down, *Experiments in Fluids*, **12**, 173-180

Meier, G. E. A., Ehrenfried, K., Grabitz, G., and Rein, M. (1991) Wave Models for Unsteady Separation and Jets, in: Kozlov, V. V. and Dovgal, A. V. (eds.), *Separated Flows and Jets*, IUTAM Symp. Novosibirsk/USSR 1990, Berlin, Heidelberg, 503-512

Melosh, H. J. (1989) *Impact cratering: a geologic process*, Oxford University Press, New York NY

Oğuz, H. N. and Prosperetti, A. (1990) Bubble entrainment by the impact of drops on liquid surfaces, *J. Fluid Mech.*, **219**, 143-179

Peregrine, D. H. (1981) The fascination of fluid mechanics, *J. Fluid Mech.*, **106**, 59-80

Prosperetti, A. and Oğuz, H. N. (1993) The impact of drops on liquid surfaces and the underwater noise of rain, *Ann. Rev. Fluid Mech.*, **25**, 577-602

Pumphrey, H. C. and Elmore, P. A. (1990) The entrainment of bubbles by drop impacts, *J. Fluid Mech.*, **220**, 539-567

Pumphrey, H. C. and Walton, A. J. (1988) Experimental study of the sound emitted by water drops impacting on a water surface, *Euro. J. Phys.*, **9**, 225-231

Rayleigh, J. W. S. (1882) Further observations upon liquid jets, in continuation of those recorded in the Royal Society's 'Proceedings' for March and May, 1879, *Proc. R. Soc. London*, Vol. 34, 130-145

Rein, M. (1993) Phenomena of liquid drop impact on solid and liquid surfaces, *Fluid Dyn. Res.* **12**, 61-93

Rein, M. (1994) Maximum pressures during hypervelocity liquid-liquid impact, in: Morioka, S. and van Wijngaarden, L. (eds.), *Waves in liquid/gas and liquid/vapor two-phase system*, IUTAM Symp. Kyoto/Japan 1994, Kluwer Academic Publishers

Rein, M. and Meier, G. E. A. (1988) Numerical Simulation of the Dynamics of Cavitation Bubble Fields Generated by Strong Expansion Waves, in: Furuya, O. (ed.): *Cavitation and Multiphase Flow Forum - 1988*, ASME, New York, 40-44

Rein, M. and Meier, G. E. A. (1990a) On the Dynamics of Heterogeneous Shock Cavitation, *Acustica*, **71**, 1-13

Rein, M. and Meier, G. E. A. (1990b) On the influence of different parameters on heterogeneous shock cavitation, *J. Acoust. Soc. Am.*, **88**, 1921-1928

Reynolds O. (1881) On the floating of drops on the surface of water depending only on the purity of the surface, *Proc. Manchester Lit. Phil. Soc.*, Vol. 21, (also in: Reynolds, O. (1900) *Papers on Mechanical and Physical Subjects*, Vol. 1, 413-414, Cambridge)

Rodriguez, F. and Mesler, R. (1985) Some Drops Don't Splash, *J. Colloid Interface Sci.*, **106**, 347-352

Sigler, J. and Mesler, R. (1990) The Behavior of the Gas Film Formed upon Drop Impact with a Liquid Surface, *J. Colloid Interface Sci.*, Vol. 134 (2), 459-474

Skogan, N. (1956) Inverted Soap Bubbles - A Surface Phenomenon, *American J. Physics*, **24**, 239-241

Stong, C. L. (1974) The Amateur Scientist: Curious bubbles in which a gas encloses a liquid instead the other way around, *Scientific American*, **230**, 116-121

Taylor, G. I. (1959) The dynamics of thin sheets of fluid II. Waves on fluid sheets, *Proc. R. Soc. Lond. A*, **253**, 296-312

van Wijngaarden, L. (1968) On the equations of motion for mixtures of liquid and gas bubbles, *J. Fluid Mech.*, **33**, 465-474

van Wijngaarden, L. (1972) One-dimensional flow of liquids containing small gas bubbles, *Ann. Rev. Fluid Mech.*, **4**, 369-396

Walker, J. (1978) The Amateur Scientist: Drops of liquid can be made to float on the liquid. What enables them to do so?, *Scientific American*, **238**, 123-129

Walsh, J. M., Shreffler, R. G. and Willig, F. J. (1953) Limiting Conditions for Jet Formation in High Velocity Collisions, *J. Appl. Phys.*, **24**, 349-359

Weiss, D. A. (1993) Periodischer Aufprall monodisperser Tropfen gleicher Geschwindigkeit auf feste Oberflächen, Mitt. Max-Planck-Inst. f. Strömungsforschung, Nr. **112**, Göttingen

Worthington, A. M. (1876) On the Forms assumed by Drops of Liquids falling vertically on a horizontal Plate, *Proc. R. Soc. London*, **25**, 261-271

Worthington, A. M. (1877) A Second Paper on the Forms assumed by Drops of Liquids falling vertically on a horizontal Plate, *Proc. R. Soc. London*, **25**, 498-503

Yarin, A. L. and Weiss, D. A. (1994) Impact of drops on solid surfaces: self-similar capillary waves and splashing as a new type of kinematic discontinuity, *J. Fluid Mech.* (accepted for publication)

Zhbankova, S. L. and Kolpakov, A. V. (1990) Collision of water drops with a plane water surface, *Fluid Dynamics*, **25**, 470-473

MAXIMUM PRESSURES DURING HYPERVELOCITY LIQUID-LIQUID IMPACT

Martin Rein*)
Max-Planck-Institut für Strömungsforschung
Bunsenstraße 10, 37073 Göttingen, Germany

Summary. The initial phase of hypervelocity liquid-liquid impact is considered taking into account the geometry of the colliding liquid surfaces. Right after impact shock waves are formed in both liquids. Initially, the shocks remain attached to the contact edge, i.e., to the circumference of the contact zone between the two liquids. The shocks enclose a high pressure region. When the conditions for the shocks to remain attached break down the shocks move ahead of the contact edge and lateral jetting begins. Expansion waves then decrease the pressure in the liquids. Maximum pressures that well exceed the pressure obtained by a one-dimensional approximation, are reached at the contact edge right before shock detachment. At this moment one of the shocks satisfies the strong solution. The peak pressures occuring during such events are determined. The dependence of the maximum pressures on the impact velocity and impact angle are discussed. When the impact velocity is constant oblique impacts may result in higher pressures than normal impacts.

1. Introduction

There are many examples for high-speed collisions between two liquids. In steam turbines droplets may impinge on rotating turbine blades. The surfaces of the blades are sometimes rippled in order to enhance the formation of a liquid film and, thus, reduce the impact pressure. Another example concerning erosion is the collapse of a cavitation bubble near a solid wall. During its collapse a liquid jet is formed that penetrates the cavity towards the wall with velocities up to $500 m/s$. Its impact on the opposite wall of the cavity leads to the formation of high pressures that may cause damage to the neighboring solid wall. A collapse of cavitation bubbles with jet formation can also be induced by shock waves passing the bubbles (Dear and Field, 1988, Yamada and Takayama, 1994). The collision of micrometeoroids with space vehicles may also be considered as a hypervelocity liquid-liquid impact (Schneider and Stilp, 1987). Here, the pressures may be that high that the strength of the solids can be neglected and the materials behave hydrodynamically. On a larger scale this process occurs during planetary impacts (Melosh, 1989). It obtained much attention in planetary sciences. Meteoroid impacts have been simulated numerically. In numerical computations, however, the detailed resolution of the brief initial stage of liquid impact often suffers from

*) *present address: Institut für Angewandte Mechanik und Strömungsphysik, Universität Göttingen, Bunsenstraße 10, 37073 Göttingen, Germany*

S. Morioka and L. van Wijngaarden (eds.),
IUTAM Symposium on Waves in Liquid/Gas and Liquid/Vapour Two-Phase Systems, 191-200.

artificial viscosity (Lesser and Field, 1983). It is, therefore, of great interest to obtain analytical solutions for the maximum pressures that occur during the brief initial stage. A first step in this direction was made by Kieffer (1977). She considered symmetrical impacts of planar surfaces colliding at different angles, using a shock polars approach. The results clearly demonstrate the importance of the specific geometry of the impact process. Further, the transition from jetless to jetting conditions (see below) is discussed. In order to determine the impact pressures of droplets impinging normally on solid walls Lesser (1981) introduced a somewhat different procedure. His approach leads to a single equation for the pressure at the circumference of the contact zone between the liquid and the wall. Besides depending on the impact velocity the pressure depends also on the contact angle between the drop's surface and the wall. Faddeyev *et al.* (1988) later extended Lesser's (1981) approach to include oblique impacts.

In this paper the basic physical processes of the initial stage of hypervelocity impacts between two liquid masses will be exemplified by considering the collision of a liquid sphere with a planar liquid surface. The collision of two arbitrarily shaped bodies can be treated analogously. Right after impact shock waves propagate into the liquids increasing the pressure to high values. The shock waves are oblique and the geometry of the collision process is of great importance for the pressure rise across the shocks. Two early stages of compressible droplet impact on a flat liquid surface are sketched in Fig. 1. On the l.h.s. both shocks adhere to the circumference of the interface between the two liquids (the circumference is also called a contact edge hereafter) and the shocks enclose the high pressure region. At a later time, sketched on the r.h.s. of Fig. 1, the shocks detach from the contact edge. The high pressure region is no longer fully enclosed by shocks and expansion waves release the pressure. This results in lateral jetting. The transition from attached to detached shocks is caused by the increase with time of the contact angle between the surfaces of the drop and the target liquid. Initially, this angle is extremely small and the contact edge moves radially with very high velocities. With increasing time the contact angle increases and the velocity of the contact edge decreases. Eventually, the shocks pass the contact edge. The detachment from the contact edge of the shocks will be explained in terms of a reversed wedge flow with oblique shocks in chapter 5. First, however, an equation of state is introduced for the liquids that may also represent rocky materials at high pressures. In this manner planetary impacts may be considered as well. Since a meteoroid is no droplet we will also use the general terms projectile and target liquid. – Shock relations based on the particular equation of state, are then derived and the physical mechanisms leading to the transition from attached to detached shocks are discussed in shock polar diagrams. It will be seen that one of the shocks satisfies a strong solution right before shock detachment. Finally, in chapter 6, an equation for the pressure at the contact edge is derived using the approach of Lesser (1981). Based on this equation the maximum pressures occuring during normal and oblique impacts as well as the time dependence of the edge pressure will be discussed.

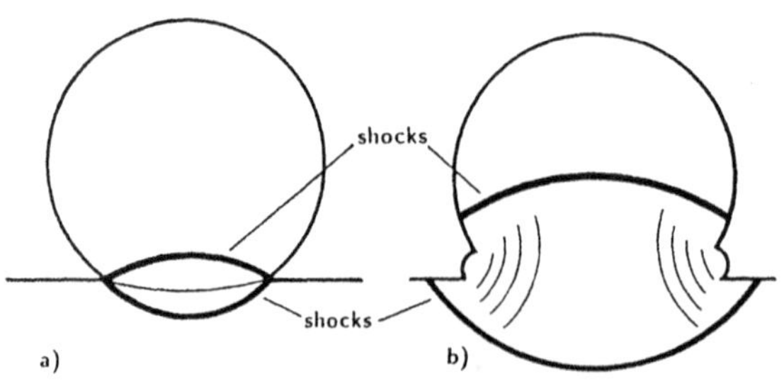

Fig. 1: Initial stages of compressible droplet impact on a planar liquid surface:
a) attached shocks enclosing a high pressure zone.
b) detached shocks, expansion waves and jetting commencement.

2. Equation of state

The physical mechanisms of high-speed liquid-liquid impact are of our main interest. Therefore, all considerations are based on a simple barotropic equation of state for both the projectile and the target liquid. High-pressure phase transitions that may occur in shocked solids will not be considered. Here, we take a Tait form of equation of state. In the context of planetary impacts this form is also called a Murnaghan equation of state. It reads

$$\frac{p+B}{p_0+B}=\left(\frac{\rho}{\rho_0}\right)^A \quad , \tag{2.1}$$

where the constants A and B need to be determined from experiments or from calculations based on solid-state physics. In the following B will be replaced by the sound speed c_0 at a reference pressure and density, p_0 and ρ_0, using $c_0^2=(\partial p/\partial\rho)_{\rho_0}$. This yields

$$\frac{p-p_0}{\rho_0 c_0^2}=\frac{1}{A}\left[\left(\frac{\rho}{\rho_0}\right)^A-1\right] \quad . \tag{2.2}$$

The sound speed scaled by c_0 is then given by

$$\left(\frac{c}{c_0}\right)^2=\left(\frac{\rho}{\rho_0}\right)^{A-1} \quad . \tag{2.3}$$

Later, the sound speed and density at the reference state will be used for scaling and, therefore, do not explicitly appear in the calculations below. However, when collisions of different liquids are considered, besides the exponent A, the ratios of the sound speeds and of the impedances of the drop and the target liquid at reference conditions need to be specified. Later we will consider the impact of water on water, and of an iron projectile on gabbroic anorthosite (GA), the latter material being characteristic of, e.g., lunar surfaces (Ahrens and O'Keefe, 1977). In the case of water we take $A=7$ as usual. The values of A and the ratios of the sound speeds and impedances for the case of iron impacting on gabbroic anorthosite are infered from data provided in Kieffer and Simonds (1980): $A=5.32$ for iron and $A=4.0\cdots5.5$ for gabbroic anorthosite in the high-pressure phase. Here we choose $A=5.32$, i.e., the same value as in the case of iron. The ratios of the sound speeds and impedances are $c_0^p/c_0^t=0.498$ and $\rho_0^p\,c_0^p/\rho_0^t\,c_0^t=0.965$ where the superscripts p and t denote the iron projectile (drop) and target liquid (gabbroic anorthosite), respectively.

3. Normal and oblique shock waves

The relations describing the change in the variables across normal and oblique shock waves in liquids can be derived analogously to the case of ideal gases. A detailed derivation was given by Ridah (1988) for the case of water. Here, we will only derive those relations that are subsequently used. These relations are valid for any liquid that obeys a Tait equation of state.

Hereafter all variables are non-dimensional. The velocity, density and pressure are scaled by c_0, ρ_0 and $\rho_0 c_0^2$, respectively. The conservation of mass and momentum as well as the pressure jump across a normal shock wave moving with velocity w are then given by

$$(w-u)\rho_1 = (w-v)\rho_2$$
$$(3.1) \qquad (w-u)^2\rho_1 + p_1 = (w-v)^2\rho_2 + p_2$$
$$\Delta p = p_2 - p_1 = \frac{1}{A}(\rho_2^A - \rho_1^A) \quad .$$

The velocities of the liquid ahead and after the shock are u and v, and the state ahead and after the shock is denoted by the subscripts 1 and 2, respectively. The density and velocity after the shock can be expressed in terms of the pressure jump:

$$\frac{\rho_2}{\rho_1} = \left(1 + A\frac{1}{\rho_1^A}\Delta p\right)^{\frac{1}{A}}$$
$$(3.2) \qquad (w-u)^2 = \frac{\Delta p}{\rho_1}\left(1 - (\frac{\rho_2}{\rho_1})^{-1}\right)^{-1} = F(\Delta p, \rho_1)^{-1}$$
$$(u-v)^2 = \frac{\Delta p}{\rho_1}\left(1 - (\frac{\rho_2}{\rho_1})^{-1}\right) = \left(\frac{\Delta p}{\rho_1}\right)^2 F(\Delta p, \rho_1) \quad .$$

The function $F(\Delta p, \rho_1)$ that was introduced by Lesser (1981), is used for brevity. We will now consider the reference frame of the shock, i.e., $w = 0$. Using eqs. (2.3) and (3.2), and introducing the Mach number ahead of the shock wave, $M_1 = u/c_1$, eqs. (3.1) can be rearranged, yielding

$$\left(\frac{u}{v}\right)^{A+1} - (AM_1^2 + 1)\frac{u}{v} + AM_1^2 = 0$$
$$(3.3)$$
$$AM_1^2\left[1 - \left(1 + A\frac{1}{\rho_1^A}\Delta p\right)^{-\frac{1}{A}}\right] - A\frac{1}{\rho_1^A}\Delta p = 0 \quad .$$

Let us now consider oblique shock waves. In order to derive the equations for shock polars we proceed in the usual way. Fig. 2 shows different velocities at a shock forming an angle Θ with the flow ahead of it. At the shock the flow is deflected by an angle ϑ. The velocity u ahead of the shock can be decomposed into components normal and parallel to the shock, u_n and u_t. Similarly, the normal component of the velocity v after the shock is v_n. The meaning of the velocity components v_a and v_b is evident from Fig. 2. The following relations hold between the various components of the velocities and the angles Θ and ϑ:

$$u_t = u\cos\Theta \quad , \quad u_n = u\sin\Theta \quad ,$$
$$(3.4) \qquad v_n = u\sin\Theta - \frac{v_b}{\cos\Theta} \quad ,$$
$$\tan\vartheta = \frac{v_b}{v_a} \quad , \quad \tan\Theta = \frac{u - v_a}{v_b} \quad .$$

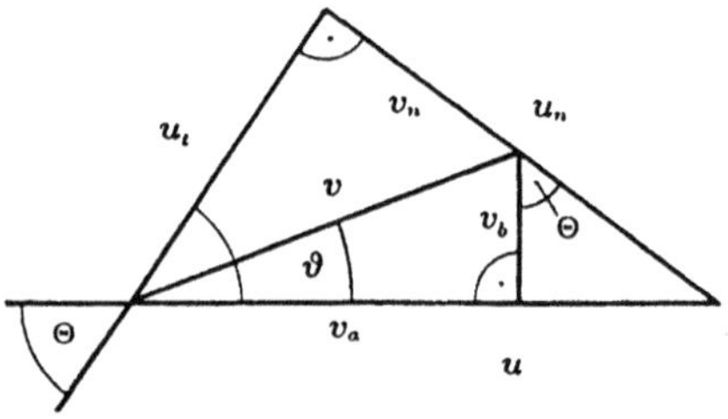

Fig. 2: Velocities at oblique shock.

The velocity component u_t parallel to the shock is not affected on crossing the shock while the shock relations for a normal shock hold for the normal component u_n. Thus, replacing M_1 by $M_1 \sin\Theta$ in the first of eqs. (3.3) and making use of eq. (3.4) yields a relation between ϑ, Θ and M_1:

$$G(\vartheta,\Theta,M_1) = 1-(AM_1^2\sin^2\Theta+1)\lambda^A AM_1^2\sin^2\Theta\lambda^{A+1} = 0 \quad ,$$

(3.5)

$$\text{were } \lambda = \frac{v_n}{u_n} = 1 - \frac{1}{\sin^2\Theta + \cot\vartheta\sin\Theta\cos\Theta} \quad .$$

Similarly, from the second of eqs. (3.3),

$$(3.6) \qquad P(\Delta p,\Theta,M_1) = AM_1^2\sin^2\Theta\left[1-\left(1+A\frac{1}{\rho_1^A}\Delta p\right)^{-\frac{1}{A}}\right] - A\frac{1}{\rho_1^A}\Delta p = 0 \quad .$$

For any constant Mach number M_1 ahead of a shock eqs. (3.5) and (3.6) provide the states possible after an oblique shock in the Θ-ϑ plane and in the Δp-ϑ plane, respectively. As in the case of an ideal gas a particular deflection may be obtained either via a weak or via a strong shock. Further, there exists a maximum deflection angle beyond which the flow cannot be turned by the shock. This angle may be determined, e.g., from eq. (3.5) using the additional constraint $\partial G/\partial\Theta = 0$.

4. Planar impact approximation

The nonlinear pressure rise during the collision of a meteoroid with a lunar or terrestical surface was estimated by Gault and Heitowit (1963) using a one-dimensional calculation by considering the normal impact of two semi-infinite bodies along planar surfaces. This is called the planar impact approximation.

The projectile approaches the stationary target with an impact velocity u_i, the corresponding Mach number is M_i. Hereafter, the reference state taken for scaling is the state in the unshocked projectile, hence $M_i = u^p$ where the superscripts p and, later, t, denote the projectile and target liquid. After impact two planar shocks propagate into the two liquids. This is sketched in Fig. 3. At the interface the velocities and pressures are the same in both liquids. The pressure increase across the shock waves is obtained from the last of eqs. (3.2):

$$(4.1) \qquad \begin{aligned} (v^p - M_i)^2 &= \frac{\Delta p^p}{\rho_1^p}\left(1-\left(\frac{\rho_2^p}{\rho_1^p}\right)^{-1}\right) \\ v^{t^2} = v^{p^2} &= \frac{\Delta p^t}{\rho_1^t}\left(1-\left(\frac{\rho_2^t}{\rho_1^t}\right)^{-1}\right) \quad . \end{aligned}$$

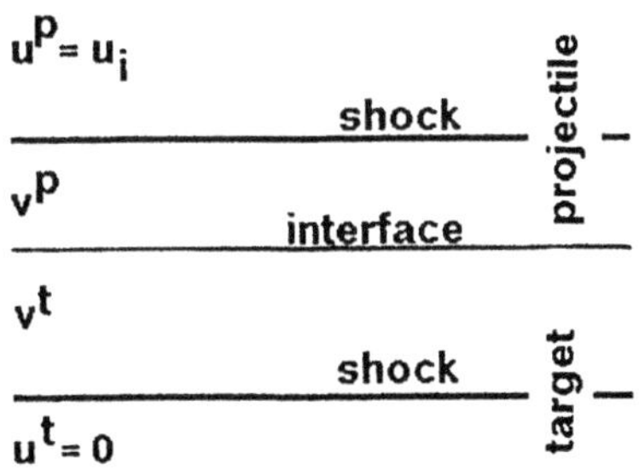

Fig. 3: Planar impact approximation.

The initial pressures in the projectile and target may be neglected with respect to the pressure increase, hence $\Delta p^p = p_2^p = p_2^t = \Delta p^t$. The second equation of eq. (4.1) can then be substituted into the first one yielding an equation for the pressure in the high pressure region between the shocks.

5. System of two oblique shock waves at contact edge

The impact of a spherical liquid mass with a plane surface is different form that of two colliding plane surfaces that was considered in the planar impact approximation. The shocks are no longer normal. At the contact edge the flow and shock system is similar to a reversed wedge flow with different fluids and velocities on the two faces of the wedge. This is most easily seen in a reference system in which the contact edge is stationary (cf. Fig. 4). In this system ahead of the shocks the velocities of the projectile and target liquid, u_e^p and u_e^t, are parallel to the respective free surfaces. Their absolute values depend on the contact angle β and the impact velocity u_i: $u_e^p = u_i/\sin\beta$ and $u_e^t = u_i/\tan\beta$. The dependence on time of β and, hence, of u_e^p and u_e^t, is determined by the geometry of the colliding liquids and by the impact velocity. The state after the shocks can be obtained via shock polars in the usual way. The Mach numbers corresponding with u_e^p and u_e^t are M_1^p and M_1^t, respectively. Shock polars are calculated from eqs. (3.5) and (3.6) using the appropiate densities and exponents A of the equation of state. In Fig. 5 shock polars of the two shocks are plotted showing the deflection angles ϑ^p and ϑ^t and the pressure jump across the shocks. Since β equals the combined deflection, $\beta = |\vartheta^p| + |\vartheta^t|$, the shock polars are shifted by this angle. Behind the shocks the pressures need to be the same in the projectile and target, hence, the intersection of the polars provides the pressure (again, the initial pressures ahead of the shocks are neglected). Two solutions are possible, here, the one with the smaller pressure is appropiate. Both shocks are then weak shocks. With increasing β the origins of the polars move apart. Further, the Mach numbers M_1^p and M_1^t decrease, i.e., the extension in ϑ-direction of the polars becomes smaller. Hence, there exists a time at which β is so large that the polars no longer intersect. Physically this means that the edge configuration with two attached shocks breaks down. The shocks detach, move upstream and lateral jetting becomes possible. Shortly before this happens the intersection of the shock polars is situated so that one of the shocks is a strong shock. This is depicted in Fig. 6 for the case of an iron projectile impacting on gabbroic anorthosite (GA). Here, the different equations of state lead to clearly different forms of the shock polars. But also with the same liquids in both, the projectile and the target, a strong shock appears. This is due to $u_e^p > u_e^t$ and, hence, the shock polars have always a slighly different extension. It is interesting that the appearance of a strong shock is caused by the flow configuration upstream of the shocks. Strong shocks are normally due to some cause downstream of a shock. In the case of wedge flows strong attached shocks are known to exist when the geometry of the wedge downstream of the shocks is carefully changed (Zebel 1950, Busemann 1953, Guderley 1957).

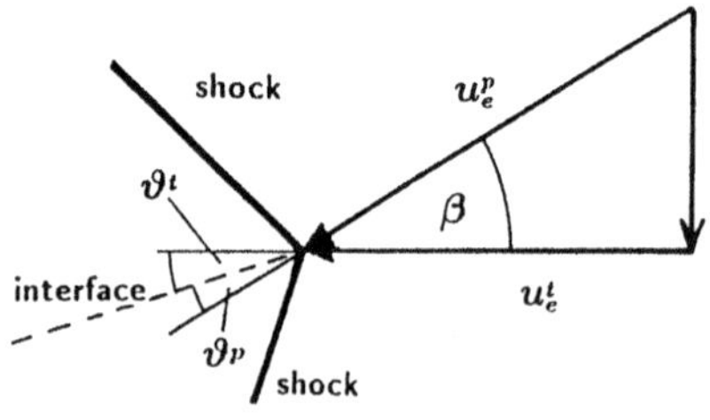

Fig. 4: Shock configuration at contact edge.

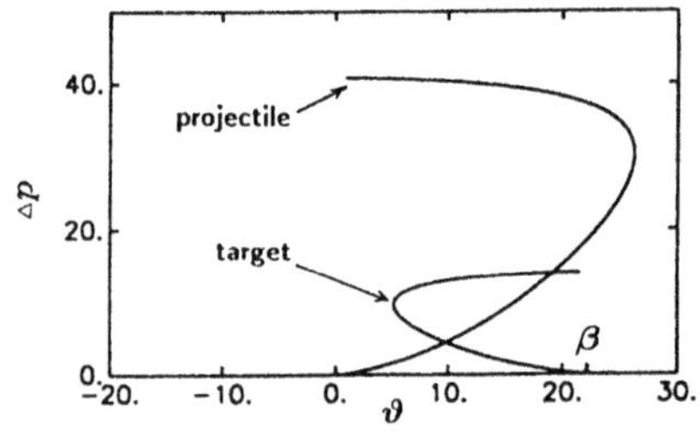

Fig. 5: Shock polars of shocks at contact edge (iron on GA, $u_i = 3$, $\beta = 22.0$).

For any particular combination of Mach numbers M_1^p and M_1^t ahead of the shocks there exists a maximum combined deflection, β_m, beyond which a solution with attached shocks is not possible. A combined deflection of β_m can only be achieved in one way and the pressure p_m connected with β_m is single valued. In order to clarify the situation the necessary branches of the two shock polars have been plotted starting at the same origin in Fig. 7. The shocks detach from the contact edge when $\beta = \beta_m$. The angle $\beta = |\vartheta^p| + |\vartheta^t|$ is a function of several variables: $\beta = \beta(\vartheta^p, \Theta^p, \vartheta^t, \Theta^t, \Delta p)$. This function is subject to the constraints $G^p(\vartheta^p, \Theta^p, M_1^p) = 0$, $P^p(\Delta p, \Theta^p, M_1^p) = 0$, $G^t(\vartheta^t, \Theta^t, M_1^t) = 0$ and $P^t(\Delta p, \Theta^t, M_1^t) = 0$. For a particular combination of M_1^p and M_1^t the maximum combined deflection β_m and, e.g., the corresponding pressure p_m can thus be determined via Lagrange multipliers.

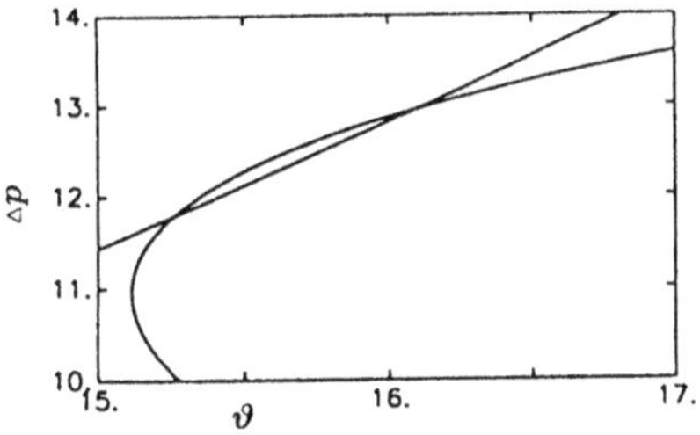

Fig. 6: Shocks polars (section): strong shock shortly before shock detachment (iron on GA, $u_i = 5$, $\beta = 32.6$).

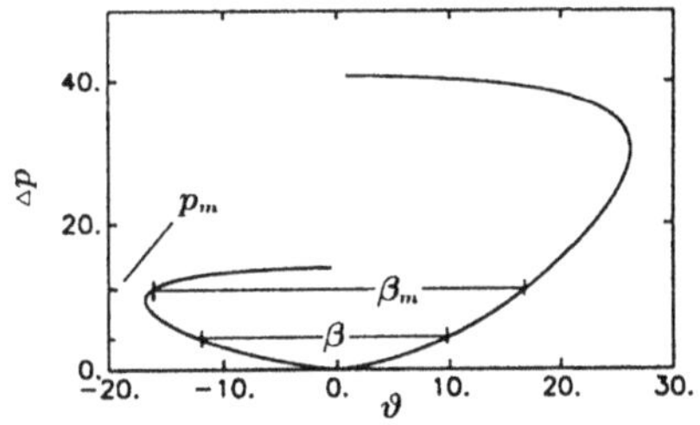

Fig. 7: Maximum combined deflection angle β_m, corresponding pressure p_m, as well as β (data as in Fig. 5).

6. Pressure at contact edge

In this section an equation for the pressure at the contact edge will be derived, following the approach of Lesser (1981). First, let us introduce the notation used. We will distinguish between different reference systems: the frames of the unshocked projectile and target, and of the stationary contact edge. These reference frames are denoted by <u>sub</u>scripts p, t, and e, respectively. The different fluids are marked by <u>super</u>scripts p and t, as in the preceeding chapters. Vectors are characterized by underlined letters. The general case of oblique impacts is considered. Fig. 8 shows the impact angle δ and impact velocity u_i as well as its components normal (u_{in}) and tangential (u_{it}) to the planar liquid surface. Further, the component of u_{it} in the direction of an azimuthal

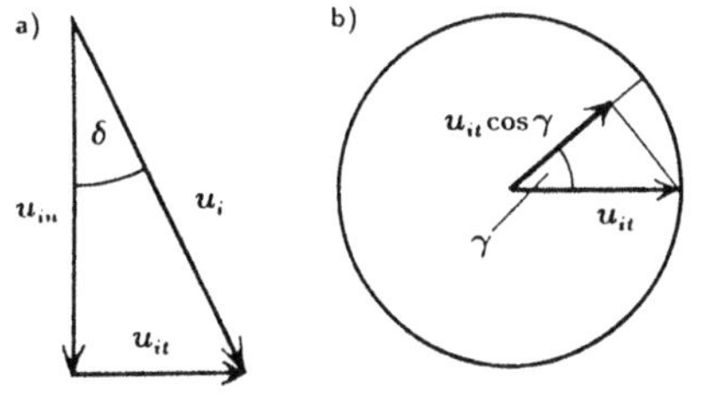

Fig. 8: Impact velocity (a) side-view: impact angle, normal and tangential velocity components, (b) top-view: azimuthal angle.

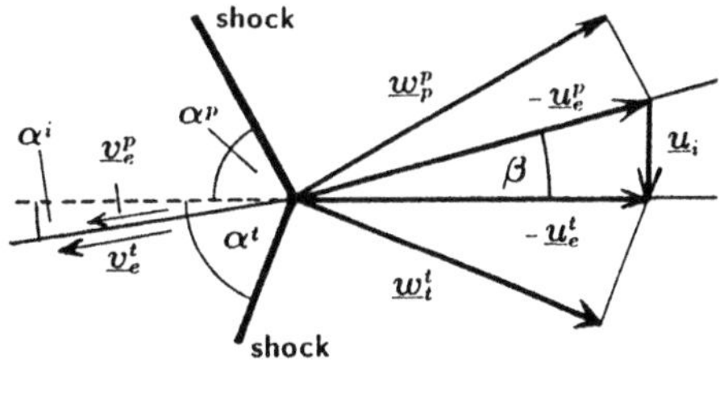

Fig. 9: sketch of contact edge with velocity vectors and projections of these vectors.

angle γ is illustrated. The different velocities at the contact edge are depicted in Fig. 9. Due to the obliquity of the impact the effective normal component of the impact velocity, $u_{i\gamma}$, depends on δ and γ. It can be seen that

$$(6.1)\quad u_e^t = u_i \cos\delta \,\frac{1}{\tan\beta} + u_i \sin\delta\,\cos\gamma \;,\quad u_e^p = u_e^t\,\frac{1}{\cos\beta} \;,\quad u_{i\gamma} = u_e^t\,\tan\beta \;.$$

In order for the shocks to remain attached to the contact edge it is necessary that the projection of the velocities $-\underline{u}_e^p$ and $-\underline{u}_e^t$ onto the direction of the shock velocities, $\underline{n}_p^p$ and $\underline{n}_t^t$, just equal the shock speeds w_p^p and w_t^t. Here, $\underline{n}_p^p = n_{pr}^p\ \underline{e}_r + n_{pz}^p\ \underline{e}_z$ and $\underline{n}_t^t = n_{tr}^t\ \underline{e}_r + n_{tz}^t\ \underline{e}_z$ are the unit vectors in the direction of the shock velocity in the projectile and target system ($\underline{e}_r$ and $\underline{e}_z$ are unit vectors in radial and axial direction). Further, after the shocks the flow velocities are in the direction of the shock velocities (in the preshock reference systems). This yields:

$$(6.2)\quad \begin{array}{lll} \text{(a)}\ \underline{u}_e^p\,\underline{n}_p^p = w_p^p \;, & \text{(b)}\ \underline{v}_p^p = v_p^p\,\underline{n}_p^p \;, & \text{(c)}\ (n_{pr}^p)^2 + (n_{pz}^p)^2 = 1 \;, \\ \text{(d)}\ \underline{u}_e^t\,\underline{n}_t^t = w_t^t \;, & \text{(e)}\ \underline{v}_t^t = v_t^t\,\underline{n}_t^t \;, & \text{(f)}\ (n_{tr}^t)^2 + (n_{tz}^t)^2 = 1 \;. \end{array}$$

We now solve for the components of $\underline{n}_p^p$ and $\underline{n}_t^t$. From eqs. (6.1) and (6.2a/d):

$$(6.3)\qquad n_{pr}^p = \frac{1}{u_e^t}(w_p^p - u_{i\gamma} n_{pz}^p) \;,\quad n_{tr}^t = \frac{w_t^t}{u_e^t} \;.$$

Together with eqs. (6.2c) and (6.2f) this yields:

(6.4)

$$n_{pz}^p = \frac{u_{i\gamma}}{u_{i\gamma}^2 + (u_e^t)^2}\left[w_p^p + \frac{u_e^t}{u_{i\gamma}}\left(u_{i\gamma}^2 + (u_e^t)^2 - (w_p^p)^2\right)^{\frac{1}{2}}\right] \;,\quad n_{tz}^t = -\left[1 - \left(\frac{w_t^t}{u_e^t}\right)^2\right]^{\frac{1}{2}} \;.$$

In taking the roots the sign was taken so that $n_{pz}^p > 0$ and $n_{tz}^t < 0$ for $\beta \to 0$. The velocities w_p^p, w_t^t, v_p^p and v_t^t are given by the shock relations (3.2):

$$(6.5)\quad \begin{array}{ll} w_p^p = F^p\,(\Delta p^p, 1)^{-1/2} \;, & v_p^p = \Delta p^p\ F^p(\Delta p^p, 1)^{1/2} \;, \\ w_t^t = F^t\,(\Delta p^t, \rho_1^t)^{-1/2} \;, & v_t^t = \Delta p^t\ F^t(\Delta p^t, \rho_1^t)^{1/2} \end{array}$$

At the interface between the projectile and target liquid the pressure is the same in both liquids. As before, we neglect the initial pressures and take $\Delta p^p \approx \Delta p^t = \Delta p$. Further, at the interface the flow velocities are parallel. In the reference system of the contact edge these velocities are:

$$(6.6)\quad \underline{v}_e^p = v_p^p\,\underline{n}_p^p - \underline{u}_e^p = v_{er}^p\ \underline{e}_r + v_{ez}^p\ \underline{e}_z \;,\quad \underline{v}_e^t = v_t^t\,\underline{n}_t^t - \underline{u}_e^t = v_{er}^t\ \underline{e}_r + v_{ez}^t\ \underline{e}_z \;.$$

The coefficients of the velocity vectors $\underline{v}_e^p$ and $\underline{v}_e^t$ have all been determined above. The constraint of these vectors being parallel can be written

$$(6.7)\qquad v_{ez}^p\ v_{er}^t - v_{ez}^t\ v_{er}^p = 0 \;.$$

The only unknown in this equation is the pressure behind the shocks. Eq. (6.7) is thus a relation for the edge pressure at a certain edge angle β. Since the time dependence of β is known this relation provides the edge pressure as a function of time. When the shocks detach it is no longer valid. Then, the solution of eq. (6.7) breaks down.

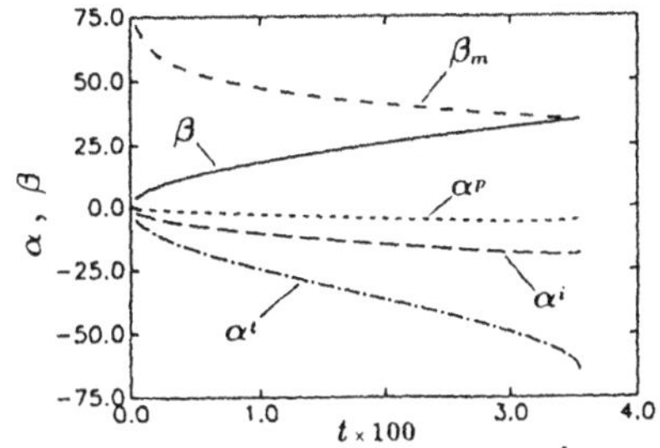

Fig. 10: β, β_m, α^p, α^t *and* α^i *vs. time (iron on GA,* $u_i = 5$, $\delta = 40°$, $\gamma = 0°$*).*

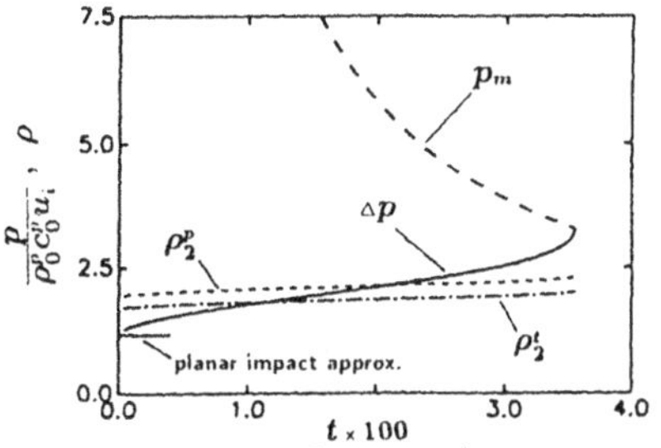

Fig. 11: Δp, p_m, ρ_2^p *and* ρ_2^t *vs. time (iron on GA,* $u_i = 5$, $\delta = 40°$, $\gamma = 0°$*).*

The dependence on time of the angles β and β_m, and of the angles of the shocks and the interface with respect to the undisturbed target surface (cf. Fig. 9), are shown in Fig. 10 for the case of an iron projectile impacting on a GA target. The time is scaled by R/c_0 where R is the radius of the projectile. When $\beta = \beta_m$ the solution breaks down and jetting begins. Then $\Delta p = p_m$ as can be seen in Fig. 11 in which the pressures Δp and p_m, and the densities in the shocked liquids are plotted versus time. Note that actually $\Delta p/\rho_0^p c_0^p u_i$ and $p_m/\rho_0^p c_0^p u_i$ are plotted, i.e., the pressure is now scaled by the linear pressure rise in the case of a planar impact. A characteristic feature is the strong pressure rise shortly before shock detachment. Right after impact (at $t = 0$) the edge pressure equals the one obtained from the planar impact approximation. The maximum pressure rise Δp_{max} at the contact edge that is reached right before shock detachment, depends on the impact velocity u_i, the impact angle δ and the azimuthal angle γ. This is shown in Fig. 12 where $\Delta p_{max}/\rho_0^p c_0^p u_i$ is plotted in a polar diagram for the case of $u_i = 5$ and different impact angles δ. It is evident that for every δ the peak pressure (with $u_i = 5 = const$) is obtained in the direction of $\gamma = 0$. Further, a maximum pressure occurs at an oblique impact. This is more obvious in Fig. 13. Here, the dependence on δ of Δp_{max} is shown for different u_i and γ. The maximum of Δp_{max} occuring at an oblique impact is due to the effective normal component of the impact velocity u_{i_γ} depending on δ and β. In Fig. 14 the dependence on the impact velocity of Δp_{max} is plotted, this time for different δ and γ. For a constant δ and γ, $\Delta p_{max}/\rho_0^p c_0^p u_i$ increases almost linearly with u_i. However, this is no longer true for small u_i where the deviation from linear theory increases strongly with decreasing u_i. At small u_i the influence of the edge geometry is very important. It should be noted that the absolute pressure approaches zero with $u_i \to 0$. This is shown in Fig. 15 for the case of water, since neither iron nor GA will reach the hydrodynamic regime at such low impact speeds.

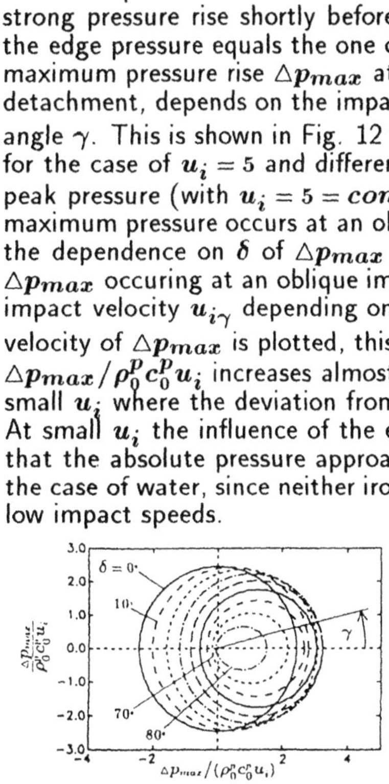

Fig. 12: Δp_{max} *in a polar diagram (iron on GA,* $u_i = 5$, $\delta = 0°, 10°, \cdots, 80°$*).*

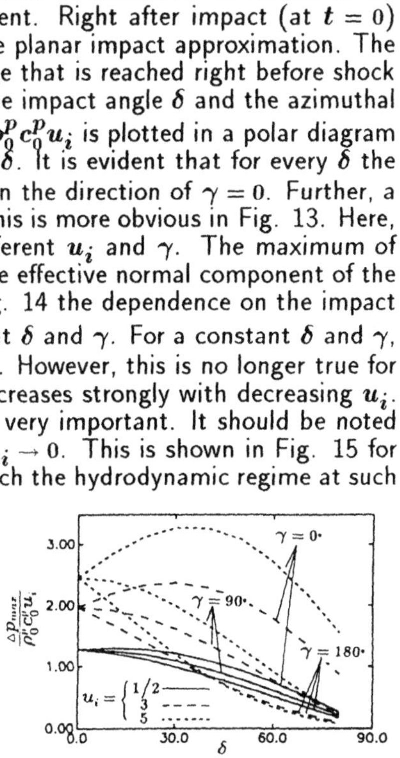

Fig. 13: Δp_{max} *vs.* δ *(iron on GA,* $u_i = 1/2, 3, 5$, $\gamma = 0°, 90°, 180°$*).*

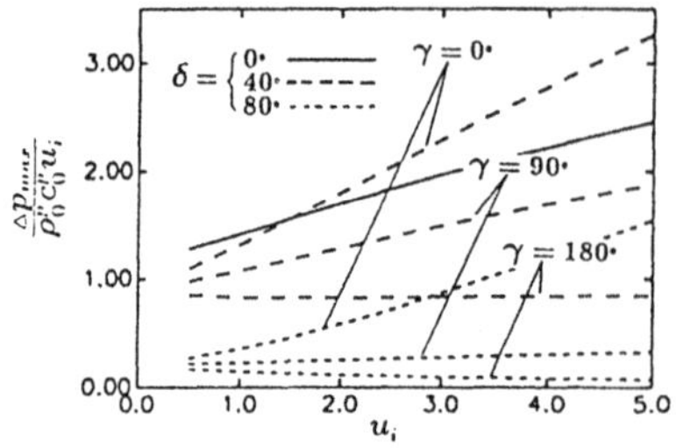

Fig. 14: Δp_{max} vs. u_i (iron on GA, $\delta = 0°, 40°, 80°, \gamma = 0°, 90°, 180°$).

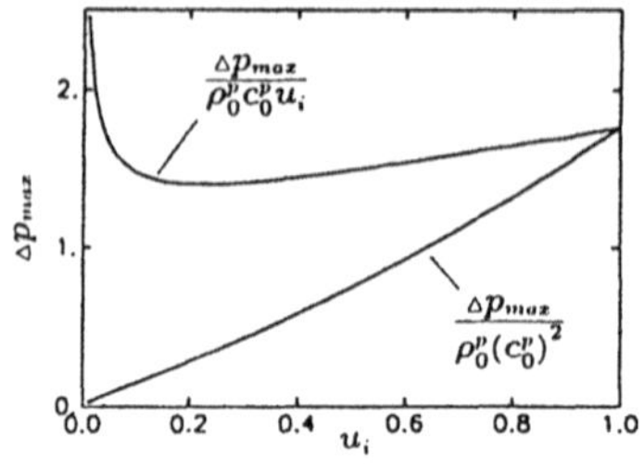

Fig. 15: Normal impact of water on water. Δp_{max} vs. u_i.

7. Conclusions

The maximum pressures occuring during the impact of a spherical liquid mass on a plane liquid surface were calculated. The peak pressure was observed to be always greater than that obtained from the planar impact approximation. Further, at small impact velocities the peak pressure is much greater than that given by a linear one-dimensional estimate. It is thus very important to take the geometry at the contact edge into proper account. In the case of oblique impacts with a constant absolute impact velocity the highest pressure is reached at a finite impact angle. The case of colliding liquid bodies of different geometries may be treated accordingly as long as the contact angle β is known. During high-speed impacts air bubbles may be entrained (cf. Shi *et al.*, 1993). When the geometry of the colliding surfaces is known this case can again be treated as shown above. – Further, a strong oblique shock that is caused by upstream flow conditions was shown to exist shortly before shock detachment.

8. References

Ahrens, T. J. and O'Keefe, J. D. (1977) in: Roddy, D. J., Pepin, R. O. , and Merril, R. B. (eds.), *Impact and explosion cratering*, Pergamon Press, NY, 639-656

Busemann, A., in: Shapiro, A. H. (1953) *The dynamics and thermodynamics of compressible fluid flow*, **1**, The Ronald Press Co., NY, 547

Dear, J. P. and Field, J. E. (1988) *J. Fluid Mech.*, **190**, 409-425

Faddeyev, I. P., Khaviya, S. and Mosenzhnik, B. (1988) *Fluid Mech. – Sov. Res.*, **17**, 38-44

Gault, D. E. and Heitowit, E. D. (1963) *Proc. Sixth Int. Hypervelocity Impact Symp.*, Cleveland, Ohio, 2, 419-456

Guderley, K. G. (1957) *Theorie schallnaher Strömungen*, Springer Verlag, Berlin, 164-168

Kieffer, S. W. (1977) in: Roddy, D. J., Pepin, R. O. , and Merril, R. B. (eds.), *Impact and explosion cratering*, Pergamon Press, NY, 751-769

Kieffer, S. W. and Simonds, C. H. (1980) *Rev. Geophys. Space Phys.*, **18** (1), 143-181

Lesser, M. B. (1981) *Proc. R. Soc. Lond. A*, **377**, 289-308

Lesser, M. B. and Field, J. E. (1983) *Ann. Rev. Fluid Mech.*, **15**, 97-122

Melosh, H. J. (1989) *Impact cratering: a geologic process*, Oxford University Press, New York

Ridah, S. (1988) *J. Appl. Phys.*, **64**, 152-159

Schneider, E. and Stilp, A. (1987) *Int. J. Impact Engng.*, **5**, 561-568

Shi, H.-H., Field, J. E., and Bourne, N. K. (1993) *Fluid. Dyn. Res.*, **11**, 79-83

Yamada, K. and Takayama, K. (1994) in: Morioka, S. and van Wijngaarden, L. (eds.), *Proc. IUTAM Symposium Waves in liquid/gas and liquid/vapor two-phase system*, Kluwer Academic Publishers

Zebel, G. (1950) *Über die Eindeutigkeit der anliegenden Kopfwelle an Keil und Kegel*, Diploma Thesis, University of Göttingen

On Travelling Waves in Two-Phase Systems

Gerd E.A. Meier
Institut für Strömungsmechanik der
Deutschen Forschungsanstalt für Luft- und Raumfahrt e.V. (DLR)
Göttingen, Deutschland

Abstract

The propagation of travelling velocity waves in multiphase or disperse media results in concentration fluctuations and/or density variations. Peaks may occur connected with the generation of catastrophic events like collision shocks. Without any special interaction (forces) between the droplets or particles of the medium also steep wave fronts may develop. The exact solution of a one dimensional nonlinear equation of motion for such an interaction free multiphase medium explains many observations concerning concentration waves in two phase flow. Contrary to the exponential growth of instability waves in linearized flow the amplitudes of these concentration waves grow in such a manner that at a finite propagation distance infinite amplitudes occur for an initially weak sinusoidal disurbance. Influence of gravity or other volume forces can also be included in the equations and in the exact solution. This influence results in wave length variations during the propagation.

As the described phenomena may occur in most two phase flows more or less pronounced these basic considerations are of some principal importance. Related experiments have been performed with a velocity modulated two phase spray of water droplets in air. The corresponding system was also treated theoretically. The exact solution of the one dimensional set of nonlinear equations of motion provides an insight into the kinematics of the travelling concentration waves. To illustrate the wave propagation for certain initial conditions Fig. 1 provides a visualisation of a computational simulation.

Fundamental Equations

For a one dimensional spray flow with the velocity $u(x,t)$ and the number density of the spray particles $m(x,t)$, one has for the conservation equations of mass and momentum

S. Morioka and L. van Wijngaarden (eds.),
IUTAM Symposium on Waves in Liquid/Gas and Liquid/Vapour Two-Phase Systems, 201-204.

$$\frac{\partial m}{\partial t} + u\frac{\partial m}{\partial x} + m\frac{\partial u}{\partial x} = 0 \tag{1}$$

$$\frac{\partial u}{\partial t} + u\frac{\partial u}{\partial x} = 0. \tag{2}$$

An additional acceleration by a uniform volume force (e.g. gravitation f) can be included:

$$\frac{\partial u}{\partial t} + u\frac{\partial u}{\partial x} = f \tag{3}$$

Conditions at Start (Nozzle)

The initial and boundary conditions for an arbitrary location ($x = 0$) in the spray flow are for the velocity

$$u(0\,,t) = u_0(t) \tag{4}$$

and for the number density of particles

$$m(0\,,t) = m_0(t). \tag{5}$$

Lagrangian Transform

Introducing a Lagrangian co-ordinate transform, with T_0 being the starting time of the bulk flow particles at $x = 0$ and T the actual time

$$t(T,T_0) = T \tag{6}$$

$$x(T,T_0) = u_0(T_0)\,(T - T_0) \tag{7}$$

we get the following solutions of the fundamental equations [1].

Exact Solutions

For the number density from eq. (1) and (2) results

$$m(T,T_0) = \frac{m_0(T_0)\,u_0(T_0)}{\left|u_0(T_0) - (T - T_0)\dfrac{d\,u_0(T_0)}{d\,T_0}\right|} \tag{8}$$

and for the velocity of the particles

$$u(T,T_0) = u_0(T_0) \tag{9}$$

Presence of a Volume Force

The solutions are only slightly modified for the case of the presence of a volume force (accelleration f in eq. (3)) on the particles. We get for the number density

$$m(T,T_0) = \frac{m_0(T_0)\, u_0(T_0)}{\left| u_0(T_0) - (T-T_0)\left(\frac{d\,u_0(T_0)}{d\,T_0} - f\right)\right|} \tag{10}$$

and for the velocity

$$u(T,T_0) = u_0(T_0) + f(T-T_0) \tag{11}$$

Example of a Distorted Flow

We now introduce as an example a sinussoidal velocity distortion with the amplitude ε and the frequency ω

$$u_0(T_0) = (1 + \varepsilon sin(\omega T_0)). \tag{12}$$

The solution of eq. (9) becomes

$$\frac{m(T,T_0)}{m(T_0)} = \frac{1 + \varepsilon \sin(\omega T_0)}{\left|1 + \varepsilon sin(\omega T_0) - (T-T_0)\,\varepsilon\,\omega\cos\omega T_0\right|} \tag{13}$$

where the location of the particles is defined by eq. (8):

$$x(T,T_0) = U(1 + \varepsilon sin(\omega T_0))\,(T-T_0) \tag{14}$$

Considerations and Applications

The main question with respect to a model concerning an interaction-free disperse medium is if there are any real systems which fulfill this assumption and if not whether one can learn something also for the case of a weakly interacting medium.

The first part of the question can be answered in such a way that in addition to a neutron flow or a molekular beam of neutral particles, some dust or droplet flows fulfill the assumption of no interaction, too, provided there is no dense carrier gas or collision of particles. As the number density *m* can be varying in time one can study the development of very complicated distorted particle flows.

With respect to the second part of the above formulated question it can be stated that also for weakly interacting structured media for which a local density and the velocity of the structures are defined in certain regions of a one dimensional stream or flow the future and local nonlinear development of the considered structure can be predicted. Especially the development of cusps and triple (or more) valued regimes of the solution shows an interesting aspect of travelling waves in dilute particle flows.

Literature

[1] G. Grabitz, G.E.A. Meier:
Über Laufzeitinstabilitäten in Flüssigkeitsstrahlen.
ZAMM 63, 1983, T 255 - T 257

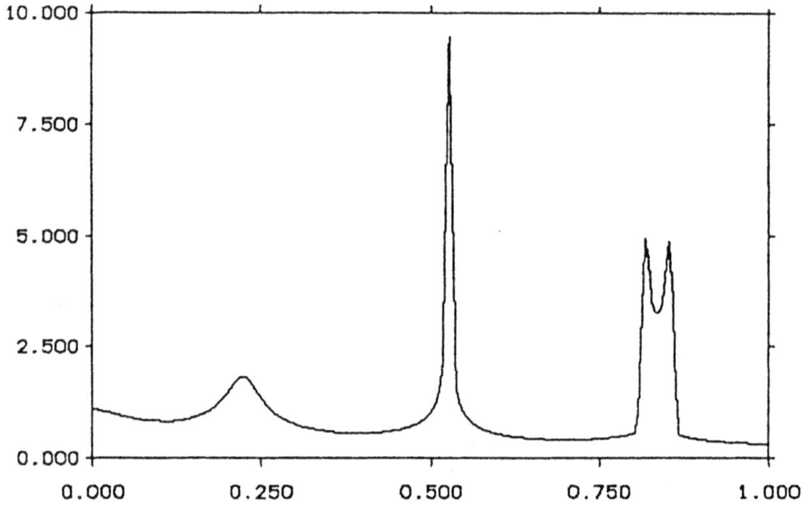

Figure 1. Typical density distribution over propagation distance of a sinussoidally distorted noninteracting particle stream.

Chaotic Behavior of Waves in Two-Phase System

Takuji Kawahara
Department of Physics, Faculty of Science
Kyoto University, Kyoto 606 Japan

Abstract

A problem concerning the possibility of chaotic wave behavior in two-phase systems is discussed. It is shown how a simple nonlinear evolution equation like the Benney equation is derived from a genaral system of equations governing the voidage waves in a variety of two-phase systems. Some properties of numerical solutions for the initial value problem of the Benney equation are exhibited briefly.

1 Introduction

Recent investigations on nonlinear evolution equations have clarified a variety of fascinating wave dynamics involving soliton and chaos phenomena. Roles of solitons as fundamental entities in the evolutions of purely dispersive nonlinear waves are now well-known and mathematically rigorous treatments are established for soliton equations such as the Korteweg-de Vries(KdV) equation and the nonlinear Schrödinger equation. Meanwhile, simultaneous inclusions of instability and dissipation mechanisms yield irregular spatio-temporal variations or chaotic behaviors in wave evolutions. Typical examples of such evolution equations are the Benney equation(KdV equation with Kuramoto-Sivashinsky terms) for unstable long waves or the complex Ginzburg-Landau equation governing nonlinear wave modulations of a quasi-monochromatic wave near a critically unstable point.

As to the long nonlinear waves in two-phase systems, for example, the propagation of pressure waves in a bubbly liquid was discussed by van Wijngaarden[1] on the basis of the KdV equation. The dissipative effect was also considered in terms of the KdV-Burgers equation to describe shock wave propagation [2][3][4]. On the other hand, in the investigations of voidage waves, the KdV-Burgers equation was also introduced [5][6]. Recently, the Benney equation, first derived by Benney[7] for unstable long

S. Morioka and L. van Wijngaarden (eds.),
IUTAM Symposium on Waves in Liquid/Gas and Liquid/Vapour Two-Phase Systems, 205-224.

waves on a falling liquid film, has been found to describe unstable voidage waves in two-phase systems such as in fluidized beds[8][9][10], in bubbly liquid[11][12], etc. The governing set of partial differential equations for these two-phase systems(the equations of continuity and the momentum conservation equations for each phase component) take mathematically quite similar forms except for detailed differences in stresses, interaction forces, or body forces. Thus it is expected that the Benney type equation becomes a ubiquitous approximate nonlinear evolution equation for unstable long voidage waves in two-phase systems.

In this paper, we focus our attention to the possibility of chaotic wave behavior in two-phase systems and discuss how such simple evolution equations of Benney type are derived. For simplicity, attention is paid to a general one-dimensional simplified model equations and a relation between the instability of voidage waves and the wave hierarchy interpretation will be discussed based on the simple model equations. It can be concluded that the Benney type equation governs unstable long voidage waves in a variety of two-phase systems when the effects of inertia and of suitable dissipation are taken into account.

Then, nonlinear wave phenomena including soliton and chaos are reviewed based on numerical solutions of the initial value problem for some typical simple nonlinear evolution equations like the Benney equation or the unstable KdV-Burgers equation paying attention to the roles of the spatially localized structures. References are given to show that solutions of the initial value problem can be described qualitatively by a superposition of spatially localized pulses and transition from regular(soliton) to irregular(chaos) state can be explained in terms of the soliton lattice model. Furthermore, another possible source of chaotic behavior of the fifth order KdV equation and an extension to "two-dimensional problem" are discussed briefly. Some conjectures available from the theoretical analysis of simple approximate nonlinear evolution equations will be presented for the nonlinear waves in the two-phase systems.

2 Basic equations for two-phase systems

We restrict the discussion to the two-fluid model and consider the equations for mass conservation

$$\partial_t \rho^{(i)} + \nabla \cdot (\rho^{(i)} \boldsymbol{v}^{(i)}) = 0, \tag{2.1}$$

and momentum conservation

$$\rho^{(i)}\{\partial_t \boldsymbol{v}^{(i)} + (\boldsymbol{v}^{(i)} \cdot \nabla)\boldsymbol{v}^{(i)}\} = \nabla \cdot \boldsymbol{\sigma}^{(i)} + \boldsymbol{I}^{(i)} + \rho^{(i)}\boldsymbol{F}^{(i)}, \tag{2.2}$$

where $\rho^{(i)}, \boldsymbol{v}^{(i)}$, $\boldsymbol{\sigma}^{(i)}, \boldsymbol{I}^{(i)}$ and $\boldsymbol{F}^{(i)}$ denote density, velocity, stress, interaction force and body force of the i-th phase(i=1,2), respectively. The interaction force $\boldsymbol{I}$ includes the mutual drag and the voidage gradient force and the body force $\boldsymbol{F}$ includes the buoyancy force. This set of equations governs voidage waves in a variety of two-phase systems, such as magma migrations or flows in porous media for solid-liquid system, fluidized beds for solid-gas system, and bubbly flow or liquid drop flow for liquid-gas system, except for detailed differences in the forces in the momemtum conservation equations.

2.1 Model equations for magma

In order to make clear the mathematical structures of existing approximation procedures, we take up the case of magma. Equations for magama dynamics proposed by McKenzie[13][14] are

$$\partial_t \phi + \nabla \cdot (\phi \boldsymbol{u}) = 0, \qquad \text{(melt)} \tag{2.3}$$

$$\partial_t (1-\phi) + \nabla \cdot \{(1-\phi)\boldsymbol{v}\} = 0, \qquad \text{(matrix)} \tag{2.4}$$

$$\rho_m\{\partial_t \boldsymbol{u} + (\boldsymbol{u}\cdot\nabla)\boldsymbol{u}\} = -\nabla P - \mu\phi K^{-1}(\boldsymbol{u}-\boldsymbol{v}) \qquad \text{(melt)}, \tag{2.5}$$

$$\begin{aligned}\rho_s\{\partial_t \boldsymbol{v} + (\boldsymbol{v}\cdot\nabla)\boldsymbol{v}\} = -\nabla P - g\rho_{dif}\cdot\boldsymbol{e}_k + (1-\phi)^{-1}\{\eta\nabla^2\boldsymbol{v} + (\xi+\eta/3)\nabla(\nabla\cdot\boldsymbol{v})\}\\ +\mu\phi^2(1-\phi)^{-1}K^{-1}(\boldsymbol{u}-\boldsymbol{v}) \qquad \text{(matrix)},\end{aligned} \tag{2.6}$$

where $\rho_{dif} \equiv \rho_s - \rho_m$, $K \equiv K_0\phi^n$ $(n \cong 3)$, ξ, η are viscosity coefficients, $\boldsymbol{e}_k$ denotes the direction of gravity, and ∇ is the gradient operator. Equations (2.3) and (2.4) give the divergence-free condition for the mean velocity

$$\nabla \cdot \{\phi\boldsymbol{u} + (1-\phi)\boldsymbol{v}\} = 0, \tag{2.7}$$

which means

$$\phi\boldsymbol{u} + (1-\phi)\boldsymbol{v} = \boldsymbol{U}(t). \tag{2.8}$$

One of the simplifications often introduced is to ignore the inertia of the light phase (melt). Then eq.(2.5) reduces to

$$\phi(\boldsymbol{u}-\boldsymbol{v}) + \mu^{-1}K\nabla P = 0, \tag{2.9}$$

which is nothing but Darcy's law. In terms of (2.8) and (2.9), the set of equations (2.3)–(2.6) is reduced to two coupled evolution equations for the concentration (ϕ) and the velocity ($\boldsymbol{v}$). Even when the inertia of the light phase is included, the set of equations can be reduced to two coupled evolution equations, since ∇P is eliminated by using (2.5),(2.6) and then $\boldsymbol{u}$ can be eliminated from the resulting equation by using (2.8).

Further simplification can be made by ignoring the inertia of heavy phase (matrix), i.e., the left hand side of (2.6). Then we obtain a single evolution equation for the dimensionless voidage ϕ:

$$\partial_t\phi - \phi^3\nabla^2\partial_t\phi - 3\phi^2\nabla\phi\cdot\nabla\partial_t\phi + 3\phi^2\partial_x\phi = 0. \tag{2.10}$$

which was treated in details by Barcilon & Lovera[15]. Weakly nonlinear approximation of eq.(2.10) leads to the regularized-long-wave (RLW) equation

$$(1-\nabla^2)\partial_t\tilde{\phi} + 3\partial_x\tilde{\phi} + 6\tilde{\phi}\partial_x\tilde{\phi} = 0, \tag{2.11}$$

or the Zakharov-Kuznetsov (ZK) equation

$$\partial_t\tilde{\phi} + 3\nabla^2\partial_x\tilde{\phi} + 3\partial_x\tilde{\phi} + 6\tilde{\phi}\partial_x\tilde{\phi} = 0, \tag{2.12}$$

where $\phi = 1 + \tilde{\phi}$. The linear part of (2.11) gives the linear dispersion relation

$$\omega = 3k_x/(1 + k_x^2 + k_y^2 + k_z^2). \tag{2.13}$$

Expansion for small wave number reduces (2.13) to $\omega = 3k_x(1 - k_x^2 - k_y^2 - k_z^2)$, which is the linear dispersion relation for the ZK equation (2.12).

2.2 Model equations for fluidized beds

The set of dimensionless equations in what follows is a model for fluidized beds proposed by Göz[16].

$$\partial_t\phi + \nabla\cdot(\phi\boldsymbol{u}) = 0, \qquad \text{(fluid)} \tag{2.14}$$

$$\partial_t(1-\phi) + \nabla\cdot\{(1-\phi)\boldsymbol{v}\} = 0, \qquad \text{(particle)} \tag{2.15}$$

$$F\varrho\phi\{\partial_t\boldsymbol{u} + (\boldsymbol{u}\cdot\nabla)\boldsymbol{u}\} = -\phi\nabla P - \varrho\phi\boldsymbol{e}_k - B(\phi)(\boldsymbol{u}-\boldsymbol{v})$$

$$+FR^{-1}(\nabla^2+\kappa'\nabla\nabla\cdot)\boldsymbol{u}, \qquad \text{(fluid)} \tag{2.16}$$

$$F(1-\phi)\{\partial_t\boldsymbol{v}+(\boldsymbol{v}\cdot\nabla)\boldsymbol{v}\}=-(1-\phi)\nabla P-(1-\phi)\boldsymbol{e}_k+B(\phi)(\boldsymbol{u}-\boldsymbol{v})$$
$$-G(\phi)\nabla\phi+FR^{-1}(\nabla^2+\kappa\nabla\nabla\cdot)\boldsymbol{v}, \qquad \text{(particle)} \tag{2.17}$$

where ϕ denotes the volume fraction of fluid(gas), F is the Froude number, R the Reynolds number, ϱ is the density ratio of fluid to particle (ρ_f/ρ_s), $B(\phi)$ the coefficient of drag force and $G(\phi)$ represents the effective pressure due to particle-particle interaction. Except for the viscous stress term in (2.16), the set of equations (2.14)–(2.17) are similar to that for magma dynamics (2.3)–(2.6), so that approximation procedures analogous to the case of magma are also valid.(We do not go into detailes here.)

2.3 One-dimensional model equations for liquid drops

General formulations for bubbly flow or liquid drops are also available in the three-dimensional case and reductions similar to the above-mentioned two models are possible. However, for simplicity, we shall focus our discussion on one-dimensional case from now on. The following set of equations is a model for flow of liquid drops in a gas[17].

$$\partial_t\phi+\partial_x(\phi u)=0, \qquad \text{(gas)} \tag{2.18}$$

$$\partial_t(1-\phi)+\partial_x\{(1-\phi)v\}=0, \qquad \text{(liquid)} \tag{2.19}$$

$$-\partial_x(\phi p)-\rho_s(1-\phi)R(u-v)=0, \qquad \text{(gas)} \tag{2.20}$$

$$\rho_s(1-\phi)(\partial_t v+v\partial_x v)=-\partial_x p-\rho_s g(1-\phi)+\rho_s(1-\phi)R(u-v)$$
$$+(4p/3\nu)\partial_x^2 v, \qquad \text{(liquid)} \tag{2.21}$$

$$p=(\kappa T_p/v_p)(1-\phi),$$

where the assumptions $4p/3\nu$ =const. and $\kappa T_p/v_p$ =const. have been introduced and the inertia of light phase(gas) has been omitted from the outset in (2.20). Then the one-dimensional reduced equations in suitably normalized forms are given by

$$\partial_t\phi+\partial_x(\phi u)=0, \qquad \text{(gas)} \tag{2.22}$$

$$\partial_t(1-\phi)+\partial_x\{(1-\phi)v\}=0, \qquad \text{(liquid)} \tag{2.23}$$

$$(1-\phi)(\partial_t v+v\partial_x v)=\mu\partial_x\phi+\nu\partial_x^2 v-g(1-\phi)$$
$$+R(1-\phi)(u-v). \qquad \text{(liquid)} \tag{2.24}$$

It is easy to derive one-dimensional equations from the sets of equations for magma and for fluidized beds by neglecting the inertia of the light phase. They are given as follows.
For magma:

$$\partial_t\phi+\partial_x(\phi u)=0, \qquad \text{(magma)} \tag{2.25}$$

$$\partial_t(1-\phi)+\partial_x\{(1-\phi)v\}=0, \qquad \text{(matrix)} \tag{2.26}$$

$$(1-\phi)(\partial_t v+v\partial_x v)=\nu\partial_x^2 v-g(1-\phi)$$
$$+C(\phi)(1-\phi)(u-v). \qquad \text{(matrix)} \tag{2.27}$$

For fluidized beds:

$$\partial_t\phi+\partial_x(\phi u)=0, \qquad \text{(fluid)} \tag{2.28}$$

$$\partial_t(1-\phi)+\partial_x\{(1-\phi)v\}=0, \qquad \text{(liquid)} \tag{2.29}$$

$$F(1-\phi)(\partial_t v+v\partial_x v)=-G(\phi)\partial_x\phi+FR^{-1}\kappa\partial_x^2 v-(1-\phi)$$
$$+B(\phi)\phi^{-1}(u-v). \qquad \text{(liquid)} \tag{2.30}$$

3 General one-dimensional model

The sets of equations given above for liquid drops((2.22)–(2.24)), magma((2.25)–(2.27)), and fluidized beds((2.28)–(2.30)) are quite similar, so that we consider hereafter the following set of equations.

$$\partial_t\phi+\partial_x(\phi u)=0, \qquad \text{(light phase)} \tag{3.1}$$

$$\partial_t(1-\phi)+\partial_x\{(1-\phi)v\}=0, \qquad \text{(heavy phase)} \tag{3.2}$$

$$(1-\phi)(\partial_t v + v\partial_x v) = -\mu\partial_x\{G(\phi)\} + \nu\partial_x^2 v - \lambda(1-\phi)$$
$$+\kappa B(\phi)(1-\phi)(u-v). \qquad \text{(heavy phase)} \tag{3.3}$$

3.1 Linearization

Now we consider the linearized problem for the general one-dimensional model. The steady state when the light phase has a constant velocity u_0 and the heavy one has v_0 is given by

$$\phi = \phi_0, \quad u = u_0 = U + \lambda(1-\phi_0)/\kappa B(\phi_0), \quad v = v_0 = U - \lambda\phi_0/\kappa B(\phi_0),$$
$$U = \phi_0 u_0 + (1-\phi_0)v_0, \tag{3.4}$$

where the relation $\phi u + (1-\phi)v = U$ has been utilized. Linearizing eqs.(3.1)-(3.3) around the steady state, we obtain

$$(\partial_t + v_0\partial_x)\phi - (1-\phi_0)\partial_x v = 0, \tag{3.5}$$

$$(1-\phi_0)(\partial_t + v_0\partial_x)v + \mu G'(\phi_0)\partial_x\phi - \nu\partial_x^2 v + \lambda(1-\phi_0)\phi_0^{-1}\{1-\phi_0 B'(\phi_0)/B(\phi_0)\}\phi$$
$$+\kappa(1-\phi_0)\phi_0^{-1}B(\phi_0)v = 0, \tag{3.6}$$

where the prime denotes a differentiation. Eliminating ϕ from the above equations, we get a linearized equation for v :

$$(\partial_t + v_0\partial_x)^2 v + \kappa\phi_0^{-1}B(\phi_0)(\partial_t + v_0\partial_x)v - \nu(1-\phi_0)^{-1}(\partial_t + v_0\partial_x)\partial_x^2 v$$
$$+\lambda(1-\phi_0)\phi_0^{-1}\{1-\phi_0 B'(\phi_0)/B(\phi_0)\}\partial_x v + \mu G'(\phi_0)\partial_x^2 v = 0. \tag{3.7}$$

For simplicity, we rewrite eq.(3.7) into

$$\zeta\partial_t^2 v + \beta\partial_t v - \nu\partial_t\partial_x^2 v + \alpha\partial_x v + \gamma\partial_x^2 v = 0. \tag{3.8}$$

Here the positive parameter ζ has been introduced to signify the importance of the inertia of the heavy phase and $\partial_t + v_0\partial_x$ has been rewritten into ∂_t. Note that this replacement means a mere shift of the wave velocity by v_0. It will be shown that the inclusion of the inertia term signified by ζ is essential for the derivation of the Benney type equation.

3.2 Linear dispersion relation

Assuming $v \propto \exp(ikx - i\omega t)$ with $\omega = \omega_r + i\omega_i$, we have the linear dispersion relation from eq.(3.8)

$$\zeta\omega^2 + i(\beta + \nu k^2)\omega - i\alpha k + \gamma k^2 = 0. \tag{3.9}$$

When $\zeta \neq 0$, the linear dispersion equation is quadratic in ω, and it is readily solved to give

$$\begin{aligned} \omega_i &= -[\beta + \nu k^2 \mp \{(\beta + \nu k^2)^2 + 4X\}^{1/2}]/2\zeta, \qquad (3.10)\\ X &= -\{A \mp (A^2 + 16B)^{1/2}\}/8, \\ A &= (\beta + \nu k^2)^2 - 4\zeta\gamma k^2, \\ B &= \zeta\gamma k^2(\beta + \nu k^2)^2 + \zeta^2\alpha^2 k^2, \\ \omega_r &= \alpha k\{2\zeta\omega_i + (\beta + \nu k^2)\}^{-1}. \qquad (3.11) \end{aligned}$$

The linear wave is unstable when $\omega_i > 0$ and stable when $\omega_i < 0$.

We now consider the limiting cases of the dispersion relation for small and large wave numbers. Asymptotic relations for two modes are given as follows:

(1)

$$\begin{aligned} &\omega_i \approx -(1/2)\beta\zeta^{-1}, && \omega_r \approx \alpha\nu^{-1}k^{-1} && \text{for small } k \\ &\omega_i \approx -(1/2)\nu\zeta^{-1}k^2, && \omega_r \approx \alpha\beta^{-1}k && \text{for large } k \end{aligned}$$

(2)

$$\begin{aligned} &\omega_i \approx (\gamma\beta^2 + \zeta\alpha^2)\beta^{-3}k^2, && \omega_r \approx \alpha\beta^{-1}k && \text{for small } k \\ &\omega_i \approx \gamma\nu^{-1}, && \omega_r \approx \alpha\nu^{-1}k^{-1} && \text{for large } k. \end{aligned}$$

The parameters β, ν are positive but α, γ can take either signs. The mode (1) is always stable(damping) because $\beta, \nu > 0$. The mode (2) is stable (damping) for large k if $\gamma = \mu G'(\phi_0) < 0$ but becomes unstable for small k if $\gamma\beta^2 + \zeta\alpha^2 > 0$. This unstable case becomes possible when the effect of inertia ($\zeta \neq 0$) is included.

For the case when the linear dispersion relation gives growth for small wave number but damping for large wave number, it is possible to derive an approximate nonlinear evolution equation of the Benney type. We consider the long wave approximation and expand the linear dispersion relation for the mode (2) into a series of small k. Then we have

$$\begin{aligned} \omega = {} & \alpha\beta^{-1}k + i(\gamma\beta^2 + \zeta\alpha^2)\beta^{-3}k^2 - \alpha\{2\zeta(\gamma\beta^2 + \zeta\alpha^2) + \nu\beta^3\}\beta^{-5}k^3 \\ & -i\{\zeta(\gamma\beta^2 + \zeta\alpha^2)(\gamma\beta^2 + 5\zeta\alpha^2) + (\gamma\beta^2 + 3\zeta\alpha^2)\nu\beta^3\}\beta^{-7}k^4 + \cdots \end{aligned}$$

$$\equiv c_1 k + i c_2 k^2 - c_3 k^3 - i c_4 k^4 + \cdots. \tag{3.12}$$

The following Benney equation can be derived by carrying out a suitable perturbation expansion to include the lowest order effect of the nonlinearity:

$$(\partial_T + c_1 \partial_X)\Phi + c_2 \partial_Z^2 \Phi + \epsilon c_3 \partial_Z^3 \Phi + \epsilon^2 c_4 \partial_Z^4 \Phi + \epsilon^2 c_0 \Phi \partial_Z \Phi = 0, \tag{3.13}$$

where

$$\phi = \epsilon^3 \Phi + \cdots, \quad T \equiv \epsilon^2 t, \quad X \equiv \epsilon^2 x, \quad Z \equiv \epsilon(x - c_1 t).$$

We consider the limiting case of $\zeta \to 0$. The linear dispersion relation (3.9) is reduced to

$$\omega = (\alpha k + i\gamma k^2)/(\beta + \nu k^2), \tag{3.14}$$

and the asymptotic behaviors are

$$\begin{array}{lll} \omega_i \approx \gamma\beta^{-1}k^2, & \omega_r \approx \alpha\beta^{-1}k & \text{for small } k \\ \omega_i \approx \gamma\nu^{-1}, & \omega_r \approx \alpha\nu^{-1}k^{-1} & \text{for large } k. \end{array}$$

Small k expansion of (3.14) gives

$$\omega = \alpha\beta^{-1}k + i\gamma\beta^{-1}k^2 - \alpha\nu\beta^{-2}k^3 - i\gamma\nu\beta^{-2}k^4. \tag{3.15}$$

It follows from (3.14) that the wave is unstable for all wave numbers when $\gamma > 0$ and stable when $\gamma < 0$. On the other hand, small k expansion (3.15) indicates that the term of k^2 represents damping when $\gamma < 0$ but the term of k^4 represents growth, and vice versa for $\gamma > 0$. Thus the asymptotic expansion (3.15) for small k contradicts the dispersion relation (3.14). The linear equation corresponding to the linear dispersion relation (3.14) for the case of $\zeta = 0$ is provided by the RLW type equation:

$$\beta \partial_t v - \nu \partial_t \partial_x^2 v + \alpha \partial_x v + \gamma \partial_x^2 v = 0. \tag{3.16}$$

It should be noted that the perturbation expansion in terms of small k might give a Benney type equation for the case of $\zeta = 0$ but it provides an erroneous equation which is inconsistent with the linear dispersion relation (3.14) of the original system. In such a case, the RLW type equation whose linear part is given by (3.16) should be an appropriate equation.

3.3 Wave-hierarchy interpretation

Now we touch upon the wave-hierarchy interpretation of the linearized equation (3.8). The lower-order part $\beta\partial_t v + \alpha\partial_x v = 0$ gives a characteristic wave speed $c_1 = \alpha\beta^{-1}$. The higher-order characteristic wave speeds are given as $c_\pm = \pm\sqrt{-\gamma/\zeta}$ from $\zeta\partial_t^2 v + \gamma\partial_x^2 v = 0$ for $\gamma < 0$. The stability criterion for the solutions to (3.8) is $c_- < c_1 < c_+$, that is, $\gamma\beta^2 + \zeta\alpha^2 < 0$. On the other hand, waves become unstable when $\gamma\beta^2 + \zeta\alpha^2 > 0$ and the higher-order part of the linear equation gives negative diffusion as pointed out by Whitham[18], Liu[19], and others[20].

This criterion, $\gamma\beta^2 + \zeta\alpha^2 > 0$, coincides exactly with the growing case of the coefficient of the k^2 term in the expanded linear dispersion relation (3.12). Consequently, the case when the higher-order wave gives instability(or negative diffusion term) corresponds to the case when the Benney type equation should be introduced. Therefore, as noted already, inclusions of the inertia and the dissipation are necessary in order to have asymptotically stable wave behaviors.

At the conclusion of this section, we may say that the inclusions of inertia term together with the viscous effects are necessary for deriving the Benney type equation. An important point is the existence of growth for small wave numbers and damping for large wave numbers. In such a case, asymptotically valid nonlinear evolution equation is available because long wavelength instability can be stabilized by the diffusion effects for high wave numbers, which assures an energetic balance in wave evolutions. Although the discussion here is restricted to the one-space dimensional case, the fundamental concept can be equally applied to higher dimensional cases.

4 Solutions of the Benney equation

In this section we briefly review numerical solutions of the initial value problem of the Benney equation and related equations. Outlines of numerical solutions are exhibited both for the one-space dimensional case and two-space dimensinal case. Also touched upon briefly are other types of evolution equations showing chaotic wave behavior.

4.1 One-dimensional Benney equation

The Benney equation is given by

$$\partial_t u + u\partial_x u + \delta\partial_x^3 u + \partial_x^2 u + \partial_x^4 u = 0, \tag{4.1}$$

where the parameter δ represents the relative importance of the dispersion with respect to the instability and the dissipation terms. The linear dispersion relation for this case is

$$\omega = \omega_r + i\omega_i = -\delta k^3 + i(k^2 - k^4), \tag{4.2}$$

so that the wave is linearly unstable for $k < 1$ and stable otherwise. The maximum growth rate is given at the finite wave number $k = 1/\sqrt{2}$. Initial value problems under periodic boundary condition are solved by pseudo-spectral method for spatial integration and the usual Runge-Kutta-Gill method for temporal integration. Examples of numerical solutions for different values of δ are shown in Figs.1–2. Figure 1 shows the evolutions of wave forms for $\delta = 0, \delta = 2.15$ and $\delta = 8.8$. Note that the case $\delta = 0$ corresponds to the Kuramoto-Sivashinsky chaos. For strongly dispersive cases($\delta = 2.15, \delta = 8.8$), the asymptotic states become a law of soliton-like pulses with equal amplitude. Figure 2 shows the results around the value of δ where a transition from regular to irregular behaviors takes place. Trajectories of pulse peaks, an instantaneous wave form, inter-pulse distances(s), pulse amplitudes(u), and the shapes of steady progressive pulse solutions are exhibited. Note that the pulse-like solutions are discernible even when the wave forms appear to be quite chaotic.

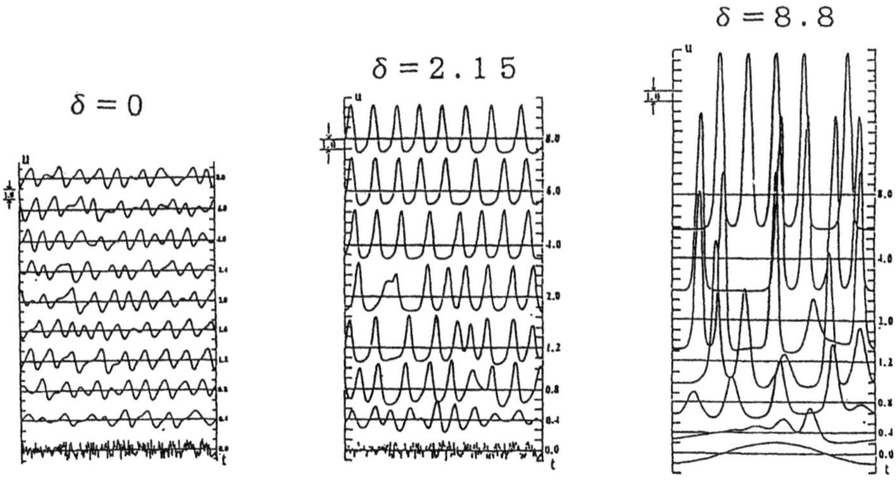

Fig.1: Numerical solutions of the initial value problem for eq.(4.1). (see Ref.[21])

The results are summarized as follows:(For details see [21]–[27])

(1) The ultimate state for strongly dispersive case(large δ) is a row of soliton-like pulses with the same amplitude.

(2) For weakly dispersive cases(small δ), wave evolutions are chaotic and tend to the Kuramoto-Sivashinsky chaos in the limit of δ=0.

(3) Transitions from regular to irregular behavior take place around $\delta \approx 0.2$ [24] .

(4) The steady pulse solutions play essential roles in the evolutions of initial value problems even in chaotic cases [27].

(5) Qualitative features of wave evolutions can be described well by a superposition of the steady pulse like solutions. The concept of soliton lattice has been introduced based on this fact [25] [26].

It should be noted that in case of the Korteweg-de Vries equation($\delta = \infty$), the recurrence of the initial state occurs after interactions of generated solitons with different amplitudes as is well-known from the famous numerical calculations due to Zabusky and Kruskal. However, the ultimate state is a row of similar amplitude pulses when δ is finite. The scale of pulses or the inter-pulse distances are, roughly speaking, determined by the scale corresponding to the wave number of the maximum growth rate.

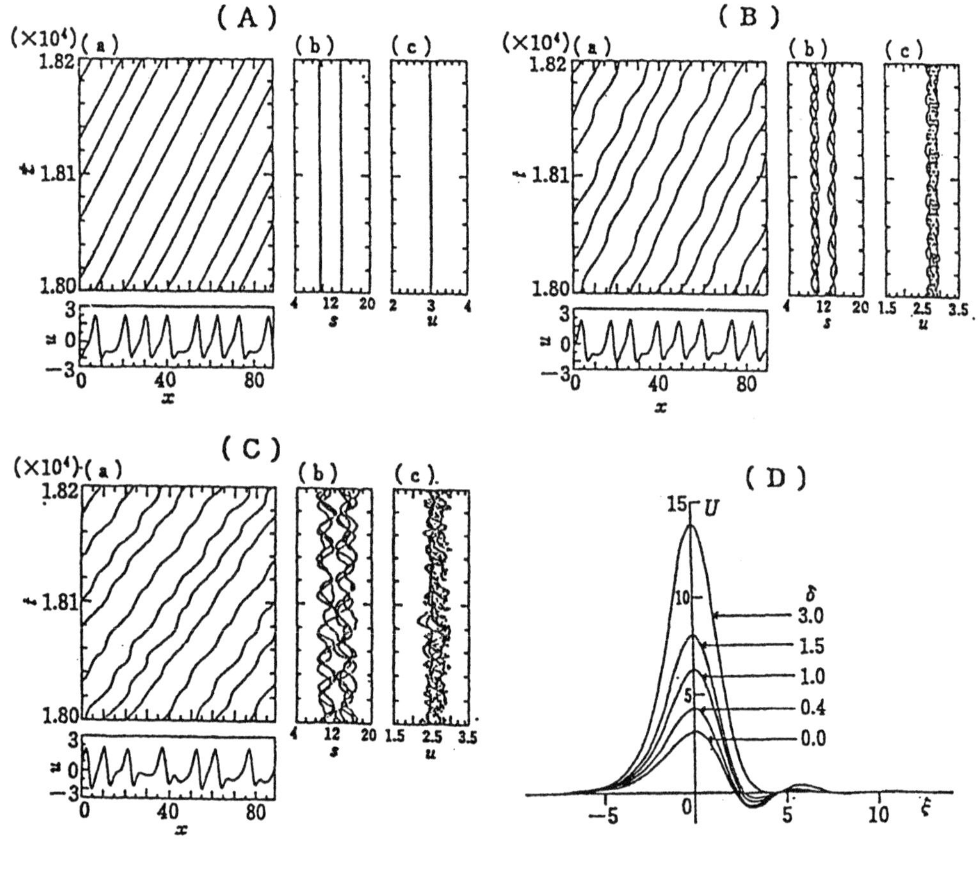

Fig.2: Regular to irregular wave evolutions. (A) regular behavior($\delta = 0.4$). (B) periodic behavior($\delta = 0.3$). (C) irregular behavior($\delta = 0.15$). (D) steady pulse solutions. (a) peak trajectories, (b) inter-pulse distance, (c) peak amplitude, and waveform at t=18,000. (see Ref.[24])

4.2 Two-dimensional Benney equation

Let us consider an extension to the two-dimensional Benney equation which governs the long waves on the liquid film flowing down an inclined plane. We consider

$$\partial_t u + u\partial_x u + \delta\Delta\partial_x u + \partial_x^2 u + \Delta^2 u = 0, \tag{4.3}$$

where $\Delta \equiv \partial_x^2 + \partial_y^2$ is the two-dimensional Laplacian. The linear dispersion relation for (4.3) is

$$\omega = \omega_r + i\omega_i = -\delta k_x(k_x^2 + k_y^2) + i\{k_x^2 - (k_x^2 + k_y^2)^2\}. \tag{4.4}$$

In the strongly dispersive limit($\delta \to \infty$), eq.(4.3) reduces after renormalization to the Zakharov-Kuznetsov equation

$$\partial_t u + u\partial_x u + \partial_x^3 u + \partial_x\partial_y^2 u = 0. \tag{4.5}$$

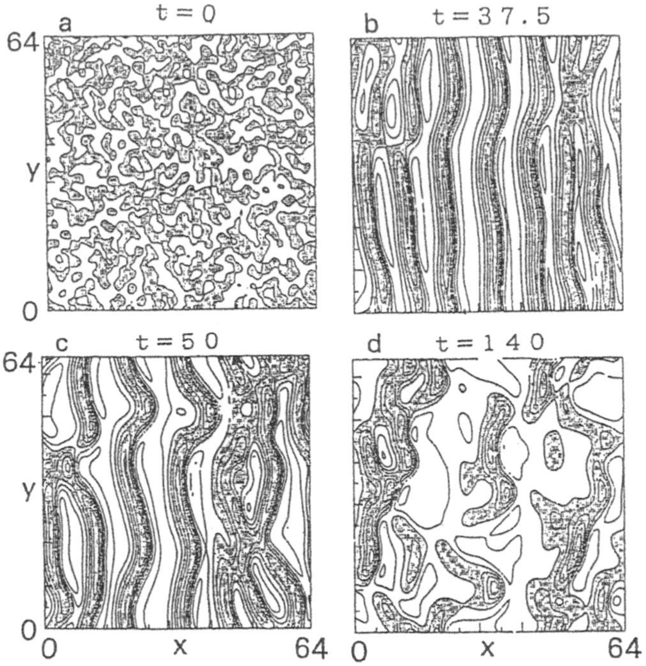

Fig.3: Evolution of solution to eq.(4.3) for $\delta = 0.8$ from random initial condition. Contour plots show irregular behavior but horse-shoe patterns can be observed in d at $t = 140$. (see Ref.[28])

Numerical solutions of eq.(4.3) show behaviors qualitatively similar to the one-dimensional case as summarized as follows:

(1) Wave evolutions are regular for strongly dispersive case and irregular for weakly dispersive case.

(2) Localised pulse structures are generated and play essential roles in the wave evolutions.

(3) Horse-shoe like patterns are observed in the irregular wave patterns for the case of weak dispersion(small δ).

(4) A quasi-stationary lattice consisting of a row of cylindrical bell-shaped pulses is formed ultimately for the case of large δ. (For details see [28][29]).

Figures 3 and 4 show examples of numerical results for the two-dimensional Benney equation. Figure 3 shows an example for relatively weak dispersion when the wave behavior is irregular. A pattern similar to a horse-shoe can be found in irregular wave patterns. Figure 4 shows the case with strong dispersion when the localized bell-shaped pulses are generated and asymptotically form a quasi-stationary lattice like structure.

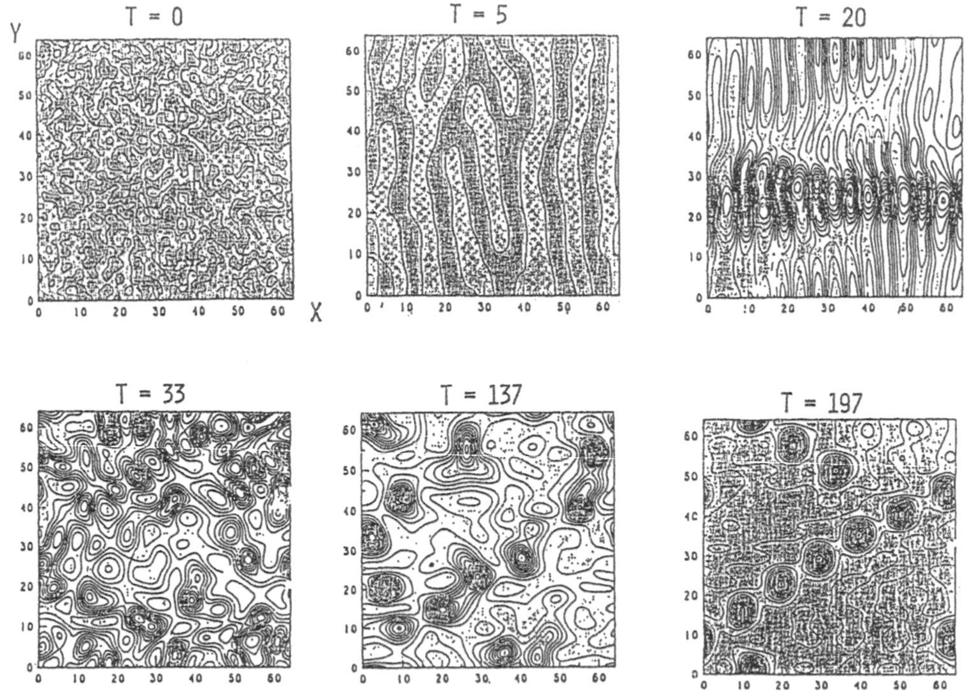

Fig.4: Formation of quasi-stationary lattice ($\delta = 25$). Generation of 1-D structure ($T = 5$), 2-D instability ($T = 20$), generation of 2-D pulses with similar amplitude ($T = 137$) and lattice configuration($T = 197$). (see Ref.[28])

4.3 Other evolution equations related to chaos

The following equations are typical examples of simple nonlinear evolution equations involving chaotic behaviors:

(A) Unstable KdV-Burgers equation:

$$\partial_t u + u\partial_x u + \partial_x^3 u - \eta(u + \partial_x^2 u) = 0. \tag{4.6}$$

(B) Complex Ginzburg-Landau equation:

$$i\partial_t a + (p_r + ip_i)\partial_x^2 a + (q_r + iq_i)|a|^2 a - i\gamma a = 0. \tag{4.7}$$

(C) 5th order KdV equation:

$$\partial_t u + u\partial_x u + \partial_x^3 u + \mu\partial_x^5 u = 0. \tag{4.8}$$

The growth and the damping terms of the unstable KdV-Burgers equation take different forms, that is, uniform excitation and usual diffusion. The maximum growth rate is given in this case at k=0, so that the longest wave component becomes most unstable. This difference brings about qualitatively quite different features for the overall wave evolutions. The overall evolutions become irregular for strongly dispersive case and regular for weakly dispersive case in contrast with those for the Benney equation[30].

(1) For strongly dispersive case, groupings of soliton like pulses occur in the largest scale of numerical calculation(i.e., the scale corresponding to the lowest wave number admitted in numerical calculations.). The pulses interact irregularly and the energy spectra exhibit irregular variations in time.

(2) For weakly dispersive case, steady pulse like solutions arise at initial stage but the final state is a shock with dispersive effect at the peak of shock wave. Thus the steady pulse solutions do not play fundamental roles in this case.

Figure 5 shows the wave evolution for strongly dispersive case. Initial condition is a sinusoidal wave and some pulse-like structures are formed at the early stage. They are modulationally unstable and interact irregularly forming groups in the largest available scale. Figure 6 shows weakly dispersive case when the asymptotic state is a shock with dispersive effect at the shock front. Although pulse-like structures arise also in this case at the early stage, they are unstable and eventually shrink into oscillatory structures at the shock front. Thus the overall wave evolutions are irregular for strongly dispersive case and regular for weakly dispersive case for the unstable KdV-Burgers equation (4.6) in contrast to the Benney equation. Note that this difference comes from the differences in the form of linear growth rate.

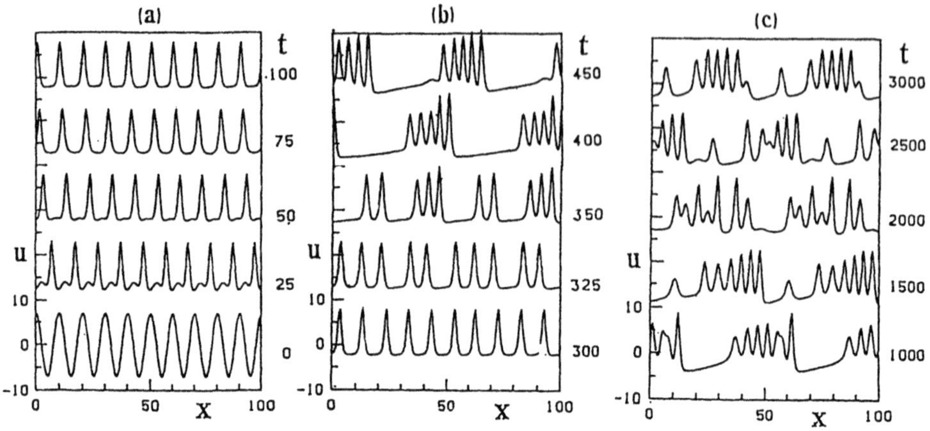

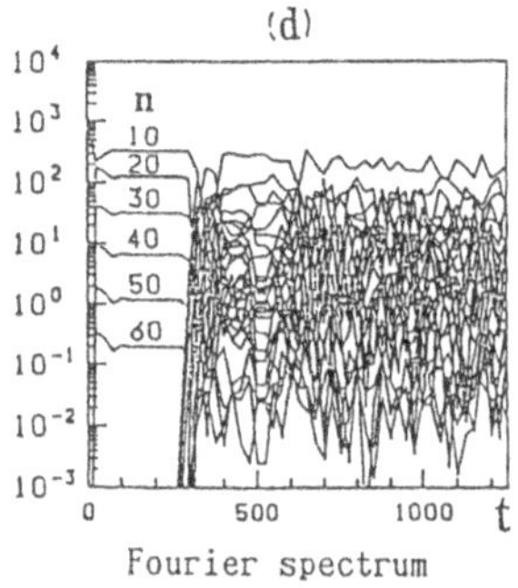

Fig.5: Irregular behavior for strongly dispersive case($\eta = 0.1$). Initial condition: $u(x,0) = 7\cos 0.2\pi x$. (a) arrangement of steady pulses, (b) modulational instability, (c) irregular grouping of pulses, (d) Fourier spectrum showing irregularity.

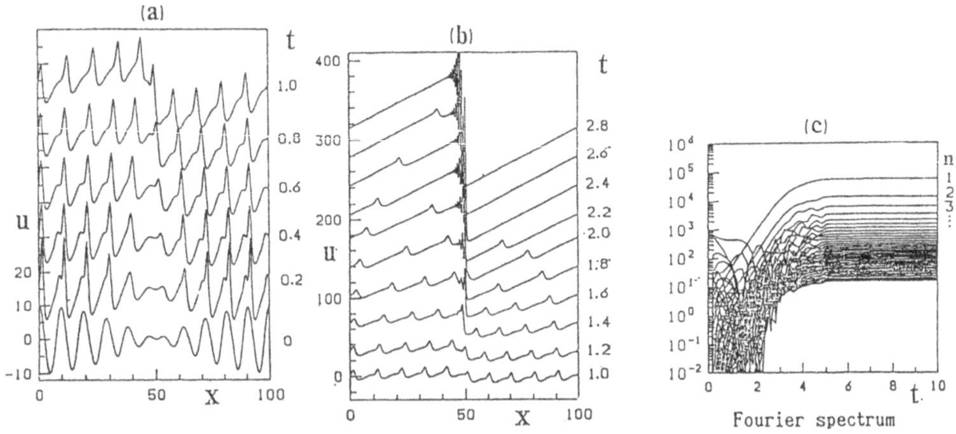

Fig.6: Shock formation for weakly dispersive case($\eta = 1.5$). Initial condition: $u(x,0) = 5(\cos 0.2\pi x + \cos 0.22\pi x)$. (a) formation of steady pulse, (b) development of largest scale shock, (c) Fourier spectrum indicating regular shock.

The complex Ginzburg-Landau equation (4.7) is one of the most well-known examples of nonlinear evolution equations involving chaotic behaviors. This equation describes the modulation of the complex amplitude a of a quasi-monochromatic wave with finite wave number near the critically unstable point. Because of its ubiquity a number of investigations have been done concerning this equation (see for example [31]–[34]). Transitions to chaos, spatio-temporal evolutions of chaos, etc. are explored extensively. This type of equation will arize if the instability takes place at a finite wave number for two phase systems.

The 5th order KdV equation is a purely dispersive equation, but it also exhibits irregular behaviors ([35][36]). The possibility of this equation for two-phase system is anticipated for the simplest model in which the inertia of heavy phase is ignored but some dissipative effects are included. We can derive the RLW type equation, if the Darcy's law with dissipative effects is used in the equation of continuity. In this approximation, the RLW equation involves a higher order dispersion originated from the 'dissipation term'.

5 Concluding remarks

It has been shown that the voidage waves in a variety of two-phase systems can be descrived by Benney's equation. On the basis of the two-fluid model, it is shown by using a general one-dimensional model that the inclusions of the inertia and the dissipative effect lead to Benney's equation. In the existing investigations, voidage waves were thought to be unstable when the stability criterion due to the wave-hierarchy interpretation is violated. However, even for that case, asymptotically stable wave evolutions are possible if we take into account suitable higher-order dissipation effects. Chaotic wave behavior described by the Benney type equations is thus expected for voidage waves in a variety of two-phase systems.

In terms of the existing results for numerical solutions of simple nonlinear evolution equations, we point out here some remarks that should be taken into account in the investigations of nonlinear waves in actual physical systems. Solutions of the nonlinear evolution equations sensitively depend on the form of included terms and even a slight difference in linear terms may cause drastic differences in overall wave evolutions. Therefore, deep care should be taken when we are concerned with the correspondence between actual physical phenomena and those described by simple approximate equations. One-dimensional solutions often become unstable when extended to higher space dimensional case, so that extensions of asymptotic approximations to higher space dimensional cases are highly required to deal with actual physical phenomena much more correctly.

Important points to be investigated further are as follows:

(1) Whether such wave behaviors described by the Benney equation as soliton lattice, irregular pulse low, chaos can be observed in actual two-phase systems? Pulse

amplitudes will tend to a constant value, if the actual wave evolutions are governed by the Benney type equation.
(2) Extensions to higher space dimension are required to access further to actual physical phenomena, because one-dimensionally stable structures often become unstable in higher dimensional cases.
(3) Is there the possibility of other types of chaotic behavior described by the complex Ginzburg-Landau equation or the unstable KdV-Burgers equation, etc.?
(4) Although the one-dimensional model system dealt with here includes basic factors essential to the two-phase systems in a general form, it is an extremely simplistic mathematical model. Elaborate modellings and detailed investigations of them are required to discuss individual physical phenomena much more definitely.

References

[1] L. van Wijngaarden: *On the equations of motion for mixtures of liquid and gas bubbles*, J.Fluid Mech. 33 (1968) 465–474.

[2] L. van Wijngaarden: *One-dimensional flow of liquids containing small gas bubbles*, Ann. Rev. Fluid Mech. 4 (1972) 369–396.

[3] V.V.Kuznetsov, V.E.Nakoryakov, B.G.Pokusaev & I.R.Shreiber: *Propagation of perturbations in a gas-liquid mixture*, J. Fluid Mech. 85 (1978) 85–96.

[4] V.E.Nakoryakov, B.G.Pokusaev & I.R.Shreiber: *Wave Propagation in Gas-Liquid Media*, CRC Press (1993).

[5] S.E.Harris & D.G.Crighton: *Solitons, solitary waves, and voidage disturbances in gas-fluidized beds*, J.Fluid Mech. 266 (1994) 243 –276.

[6] A.E.Beylich & A.Gülhan: *On the structure of nonlinear waves in liquids with gas bubbles*, Phys.Fluids A2 (1990) 1412–1428.

[7] D.J.Benney: *Long waves on liquid films*, J. Math. & Phys. 45 (1966) 150–155.

[8] S.Sasa & H.Hayakawa: *Void-fraction dynamics in fluidization* , Europhys. Lett. 17 (1992) 685–689.

[9] T.S.Komatsu & H.Hayakawa: *Nonlinear waves in fluidized beds*, Phys.Lett.A183 (1993) 56–62.

[10] H.Hayakawa, T.S.Komatsu & T.Tsuzuki: *Pseudo-solitons in fluidized beds*, Physica A204 (1994) 277–289.

[11] A.Biesheuvel: *Structure and stability of void waves in bubbly flows*, Abstracts of IUTAM Symposium on "Waves in liquid/gas and liquid/vapor two-phase system".

[12] S.Morioka: *Behavior of void wave on the basis of turbulent liquid-bubbles two-phase flow model* , Abstracts of IUTAM Symposium on "Waves in liquid/gas and liquid/vapor two-phase system".

[13] D.P.McKenzie: *The generation and compaction of partially molten rock*, J.Petrol. 25 (1984) 713–765.

[14] V.Barcilon & F.M.Richter: *Nonlinear waves in compacting media*, J. Fluid Mech. 164 (1986) 429–448.

[15] V.Barcilon & O.M.Lovera: *Solitary waves in magma dynamics*, J.Fluid Mech. 204 (1989) 121–133.

[16] M.F.Göz: *Bifurcation of plane voidage waves in fluidized beds*, Physica D65 (1993) 319–351.

[17] M.J.Miksis & Lu Ting: *Effective equations for multi phase flow—Waves in a bubbly liquid*, Adv. in Appl. Mech. 28 (1992) 141–260.

[18] G.B.Whitham: *Linear and Nonlinear Waves*, Wiley (1974) Chapter 10.

[19] J.T.C.Liu: *Note on a wave-hierarchy interpretation of fluidized bed instabilities*, Proc.Roy.Soc.Lond. A380 (1982) 229–239. *Nonlinear unstable wave disturbances in fluidized beds*, Proc.Roy.Soc.Lond. A389 (1983) 331-347.

[20] A.Prosperetti & J.V.Satrape: *Stability of two-phase flow models*, D.D.Joseph & D.G.Schaeffer Eds., *Two Phase Flows and Waves*, Springer (1986) 98–117.

[21] T.Kawahara: *Formation of saturated solitons in a nonlinear dispersive system with instability and dissipation*, Phys.Rev.Lett. 51 (1983) 381–383.

[22] T.Kawahara & S.Toh: *Nonlinear dispersive periodic waves in the presence of instability and damping*, Phys. Fluids 28 (1985) 1636–1638.

[23] S.Toh & T.Kawahara: *On the stability of soliton-like pulses in a nonlinear dispersive system with instability and dissipation*, J.Phys.Soc.Jpn. 54 (1985) 1257–1269.

[24] T.Kawahara & S.Toh: *Pulse interactions in an unstable dissipative-dispersive nonlinear system*, Phys. Fluids 31 (1988) 2103–2111.

[25] T.Kawahara & M.Takaoka: *Chaotic motions in an oscillatory soliton lattice*, J.Phys.Soc.Jpn. 57 (1988) 3714–3732.

[26] T.Kawahara & M.Takaoka: *Chaotic behaviour of soliton lattice in an unstable dissipative dispersive nonlinear system*, Physica D 39 (1989) 43–58.

[27] S.Toh: *Statistical model with localized structures describing the spatio-temporal chaos of Kuramoto-Sivashinsky equation* , J.Phys.Soc.Jpn. 56 (1987) 949–962.

[28] S.Toh, H.Iwasaki & T.Kawahara: *Two-dimensionally localized pulses of a nonlinear equation with dissipation and dispersion*, Phys.Rev. A 40 (1989) 5472–5475.

[29] H.Iwasaki, S.Toh & T.Kawahara: *Cylindrical quasi-soliton of the Zakharov-Kuznetsov equation*, Physica D 43 (1990) 293–303.

[30] T.Kawahara & S.Toh: *On some properties of solutions to a nonlinear evolution equation including long wavelength instability*, A.Jeffrey Ed. *Contributions to Nonlinear Wave Motion*, Pitman Monographs and Surveys in Pure and Applied Mathematics, (1988) 95–117.

[31] M.C.Cross & P.C.Hohenberg: *Pattern formation outside of equilibrium*, Rev.Mod.Phys. 65 (1993).

[32] H.T.Moon, P.Huerre & L.G.Redekopp: *Transitions to chaos in the Ginzburg-Landau equation*, Physica D7 (1983) 135–150.

[33] K.Nozaki & N.Bekki: *Pattern selection and spatio-temporal transition to chaos in the Ginzburg-Landau equation*, Phys.Rev.Lett. 51 (1983) 2171–2174.

[34] S.Kishiba, S.Toh & T.Kawahara: *An estimate of energy spectra of the Ginzburg-Landau equation*, Physica D 54 (1991) 43–57.

[35] K.A.Gorshkov & V.V.Papko: *Dynamic and stochastic oscillations of soliton lattices*, Sov.Phys.JETP 46 (1977) 92–97.

[36] H.Nagashima: *Chaos in a nonlinear wave equation with higher order dispersion*, Phys.Lett. A105 (1984) 439–442.

EVOLUTION OF TIME PERIODIC PERTURBATIONS ON FILMS FALLING DOWN WALLS

L. A. Dávalos-Orozco[1], S. H. Davis[2], S. G. Bankoff[3]
[1]Inst. de Investigaciones en Materiales, Universidad Nacional Aut. de México, ApdoPostal 70-360, Coyoacan, 04510 México, DF.
[2]Dept. of Engineering Science and Applied Mathematics, Northwestern University, Evanston, IL 60208, USA.
[3]Dept. of Chemical Engineering, Northwestern University, Evanston, IL 60208, USA.

INTRODUCTION

The research on the stability of films falling down walls has recently been the attraction of many scientists due to its practical importance and because of the many challenging problems it still presents not only from the theoretical but also from the experimental point of view. A model non-linear evolution equation which describes the film stability is the Benney equation (BEQ) Benney (1966). The BEQ has been investigated theoretically by Gjevik (1970), Nakaya (1975), Krishna and Lin (1977), Pumir et al. (1983), Chang (1989), Nakaya (1989), Joo et al. (1991a,b), Joo and Davis (1992a,b) and Liu and Gollub (1994).
In all these papers the temporal growth of the perturbations is examined in the reference frame of the laboratory. A different point of view was presented by Deissler et al. (1991) for waves of the Kuramoto-Sivashinsky equation (KSEQ). Making a distinction between convective and absolute instability they solved numerically the KSEQ. The problem of convective and absolute instability for falling films from the point of view of the BEQ has been discussed in the paper by Joo and Davis (1992a) for falling films.
From the experimental point of view, it would be desirable to have theoretical support for the description of the phenomena as seen in the laboratory. For films falling down walls, it is necessary to describe the flow instability in the laboratory frame of reference subjected to spatial growth. Here, for the first time we describe the space-time evolution of non-linear waves on falling films as described by a periodically forced inhomogeneous BEQ. We suppose that the film surface is forced locally by a time periodic tangential shear stress (defined below Eq.9) with Gaussian (bell shaped) spatial distribution. The frequency of oscillation is imposed at $t = 0$ and a train of waves propagates behind the wave front along the film. Very far upstream from $x = 0$ and very far downstream from a distance x = phase velocity times time, we impose as boundary conditions that the surface height is uniform and that its x-derivative is zero. By "very far" we mean 50 non-dimensional spatial units or 100, depending on the proper wavelength

S. Morioka and L. van Wijngaarden (eds.),
IUTAM Symposium on Waves in Liquid/Gas and Liquid/Vapour Two-Phase Systems, 225-233.

which may arise. We present numerical results for the spatial evolution of the waves for a variety of frequencies and different values of the Reynolds number in the region between the lines of criticality and subcriticality and for values below subcriticality. For large enough Reynolds numbers, our results show the possibility of spatial chaos when use is made of the power spectrum against the wavenumber.

EQUATIONS OF MOTION AND THE FORCED BENNEY EQUATION

The fluid layer has density ρ viscosity μ and thickness d_0 and is falling down a vertical wall. The x-axis, parallel to gravity, is directed down the wall and the z-axis is perpendicular to the wall. Let $u_0 = \mu/\rho d_0$. Then, the variables are made non-dimensional as follows: $(u,w) = (u',w')/u_0$, $t = (u_0/d_0)t'$, $(x,z,h) = (x',z',h')/d_0$, $(p,F) = (p',F')/\rho u_0^2$. Where the primed are dimensional variables. Here, u and w are the x and z components of the velocity, x, z and h are the x and z coordinates and h is the height of the wave. p and F are the pressure and the time dependent stress in the x direction. $\varepsilon = d_0/l$ is small and l is proportional to the disturbance wavelength. The equations of motion and continuity for the x and z components of velocity including a small time dependent forcing term in the x direction are in non-dimensional form:

$$\varepsilon u_\tau + \varepsilon u u_\xi + w u_\zeta = -\varepsilon p_\xi + \varepsilon^2 u_{\xi\xi} + u_{\zeta\zeta} + G + \varepsilon F(\xi,\tau) \qquad (1)$$

$$\varepsilon w_\tau + \varepsilon u w_\xi + w w_\zeta = - p_\zeta + \varepsilon^2 w_{\xi\xi} + w_{\zeta\zeta} \qquad (2)$$

$$\varepsilon u_\xi + w_\zeta = 0. \qquad (3)$$

The kinematic, normal stress, tangential stress and no slip boundary conditions are, respectively:

$$\varepsilon h_\tau + \varepsilon u h_\xi - w = 0 \qquad \zeta = h \qquad (4)$$

$$p - 2[\varepsilon u_\xi(\varepsilon^2 h_\xi^2 - 1) - \varepsilon h_\xi(u_\zeta + \varepsilon w_\xi)]/N^2 = - 3\,\bar{S}h_{\xi\xi}/N^3 \qquad \zeta = h \qquad (5)$$

$$(u_\zeta + \varepsilon w_\xi)(1 - \varepsilon^2 h_\xi^2) - 4\varepsilon^2 u_\xi h_\xi = 0 \qquad \zeta = h \qquad (6)$$

$$u = w = 0 \qquad \zeta = 0 \qquad (7)$$

where $N = (1 + \varepsilon^2 h_\xi^2)^{1/2}$, $G = d_0^3 g/\nu^2$ is the Reynolds numbers , $\bar{S} = \varepsilon^2 S$, $S = \sigma_0 d_0/3\rho\nu^2$ is the capillary number. Note that a lubrication type approximation have been used in the space and time as follows: $\xi = \varepsilon\, x$, $\zeta = z$ and $\tau = \varepsilon\, t$. Now, the following expansion

$$u = u_0 + \varepsilon u_1 + \ldots\ldots$$
$$w = \varepsilon\,(w_0 + \varepsilon w_1 + \ldots\ldots) \tag{8}$$
$$p = p_0 + \varepsilon p_1 + \ldots\ldots$$

is necessary to obtain, from these equations, the inhomogeneous Benney equation for the wave amplitude h:

$$h_\tau + G\,h^2h_\xi + \varepsilon\,[\tfrac{2}{15}\,G^2h^6h_\xi + \bar{S}h^3h_{\xi\xi\xi}]_\xi = -\,\tfrac{\varepsilon}{3}\,[F(\xi,\tau)h^3]_\xi \tag{9}$$

where $F(\xi,\tau) = A\sin(\omega\tau)\exp(-a\xi^2)$ with A, $a > 0$.
The homogeneous BEQ was investigated theoretically by Gjevik (1970). He studied the growth of a linear unstable perturbation and its interaction with higher harmonics. He found steady finite-amplitude surface waves. By means of a Fourier series expansion he gave a set of coupled non-linear equations for the first three time dependent amplitude terms. From this set of approximate equations, a condition for the existence of stationary solutions was obtained. He gave a line of neutral stability or criticality and a line of subcriticality. Those lines are shown in Fig.1 as k_c and k_s in a plot of k, the wavenumber, against G for the case of a vertical wall. Between those lines stationary waves were found. However, for k's smaller than k_s no equilibration was found. Joo et al. (1991a) investigated these two areas of Fig.1 by means of a homogeneous BEQ for temporal growth.
In this paper, due to limitation of space, a numerical solution of the inhomogeneous BEQ equation is given by means of two examples corresponding to two different frequencies of oscillation of the tangential shear. We follow the curves drawn in Fig.1 to give different values of G. This will be explained presently.

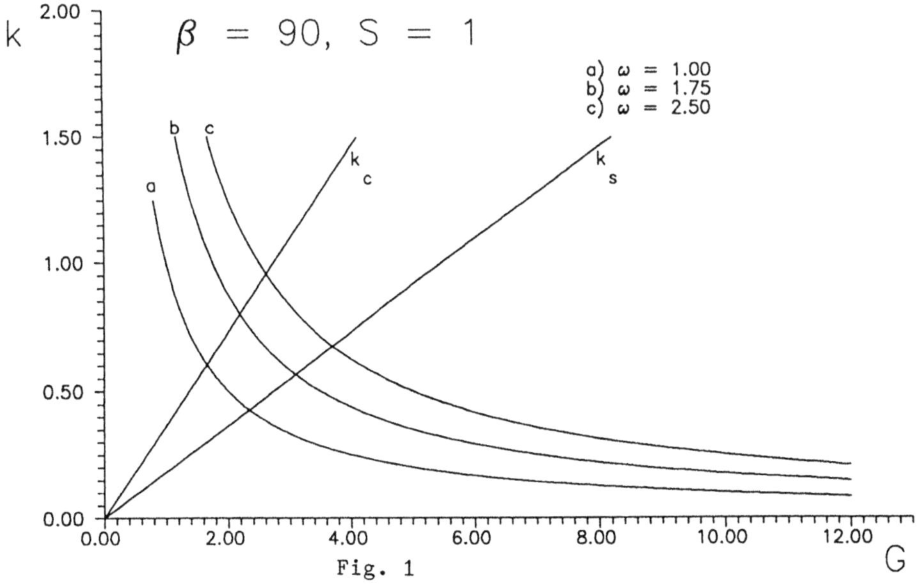

Fig. 1

NUMERICAL SOLUTION AND RESULTS

Equation (9) was solved numerically by means of an explicit method using second-order central differences. The product $G^2\Delta\tau/\Delta\xi^2$ was always made ≤ 0.1 . Following the curves $k = \omega/G$, the values of the Reynolds number were selected to investigate the spatial and temporal development of the wave trains in the equilibration region, between k_c and k_s, and the time dependent (growth) region of subcriticality, below k_s.

Fig.2 shows the case for $\omega = 1$. The values of the parameters are shown in the figure. Note that all graphs are presented in three parts. The first shows a panorama of the wave for all times considered. The second, in the middle, shows an amplification of an interesting region. The third, is the power spectrum of the first graphs. Fig.2a corresponds to $G = 1.7$, very near below to criticality, and shows a small decrease because the perturbation amplitude is larger than the natural stationary amplitude of the wave which needs a distance to present equilibration. Fig.2b, for $G_m = 1.968$ corresponding to the maximum growth rate of the linear theory, shows modulation of the wave. For $G = 2.1$, Fig2c shows, apart of modulation, a harmonic wave. Near, but below the condition for subcriticality, for $G = 2.7$, Fig.2d shows a longwave modulation, larger amplitude harmonics with growth in the wave front. For $G = 3.2$ in Fig.2e, the growth is more important and only shorter time ($\tau = 150$) calculations are possible but a new harmonic appears. Fig.2f for $G = 3.5$ shows the appearance of more harmonics when the growth is very fast. Here, a broad-band power spectrum exists, very similar that of a typical chaotic system.

It was found that even in the area between k_C and k_s growth could be found from frequencies somewhere between $\omega = 1.5$ and about 1.75. See Fig.1 for the situation of the curve $\omega = 1.75$. An example for $\omega = 2.5$ is given in Fig.3. It is clear that growth is present in the wave when $G = 2.7$ in Fig.3a. However, it has a well defined modulation wave in spite of been near criticality. This modulation is present too for larger Reynolds numbers. Here we note that the appearance of harmonics is delayed until the wavenumber is below subcriticality, as shown in Fig.3d. Due to the growth in the subcritical region the calculation time for the case $G = 4.4$ and $G = 5$ (Figs.3e and 3f) was very short. Here too, the power spectra of both figures shows a broad-band structure and exponential decay which are characteristic of a typical chaotic system.

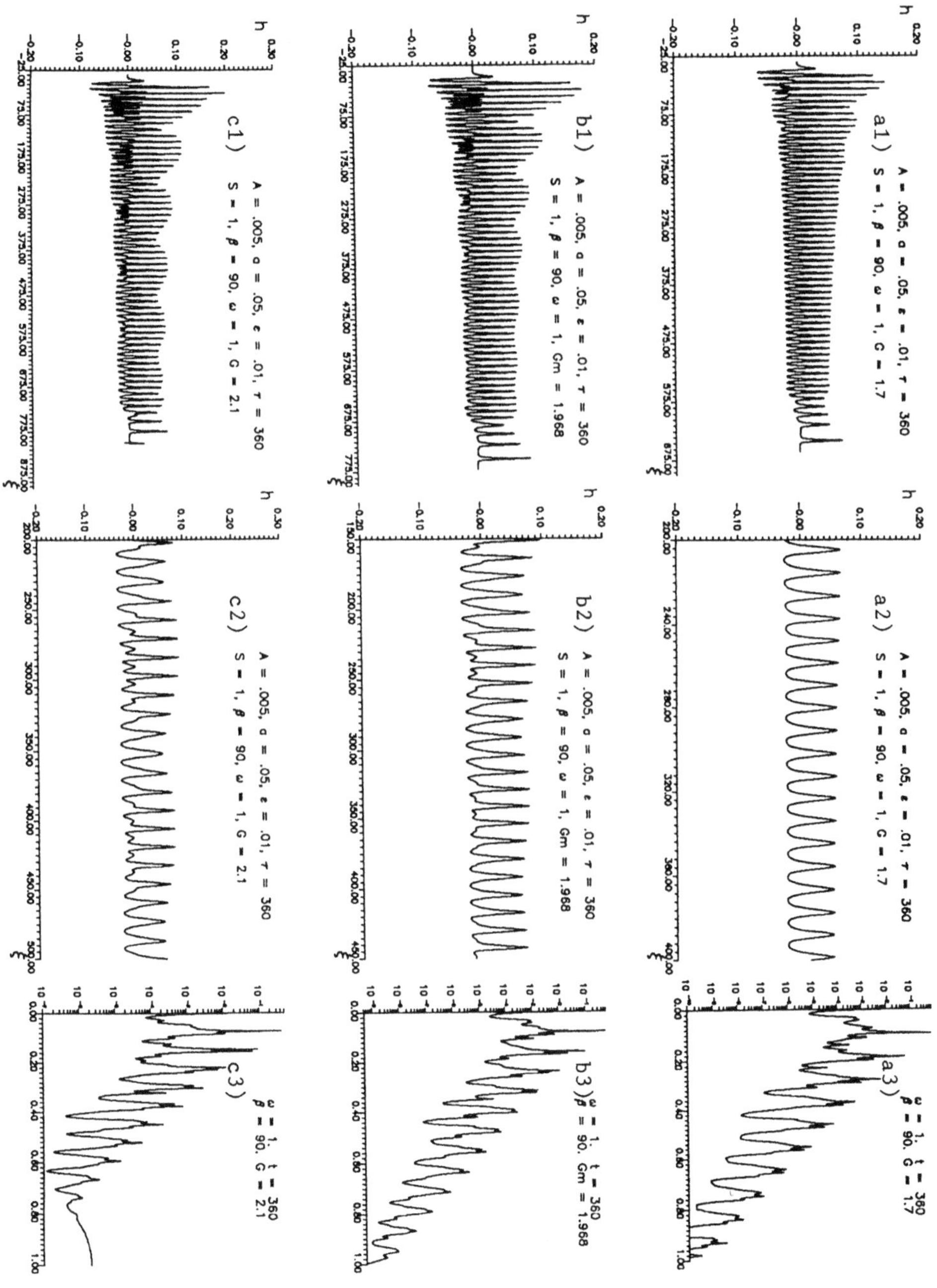
a1) A = .005, α = .05, ε = .01, τ = 360
S = 1, β = 90, ω = 1, G = 1.7
a2) A = .005, α = .05, ε = .01, τ = 360
S = 1, β = 90, ω = 1, G = 1.7
a3) ω = 1. t = 360
β = 90. G = 1.7
b1) A = .005, α = .05, ε = .01, τ = 360
S = 1, β = 90, ω = 1, Gm = 1.968
b2) A = .005, α = .05, ε = .01, τ = 360
S = 1, β = 90, ω = 1, Gm = 1.968
b3) ω = 1. t = 360
β = 90. Gm = 1.968
c1) A = .005, α = .05, ε = .01, τ = 360
S = 1, β = 90, ω = 1, G = 2.1
c2) A = .005, α = .05, ε = .01, τ = 360
S = 1, β = 90, ω = 1, G = 2.1
c3) ω = 1. t = 360
β = 90. G = 2.1
h

Fig. 2

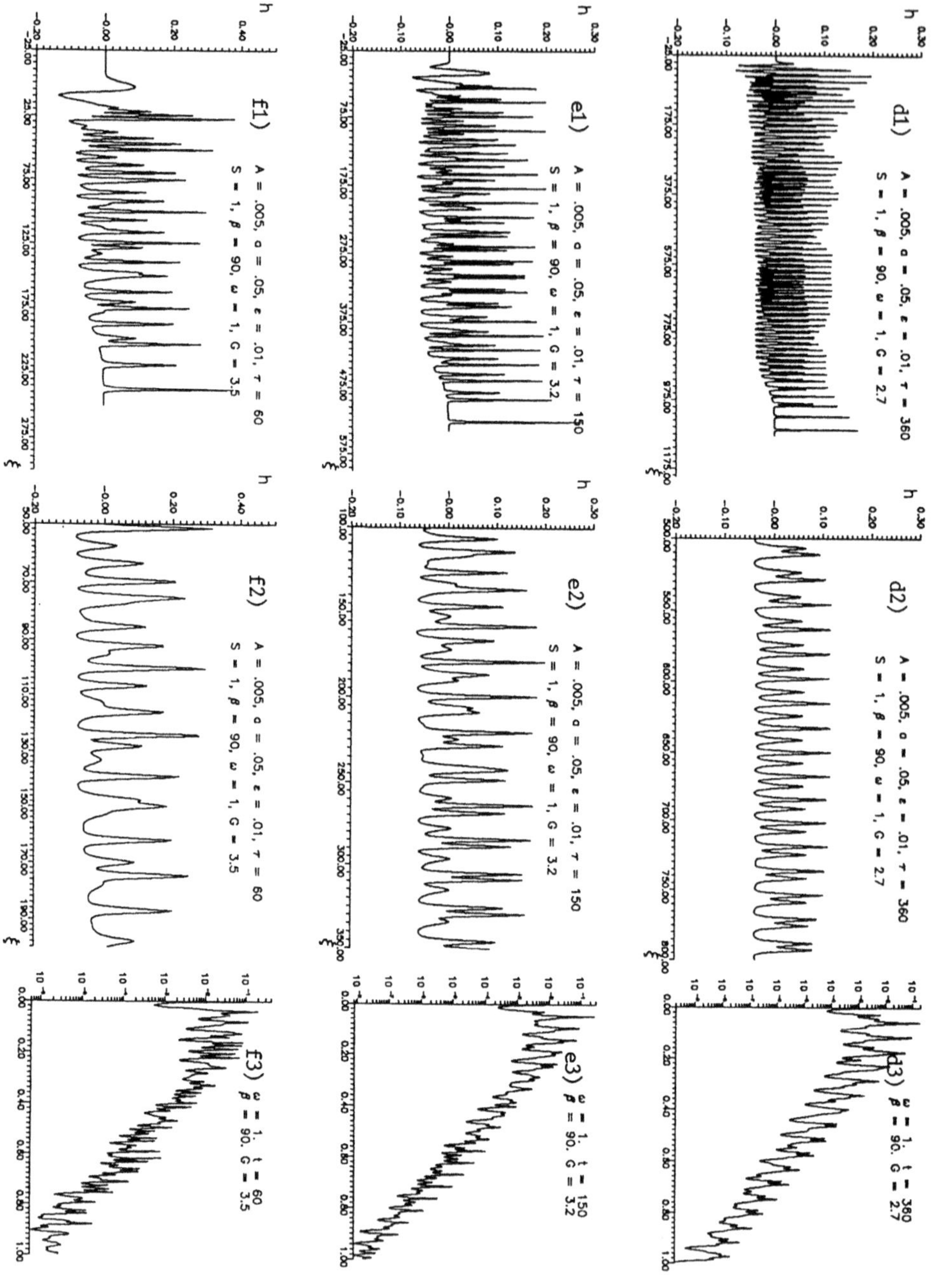
d1)
A = .005, α = .05, ε = .01, τ = 360
S = 1, β = 90, ω = 1, G = 2.7
d2)
A = .005, α = .05, ε = .01, τ = 360
S = 1, β = 90, ω = 1, G = 2.7
d3) ω = 1. t = 360
β = 90. G = 2.7
e1)
A = .005, α = .05, ε = .01, τ = 150
S = 1, β = 90, ω = 1, G = 3.2
e2)
A = .005, α = .05, ε = .01, τ = 150
S = 1, β = 90, ω = 1, G = 3.2
e3) ω = 1. t = 150
β = 90. G = 3.2
f1)
A = .005, α = .05, ε = .01, τ = 60
S = 1, β = 90, ω = 1, G = 3.5
f2)
A = .005, α = .05, ε = .01, τ = 60
S = 1, β = 90, ω = 1, G = 3.5
f3) ω = 1. t = 60
β = 90. G = 3.5

Fig. 2

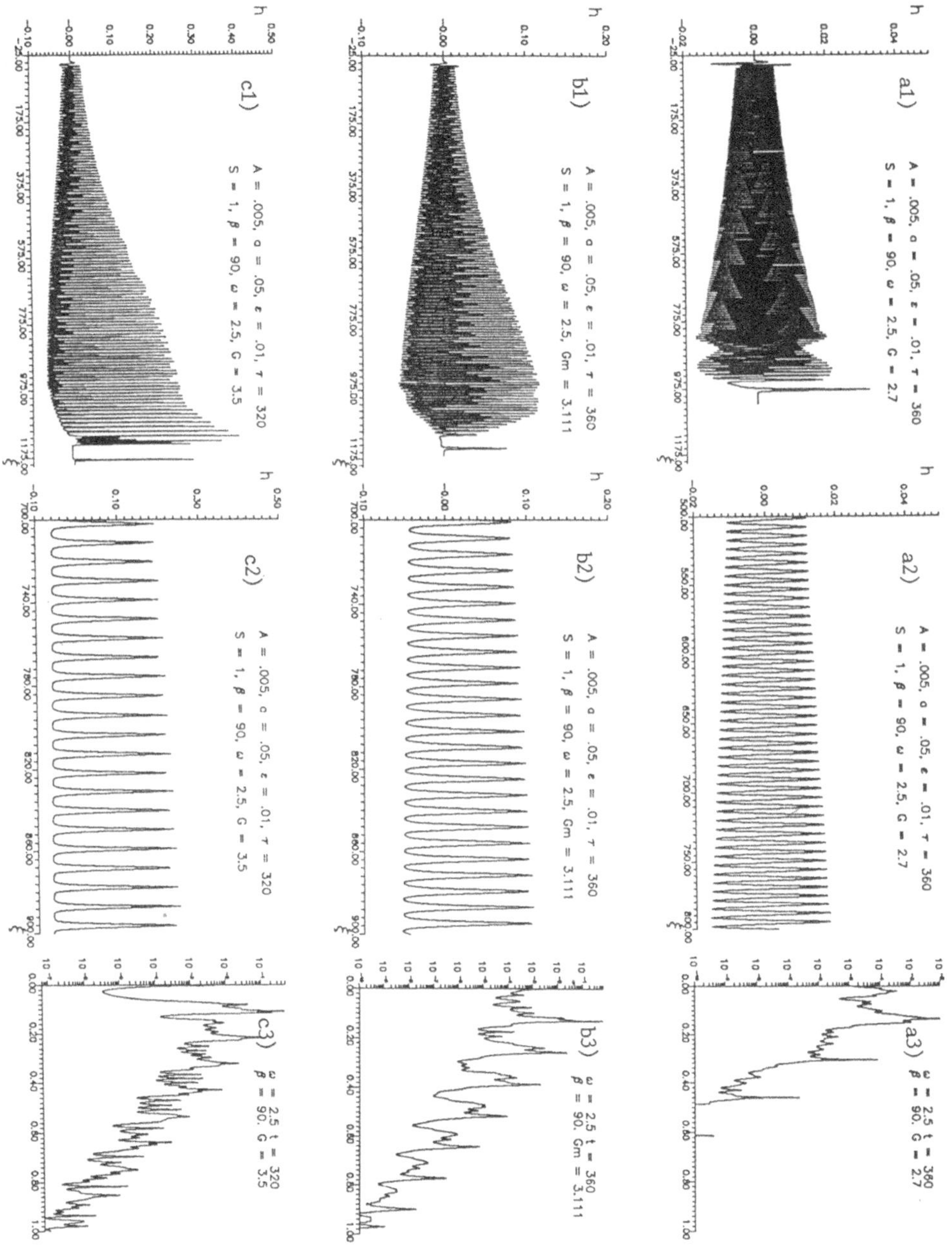
a1)
A = .005, α = .05, ε = .01, τ = 360
S = 1, β = 90, ω = 2.5, G = 2.7
a2)
A = .005, α = .05, ε = .01, τ = 360
S = 1, β = 90, ω = 2.5, G = 2.7
a3) ω = 2.5 t = 360
β = 90. G = 2.7
b1)
A = .005, α = .05, ε = .01, τ = 360
S = 1, β = 90, ω = 2.5, Gm = 3.111
b2)
A = .005, α = .05, ε = .01, τ = 360
S = 1, β = 90, ω = 2.5, Gm = 3.111
b3) ω = 2.5 t = 360
β = 90. Gm = 3.111
c1)
A = .005, α = .05, ε = .01, τ = 320
S = 1, β = 90, ω = 2.5, G = 3.5
c2)
A = .005, α = .05, ε = .01, τ = 320
S = 1, β = 90, ω = 2.5, G = 3.5
c3) ω = 2.5 t = 320
β = 90. G = 3.5
h

Fig. 3

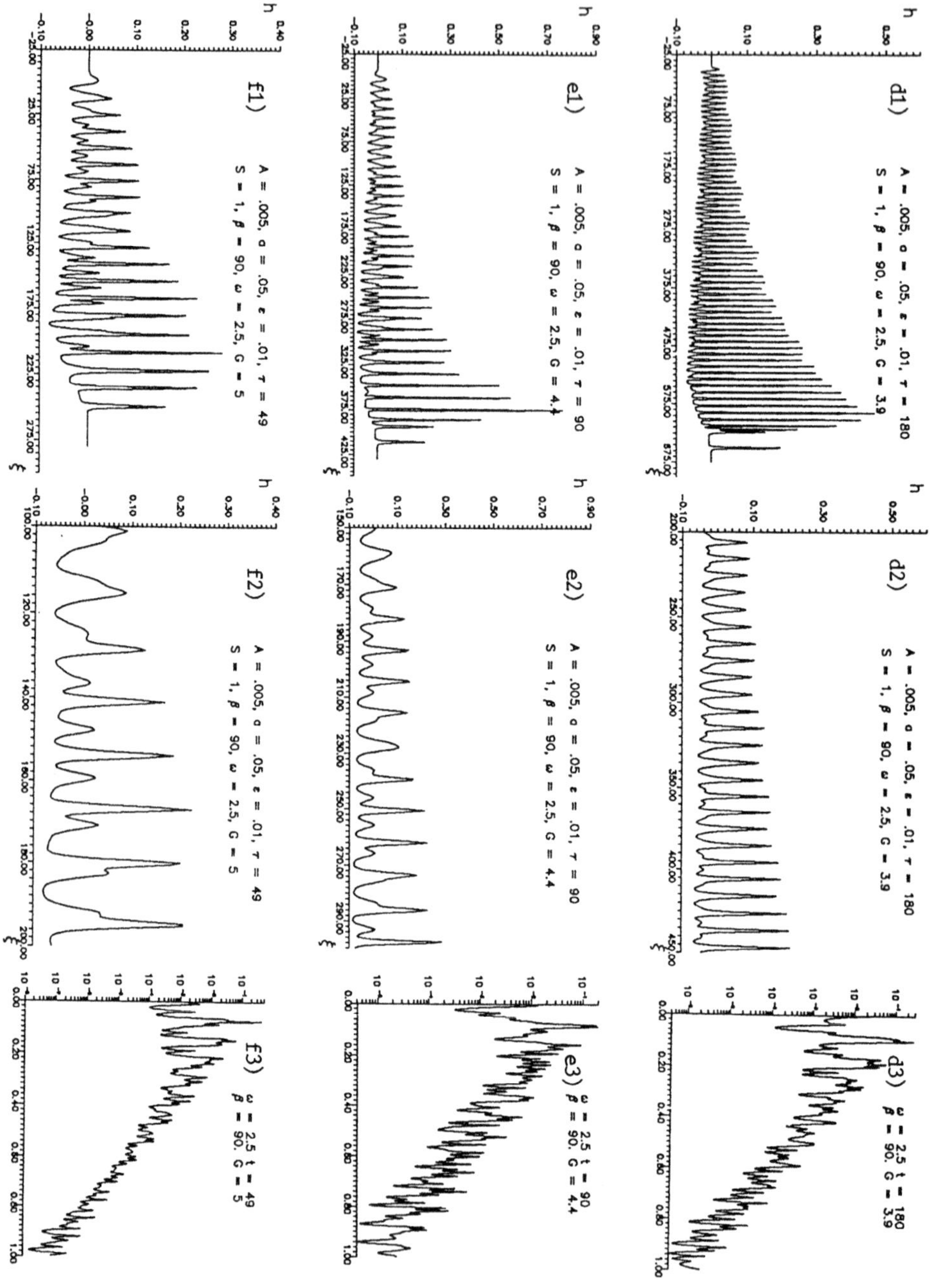
d1)
A = .005, α = .05, ε = .01, τ = 180
S = 1, β = 90, ω = 2.5, G = 3.9
d2)
A = .005, α = .05, ε = .01, τ = 180
S = 1, β = 90, ω = 2.5, G = 3.9
d3) ω = 2.5 t = 180
β = 90. G = 3.9
e1)
A = .005, α = .05, ε = .01, τ = 90
S = 1, β = 90, ω = 2.5, G = 4.4
e2)
A = .005, α = .05, ε = .01, τ = 90
S = 1, β = 90, ω = 2.5, G = 4.4
e3) ω = 2.5 t = 90
β = 90. G = 4.4
f1)
A = .005, α = .05, ε = .01, τ = 49
S = 1, β = 90, ω = 2.5, G = 5
f2)
A = .005, α = .05, ε = .01, τ = 49
S = 1, β = 90, ω = 2.5, G = 5
f3) ω = 2.5 t = 49
β = 90. G = 5

Fig. 3

CONCLUSIONS

A numerical solution for spatial growth in an inhomogeneous Benney equation has been obtained. It has been shown that equilibration between the lines of criticality and subcriticality is only attainable for values of the frequency of the oscillating tangential stress below some value between w = 1.5 and 1.75. For values larger than ω = 1.75 all waves grow with modulation, delaying the appearance of harmonics only for wavenumbers below the line of subcriticality.
It has been found that a little below subcriticality there exist the possibility of spatial chaos due to the broad-band power spectrum shown by the waves for relatively large Reynolds numbers.

REFERENCES

Benney, D. J. (1966) J. Math. & Phys. 45, 150.
Chang, H.-C., (1989), Phys. Fluids A 1, 1314.
Deissler, R. J., Oron A. & Lee, Y. C. (1991) Phys. Rev. A 43, 4558.
Gjevik, B., (1970) Phys. Fluids 13, 1918.
Joo, S. W., Davis, S. H. and Bankoff, S. G. (1991a) J. Fluid Mech. 230, 117.
Joo, S. W., Davis, S. H. and Bankoff, S. G. (1991b) Phys. Fluids A 3, 231.
Joo, S. W., Davis, S. H., (1992a) J. Fluid Mech. 242, 529.
Joo, S. W., Davis, S. H., (1992b) Chem. Eng. Comm. 118, 111.
Krishna, M. V. G. & Lin, S.-P. (1977) Phys. Fluids 20, 1039.
Liu, J. & Gollub, J. P. (1994) Phys. Fluids 6, 1702.
Nakaya, C. (1975) Phys. Fluids 18, 1407.
Nakaya, C. (1989) Phys. Fluids A 1, 1143.
Pumir A., Manneville P. & Pomeau Y. (1983) J. Fluid Mech. 135, 27.

THERMAL CONTROL OF INTERFACIAL INSTABILITIES IN MULTILAYER THIN-FILM FLOWS

Sang W. Joo
Department of Mechanical Engineering
Wayne State University
Detroit, MI 48202
U.S.A.

Abstract. Interfacial instabilities in thin-film flows are studied via nonlinear evolution equations derived using the long-wave asymptotics of Benney (1966). Linear-stability analysis, weakly-nonlinear analysis, and spectral computations are performed to understand the competing effects of viscous, elastic, and Marangoni instabilities.

1. Introduction

Liquid layers flowing down an incline are subject to various modes of hydrodynamic instabilities. When the layer is thin, so that the inertial effects are small, the instabilities occur in the form of surface waves. Near the wave inception, these waves are usually two-dimensional, and tends toward nonlinear permanent waves. Further downstream, they become three-dimensional and eventually erratic. The relevant primary and secondary instabilities and the transition to chaos are adequately described by long-wave evolution equations based on lubrication-type approximation, as reported by Joo & Davis (1992). In nonisothermal layers, the dynamics is much more intriguing. Although the buoyancy effect is negligible for thin films, the thermocapillary effect can be significant. Unstably heated films can exhibit, in addition to the behaviors of an isothermal layer, local dryout (or rupture), completely alien pattern selection, and formation of rivulets, due to the coupled surface-wave and thermocapillary instabilities.

Analogous (but much more involved) instabilities can exist in stratified systems, such as liquid/gas or liquid/liquid films between parallel plates. The linear and nonlinear flow developments are further complicated by the active participation of both phases. For nonisothermal layers, the film rupture and subsequent dry-patch formation can be affected by the imposed streamwise pressure gradient.

In the present study, we consider two liquid films extruded between parallel plates at two different temperatures. Both layers are thin, so that interfacial motion is studied via an evolution equation. The nonlinear flow developments are analyzed by using

S. Morioka and L. van Wijngaarden (eds.),
IUTAM Symposium on Waves in Liquid/Gas and Liquid/Vapour Two-Phase Systems, 235-244.

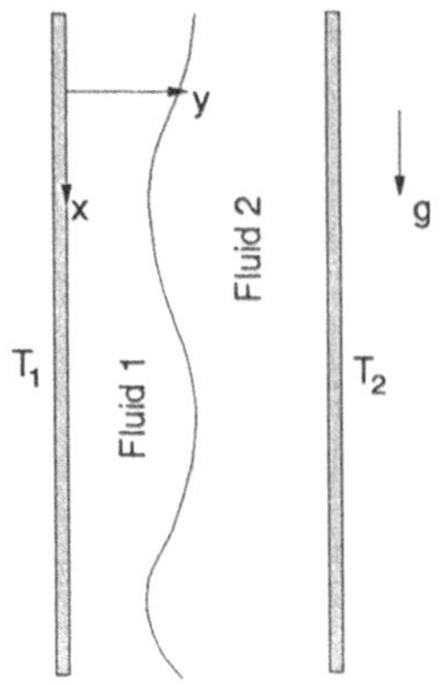

Figure 1: Two-layer flow configuation

a dynamical-system approach and by numerically integrating the fully-nonlinear evolution equation. Since an immediate application of the resulting contributions is to a thermocapillary stabilization of interfacial waves in polymer film processing, the Oldroyd four-constant model is used to model viscoelastic fluids with shear thinning (see Bird et al, (1987) for a detailed discussion of this and other viscoelatic models).

2. Formulation

Two immiscible viscoelastic fluids are flowing in a vertical parallel-plate channel as shown in Fig. 1. The flow is induced by the combined effect of gravity and pressure gradient, so that the direction of the mean flow can be either vertically upward or downward in each phase. Each fluid has constant density ρ_i, thermal conductivity k_i and elastic time-constants, where the index $i = 1$ ($i = 2$) represents fluid 1 (fluid 2), but the viscosity varies with local shear rate and temperature, as described below. The channel walls in contact with fluid 1 and 2 are maintained at constant temperatures T_1 and T_2, respectively.

We use a Cartesian coordinate sytem (x, y), measured in units of the channel gap d, where x is directed vertically downward and y is directed normal to plate 1 into the fluids. The nondimensional time t, velocity $(u^{(i)}, v^{(i)})$, and stress components are based on the viscous scales of fluid 1, d^2/ν_1, ν_1/d, and $\rho_1\nu_1^2/d^2$, respectively, where $\nu_1 = \mu_{01}/\rho_1$ and μ_{01} is the zero-shear-rate viscosity of fluid 1 at temperature T_1. The momentum equations for each phase are then written as

$$u_t^{(i)} + u^{(i)}u_x^{(i)} + v^{(i)}u_y^{(i)} = \frac{\rho_1}{\rho^{(i)}}\left[-p_x^{(i)} + \tau_x^{xx(i)} + \tau_y^{xy(i)}\right] + G \tag{1}$$

$$v_t^{(i)} + u^{(i)}v_x^{(i)} + v^{(i)}v_y^{(i)} = \frac{\rho_1}{\rho^{(i)}}\left[-p_y^{(i)} + \tau_x^{xy(i)} + \tau_y^{yy(i)}\right], \tag{2}$$

where the subscripts x, y, and t represent partial differentiation. The components of the extra-stress tensor $\tilde{\tau}$ are denoted by superscripts for convenience. The parameter

$$G = \frac{d^3 g}{\nu_1^2}$$

measures the gravity g. A parameter for the density ratio

$$\rho = \frac{\rho_2}{\rho_1}$$

appears in the momentum equations for fluid 2. The continuiy equation for each phase is

$$u_x^{(i)} + v_y^{(i)} = 0. \tag{3}$$

For both fluids we modify the Oldroyd four-constant model to include the temperature dependency of the viscosity. The Oldroyd four-constant model describes viscoelasticity with shear thinning in dynamic viscosity and in first normal-stress coefficient but zero second normal-stress coefficient (see, e.g., Bird et al. (1987) for a detailed review of the model). As we shall see later, for thin layers the flow behaviors are not sensitive to the choice of rheological models. We assume that the zero-shear-rate viscosity μ^{0i} varies linearly with temperature:

$$\mu^{0i} = \mu_{0i} \left[1 - \beta_i \left(\tilde{T} - T_1 \right) \right], \tag{4}$$

where $\tilde{T}$ is the dimensional temperature. For most common fluids the constants β_i are positive (thermal thinning). If we scale the temperature using $T = (\tilde{T} - T_1)/(T_2 - T_1)$, the model becomes

$$\tilde{\tau}^{(i)} + De_{1i} \tilde{\tau}_*^{(i)} + \bar{Z}^{(i)} \left(\mathrm{tr}\, \tilde{\tau}^{(i)} \right) \tilde{\gamma} = \left(1 - \bar{B}^{(i)} T \right) \left(\tilde{\gamma} + De_{2i} \tilde{\gamma}_* \right), \tag{5}$$

where $\tilde{\gamma} = \nabla \vec{v} + (\nabla \vec{v})^T$ is the rate-of-strain tensor,

$$()_* = \frac{\mathrm{D}()}{\mathrm{D}t} - \left[() \cdot \nabla \vec{v} + (() \cdot \nabla \vec{v})^T \right], \tag{6}$$

superscript T denotes transpose, $\mathrm{D}/\mathrm{D}t$ is the material derivative, and $\nabla = (\partial_x, \partial_y)$ is the gradient operater. The parameters

$$De_{1i} = \frac{\nu_1 \lambda_{1i}}{d^2} \qquad \text{and} \qquad De_{2i} = \frac{\nu_1 \lambda_{2i}}{d^2}$$

are dimemsionless relaxation and retardation time for each fluid, respectively, and become zero for inelastic fluids. For convenience we define the Deborah number in each phase as

$$De^{(i)} = De_{1i} - De_{2i},$$

which is always positive due to the constraint of the model constants. The nondimensional time constant

$$\bar{Z}^{(i)} = \frac{\nu_1 \eta_{0i}}{d^2}$$

and the parameter

$$\bar{B}^{(i)} = \beta_i (T_2 - T_1)$$

measure, respectively, the shear and the thermal thinning. For example, in a steady simple shear flow, the fluid viscosity would behave as

$$\mu_i = \mu^{0i} \frac{1 + \lambda_2 \eta_{0i} \dot{\gamma}^2}{1 + \lambda_1 \eta_{0i} \dot{\gamma}^2}. \tag{7}$$

The energy equations become, after nondimensionalization,

$$\frac{\nu_1}{\nu_i} P^{(i)} \left(T_t^{(i)} + u^{(i)} T_x^{(i)} + v^{(i)} T_y^{(i)} \right) = T_{xx}^{(i)} + T_{yy}^{(i)}. \tag{8}$$

The Prandtl numbers

$$P^{(i)} = \frac{\nu_i}{\kappa_i},$$

where κ_i is the thermal diffusivity for each fluid. The kinematic-viscosity ratio

$$\nu = \frac{\nu_2}{\nu_1}$$

appears in the energy equation for fluid 2.

The interface between the fluids is located at $y = h(x, t)$, where $0 < h < 1$ due to the scaling. The mean location of the interface $h_0 = d_1/(d_1 + d_2)$ is imposed as a control parameter, where d_i is the mean layer thickness in each phase, but he local shape $h(x, t)$ must be obtained as a part of the solution to the above system. The interfacial tension σ varies linearly with temperature:

$$\sigma = \sigma_0 \left[1 - \gamma \left(\tilde{T} - T_1 \right) \right], \tag{9}$$

where σ_0 is the interfacial tension at $\tilde{T} = T_1$. There is a jump in the normal stress at the interface due to the capillary force:

$$-p^{(2)} + p^{(1)} + \left[\!\left[\vec{n} \cdot \tilde{\tau}^{(i)} \cdot \vec{n} \right]\!\right] = -\frac{3}{N^3} \bar{S} h_{xx} \quad \text{at} \quad y = h, \tag{10}$$

where

$$\vec{n} = \frac{1}{N} (-h_x, 1) \tag{11}$$

is the unit normal vector at the interface directed into fluid 2,

$$\bar{S} = \frac{\bar{\sigma}_0 d}{3 \rho_1 \nu_1^2}$$

is the nondimensional mean interfacial tension, and $N = \sqrt{1 + h_x^2}$. The jump in quantities across the interface is denoted by $[\![\cdot]\!] = (\cdot)^{(1)} - (\cdot)^{(2)}$. Due to the thermo-capillarity, there is a jump in the shear stress, which is described as

$$\left[\!\left[\vec{t} \cdot \tilde{\tau}^{(i)} \cdot \vec{n} \right]\!\right] = \frac{1}{N} M \vec{t} \cdot \nabla T \quad \text{at} \quad y = h, \tag{12}$$

where

$$\vec{t} = \frac{1}{N}(1, h_x) \tag{13}$$

is the unit tangent vector at the interface. The Marangoni number, defined as

$$M = \frac{\gamma\sigma_0 d(T_2 - T_1)}{\rho_1\nu_1^2},$$

measures the thermocapillarity, and can assume a positive or negative value depending on the controlled temperatures T_1 and T_2. There is no slip at the interface, so that the normal and tangential components of the volocity vector are continuous across the interface:

$$[\![\vec{u}\cdot\vec{n}]\!] = 0 \quad \text{at} \quad y = h \tag{14}$$

$$[\![\vec{u}\cdot\vec{t}]\!] = 0 \quad \text{at} \quad y = h. \tag{15}$$

The local temperature and heat flux are also continuous across the interface, which gives

$$T^{(1)} = T^{(2)} \qquad \text{at} \quad y = h \tag{16}$$

and

$$[\![k_i\vec{n}\cdot\nabla T^{(i)}]\!] = 0 \qquad \text{at} \quad y = h. \tag{17}$$

The location of the interface is defined by the kinematic condition $v = h_t + uh_x$ at $y = h$, which is converted into

$$h_t + \left(\int_0^h u\,dy\right)_x = 0 \tag{18}$$

by consuming the normal-velocity condition (14) and introducing a nondimensional flow rate

$$Q = \int_0^1 u\,dy,$$

which is in units of $\nu_1 d$. Here Q is positive (negative) for a vertically downward (upward) flow, and its magnitude can be considered as the Reynolds number. In the single-layer limit (μ_2, $\rho_2 \to 0$), it is impossible to impose the pressure gradient or control the flow rate except by changing the mean layer thickess, and the parameter G becomes the Reynolds number.

Both walls are rigid and kept at constant temperatures. The boundary conditions are thus

$$u^{(1)} = v^{(1)} = T^{(1)} = 0 \qquad \text{at} \quad y = 0 \tag{19}$$

$$u^{(2)} = v^{(2)} = 0 \quad \text{and} \quad T^{(2)} = 1 \qquad \text{at} \quad y = 1. \tag{20}$$

3. Evolution equation

When both layers are thin and the fluids have moderate elasticity (Q, G, $De^{(i)} = O(1)$), the instability occurs in the form of interfacial waves. The shear instability, which is essentially a short-wavelength disturbance of the Tollmien-Schlichting type, and the thermocapillary instability of the Pearson type (Pearson 1958), which also occurs for disturbance wavelength comparable to the layer thickness, do not exist. The interfacial modes start to grow for long-wave disturbances, while short waves are suppressed by the capillary force. The dynamics of the flow thus can be efficiently studied by applying the long-wave asymptotics. The interface deformation can have a profound effect on the quality of the processed materials, and needs to be thoroughly examined. While the linear stability of the interface can be studied by the usual infinitesimal amplitude normal-mode analysis, as performed by Yih (1967) and Yiantsios & Higgins (1988) for isothermal Newtonian fluids and by Li (1969) and Su & Khomami (1992) for isothermal viscoelastic fluids among others, we derive an evolution equation, which enables the stability analysis and the study of highly nonlinear flow behaviors beyond the initial disturbance growth. An equivalent equation for isothermal Newtonian fluids has been reported by Tilly et al. (1994). Here, we present an equation for nonisothermal non-Newtonian fluids.

We apply the long-wave approximation of Benney (1967), we obtain, for the spatio-temporal evolution of the interface,

$$h_\tau + J_0 h_x + \left(J_1 h_x + J_2 h_{xxx} + J_3\right)_x = 0, \tag{21}$$

where the coefficients

$$\begin{aligned} J_0 &= J_0\left(h; G, Q, \rho, \mu\right) \\ J_1 &= J_1\left(h; G, Q, \rho, \mu, k, De^{(i)}, M\right) \\ J_2 &= J_2\left(h; \mu, S\right) \\ J_3 &= J_3\left(h; G, Q, \rho, \mu, k, Z^{(i)}, B^{(i)}\right). \end{aligned}$$

Derivation of these equations is given in more detail by Joo (1994). The evolution equation for isothermal fluids is obtained by taking $M = B^{(i)} = 0$ in (21). If we further set $De^{(i)} = Z^{(i)} = 0$, the evolution equation for two Newtonian fluids, derived by Tilly et al. (1994), is recovered except for the hydrostatic effect, which is absent in the present case of vertical channel. The sigle-fluid limit is achieved by setting $\mu = \rho = 0$. If we take this limit with $M = B^{(i)} = 0$ on (21), the evolution equation for a vertical draining viscoelastic film, reported by Joo (1994), is obtained. If we then set $De^{(i)} = Z^{(i)} = 0$, the evolution equation for a Newtonian draining film, first obtained by Benney (1966) is recovered.

The linear stability analysis of the stratified-flow system can be performed via the evolution equation. The basic state is a flat undisturbed interface, described by $h = h_0$, where h_0 is a constant. We superpose a simple harmonic disturbance of an infinitesimal amplitude to the basic state, and write

$$h = h_0 + \delta \left(e^{i\alpha x + \Gamma \tau} + \text{c.c.} \right), \tag{22}$$

where α and Γ are the disturbance wavenumber and the complex growth rate, respectively, c.c. represents the complex conjugate, and $\delta \ll 1$. The linear stability of the system then is determined by the sense of the real part Γ_r of the growth rate; the disturbance will grow in time if $\Gamma_r > 0$. If we substitute (22) into the evolution equation and linearize in δ, we obtain

$$\Gamma_r = \alpha^2 \left(J_1 - \alpha^2 J_2 \right). \tag{23}$$

It is obvious that $J_1 > 0$ and $J_2 > 0$ will, respectively, destablize and stabilize the flow.

The interface can become unstable due to strafications in density, viscosity, elasticity, and thermocapillarity. While other combinations of these effects are reported in many previous works, including that by Joo (1994) most recently, only those related to the thermocapillarity will be summarized here.

Even when these is no elasticity and stratification in density and viscosity, the flow can become unstable due to the thermocapillarity. The full expression for this effect is

$$J_{1T} = -\frac{k(k-1)h^2(h-1)^2\left(h^2\mu - h^2 + 2h - 1\right)}{2f_0\left[h(k-1)+1\right]^2}, \tag{24}$$

where

$$f_0\left(h;\mu\right) = h^4\mu^2 - 2h(h-1)(h^2 - h + 2)\mu + (h-1)^4. \tag{25}$$

If we want to extract the purely thermocapillary instability, the viscosity ratio μ must be set to unity, which yields

$$J_{1T} = -\frac{h^2(h-1)^2(2h-1)k(k-1)}{2\left[h(k-1)+1\right]^2}. \tag{26}$$

The interface will become unstable due to the pure themocapillarity if

$$M(2h-1)(k-1) < 0. \tag{27}$$

If the channel wall in contact with the fluid 2 is maintained at a higher temperature ($M > 0$), the interface becomes unstable if the layer 2 has lower thermal conductivety and smaller thickness or higher conductivety and larger thickness. Therefore, if one pursues the stabilizing effect of the thermocapillarity by controlling the temperature of one of the channel walls, Table 1 can be used as a guide.

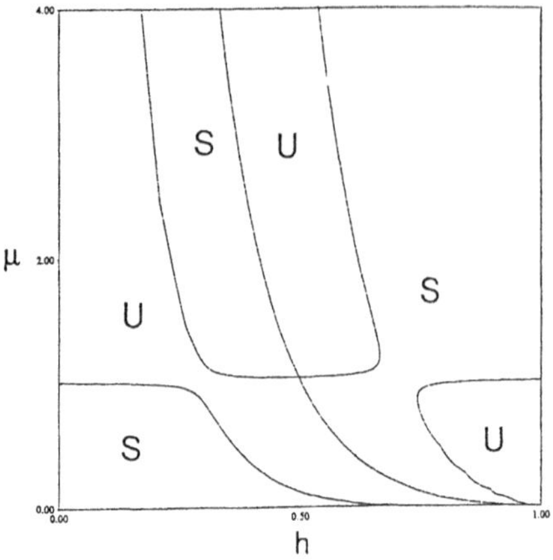

Figure 2: Instability due to elasticity and thermocapillarity when $k = 2$, $De^{(1)} = De^{(2)} = 0.1$, and $M = 0.1Q^2$.

When the viscosity stratification is also present, (24) must be considered with the effects of isothermal viscosity stratification superimposed:

$$J_1 = h^2(h-1)^2\left(h^2\mu - h^2 + 2h - 1\right)\left[Q^2(\mu - 1)E_{3V} - M\frac{k(k-1)}{2f_0\left[h(k-1)+1\right]^2}\right]. \quad (28)$$

The neutral stability branch related to isothermal viscosity stratification persists, but the flow can be stabilized or further destabilized, depending upon the heating condition M and the thermal-conductivety ratio k.

In order to examne the effect of thermocapillarity on viscoelastic fluids with viscosity stratification, the full equation for J_1 should be considered. In the absence of density stratification, each term has a common factor $(h^2\mu - h^2 + 2h - 1)$, which corresponds to the neutral branch of the purely viscous instability. This branch thus is not affected by elasticity or thermocapillarity. Other branches, on the other hand, are influenced by the interaction of elasticity, stratification of elasticity, and thermocapillarity.

Fig. 2 shows stability diagrams in (μ, h) planes when there is no elasticity stratification ($De^{(1)} = De^{(2)} = 0.1$), wall 2 is heated $M = 0.1$, and $k = 2$. Since $M(k-1) > 0$, the thermocapillarity by itself is stabilizing if $h > 1/2$. Consequently, compared to the isothermal case (see Joo 1994), the stable (unstable) region to the right (left) of the viscous neutral branch has expanded. The branch $\mu = 1$ that exists in the isothermal case has disappeared. If we enhance the thermocapillary effect, the expansion of these stable and unstable regions develop further.

In Fig. 3, the stability diagrams are shown for cases where the elasticity stratification ($De^{(1)} \neq De^{(2)}$) is also present and $M = Q^2$. In Fig. 3a, the fluid in the heated side is more elastic ($De^{(1)} = 0.1Q^2$ and $De^{(2)} = 0.2Q^2$). Two unstable regions

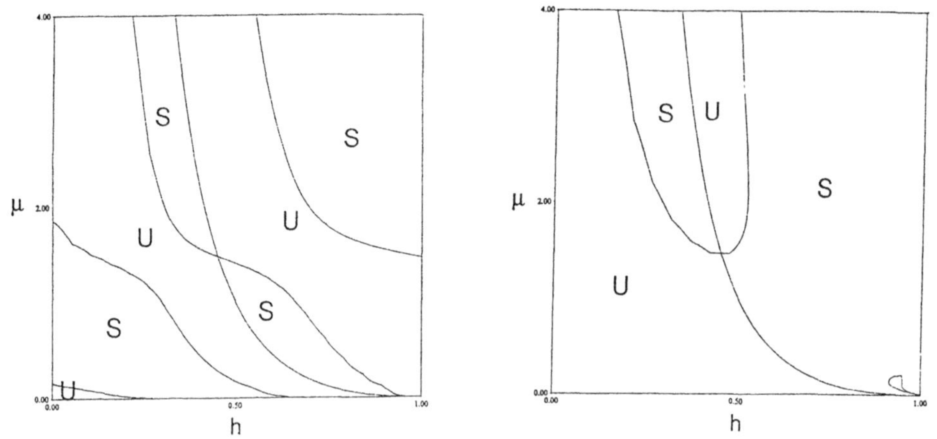

Figure 3: Instability due to thermocapillarity and stratifications in viscosity and elasticity when $k = 1$ and $M = Q^2$. (a) $De^{(1)} = 0.1$ and $De^{(2)} = 0.2$; (b) $De^{(1)} = 0.2$ and $De^{(2)} = 0.1$.

to the left of the viscous neutral branch have merged together, and a new unstable region has been created at the lower left corner of the diagram. The opposite case of the elasticity stratification ($De^{(1)} = 0.2Q^2$ and $De^{(2)} = 0.1Q^2$) is shown in Fig. 3b. Compared to the isothermal case, the stable region in the lower left corner has compeletely disappeared, whereas the unstable region in the lower right corner persists. The viscosity ratio μ for the instability to set in to the right of the viscous neutral branch has been increased. The diagrams support the rules for applying temperature difference provided in Table 1 to achieve the stability control.

5. Concluding remarks

Interfacial instabilities between two immicible viscoelastic fluids flowing through two parallel plates are studied. The interfacial evolution equation, derived by using the long-wave approximation, is examined, through linear stability analysis, weakly-nonlinear analysis, and spectral computations, to understand the flow behavors beyond criticality. In this paper, the stability results related to the thermocapillarity of viscoelastic fluids are reported.

The author acknowledges Professor S. H. Davis for his valuable guidance for most of the present study.

References

Benney, D. J. 1966 Long waves in liquid films. J. Math. & Phys., **45**, 150.

Joo, S. W. and Davis, S. H. 1992 Instabilities of three-dimensional viscous falling films. J. Fluid Mech., **242**, 529.

Joo, S. W. 1994 Interfacial instabilites in plane Poiseuille flow of two stratified viscoelastic fluids with heat transfer: part I. Evolution equation and stability analysis. Pending publication.

Li, C.-H. 1969 Stability of two superposed elasticoviscous liquids in plane Couette flow. Phys. Fluids, **12**, 531.

Su, Y.-Y. and Khomami, B. 1992a Interfacial stability of multilayer viscoelastic fluids in slit and converging channel die geometries. J. Rheol., **36**, 357.

Su, Y.-Y. and Khomami, B. 1992b Pure elastic interfacial instabilities in superposed flow of polymeric fluids. Rheol. Acta, **31**, 413.

Tilly, B. S., Davis, S. H. and Bankoff, S. G. 1994 Nonlinear long-wave stability of superposed fluids in an inclined channel. Pending publication.

Yiantsios, S. G. and Higgins, B. G. 1988 Linear stability of plane Poiseuille flow of two superposed fluids. Phys. Fluids, **31**, 3225.

Yih, C.-S. 1967 Instability due to viscosity stratification. J. Fluid Mech., **27**, 337.

HYDRODYNAMIC STABILITY OF LIQUID FILMS

From fundamental fluid mechanics to refinery

ROB MIESEN
Koninklijke/Shell-Laboratorium, Amsterdam
Postbus 38000, 1030 BN Amsterdam
The Netherlands

Abstract. In this paper I discuss the hydrodynamic stability of sheared liquid films to disturbances with wavelengths of the order of the film thickness, and a possible relation of that subject to the critical Reynolds number that has been observed for roll waves. The insight obtained is used to considerably improve a model that predicts the entrained fraction of liquid in annular dispersed flow.

1. Film stability and entrainment

The study of the generation of waves on water is one of the oldest topics in fluid mechanics. If the water is shallow the situation corresponds to the generation of waves on thin water films. Wave generation on such films is studied by considering the hydrodynamic stability [6] of the flow in the film and/or the air flowing over the film. Although such flows have been studied for a long time and considerable progress was made in the 1960's, e.g. by Miles (see [17] and references therein), there are still important questions to be answered. Examples of such questions concern the interaction of different modes of instability, the nonlinear saturation of wave growth [3], and the explanation of the critical Reynolds number for the generation of long waves on thin films [4] [9] [10].

Appart from the clear academic interest in the stability of liquid films, this topic is of importance in a technological context. Film stability plays a role in, for example, coating technology (the manufacturing of color film [24]) and the cleaning of silicon wafers. Film stability is even more important in the design and operation of process equipment in the (petro)chemical and nuclear industry (see the many AERE-reports, e.g. [8]; [22]). In such equipment gases and liquids are often present in such ratios that a flow

S. Morioka and L. van Wijngaarden (eds.),
IUTAM Symposium on Waves in Liquid/Gas and Liquid/Vapour Two-Phase Systems, 245-255.

pattern arises in which the liquid flows like a film along the wall of the equipment or pipes, and which is called annular flow [12]) (because these films are, in general, thin we can abstract from the specific geometry and describe them as films on a flat plate). Whether the film is stable or not effects markedly the performance of the equipment: film instability may lead to the entrainment of droplets from the film, i.e. to a transition to (annular) dispersed flow [11] [23]. Even if a relatively small fraction of the liquid is transported in the form of droplets this results in a considerable decrease of the residence time of the liquid in the equipment; the droplet velocity, which is approximately equal to the gas velocity, is an order of magnitude larger than the liquid velocity in the film. In many cases the liquid residence time is an important parameter in designing equipment so that the importance of knowledge about film stability is evident. From the above it is also clear that from a technological point of view one is not only interested in predicting the critical conditions for the generation of waves but also (or even more) in predicting the rate at which droplets are formed (and transported back to the film). Furthermore, interest is often not with the flow of water and air, so that the effect of the physical properties of the liquid and the gas on the critical conditions and the formation of droplets is of concern.

2. Critical conditions for short waves

The simplest scenario for film instability and droplet formation would be as follows: if the gas velocity over a liquid film is increased, the film becomes unstable (and thus wavy) at a certain critical gas velocity and if the gas velocity is increased sligthly more, droplets are formed. Although this picture is too simple the question can nevertheless be posed at which gas velocity the film becomes hydrodynamically unstable so that waves will form. This is exactly the question considered by Cohen & Hanratty [5], who study this problem both experimentally as well as theoretically.

Cohen & Hanratty [5] increase the air velocity over a film with a fixed liquid flow rate Q_l on a flat (horizontal) plate. They find that if the gas velocity is increased beyond a certain critical gas velocity U_{gc}, waves with wavelengths of the order of the film thickness (or somewhat longer) are formed. Qualitatively this can be understood as follows. The shear that the gas exerts on the (not yet wavy) film sets up an approximately linear velocity profile in the film: velocity 0 at the plate and a velocity U_{lm} at the air-liquid interface. The average liquid velocity is thus equal to $U_{lm}/2$, so that $U_{lm}d/2 \propto Q_l$ is unchanged during an experiment, with d as the film thickness. The stabilising effect of surface tension and gravity are measured by the Weber number $S = \sigma/(\rho_l U_{lm}^2 d)$ and the Froude number $F = gd/U_{lm}^2$,

σ being the coefficient of interfacial tension, ρ_l the liquid mass density, and g the gravitational acceleration. As the gas velocity increases, the shear force of the gas on the liquid also increases and thus U_{lm} increases. Therefore both $S \propto 1/(Q_l U_{lm})$ and $F \propto Q_l/U_{lm}^3$ decrease, so that at a certain gas velocity U_{gc}, destabilising inertial effects will become larger than the stabilising effect of surface tension and gravity. Cohen & Hanratty [5] furthermore find that the critical gas velocity U_{gc} decreases with increasing Q_l.

Cohen & Hanratty calculate U_{gc} by using the method suggested by Benjamin [2], coined 'devided attack' by him. First calculate the stresses that a gas flow exerts on a solid wavy plate, thus approximating the gas-liquid interface to stand still (the wave velocity is indeed much lower than the gas velocity) and not being able to relax these stresses. The stresses thus found are a function of the wavelength of the disturbance on the plate, the position with respect to the wave crest, the physical properties of the gas and the gas velocity (profile). Then use these stresses in the conditions at the air-liquid interface when solving for the liquid flow in the film. This gives at which gas velocity there is a disturbance (with an arbitrary wavelength) that is unstable, which should correspond to the critical conditions found in the experiment. Cohen & Hanratty [5] indeed find reasonably good agreement between such theoretical predictions of the critical gas velocity U_{gc} and their experiments (Table 1), although theoretical predictions are consistently too high. Some improvement in predicting U_{gc} can be made by solving the flow in the gas and the liquid simultaneously [18] (Table 1), i.e. not using the 'divided attack'.

3. More short waves

If the flow of a gas over a liquid film is solved as described in [18], it is found that a second mode of instability can lead to waves of the order of the film thickness if the Reynolds number of the liquid film, defined as

$$R = \frac{\rho_l U_t d}{\mu_l}, \qquad U_t = \frac{d\tau}{2\mu_l}, \tag{1}$$

is large enough; μ_l is the (dynamic) viscosity of the liquid and U_t is a characteristic velocity based on the shear force τ that the gas exerts on the liquid film, which can be calculated if the gas velocity (far from the interface) U_g is known [23]:

$$\tau = \frac{1}{2} c_f \rho_g U_g^2, \tag{2}$$

where c_f is the friction coefficient and ρ_g and is the mass density of the gas. This second mode of instability differs from the waves considered by

μ_l $(mPa\,s)$	d (mm)	$U_{g,exp}$	$U_{g,ch}$	$U_{g,us}$
0.929	1.89	2.72	3.02	2.70
0.933	3.54	1.99	2.24	2.20
0.925	4.91	1.44	1.93	1.76
3.89	4.48	2.81	3.98	2.81
3.95	6.41	2.00	2.28	1.97
11.2	4.67	3.80	4.48	3.54
11.2	7.32	3.15	3.69	3.08

TABLE 1. The critical values of the (bulk average) gas velocity for the seven experiments in Table 2 of the paper of Cohen & Hanratty [5]. The top three experiments were done with water (viscosity $\mu_l \approx 0.93\,mPa\,s$); the other experiments were done with water glycerine mixtures. The third column gives the measured values of the critical gas velocity, the fourth as calculated by Cohen & Hanratty, and the fifth as found by us [18].

Cohen & Hanratty [5] in the fact that the phase speed of these waves is always smaller than the liquid velocity in the liquid film, while the phase velocity of the waves studied by, for example, Cohen & Hanratty is, in general, larger. Consequently, the critical layer (i.e. the plane where the wave velocity is equal to liquid velocity) is in the liquid film instead of directly above the gas-liquid interface. The second mode discussed here is therefore called the 'internal mode' and the mode studied by Cohen & Hanratty [5] is called the 'interfacial mode'. The different position of the critical layer is directly linked to the fact that the waves obtain their energy by a different mechanism from the flow.

The internal mode is the same mode as studied asymptotically by Miles [15], who neglects all effects of the gas on the liquid flow but the effect that it exerts a constant (uniform, time-independent) shear force which is responsible for setting up the flow in the film (Yiantsios & Higgins [25] also study the internal mode, for the case that viscosities, densities, and thicknesses of two layers of liquid are of the same order of magnitude). The critical Reynolds number R_c for the internal mode in this so-called free surface approximation was found numerically to be 39.2 by Smith & Davis [20] (using the definition (1) for the Reynolds number), if the Weber number S is zero (because the Froude number F is quite small, in general, and the effect of gravity is similar to that of surface tension, the effect of F is not discussed for convenience). If S increases, R_c increases rapidly in the free surface approximation.

If the internal mode is studied by simultaneously solving the flow in the gas and the liquid [18] (as we did for the waves studied by Cohen & Hanratty), the free surface approximation is not made and all the effects of the gas on the flow in the liquid film are included. In that case R_c not only depends on S (and F), but also on the velocity profile in the gas and the ratio of the densities and the ratio of the viscosities of the gas and the liquid. For realistic values of the parameters, R_c can differ substantially from 39.2 as found in the surface approximation, and in [18] it is discussed why the free surface approximation is incorrect. It is also found that the dependence of R_c on S is much weaker than in the free surface approximation (if S is not too large). For the flow of air and water a typical critical Reynolds number can be shown to be 100.

An obvious and important question now is: can the waves that correspond to the internal mode be observed? The answer to that question is that it will be difficult, because for conditions at which the internal mode becomes unstable, the surface mode is unstable already. If critical conditions, defined as conditions for which waves are first observed, are studied in an experiment, only the surface mode will therefore be observed. I know of no direct observation of the internal mode.

4. Roll waves

For water films that are sheared by air something else, however, happens at a Reynolds number of approximately 100: if the air velocity is large enough 'roll waves' are formed [4] [9] [10]. Roll waves are very long, breaking waves that resemble waves on a beach, with wavelengths of the order hundred times the film thickness. It is thus difficult to understand that the internal mode, which corresponds to waves with a wavelength of the order of the film thickness, would have anything to do with the generation of roll waves. This possibility will nevertheless be discussed now.

The existence of roll waves on water films with a Reynolds number of, say, 200 or larger is quite adequately described by theory that uses Navier-Stokes equations that have been averaged in the direction perpendicular to the flow [1] [4]. Such theories predict, however, that roll waves should be formed for any value of the Reynolds number, if only the gas velocity is larger than a certain critical value. This is not observed in experiments: as mentioned above roll waves are formed only if the Reynolds number of the film exceeds approximately 100. An explanation of a critical Reynolds number for roll wave formation was sought for for twenty when Andreussi, Asali & Hanratty [1] showed that certain stabilising stresses at the gas-liquid interface, that should come into play if the Reynolds number drops below a critical value, could offer such an explanation; they also show that these stabilising stresses may arise if the surface roughness of the film suddenly drops, without discussing where this sudden decrease in roughness comes from. I suggest now that the change in surface roughness (or the generation of these stabilising stresses) is related to the internal mode becoming (un)stable. It should be nemtioned, however, that Bruno & McCready [4] do not find any evidence of an increase of short waves in the surface wave spectrum.

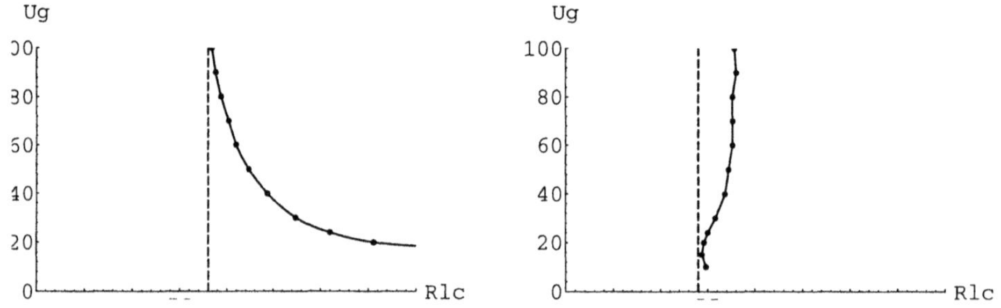

Figure 1. The gas velocity U_g (multiply by approximately 1667 to find the gas Reynolds number as used by, e.g. Andreussi at al. [1]) against the critical liquid Reynolds number R_{lc} for the internal mode of a liquid film in a channel of height $0.025\,m$. The parameters for the stability calculation have been determined as described in the Appendix. The gas velocity profile is the linear-logarithmic profile proposed by Miles [16] (see also [18]). The left figure is for the flow of water ($\rho_l = 1000\,kg/m^3$, $\mu_l = 0.001\,Pa\,s$) and air ($\rho_g = 1.2\,kg/m^3$, $\mu_g = 1.8\,10^{-5}\,Pa\,s, \sigma = 0.074\,N\,m$). The figure on the right is for a viscosity $\mu_l = 0.0025\,Pa\,s$, other parameters unchanged. The dashed lines are drawn at $R_{lc} = 90$ and 70, respectively.

In Figure 1 some more detailed calculations are presented, that show that the critical Reynolds number for roll waves on a water film ($R_{lc} \approx 100$) is close to the critical Reynolds number for the internal mode of the water film. This value is measured, however, at gas speeds of 15 to $30\,m/s$. At $30\,m/s$ we find that $R_{lc} \approx 135$ for the internal mode. This difference can have many causes. For example, the velocity profile that we have used in the gas or the liquid is only approximately correct. Furthermore, if roll waves are present on a film the Reynolds number of the film varies substantially because, e.g. the film thickness varies along the films.

We should not rule out that this agreement is a coincidence. If the critical Reynolds number for roll waves and the internal mode are, however, indeed closely linked they should show the same trend if the physical properties of the liquid (or the gas) are changed. Next to water the best documented case is the formation of roll waves on a film that is a mixture of water and glycerin with a viscosity of $0.0025\,Pa\,s$, giving a critical Reynolds number of approximately 75 [4]. For the internal mode also a critical Reynolds number of approximately 75 is found (Figure 1)! For more viscous water-glycerin mixtures roll waves have also been studied, but in less detail [1]. Moreover, for more viscous liquids, the critical Reynolds number for the internal mode becomes more sensitive to the details of the shape of the velocity profile in the gas and a comparison becomes more difficult (calculations equivalent to those in Figure 1 give critical Reynolds numbers that are about 50% too large). Another property that can be varied is the surface (or interfacial) tension of the gas-liquid system. Some experiments with kerosine ($\sigma = 0.026\, N\, m$, $\mu_l = 0.00156\, Pa\, s$, $\rho_l = 790 kg/m^3$), which were done at our laboratory, indicate that the critical Reynolds number for roll waves is approximately 55, about the same as we find for the internal mode.

From a practical point of view roll waves are very important because the transition to dispersed flow has been observed to almost coincide with the generation of roll waves [1] [8]. The mechanism of droplet formation appears to be the stripping of short waves from the crest of a roll wave [8] [21]. Thus the critical Reynolds number for roll waves (and for the internal mode?) should correspond to the critical flow rate - of liquid in the film - for droplet formation, as used in empirical models to predict the rate of entrainment (the amount of liquid per second and per square meter of film surface) from a sheared film [13]. Before coming back to that point, I would like to point out that the critical Reynolds number for roll waves has been studied for a very limited range of physical properties. If indeed the critical Reynolds for roll waves and the internal mode are the same, it is clear now that this would be of great practical importance, because the gasses and liquids that are encountered in the process industry have physical properties which are widely varying.

5. Entrainment

In process equipment, liquids and gasses are frequently flowing through pipes in such ratios that part of the liquid is in the form of a film along the pipe wall and part of the liquid (the entrained fraction) can be found in the form of droplets in the gas. Empirical relations that predict the entrained fraction of liquid are a very important ingredient of models that are used to design process equipment [19]. These relations are, however, not very successful, as they tend to predict the entrained fraction correctly only in a small parameter range. In recent years empirical models have been developed that describe the entrained fraction as a result of the competition of droplet entrainment from the film and the deposition of droplets that are present in the gas flow [14] [13]. These models, which require an expression for the rate of entrainment and an expression for the rate of depostion, have the potential to be more successful in predicting the entrained fraction over a wider range of parameters.

In a model proposed by Govan, Hewitt, Owen & Boot [7] [13] the rate of entrainment E_r is proportional to:

$$E_r \propto (\dot{m}_{lf} - \dot{m}_{lfc})^{c_1}, \tag{3}$$

where $\dot{m}_{lf}$ is the flow rate in the film (referred to the total cross-section of the channel) and $\dot{m}_{lfc}$ the critical flow rate for the onset of entrainment. This can be written as:

$$E_r \propto \left(\frac{4\mu_l}{D}\right)^{c_1} \left(\frac{\rho_l \bar{U}_l d}{\mu_l} - \frac{\rho_l \bar{U}_{lc} d}{\mu_l}\right)^{c_1}, \tag{4}$$

where D is the pipe diameter, c_1 an empirical constant, $\bar{U}_l$ the average liquid velocity in the film and $\bar{U}_{lc}$ the critical average liquid velocity for the onset of entrainment. For a horizontal (laminar) film flow the velocity profile in the film is linear and it is easily shown that $\bar{U}_l = U_{lm}/2 = U_t$, where U_{lm} is the maximum velocity in the film (attained at the interface), and U_t is the velocity defined by equation (1). Substituting this into (4) gives, together with equation (1), that

$$E_r \propto \left(\frac{4\mu_l}{D}\right)^{c_1} (R - R_c)^{c_1}. \tag{5}$$

Expression (5) shows that the Reynolds number which dominates the hydrodynamic stability of the film also determines the rate of entrainment of droplets from the film. This may be due to the mechanism of drop formation: stripping of short waves, the generation of which is dominated by R [18], from the crests of roll waves [21]. Note furthermore that it is clear now that R_c is the critical Reynolds number for roll wave formation, and possibly also the critical Reynolds number for the internal mode (in which case it would be possible to calculate R_c from 'first principles' instead of using a correlation [13].).

For a vertical film, i.e. gravity is in the direction of or opposed to the direction of the flow, the velocity profile is quadratic, and the simple relation between $\bar{U}_l$ and U_t used above is not valid. Nevertheless, (5) seems to have a sounder physical basis then the original expression (3), because R and not $\dot{m}_{lf}$ determines the stability of the film. As can be seen in Figure 2, using (5) instead of (3) leads indeed to much better results for the prediction of the entrained fraction of liquid, both in the case of upflow (liquid and gas both flow opposite to the direction of gravity) and downflow (liquid and gas both flow in the direction of gravity). Except for replacing the expression on the right-hand-side of (3), where it occurs in Hewitt & Govan's model [13], by the right-hand-side of (5), we have not changed their expression for the rate of entrainment nor their expression for the rate of deposition. The only additional change is that we have used different values for the empirical constants that appear in the expressions; these have been determined by fitting the model to a large set of experiments in which the entrained fraction is measured. The effect of the different constants is, however, minor compared to the effect of changing the expression for the rate of entrainment.

For high gas velocities, the effect of gravity on the velocity profile in the film is small (Figure 2) and the velocity profile is then approximately linear [18]. Consequently, the difference between the model of Hewitt & Govan [13] and ours is small. For lower gas velcities, where the velocity profile in the film deviates markedly from a linear profile, differences are large. Note that (for the case of upflow) at the gas speed where the stress that the gas exerts on the film is too small to drag the film upwards (this velocity follows from the calculation of the basic flow, as discussed in the Appendix), the predicted entrained fraction goes to 1, as found in the experiments [22].

Similar improvements, as shown in Figure 2, were found for other liquid flow rates, pipe diameters and different liquids (kerosine, water-glycerine). For these different liquids R_{lc} as found for the internal mode should be used in (5) (so not the correlation given in [13]) to have good agreement between the model and experiments. This is further circumstantial evidence that the critical Reynolds number for roll waves and the internal mode are related.

References

1. Andreussi, P., J.C. Asali & T.J. Hanratty, 'Initiation of roll waves in gas-liquid flow', *A.I.Ch.E.J.* **31**, 119-126 (1985).
2. Benjamin, T.B., 'Shearing flow over a wavy boundary', *J. Fluid Mech.* **6**, 161-205 (1959).
3. Blennerhassett, P.J., 'On the generation of waves by wind', *Phil. Trans. Roy. Soc. A* **298**, 451-494 (1980).
4. Bruno, K. & M.J. McCready, 'Origin of roll waves in horizontal gas-liquid flows', *A.I.Ch.E.J.* **34**, 1431-1440 (1988).
5. Cohen, L.S. & T.J. Hanratty, 'Generation of waves in the concurrent flow of air and a liquid', *A.I.Ch.E.J.* **11**, 138-144 (1965).
6. Drazin, P.G. & W.H. Reid, 'Hydrodynamic Stability', Cambridge University Press, Cambridge (1986).
7. Govan, A.H., G.F. Hewitt, D.G. Owen & T.R. Bott, 'An improved CHF modelling code', 2nd U.K. Heat Transfer Conf., Glasgow (1988).
8. Hall-Taylor, N.S. & G.F. Hewitt, 'The motion and frequency of large disturbance waves in annular two-phase flow of air-water mixtures', AERE-R 3955, Harwell, England (1962).
9. Hanratty, T.J., 'Separated flow modelling and interfacial transport phenomena', Appl. Sci. Res. **48**, 353-390 (1991).
10. Hanratty, T.J. & M.J. McCready, 'Phenomenological understanding of gas-liquid

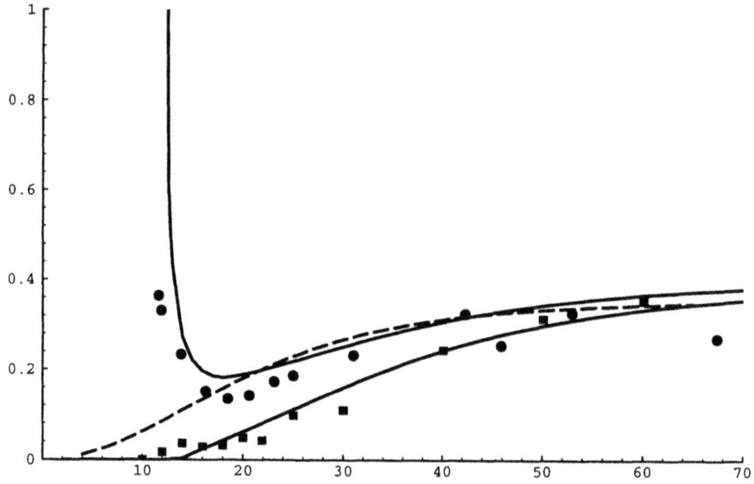

Figure 2. The entrained fraction of liquid as a function of the gas velocity. The dots are the experiments done at our laboratory [22] for upflow of air and water in a 50 mm diameter pipe and for a superficial liquid velocity (i.e. the total amount of liquid flowing per second through the pipe, divided by the liquid mass density and the cross-sectional area of the pipe) of 0.014 m/s. The squares are experiments for downflow, at the same conditions. The dashed line is found from the model by Hewitt & Govan [13] which does not discriminate between upflow and downflow. The solid lines are found for upflow and downflow if our modified expressions for the rate of entrainment are used. How to calculate, for example, R or d is explained in the Appendix.

separated flows', Proc. of Workshop on two-phase flow, London (1992).
11. Hetsroni, G., 'Handbook of Multiphase Systems', McGraw-Hill, New York (1982).
12. Hewitt, G.F. & N.S. Hall-Taylor, 'Annular Two-Phase Flow', Pergamon, Oxford (1970).
13. Hewitt, G.F. & A.H. Govan 'Phenomenological modelling of non-equilibrium flows with phase change', *Int. J. Heat Mass Transfer* **33**, 229-242 (1990).
14. Ishii, M. & K. Mishima, 'Droplet entrainment correlation in annular two-phase flow', *Int. J. Heat Mass Transfer* **32**, 1835-1845 (1989).
15. Miles, J.W., 'The hydrodynamic stability of a thin film of liquid in uniform shearing motion', *J. Fluid Mech.* **8**, 593-610 (1960).
16. Miles, J.W., 'On the generation of surface waves by shear flows, Part 4', *J. Fluid Mech.* **13**, 433-448 (1962).
17. Miles, J.W., 'Surface-wave generation revisited', *J. Fluid Mech.* **256**, 427-441 (1993).
18. Miesen, R. & B.J. Boersma, 'Stability of thin sheared liquid films', submitted for publication to *J. Fluid Mech.* (1993).
19. Oliemans, R.V.A., B.F.M. Pots & N. Trompé, 'Modelling of annular dispersed two-phase flow in vertical pipes', *Int. J. Multiphase Flow* **12**, 711-732 (1986).
20. Smith, M.K. & S.H. Davis, 'The instability of sheared liquid layers', *J. Fluid Mech.* **121**, 187-206 (1982).
21. Steen, D.A. & G.B. Wallis, 'The transition from annular to annular-mist co-current two-phase downflow', U.S. Atomic Energy Commision Report NYO-3114-2 (1964).
22. Verbeek, P.H.J., R. Miesen & C.J. Schellekens, 'Liquid entrainment in annular dispersed upflow', in *Proc. Eigth Annual European Conference on Liquid Atomisation and Spray Systems*, 33-45 (1992).
23. Wallis, G.B., 'One-dimensional Two-phase Flow', McGraw-Hill, New York (1969).

24. Wang, C.K., J.J. Seaborg & S.P. Lin, 'Instability of multi-layered liquid films', *Phys. Fluids* **21**, 1669 (1978).
25. Yiantsios, S.G. & B.G. Higgins, 'Linear stability of plane Poiseuille flow of two superposed fluids', *Phys. Fluids* **31**, 3225-3238 (1988) and Erratum *Phys. Fluids A* **1**, 897 (1989).

Appendix

1. Basic flow

To calculate critical Reynolds numbers, as presented in Figure 1, or entrained fractions, as presented in Figure 2, one needs to know how parameters like d depend on the gas velocity, i.e. the basic flow in the liquid film has to be known. Only if the basic flow is known a parameter such as the Reynolds number (see equations (1) and (5)) can be calculated.

We have determined the parameters of the basic flow in the more or less standard way as described by Hewitt in [12] or section 2.2 of [11]. We assume that the film is thin enough to be considered two-dimensional and laminar. For a film on a horizontal plate and in the absence of a pressure gradient, the velocity profile is linear. In other cases the velocity profile $U(y)$ in the film is quadratic in the distance to the wall y, and is completely determined if the thickness of the film d, the pressure gradient along the pipe and the shear stress τ of the gas flow on the film are known. The pressure gradient is approximated by $-4\tau/D$ ($-(\tau + \tau_{wall})/D$ for a channel of height D), D the pipe diameter, so that only τ and d need to be determined.

The shear stress τ of the gas on the film surface is approximately given by (2), with c_f given by [23] [11]:

$$c_f = c_{f0}\left(1 + \gamma\frac{d}{D}\right), \qquad c_{f0} = 2(2.236\log R_g - 4.639)^{-2}, \qquad R_g = \frac{\rho_g U_g D}{\mu_g}, \tag{6}$$

and $\gamma \approx 400$. The only unknown parameter, i.e. the film thickness d, is determined by the fact that the superficial film velocity U_{slf} is by definition:

$$U_{slf} = \frac{\pi D d \bar{U}_l}{\pi D^2/4} = \frac{4}{D}\int_0^d U(y)dy, \tag{7}$$

where $\bar{U}_l$ is again the average velocity in the film. Since the liquid flow rate through the pipe, and thus the superficial liquid velocity, is known, this equation gives d: if the entrained fraction E is zero, the superficial film velocity is equal to the superficial velocity through the pipe ($U_{slf} = U_{sl}$); if the entrained fraction is not zero $U_{slf} = (1 - E)U_{sl}$. In the latter case, the model that gives E (as a function of d etc.) and the equations for the basic flow have to be solved together.

INTERFACIAL WAVE STRUCTURE AND ITS EFFECT ON TRANSPORT PHENOMENA IN HORIZONTAL WAVY/ANNULAR TWO-PHASE FLOWS

Yoshimichi Hagiwara[1], Satoshi Yamaguchi[2] and Kenjiro Suzuki[3]
[1]Dept. of Mech. & System Eng., Kyoto Institute of Technology, Kyoto 606, Japan
[2]Toyota Motor Car Company, Toyota City 471, Japan
[3]Department of Mechanical Engineering, Kyoto University, Kyoto 606-01, Japan

Abstract

The measurement of time change in the liquid film thickness and the visual observation of the interface and the flow inside the liquid film were carried out for air-water wavy/annular two-phase flows in a horizontal pipe. The liquid-film time traces obtained at different peripheral locations and the images of visualized liquid flow were processed. The contribution of disturbance waves to the liquid transport downstream was found to become small for higher liquid flow rate. The wave-crest line lines of disturbance waves are found curved. The existence of a secondary flow may be suggested in the back side of disturbance waves.

1. INTRODUCTION

Interest has been growing recently in the structure of wavy/annular two-phase flows because the flows are often encountered in steam generators and vapor condensers. Various kinds of measurements have been done for understanding flow and heat transfer characteristics in different conditions of the flows. In the case of annular or wavy two-phase flow, many take such a view that the interface is usually covered with disturbance waves and ripple waves, and that disturbance waves have high propagation velocity, large amplitude and two-dimensionality. It is also known that disturbance waves play important roles in the enhancement of heat and momentum transfer, droplet entrainment, and dryout or rewetting of pipe inner surface. However, the structure of disturbance wave and its effect on the flow near the interface have not been clarified yet.

The present study aims at understanding the spatial structure of disturbance waves and its effect on the flow inside the liquid film of horizontal wavy/annular flows. For this purpose, the measurement of liquid film thickness at different peripheral locations and visual observations of the interface and liquid film are done for annular/wavy air-water two-phase flows in a horizontal pipe.

2. EXPERIMENTAL APPARATUS AND PROCEDURE

2.1 Apparatus

Figure 1 shows the experimental apparatus. It is the same as that was in HAGIWARA et al.[1989]. Air was drawn by a blower(5), passed through the wire

S. Morioka and L. van Wijngaarden (eds.),
IUTAM Symposium on Waves in Liquid/Gas and Liquid/Vapour Two-Phase Systems, 257-267.

nets(1) and the contraction nozzle(2) and was fed into the test section. The test section was made of acrylic tube of D = 49.9 mm I.D. and 5930 mm length. The air flow rate was measured with a Pitot tube(12) located at the outlet of the contraction nozzle. The axial distance, x, was measured from the outlet of the nozzle. The water in a constant–head tank(8) was supplied to an air–water mixing section(3) by a pump(9). The mixing section consisted of a core section and double annular spaces as shown in Fig. 2. Each wall contains holes to supply water to the inner spaces. Numerous small holes of the most inner wall were used to introduce a low–speed uniform injection of water into the test section. Each annular space of the mixing section was peripherally divided into eight sections and flow rate in each section was controlled independently for establishing a non–uniform distribution of the liquid phase at the mixing section: i.e., the distribution close to that to be established far downstream from the mixing section. The water flow rate was measured with an orifice(10) and a manometer. The water separation was made in a chamber(4) located at the end of the test section.

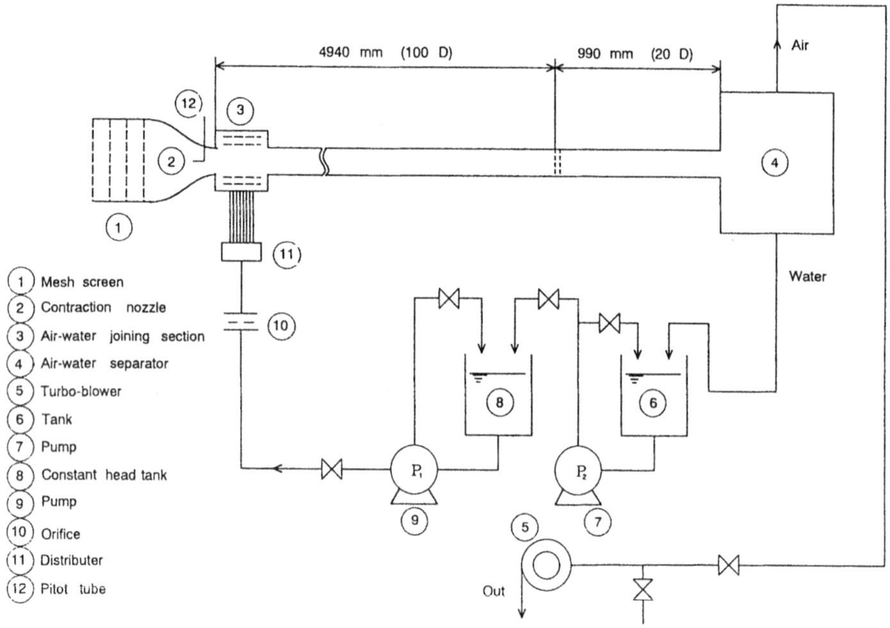

Fig.1 Experimental apparatus

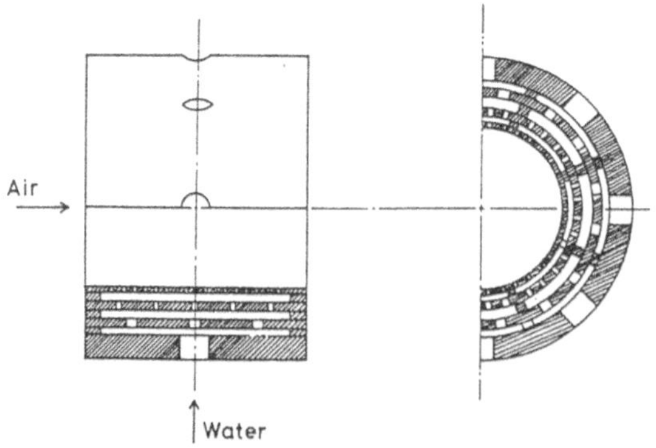

Fig. 2 Details of air–water mixing section

Conductance method was adopted for measuring time-varying liquid film thickness. The method has an advantage of higher spatial and time resolution than any other method available for measuring the thickness. The probes were mounted at eight different peripheral positions at two closely-spaced cross-section of x/D = 99 and 99.5. Each conductance probe has a pair of electrodes for measuring the conductance between them. Figure 3 shows the arrangement of the conductance probes. The parallel-wire-type probes (hereafter called PW probes) arranged at five bottom locations consist of two fine wires protruding from inner surface of the test pipe (KOSKIE et al.[1989]). The distance between the two wires of each PW probe was set 4.0 mm. On the other hand, flush-mount-type probes (called FM probes) were mounted at three top locations. FM probes consist of two wires whose edge surfaces were flush with the inner surface. The distance between the two wires of each FM probe was 3.0 mm. The probe geometry is nearly the same as that described in the book by HEWITT [1978].

An alternating electric current of 1 kHz was imposed between the two electrodes of each conductance probe to eliminate the polarizing action of water. The signal was sent to a demodulation circuit after passing through an active band-pass-filter and a low-pass filter to produce the signal proportional to the liquid film thickness (YAMAUCHI and SAWAI[1993]). Similar conductance probe was inserted into the water supplying tube to compensate the effect of water properties on the conductance probe reading. The time-varying signals of the liquid film thickness were digitized at 400 Hz. These signals were recorded simultaneously on digital magnetic tape with a TEAC DR2000 digital data recorder. The recorded signal involves errors due to the digitization and the noise. The errors were estimated to be at most 5%. The digitized data were processed using a large computer at the Data Processing Center at Kyoto University.

2.2 Calibration of the probes

Prior to the experiment, calibration had been done on the relationship between the output signal of the conductance probes and the depth of water. The depth was obtained by measuring the distance between the pipe inner surface and the needle point when the point touches to the water surface during the needle was traversed radially from the pipe axis to the interface. Since the interfacial waves cause serious error to the measurement of the water depth, air was not fed, and water was supplied calmly from the bottom part of the air-water mixing section for the calibration. A fence was attached to the upstream end of the mixing section through the calibration in order to lead water downstream. The output signal from the probe was recorded in the same manner as that mentioned above. The calibration of the probe at the bottom was thus made directly. By rotating the test section every 45 degree, the calibration of another probe was done. To obtain a thin water film flow, a plug and thin spacers were inserted into the test pipe during the calibration of FM probes. The clearance for water flow between the plug and the pipe inner surface was measured with a gauge.

The calibration data of the output voltage of each probe plotted against the liquid film thickness are shown in Fig. 4. The linear relationship was obtained between the output voltages, E, of the PW probes and the liquid film thickness, δ, ranging from 1 to 16 mm. Each calibration curve in Fig. 4 (a) was used to estimate the liquid film. The film thickness evaluated from the recorded signal with the established calibration curve was confirmed to agree with the actual thickness within about 8% accuracy. On the other hand, highly nonlinear relationship was found between E of FM probes and δ in Fig. 4(b). Based on these results, the quantitative discussion on the film thickness will be limited to the signals from PW-type probe in the present study.

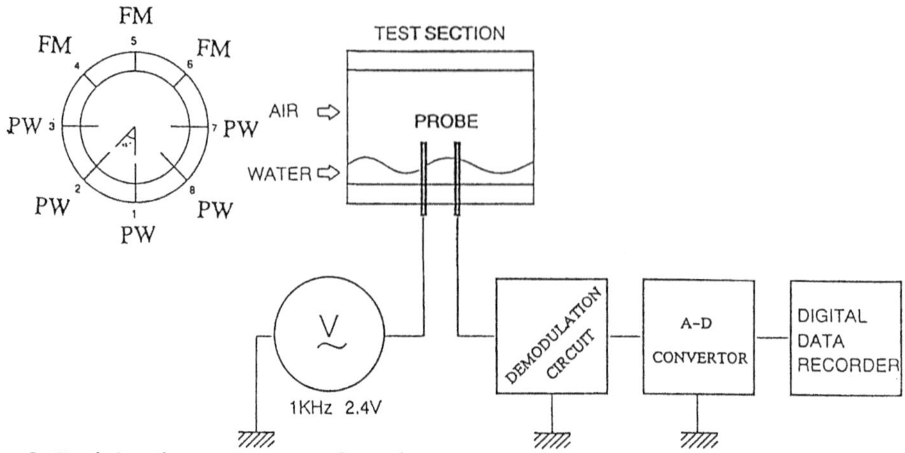

Fig. 3 Peripheral arrangement of conductance probes

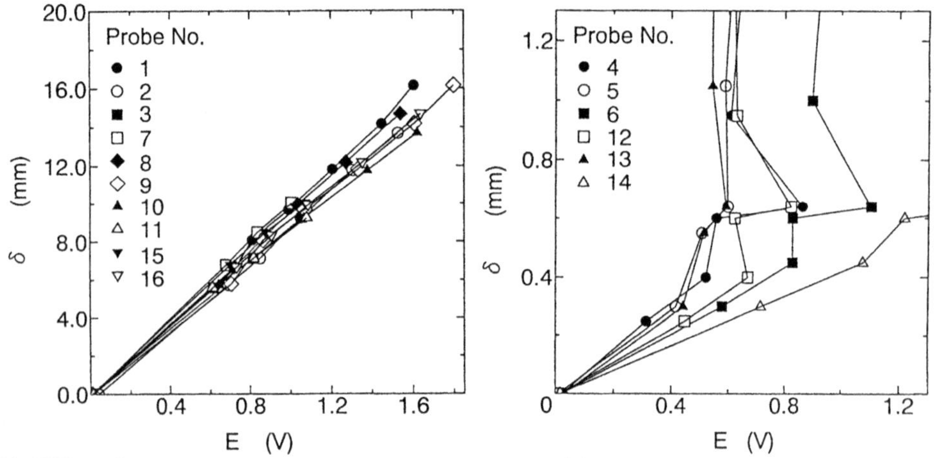

(a) PW probes (b)FM probes

Fig. 4 Calibration curve of conductance probes

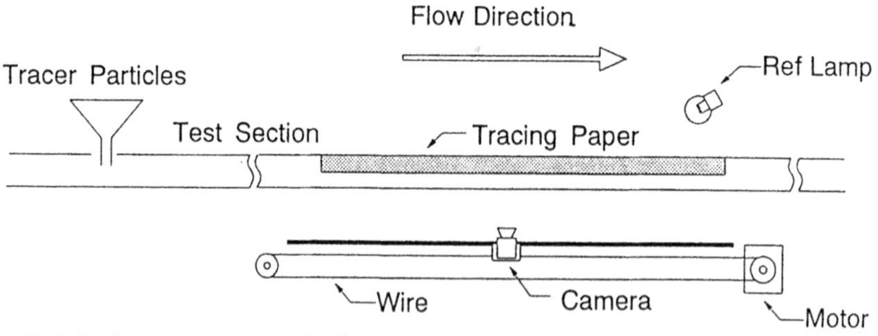

Fig. 5 Moving-camera method

2.3 Visual observation

The moving-camera method was adopted for observing the configuration of the disturbance wave. Figure 5 represents the arrangement of the camera. The small video camera-head (Sony XR77RR) moved along the test tube. The head was mounted on a small bench sliding on a rail. The rail was arranged under the test pipe. Therefore, the camera looked up the interface through the bottom of the pipe. The light from a lamp located above the test pipe was uniformly scattered by passing through a non–transparent paper covering the half top of the outer surface of the pipe. The video signals from the camera were recorded on a video tape using a video tape recorder. The spatial resolution of images recorded was estimated about 0.1 mm.

The speed of the camera was adjusted by changing the rotation speed of a motor pulling a string attached to the bench. A scale fixed at the outer surface of the test pipe along the axis was recorded with the camera. By analyzing the recorded images of the scale, it was confirmed that the camera speed agrees with each other within the error of 10% after 0.6 s has passed since the camera began to move. As the camera speed increases, the effective length to attain the constant speed becomes shorter because the rail length is limited. Therefore, the moving–camera method was adopted mainly for the case of the low propagation velocity of disturbance waves, i.e., wavy two–phase flow case. Incidentally, the visual observation was done in the upstream region of the measuring points for liquid film thickness, namely, x=60D ~ 82D.

Flow visualization was performed using tracer particles for observing fluid motion inside the liquid film. Polystyrene black particles of 0.5 mm diameter (their specific weight is 1.07) were used as the tracer particles. The images of the particles were recorded on a video tape.

2.4 Experimental conditions

The superficial gas velocity, j_G, ranged from 11 to 27 m/s (gas Reynolds number Re_G = 3.1 to 7.7 $\times$ 10^4) , which means the gas flow was turbulent. The superficial liquid velocities, j_L, lay in the range of 0.028 – 0.088 m/s (liquid Reynolds number Re_L = 1400 to 4400).

3. SIGNAL AND IMAGE PROCESSING

3.1 Detection of disturbance wave

Detection of disturbance waves was done based on the double–structured interface model (OSAKA et al.[1991]). In the model, the interfacial waves are assumed to consist of the disturbance waves and the base waves. The waves whose crest–height is higher than that of base wave, δ_{Bmax}, is regarded as disturbance waves. δ_{Bmax} is determined by the sum of the averaged thickness of base film, δ_{Bm}, and the average amplitude of the base wave, a_B. δ_{Bm} is assumed equal to the normal distance from the pipe inner surface to the position where the wave passing frequency takes its maximum. a_B is assumed to be the difference between δ_{Bm} and the minimum film thickness (see Fig. 6). The minimum film thickness is given by the normal distance from the inner surface to the point where the time fraction of the existence of water, which is calculated from the liquid film traces, is 0.99.

3.2 Statistically averaged quantities of disturbance waves

3.2.1 Height of disturbance wave

The averaged height of disturbance waves, H_D, was determined from the height of each disturbance wave, H. H was defined by subtracting the averaged base film thickness, δ_{Bm}, from the maximum film thickness, h_{max}, of the disturbance wave.

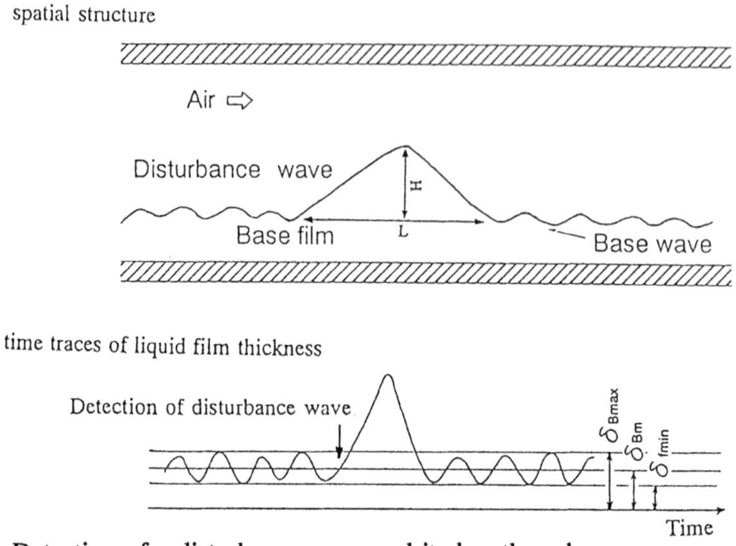

Fig. 6 Detection of a disturbance wave and its length scale

3.2.2 Streamwise length scale

The averaged value of the streamwise length scale, L_D, was calculated from the streamwise length scale of each disturbance wave, L. L was assumed to be determined from multiplying its propagation velocity, c, by its duration, t_D. c was estimated from the distance between an upstream conductance probe and a downstream one, and the time difference for the maximum values in two traces obtained by the two probes. t_D was given by the period during the liquid film thickness exceeds δ_{Bm} for the wave.

3.2.3 Volumetric flow rate transported downstream by disturbance waves

The volumetric flow rate transported downstream by disturbance waves, Q_D, was defined by multiplying the averaged passing frequency, F_D, by the averaged volume carried by a disturbance wave, Vol_D. Vol_D was obtained from the volume carried by each disturbance wave, Vol. It was assumed that the section of a disturbance wave can be expressed by a triangle formed by L and H. Vol was, thus, calculated by the peripheral integration of the triangle as follows:

$$Vol = \frac{1}{2}\int_{-\pi}^{\pi} LH\frac{D}{2}d\theta ,$$

where the increment of the peripheral distance was approximated with $Dd\theta/2$.

3.3 Image processing of visualized base-film flow

The images of visualized flow were recorded by a video tape recorder at a rate of 30 frames per second. The images were precessed with the same technique as that by TAKADA et al.[1991]: the x and y coordinates of the tracer particles on two frames obtained at different time were stored on a floppy disk using a personal computer. From the recorded coordinates in the two images, the velocity vector field was calculated from the movement of each particle. About 70 velocity vectors could be extracted from the particles appearing in the two sets of images. In order to obtain the velocity field inside the liquid film, each image was divided into a group of small rectangle cells. The velocity vectors contained within each cell were vector- averaged and the resultant vector was placed at the center of the cell.

4. RESULTS AND DISCUSSIONS

4.1 Base film thickness and wave height

Figure 7 shows the typical examples of the time traces of liquid film thickness. The upper solid line indicates the trace measured at the bottom, the middle solid and the broken lines show the traces at ± 45 degree, respectively, and the lower solid line and the dotted line at ± 90 degree, respectively. Figure 8 shows δ_{Bm} as a function of j_L. δ_{Bm} is found to increase with j_L especially at the bottom.

The averaged height of disturbance waves, H_D, is represented as a function of j_L in Fig. 9. H_D increases remarkably with j_L at the bottom and the positions of 45 degree from the bottom. On the other hand, H_D at 90 degree does not vary with a change in j_L.

4.2 Volumetric flow rate carried by disturbance waves

Figure 10 shows the volumetric flow rate carried by disturbance waves, Q_D, as a function of j_L. As j_L increases, Q_D becomes first larger, then takes its maximum value, and finally decreases slightly. This tendency comes from the fact that H_D and F_D decrease with an increase of j_L, though L_D increases with j_L (see Figs 9, 12, 11, respectively). These facts agree qualitatively with the measurement by OSAKA et al. [1992].

The fraction, Q_D/Q_{all}, is shown in Fig. 13. Q_{all} means the total liquid flow rate. A remarkable decrease in the fraction with an increase of j_L is obtained. This clearly means that the contribution of disturbance waves to the transportation of liquid downstream becomes smaller for higher liquid flow rate. This is consistent with the fact that the increase in liquid flow rate contributes to the thickening of base film at the bottom of pipe mentioned above. It is worthwhile noting that the dependency of Q_D/Q_{all} on j_L is different from that in the case of vertical flow measured by SEKOGUCHI et al.[1973].

4.3 Spatial structure of disturbance waves

The statistically-averaged configuration of disturbance waves can be made with the averaged quantities mentioned above. Figure 14 depicts the configuration. The base plane of the figure represents the pipe inner surface. The abscissa means the streamwise distance from the cross section which includes the disturbance wave crest at the bottom. The ordinate indicates the liquid film thickness. The dotted lines and the solid lines illustrate the base film and the sections of disturbance waves. The bold lines represent the projected line connecting the crests of the disturbance wave. These lines are no longer a straight line but curved; the lines are shifted upstream as the location changes from the bottom to the top of pipe. The images of disturbance waves recorded on a video tape shows the similar crest-lines[1].

4.4 Secondary flow inside the liquid film

The trajectories of the tracer particles obtained from two frames is shown in Fig. 15 in the case of wavy two-phase flow. The horizontal line shows the axis. The bold line indicates the wave-crest line. The trajectories indicate the relative motion of the particles to the disturbance waves towards upstream (from left to right in the figure). The particles are found slightly away from the axis in the back side of the disturbance

---

[1] The wave-crest line was visible in the images of interface because light was concentrated by passing the convex interface around the crest of disturbance waves.

waves. This may suggest the existence of a secondary flow near the disturbance waves inside the liquid film. The reproduced vector field from the trajectories shown in Fig. 16 may also suggest the secondary flow.

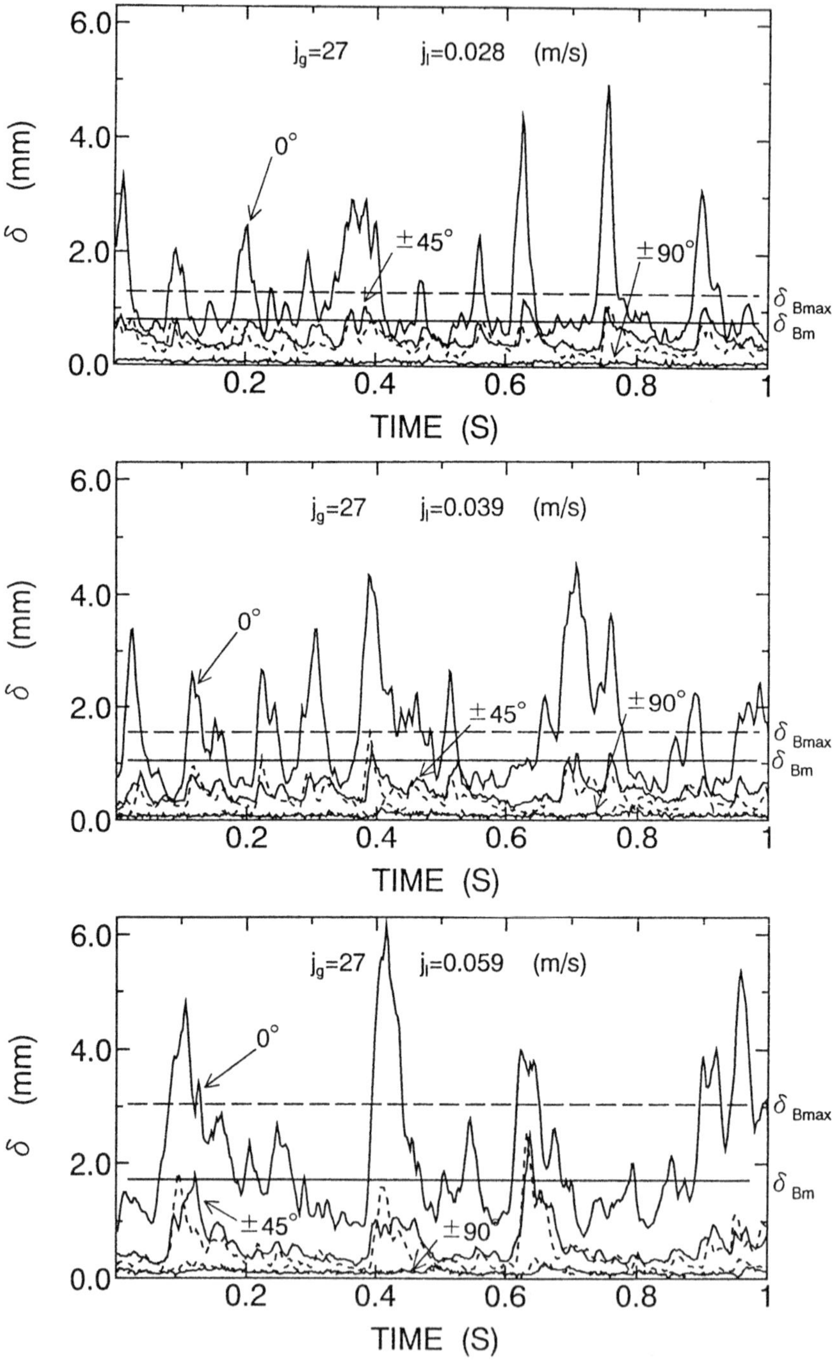

Fig. 7 Typical examples of the time traces of liquid film thickness

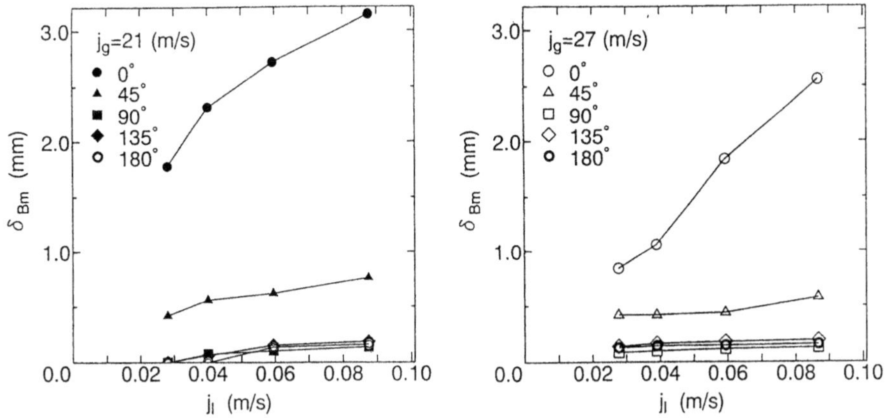

Fig. 8 Averaged base film thickness as a function of j_L

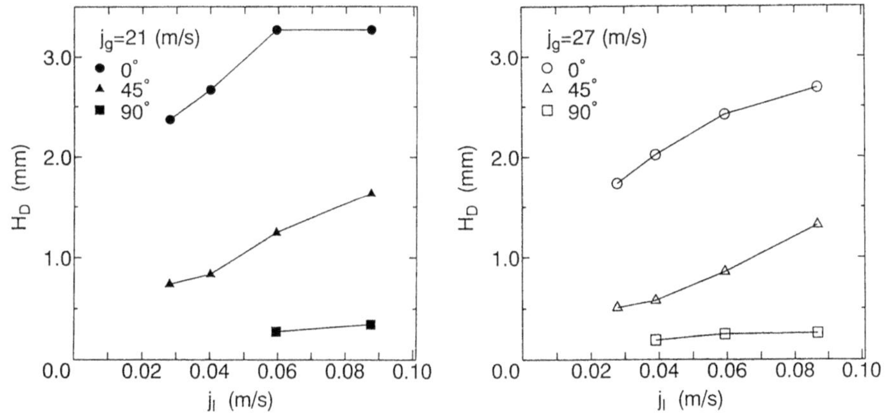

Fig. 9 Averaged height of disturbance waves as a function of j_L

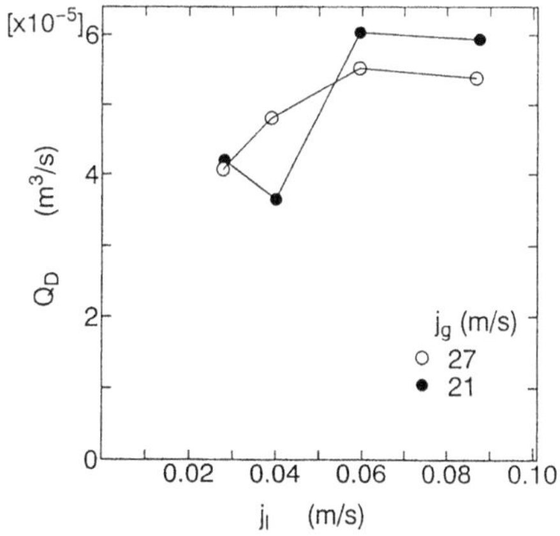

Fig. 10 Volumetric flow rate carried by disturbance waves

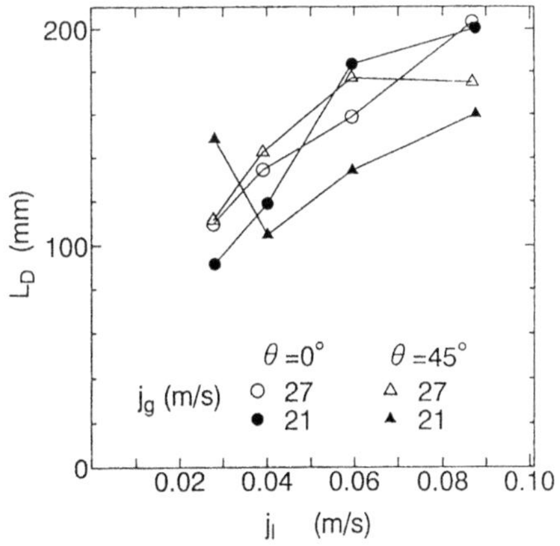

Fig. 11 Averaged streamwise length scale of disturbance waves as a function of j_L

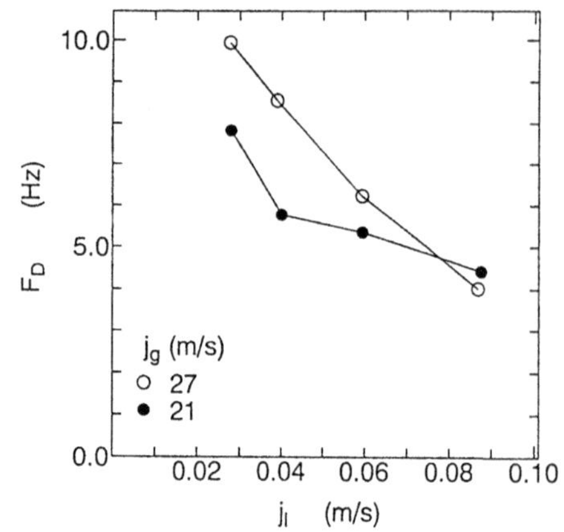

Fig. 12 Passing frequency of disturbance wave

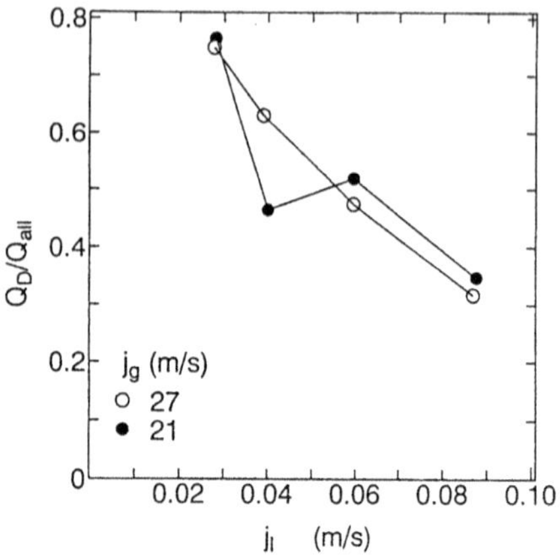

Fig. 13 Fraction of the volumetric flow rate to the total liquid flow rate

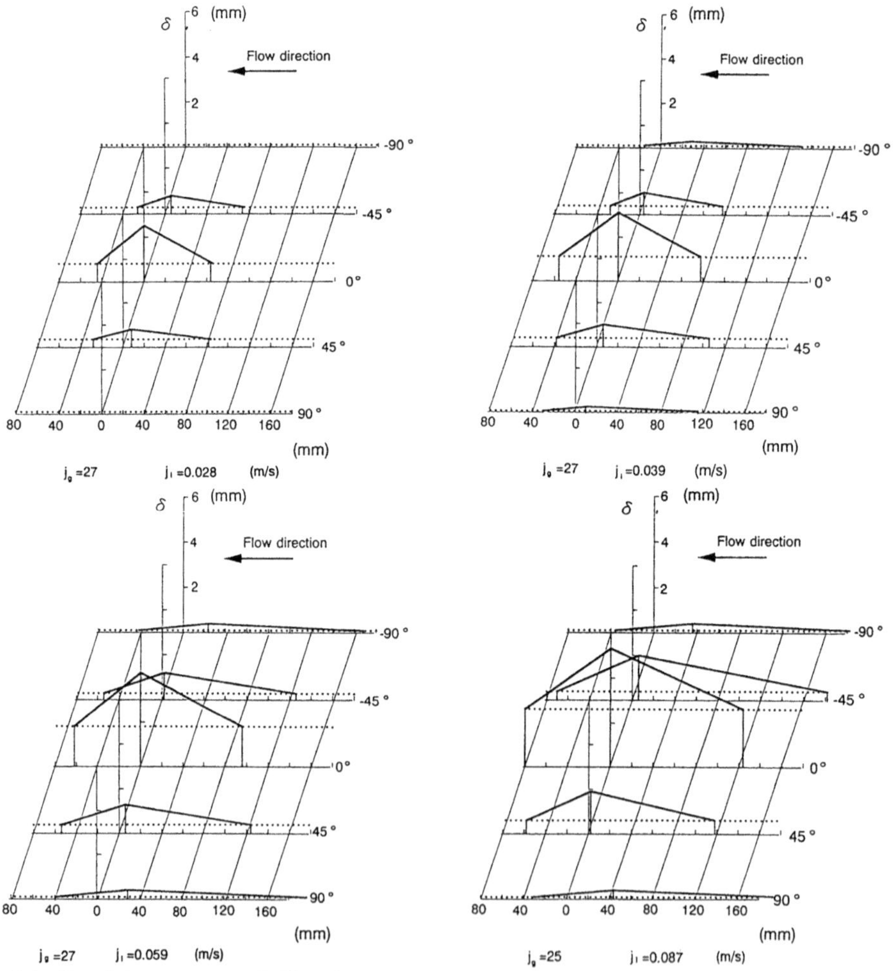

Fig. 14 Spatial structure of disturbance waves

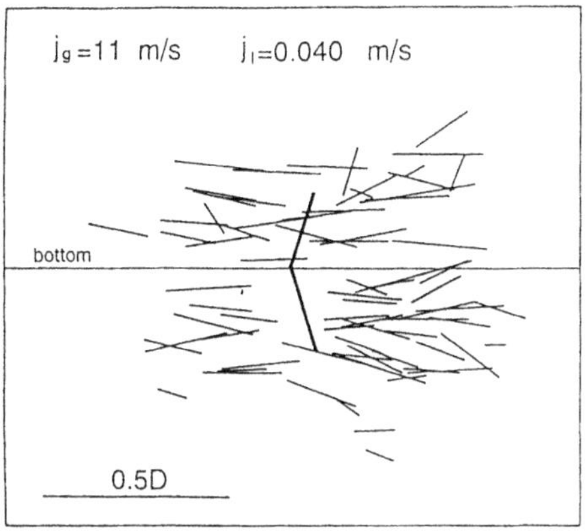

Fig. 15 Trajectories of tracer particles

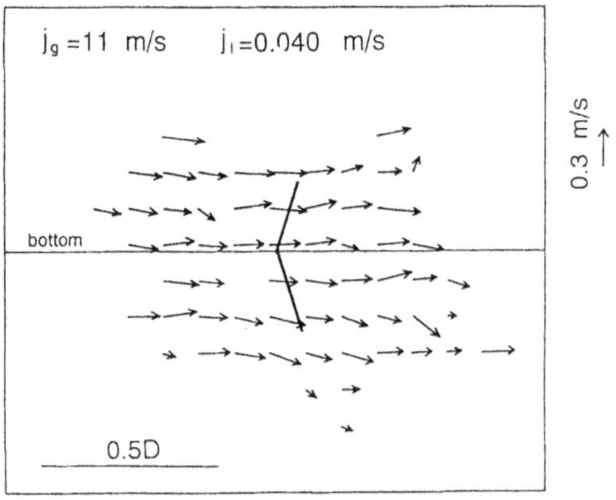

Fig. 16 Velocity vector of the base film flow

5. CONCLUSIONS

The measurement of the time change of the liquid film thickness and the visual observation of the interface and the secondary flow inside the liquid film were carried out for air–water wavy/annular two–phase flows in a horizontal pipe. The liquid–film time traces obtained at different locations and the images of visualized flow were processed. The main results obtained are as follows: (1)It was found that the contribution of disturbance waves to the liquid transport downstream becomes smaller for higher liquid flow rate. (2)The wave–crest line of disturbance waves were curved. (3)The existence of a secondary flow was suggested inside the liquid film.

REFERENCES

HAGIWARA, Y. et al., 1989, Simultaneous measurement of liquid film thickness, wall shear stress and gas flow turbulence of horizontal wavy two–phase flow, Int. J. Multiphase Flow, Vol. 15, pp.421–431.

HEWITT, G. F., 1978, Measurement of two–phase flow parameters. Academic Press.

KOSKIE, J. E. et al., 1989, Parallel–wire probes for measurement of thick liquid films, Int. J. Multiphase Flow, Vol. 15, pp.521–530.

OSAKA, A. et al., 1991, Air–water annular two–phase flow in horizontal and near horizontal tubes – circumferential variations of film thickness parameters – (in Japanese), Japanese J. Multiphase Flow, Vol.5, pp.332–342.

OSAKA, A. et al., 1992, Air–water annular two–phase flow in horizontal and near horizontal tubes – disturbance wave characteristics and liquid transportation – (in Japanese), Japanese J. Multiphase Flow, Vol.6, pp.80–87.

SEKOGUCHI, K. et al, 1973, Studies on liquid film in gas–liquid annular two–phase flows (in Japanese), Trans. of Japan Soc. of Mech. Eng., Vol. 39, pp.313–323.

TAKADA, K. et al., 1991, Analysis of fluid flow surrounding a growing, detaching and rising air bubble using image processing, Proc. of the International Conference on Multiphase Flows, '91–Tsukuba, Vol. 1, pp.381–384.

YAMAUCHI, S. and SAWAI, M., 1993, Experimental techniques on the behaviour of liquid film in gas–liquid co–current film flows and boiling two–phase flows in pipes (in Japanese), Note for the meeting of Japan Soc. of Mech. Eng.

Pressure wave propagation in a separated liquid-gas layer in a horizontal duct

T. Koshinaka, R. van der Woude[†] and S. Morioka
Dept. of Aeronautical Engineering
Kyoto University, 606-1, Japan
[†]University of Twente, P.O. Box 217
AE Enschede, The Netherlands

1 Introduction

A weak pressure wave propagating in a separated liquid-gas stratified system in a horizontally placed duct was investigated theoretically[1]. Such a pressure wave consisted of infinite modes and each mode exhibited a peculiar dispersion property, due to the geometrical dispersion, and a weak nonlinear wave was described by Korteweg-de Vries equation.

However, if we intend to prove them experimentally by a shock tube and using air and water, we meet with some difficulties[2,3]. The instantaneous removal of a diaphragm in water is difficult, and the ratio of gas density to liquid density is extremely small at the atmospheric pressure level. Further a weak plane shock wave is difficult to make in a shock tube.

In this paper, we consider a shock tube with a step, in order to remove the diaphragm instantaneously and to make a clear plane shock wave.[3] First the shock tube problem with air and water is solved analytically for a weak initial pressure ratio, then it is extended to the case of stronger initial pressure ratio. The dispersion property and the associated flow stability are discussed. Finally, an experiment is performed by a shock tube with a step, and the experimental result is compared with the theoretical one.

2 Solution for weak pressure wave

A horizontally placed duct has rectangular cross-sections with a constant width but different height in the left hand side and the right hand side, that means the duct has a step, as in Fig.1. Water is filled up to the step level in the right hand side of the duct. The diaphragm is set at the location of the step. The left hand side of the duct is set up a high pressure, and the right hand side of the duct a low pressure. When the diaphragm is broken down, a compression wave propagates into the separated liquid-gas stratified layers, and an expansion wave propagates into the gas side. The wave motion is assumed to be two-dimensional and no variation is in the direction perpendicular to the sheet.

S. Morioka and L. van Wijngaarden (eds.),
IUTAM Symposium on Waves in Liquid/Gas and Liquid/Vapour Two-Phase Systems, 269-279.

The equations governing the linear pressure wave are those of sound wave propagating in gas and liquid layers. They are provided as follows, in the nondimensional form,

$$\frac{\partial^2 p_1'}{\partial t^2} - \left(\frac{\partial^2 p_1'}{\partial x^2} + \frac{\partial^2 p_1'}{\partial y^2}\right) = 0, \tag{1}$$

$$\frac{\partial^2 p_2'}{\partial t^2} - c_*^2 \left(\frac{\partial^2 p_2'}{\partial x^2} + \frac{\partial^2 p_2'}{\partial y^2}\right) = 0, \tag{2}$$

where p' is the perturbed pressure and is made nondimensional by the initial value p_0 in the lower pressure side. The suffixes 1 and 2 denote the quantities related to gas and liqid, respectively. The time t and the space coordinates x, y are made nondimensional by L/c and L, respectively, where L is the duct height on the right hand side and c the speed of sound in gas at rest. c_* is the ratio of the speed of sound in liquid to that in gas.

The initial and boundary conditions corresponding to the above configuration are provided as follows,

$$p_1' = H(-x), \quad \frac{\partial p_1'}{\partial t} = 0, \quad p_2' = \frac{\partial p_2'}{\partial t} = 0, \quad \text{at} \quad t = 0, \tag{3}$$

$$\frac{\partial p_1'}{\partial y} = 0, \quad \text{at} \quad y = h, \quad -\infty < x < \infty, \tag{4}$$

$$\frac{\partial p_1'}{\partial y} = 0, \quad \text{at} \quad y = 0, \quad -\infty < x < 0, \tag{5}$$

$$p_1' = p_2', \quad \frac{\partial p_1'}{\partial y} = \frac{1}{\rho_*}\frac{\partial p_2'}{\partial y}, \quad \text{at} \quad y = 0, \quad 0 < x < \infty, \tag{6}$$

$$\frac{\partial p_2'}{\partial y} = 0, \quad \text{at} \quad y = -(1-h), \quad 0 < x < \infty, \tag{7}$$

$$\frac{\partial p_2'}{\partial x} = 0, \quad \text{at} \quad x = 0, \quad -(1-h) < y < 0, \tag{8}$$

where ρ_* denotes the ratio of liquid density to gas density.

Seeking for the solution of the above initial-boundary value problem in the perturbation expansion form with respect to small parameter $1/\rho_*$:

$$p_1' = p_1^{(0)} + \frac{1}{\rho_*} p_1^{(1)} + \cdots, \quad p_2' = p_2^{(0)} + \frac{1}{\rho_*} p_2^{(1)} + \cdots, \tag{9}$$

we can find the zeroth-order terms $p_1^{(0)}$ and $p_2^{(0)}$ separately.

First, $p_1^{(0)}$ can be found by the following equations :

$$\frac{\partial^2 p_1^{(0)}}{\partial t^2} - \left(\frac{\partial^2 p_1^{(0)}}{\partial x^2} + \frac{\partial^2 p_1^{(0)}}{\partial y^2}\right) = 0 \tag{10}$$

$$p_1^{(0)} = H(-x), \quad \frac{\partial p_1^{(0)}}{\partial t} = 0, \quad \text{at} \quad t = 0, \tag{11}$$

$$\frac{\partial p_1^{(0)}}{\partial y} = 0, \quad \text{at} \quad y = h, \quad -\infty < x < \infty, \tag{12}$$

$$\frac{\partial p_1^{(0)}}{\partial y} = 0, \quad \text{at} \quad y = 0, \quad -\infty < x < \infty, \tag{13}$$

Then, using the above $p_1^{(0)}$ into the boundary condition (16), $p_2^{(0)}$ can be found by the following equations :

$$\frac{\partial^2 p_2^{(0)}}{\partial t^2} - c_*^2 \left(\frac{\partial^2 p_2^{(0)}}{\partial x^2} + \frac{\partial^2 p_2^{(0)}}{\partial y^2} \right) = 0, \tag{14}$$

$$p_2^{(0)} = \frac{\partial p_2^{(0)}}{\partial t} = 0, \quad \text{at} \quad t = 0, \tag{15}$$

$$p_2^{(0)} = p_1^{(0)}, \quad \text{at} \quad y = 0, \quad x > 0, \tag{16}$$

$$\frac{\partial p_2^{(0)}}{\partial y} = 0, \quad \text{at} \quad y = -(1-h), \quad x > 0, \tag{17}$$

$$\frac{\partial p_2^{(0)}}{\partial x} = 0, \quad \text{at} \quad x = 0, \quad -(1-h) < y < 0, \tag{18}$$

Solving these linear equations, we obtain

$$p_1^{(0)} = \frac{1}{2} - \frac{1}{4}[\operatorname{sgn}(x-t) + \operatorname{sgn}(x+t)], \tag{19}$$

$$\begin{aligned} p_2^{(0)} = &\frac{1}{\pi} \int_0^\infty \frac{\operatorname{ch}[\mu(y+1-h)k]}{\operatorname{ch}[\mu(1-h)k]} \sin(tk)\cos(xk)\frac{\mathrm{d}k}{k} \\ &- 2\sum_{n=0}^{\infty}(-1)^n \int_0^\infty \frac{\left(n+\frac{1}{2}\right)(1-h)\cos\left[\left(n+\frac{1}{2}\right)\pi(y+1-h)/(1-h)\right]}{\left[\mu^2(1-h)^2k^2 + \left(n+\frac{1}{2}\right)^2\pi^2\right]\omega_n} \\ &\times \sin(t\omega_n)\cos(xk)\mathrm{d}k, \end{aligned} \tag{22}$$

where

$$\mu^2 = 1 - \frac{1}{c_*^2}, \tag{23}$$

$$\omega_n^2 = c_*^2 \left[k^2 + \left(n + \frac{1}{2}\right)^2 \pi^2 (1-h)^{-2} \right]. \tag{24}$$

Particularly, at the duct bottom, $y = -(1-h)$, we have

$$\begin{aligned} p_2^{(0)} = &\frac{1}{\pi} \int_0^\infty \frac{\sin(tk)\cos(xk)}{\operatorname{ch}[\mu(1-h)k]} \frac{\mathrm{d}k}{k} \\ &- 2\sum_{n=0}^{\infty}(-1)^n \int_0^\infty \frac{\left(n+\frac{1}{2}\right)(1-h)\sin(t\omega_n)\cos(xk)\mathrm{d}k}{\left[\mu^2(1-h)^2k^2 + \left(n+\frac{1}{2}\right)^2\pi^2\right]\omega_n}. \end{aligned} \tag{25}$$

3 Aspect of dispersion

When the ratio of gas density to liquid density is extremely small, the aspect of the dispersive pressure wave can be observed only in the liquid layer. The first integral in the right hand side in eqs.(22) and (25) corresponds to the main wave (the zeroth-mode) propagating with the speed of sound in gas, and the sum of integrals in the second line corresponds to the precursor wave whose front propagates with the speed of sound in liquid. The factor $(-1)^n$ suggests the reflection process between the duct bottom and the liquid-gas interface. Equation (24) denotes the dispersion relation of the precursor wave, and shows that it consists of infinite modes and their phase and group velocities are given by

$$\frac{\omega_n}{k} = c_* \left[1 + \frac{\left(n+\frac{1}{2}\right)^2 \pi^2}{(1-h)^2 k^2} \right], \tag{26}$$

$$\frac{\partial \omega_n}{\partial k} = \frac{c_* k}{\left[k^2 + \frac{\left(n+\frac{1}{2}\right)^2 \pi^2}{(1-h)^2} \right]^{\frac{1}{2}}}. \tag{27}$$

As the wavenumber k increases from 0 to ∞, the phase velocity decreases from ∞ to c_*, while the group velocity increases from 0 to c_*. We may notice that these are the same form as the dispersion relation for Klein-Gordon equation[3,4]. These exhibit the geometrical dispersion characteristics of the pressure wave in the liquid layer, depending on the depth of liquid layer $(1-h)$.

Figure 2 shows the temporal variation of the pressure distribution on the duct bottom, for the ratio of liquid density to gas density $\rho_* = 10^3$, and for the ratio of the speed of sound in liquid to that in gas $c_* = 4$. We can find out the diffusive structure of the main wave propagating with the speed of sound in gas and the dispersive structure of the weak precursor wave whose front propagates with the speed of sound in liquid. As the depth of liquid layer increases the diffusive structure of the main wave and the oscillatory structure of the precursor wave appear more remarkably.

4 Extension to stronger wave

As the initial pressure ratio of the shock tube increases, the pressure wave in gas layer developes to a shock wave. On the other hand, the wave in liquid layer can be still dealt with the linear theory. Now we shall discuss its validity.

The largest variation in the liquid layer is considered to be about the base of the shock wave traveling in the gas layer (see Fig.3). The time Δt in which the shock wave pass through the liquid region of the length Δx can be estimated as $\Delta t \sim \Delta x/c$, where c is the propagation velocity of shock wave in the gas phase. Let the change of the x-component of the liquid velocity within the time Δt is Δu, then $\Delta u/\Delta t \sim c\Delta u/\Delta x$ must be balanced by $\Delta p/\rho\Delta x$, according to the linearized equation of motion, where ρ is the liquid density and Δp is the pressure difference across the shock wave. Thus, we have $\Delta u \sim \Delta p/\rho c$. If we assume $\Delta p \sim 10^6$ Pa, $\rho = 10^3$kg/m^3, $c = 3.3\times10^2$m/s, $\Delta x = 10^{-2}$m, we have $\Delta u \sim 0.3$m/s. For this case, the nonlinear term $(\Delta u)^2/\Delta x$ is in the order of 10 m/s^2 and it is 10^{-3} times as small as the inertia terms of 10^4.

On the other hand, from the continuity equation or the y-component of the equation of motion, we can estimate as $\Delta v \sim \Delta u$ assuming $\Delta y \sim \Delta x$. Hence the displacement ΔY of the interface in the passing time of the shock wave has the order of magnitude $\Delta Y \sim \Delta v \Delta t \sim \Delta v \Delta x/c$. This is 10^{-3} times as small as Δx. The above discussions show that the pressure wave propagation in the liquid layer can be described precisely by the linearized equations as long as the initial pressure ratio in the shock tube is not so large.

Thus, the nonlinear effect is limited to the propergation velocity of the shock wave in gas over the speed of sound. The influence of this nonlinear effect on the pressure wave in the liquid layer can be found only by modifying the boundary condition on the liquid-gas interface in the previous initial-boundary value problem. In fact, it is given by modifying as

$$p_2^{(0)} = \frac{1}{2} H\left(t - \frac{x}{a_*}\right) \tag{28}$$

in eq.(14), where a_* denotes the ratio of the shock speed to the sound speed. Then the pressure distribution in the liquid layer is modified as follow,

$$\begin{aligned} p_2^{(0)} = {} & \frac{1}{\pi}\int_0^\infty \frac{\mathrm{ch}[\nu(y+1-h)k]}{\mathrm{ch}[\nu(1-h)k]}\sin(a_* tk)\cos(xk)\frac{\mathrm{d}k}{k} \\ & - 2a_* \sum_{n=0}^{\infty}(-1)^n \int_0^\infty \frac{\left(n+\frac{1}{2}\right)(1-h)\cos\left[\left(n+\frac{1}{2}\right)\pi(y+1-h)/(1-h)\right]}{\left[\nu^2(1-h)^2k^2 + \left(n+\frac{1}{2}\right)^2\pi^2\right]\omega_n} \\ & \times \sin(t\omega_n)\cos(xk)\mathrm{d}k, \end{aligned} \tag{29}$$

where

$$\nu^2 = 1 - \frac{a_*^2}{c_*^2}. \tag{30}$$

At the duct bottom, $y = -(1-h)$,

$$\begin{aligned} p_2^{(0)} = {} & \frac{1}{\pi}\int_0^\infty \frac{\sin(a_* tk)\cos(xk)}{\mathrm{ch}[\nu(1-h)k]}\frac{\mathrm{d}k}{k} \\ & - 2a_* \sum_{n=0}^{\infty}(-1)^n \int_0^\infty \frac{\left(n+\frac{1}{2}\right)(1-h)\sin(t\omega_n)\cos(xk)\mathrm{d}k}{\left[\nu^2(1-h)^2k^2 + \left(n+\frac{1}{2}\right)^2\pi^2\right]\omega_n}, \end{aligned} \tag{31}$$

where ω_n is given by eq.(24) again.

Figure 4 shows the temporal variation of the pressure distribution at the duct bottom, that is the graph of eq.(31). The nonlinear effect of the wave in gas layer is to increase the amplitude of the oscillatory structure in the precursor wave, as well as the speedup of the main wave.

5 Stability of associated flow

When the shock wave in the gas layer becomes stronger and hence the flow velocity behind the shock wave is faster, the Kelvin-Helmholtz instability must be taken into account. Here we shall discuss such a possibility on the basis of a simple flow of incompressible inviscid fluid. If two uniform flows of such fluids contact on a horizontal plane $y = 0$ in the duct, $h \geq y \geq -(1-h)$, where the density and the flow velocity of the upper fluid are assumed ρ_1 and U_1 and those of the lower fluid ρ_2 and U_2, respectively, then the unstable condition for the wave of the wavelength Δx can be presented approximately by

$$U_1^2 > \Delta x \cdot g \cdot \frac{\rho_2}{\rho_1} \cdot \mathrm{th}(\frac{hL}{\Delta x}), \tag{32}$$

assumeing as $\rho_1/\rho_2 \ll 1$, $U_2/U_1 \ll 1$. If we assume $\Delta x \sim 10^{-2}$m, then $U_1 \sim 10$m/s. In this discussion, the Kelvin-Helmholtz instability will be promoted with decreasing thickness of the gas layer. Here the estimation has been made for the wavelength of Δx, though the generated waves may consist of various wavelengths in general. The wave with shorter wavelength will be unstable for the lower fow velocity.

On the other hand, from the jump condition across a normal shock wave, the flow velocity U_1 behind the shock wave can be estimated as

$$U_1 = \frac{2(M^2-1)c}{(\gamma-1)M}, \tag{33}$$

where c is the speed of sound in gas at rest and M the shock Mach number. Hence the wave whos wavelength is 10^{-2}m becomes unstable for $M > 1.02$. Thus the Kelvin-Helmholtz instability is possible to begin at weak shock wave. As well known, Kelvin-Helmholtz instability is the unstable gravity wave and the propagation velocity is extremely slow compared with that of the pressure wave. Thus the instability will develope toward downstream from the step on the liquid-gas interface.

6 Experimental apparatus and instrumentation

Figure 5 shows the schematic diagram of our experimental apparatus and instrumentation. The shock tube consists of a high pressure section, an adjustment section and observation sections 1 and 2. The high pressure section corrsponds to high pressure chamber and the remaining sections to low pressure chamber. The high pressure chamber consists of a cylinder of circular crosssection with inner diameter 70 mm, and the low pressure chamber consists of a cylinder of square cross-section with side 50 mm. A spacer of stainless with length 600 mm is adhered in the adjustment section adjacent to high pressure section. The spacer serves as setting of liquid layer thickness, stoper for water with exchanging diaphragm, and generating a clear step shock wave. The diaphragm is selected among several sorts, such as 25μm rumiler, depending on the initial pressure ratio.

First, we wipe and dry up the inside wall in order that there is no water droplet. Then we fill up water to the setting level in low pressure hamber, that is the hight of spacer. We must take care of removing bubbls on the wall and particularly on the transducers, because they will induce a exclusive signal due to the bubble oscillation. Then we set the initial pressure ratio by pressurizing the high pressure chamber and by depressurizing the low pressure chamber.

The pressure has been measured by piezo transducers (Kisler 601A). The measuring points are located at upper wall (CH1, CH2) and lower wall (CH3, CH4) downstream of 1900 mm and 2000 mm from the diaphragm. The voltage signals from the transducers are amplified by charge amplifier, and are indicated on a oscilloscope (Yokokawa Denki DL1200E). Then we can observe the data memorized before the shock wave has arrived, by using a time laterder. We have also taken the photograph at the location of the transducers. The photograph has been taken by flashing a stroboscope (Sugahara MS-230A, single flash time 1μs) by making use of the above trigger system.

7 Experimantal result and discussion

The detected pressure profiles are shown in Fig.6 for six different conditions of the initial pressure ratio and the liquid layer thickness as indicated in Table 1. The four wave profiles in each figure corresponds to those at CH1, CH2, CH3 and CH4, from top to bottom. The main data measured in these conditions are shown in Table 2.

As for the precursor wave along the bottom of liquid layer, we can recognize the existence, but the oscillatory structure as predicted by the theory cannot be identified by interference of irregular signal. The pressure wave penetrating from gas to liqid may be reduced proportional to the ratio of gas density to liquid density in the present configuration.

The main wave at the bottom of liquid layer propagates with the shock speed in gas and the diffusive structure is enlarged with increasing thickness of liquid layer, consistent with the theoretical prediction. For all the data, we can find a pressure reduction behind the shock wave. Fig.7 shows the variation of the shock speed and the shock strength by the initial pressure ratio for two different thickness of liquid layer. Such a pressure reduction behind the shock wave becomes remarkable with increasing thickness of liquid layer. As one of the causes, we can consider the effect of flusing water due to Kelvin-Helmholtz instability. In fact, we can observe such an instability from the photographs (Fig.8) of liquid-gas interface taken after the shock wave has passed over.

References

1. Morioka, S & Matsui, G. 1975 J. Fluid Mech. 70, 721.
2. Matsui, G. 1975 Proc. of Non-equilibrium two-phase flow, ASME.
3. Stone, T.W. 1983 Thesis, McMaster Univ. Canada.
4. Koshinaka, T & Morioka, S 1993 Fluid Dynamics Research 12, 323.

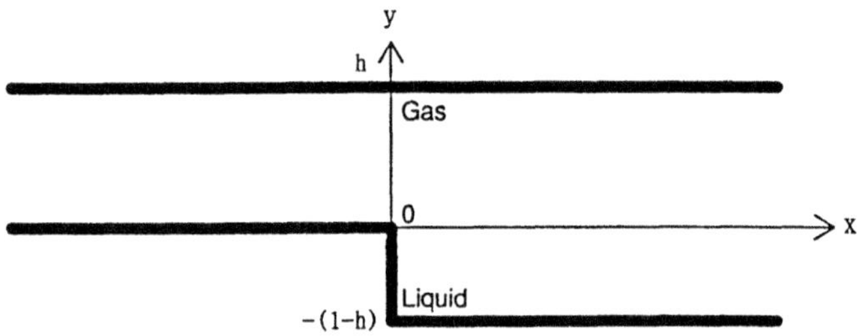

Fig. 1. Liquid-gas configuration in a duct with a step

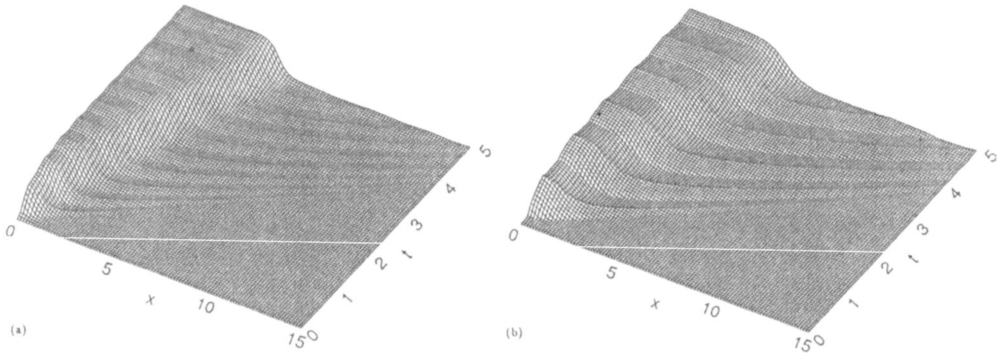

Fig. 2. Temporal variation of pressure distribution at duct bottom for a small initial pressure ratio : (a) 1-h=0.5, (b) 1-h=0.9 ; c_*=4.

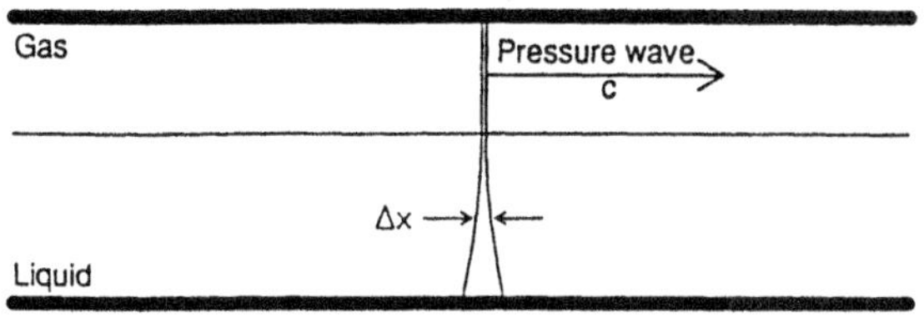

Fig.3. Acceleration of liquid with passing shock wave.

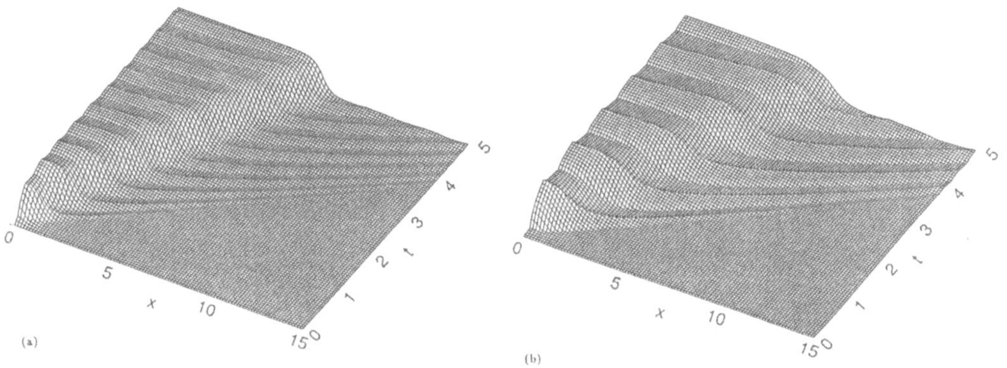

Fig.4. Temporal variation of pressure distribution at duct bottom for a large initial pressure ratio: (a) 1-h=0.5, (b) 1-h=0.9; $c_*=4$, $a_*=1.5$.

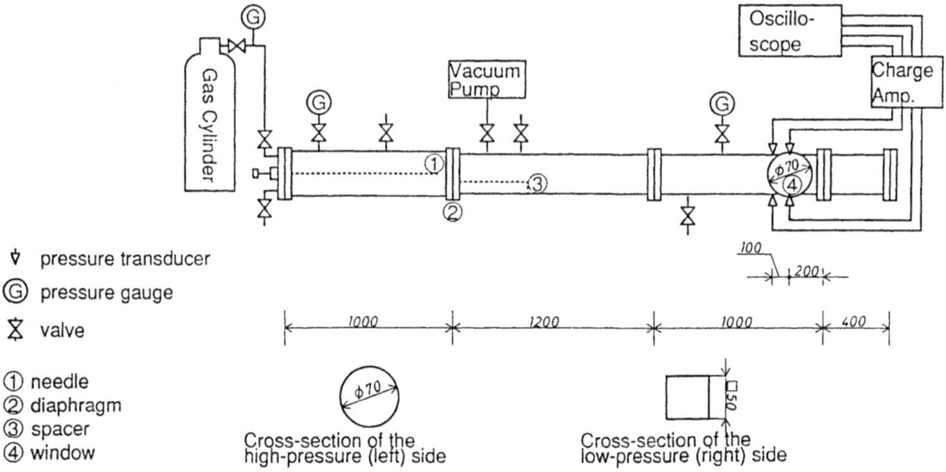

Fig.5. Schematic diagram of experimental apparatus and instrumentation.

	h (cm)	p_4/p_1	p_4 (atm)	p_1 (Torr)
Ex.1-1	2.0	1.4	1.4	760
Ex.1-2		3.4	2.7	600
Ex.1-3		5.1	4.0	600
Ex.1-4	4.0	1.4	1.4	760
Ex.1-5		3.4	2.7	600
Ex.1-6		5.1	4.0	600

Table 1. Summary of test conditions

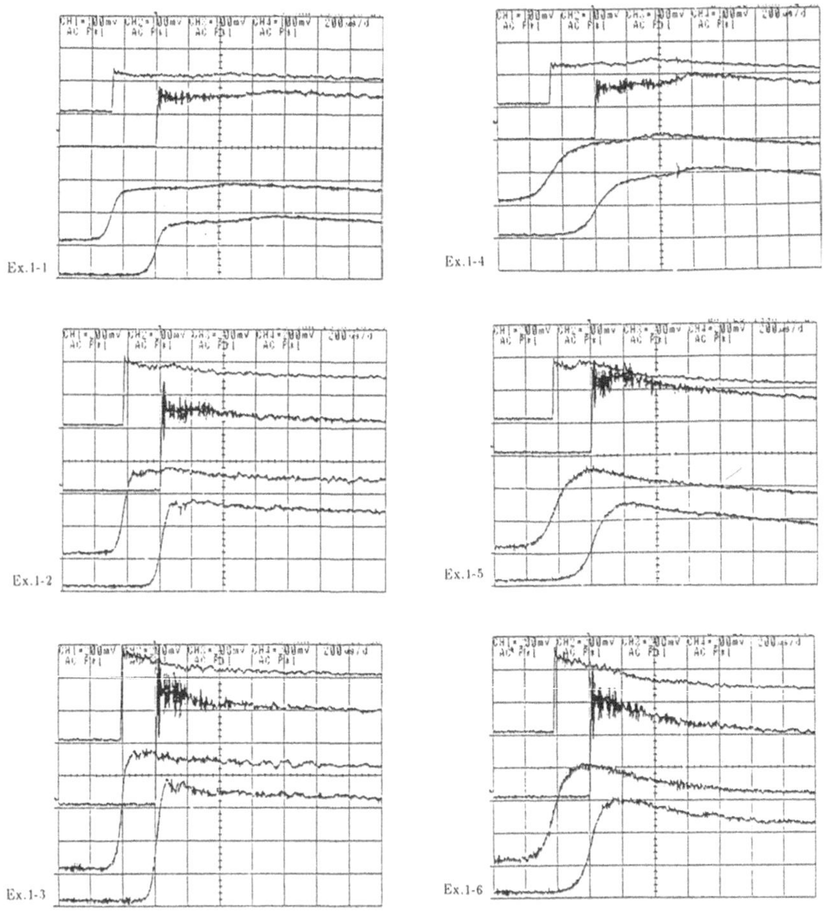

Fig.6. Detected pressure profiles for test conditions in Table 1: Four profiles from top to bottom in each figure correspond to those at CH1, CH2, CH3 and CH4, respectively.

	h (cm)	p_4/p_1	M_s	p_2/p_1	$\Delta p/p_1$
Ex.1-1	2.0	1.4	1.08	1.24	0.019
Ex.1-2		3.4	1.28	1.95	0.25
Ex.1-3		5.1	1.38	2.33	0.34
Ex.1-4	4.0	1.4	1.08	1.25	0.012
Ex.1-5		3.4	1.26	1.86	0.31
Ex.1-6		5.1	1.35	2.19	0.51

Table 2. Shock Mach number M_s, shock strength p_2/p_1 and shock attenuation $\Delta p/p_1$ vs initial pressure ratio p_4/p_1 for six test conditions.

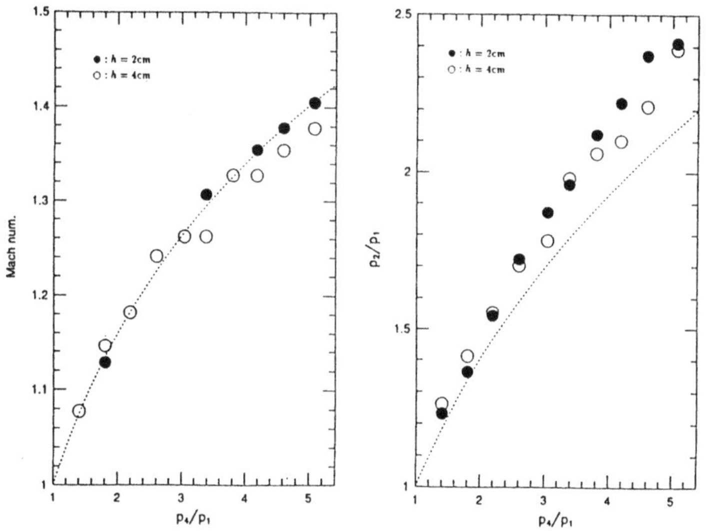

Fig.7. Shock Mach unmber and shock strength vs initial pressure ratio.

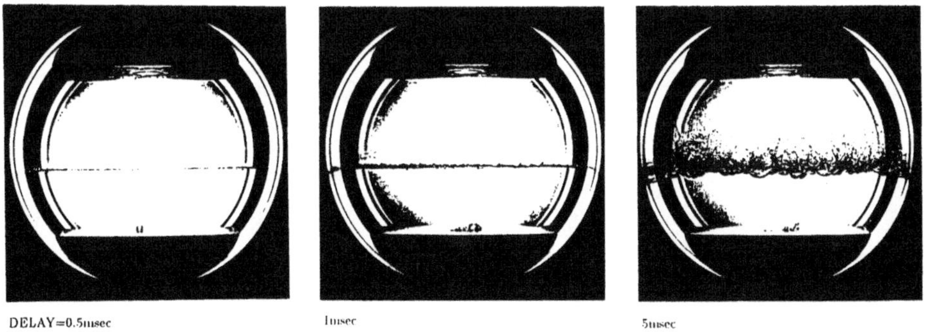

Fig.8. Photograph of liquid-gas interface after shock has passed over.

Dynamics of a Single Reactive Gas Bubble

B. Bruckert, D. L. Frost,
A. N. Meidani, and R. Chue

Mechanical Engineering Department
McGill University, Montreal, Canada

M. Brouillette

Département de génie mécanique,
Université de Sherbrooke,
Sherbrooke, Canada

Summary

The dynamics of a single combustible stoichiometric oxyacetylene gas bubble immersed in glycerine subjected to an impulsive pressure wave has been studied experimentally and theoretically. Emphasis is placed on determining the range of bubble sizes over which ignition of the gas occurs after the passage of a pressure pulse generated by the impact of a projectile with a piston. Bubbles with diameters ranging from 0.3 to 7 mm were subjected to pressure pulses with peak pressures ranging from 2.5 to 13 MPa with a typical pulse width of 50 μs. Ignition of the gas was observed only for bubbles with a diameter greater than 0.7 mm. The minimum bubble size for explosion increased as the peak pressure and impulse of the applied pressure wave decreased. A simple model has been developed which reproduces the general features of the observed bubble dynamics as well as the limiting bubble size for explosion. The minimum bubble size for explosion may be governed largely by either heat transfer from the bubble or surface tension and viscosity effects, depending on the amplitude and shape of the applied pressure pulse. For the largest bubbles tested, asymmetric collapse of the bubble in some cases prevented the ignition of the bubble.

1 Introduction

It is well known that the compression of a bubble containing a chemically reactive gas by a shock wave of sufficient strength can cause the bubble to explode [1]. In a linear array of combustible gas bubbles, Hasegawa and Fujiwara [2] have shown that the pressure disturbance generated by the explosion of one bubble can trigger the ignition of an adjacent bubble. They observed that the bubbles explode sequentially, although the propagation velocity associated with the explosions decayed with time [3]. If a shock wave is transmitted into a liquid containing multiple reactive gas bubbles, a high-pressure reactive wave may separate from the shock and propagate at a high speed through the bubbly liquid. The existence of such a self-sustained reactive wave was first demonstrated by Sychev [4] and extensive experimental work in the same laboratory [5,6] has shown that the propagation of the wave is independent of the initiation conditions and that the wave velocity is a function of the void fraction of the bubbly mixture, the energetics of the gas mixture, and the liquid viscosity.

S. Morioka and L. van Wijngaarden (eds.),
IUTAM Symposium on Waves in Liquid/Gas and Liquid/Vapour Two-Phase Systems, 281-291.

A number of researchers have investigated the propagation mechanism of a reactive wave in a bubbly liquid [6–10]. Although the reactive wave is often referred to as a "bubble detonation", it has little in common with classical Chapman-Jouguet detonation waves in a reactive gas which are driven by equilibrium thermodynamics. The reactive wave in a bubbly liquid is governed by nonequilibrium bubble dynamics, and involves a balance between the chemical energy release and the dissipation in the medium. At least three physical length scales, denoted by the terms microscale, mezoscale and macroscale in [9], corresponding to the bubble diameter, interbubble distance which is related to the void fraction and an integral scale such as tube diameter, play a role in the propagation of the wave. Although some information is available on the dependence of the wave velocity on these length scales, e.g., the influence of void fraction [4] or the influence of bubble diameter [11], little information is available on dynamic parameters such as critical initiation energy and the propagation limits. The study of the dynamics of a single reactive gas bubble is relevant to the understanding of the micro-explosions that occur within the propagating wave. However, since the propagation of a reactive wave in a bubbly liquid involves the coherent explosion of a collection of bubbles, the study of a single reactive gas bubble will not give any information on the coupling between bubbles or on the macroscopic wave phenomenon itself. Nevertheless, a detailed study of the explosion limits of a single bubble will provide important information on the limits of propagation of the reactive wave.

The present experiments focus on the dynamics of a single combustible gas bubble in a liquid, with emphasis on the limits which govern the explosion of the bubble. Stoichiometric oxyacetylene bubbles ranging in size from 0.3 to 7 mm are injected into a vertical tube filled with glycerine and subjected to shock loading. The pressure pulse that is used to compress and ignite the bubbles is generated by the impact of a projectile with a piston placed at the free surface of the liquid. In this way a short duration pressure pulse is generated which simulates the pressure field experienced by a bubble within a propagating reactive wave. Two explosion limits are being investigated: the small bubble limit where heat losses from the bubble surface are large enough to prevent the bubble temperature from reaching the autoignition temperature, and the large bubble limit, where the bubble collapse becomes highly asymmetric, also preventing a sufficient temperature rise.

In the present paper, a selection of the experimental results that illustrates the dynamics of a single reactive bubble and explosion limits will first be presented. This will be followed by a discussion of a simple model that reproduces the general features of the results and finally some suggestions for future work will be given.

2. Experimental

A schematic of the experimental apparatus is shown in Fig.1. The experiments were carried out in a stainless steel cylindrical vessel (5.1 cm dia, 20.3 cm long) filled with glycerine. A microliter syringe is used to extract a small amount of premixed stoichiometric acetylene-oxygen mixture from a mixing chamber. A bubble is injected into a narrow flexible tube filled with glycerine which is then injected into the chamber via a hypodermic needle inserted into the base of the vessel through a septum-lined valve. Due to the large viscosity of the glycerine, the bubble rises slowly allowing it to preserve its spherical shape. The compression wave that ignites the bubble is generated by the impact of a projectile accelerated by compressed air with a steel piston placed at the upper surface of the glycerine. By placing a thin lead sheet on the piston surface and adjusting the projectile velocity, a pressure profile was obtained that simulates the average pressure field (i.e., Δp ~ 5–10 MPa, pulse width 50–100 μs) associated with a self-sustained reactive wave in a bubbly liquid. Figure 2 shows the characteristic pressure pulse generated within the vessel with no bubble present as recorded by a piezoelectric pressure transducer flush-mounted on the wall of the vessel. A layer of closed cell foam was placed inside the vessel at the base to reduce the strength of the reflected pressure wave. The dynamics of the bubble motion were recorded through a window in the bottom half of the vessel with a Cordin Dynafax rotating drum camera operating at 20,000 frames/s together with a xenon flash lamp. In the absence of the flash used for lighting, the emission of bluish light generated by the combustion of the gas in the bubble was visible by eye unaided. To obtain a quantitative measure of the light emitted, a photodiode was mounted within the vessel and a conventional video camera was also used to record the light emission and to provide an estimate of the initial bubble diameter.

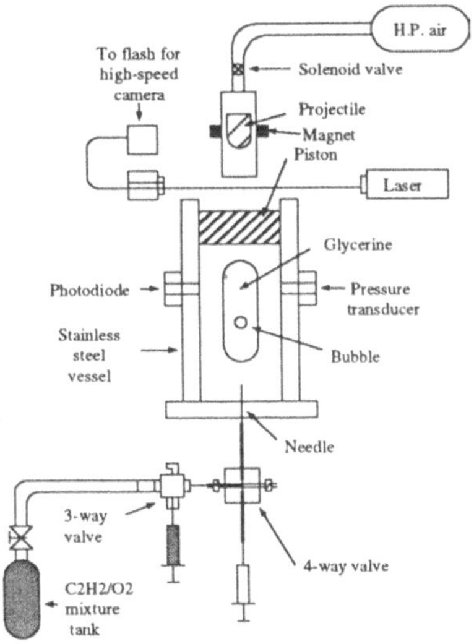

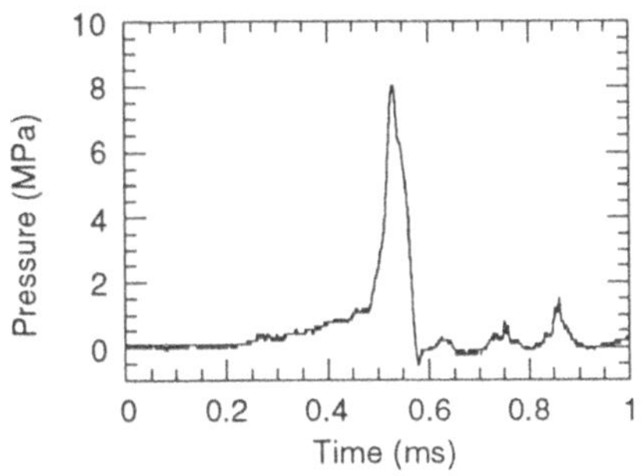

Fig. 1 Schematic of the experimental apparatus (left)

Fig. 2 Pressure recorded with no bubble present by transducer flush-mounted inside vessel (above)

3 Experimental Results

Figure 3 shows a series of single frames from the film record of a 3 mm diameter bubble subjected to a pressure pulse similar to that shown in Fig. 1, travelling from top to bottom. The photographs illustrate the characteristic compression of the bubble, ignition and combustion of the gas mixture and the subsequent bubble oscillations that occur following the interaction of the shock wave with the bubble. The bubble collapses symmetrically in a time of about 300 μs although since the time between frames is about 50 μs, the minimum diameter is not necessarily captured on film. The photodiode measurements indicate that the combustion takes place in a few microseconds. The high pressure generated by the combustion of the gas causes the bubble to expand rapidly to a maximum diameter of about 2.7 times the initial diameter in several hundred microseconds. As the bubble collapses for the second time, the shape of the bubble deforms producing a jet structure (e.g., see the photograph of the bubble at t = 924 µs) oriented in the direction of propagation of the initial pressure wave. Similar behavior has been observed by earlier investigators (e.g., [2,12]).

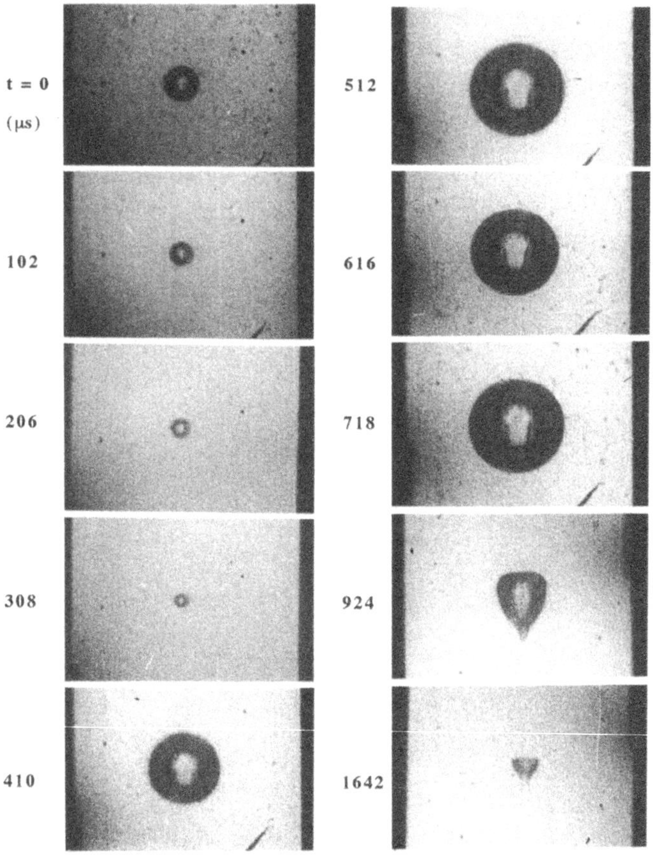

Fig. 3 Single frames from high-speed film showing collapse and explosion of a 3 mm stoichiometric oxyacetylene bubble subjected to a pressure pulse similar to that shown in Fig. 1. Applied pressure pulse propagates from top to bottom.

The history of the bubble dynamics for the bubble shown in Fig. 3 is shown in Fig. 4 in comparison to the bubble dynamics of a 3.8 mm dia noncombustible air bubble subject to a pressure pulse with a similar shape. In contrast with the reactive bubble, the air bubble rebounds to a maximum diameter of only about 1.7 times the initial diameter and the subsequent bubble oscillations decay due to viscous dissipation.

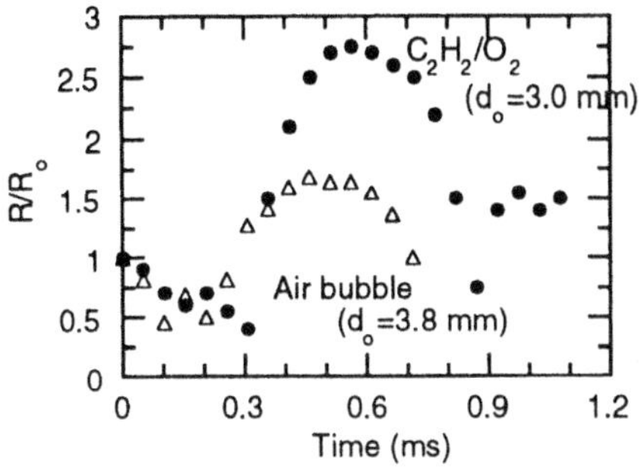

Fig. 4 Bubble dynamics for bubble shown in Fig. 3 compared with nonreactive air bubble

Since the time for bubble collapse is of the same order as the width of the applied pressure profile, small changes in the detailed shape of the applied pressure (including the strength of the reflected waves) from one experiment to the next cause significant variations in the bubble dynamics. The pressure pulse often rises relatively slowly for several hundred milliseconds before a sharp rise in the pressure. During the early stages of the applied pressure, the bubble may undergo several minor bubble oscillations before reaching the minimum diameter and exploding. An example where the bubble explosion appears to be delayed and occurs only after several small bubble oscillations is shown in Fig. 5 for a bubble with an initial diameter of 2.1 mm. The sudden increase in the bubble radius to a size of about $R/R_0 = 2.8$ after 1 ms suggests that ignition occurred at about 0.85 ms. Also shown on the same figure is the pressure recorded during this trial. The small pressure fluctuations that are recorded from 0.4 to 0.9 ms are probably due to reflected waves within the vessel and are also recorded with no bubble present (see Fig. 2). However, the sharp pressure spike recorded near 1.3 ms is probably generated by the collapse of the bubble following the large bubble oscillation.

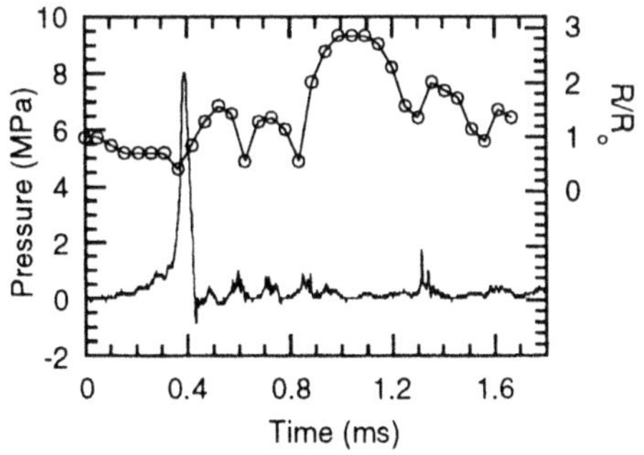

Fig. 5 Applied pressure pulse (lower trace) and bubble dynamics (upper trace) for 2.1 mm dia bubble illustrating small bubble oscillations and delayed ignition of the bubble that occurs after about 0.9 ms.

3.1 Small Bubble Explosion Limit. By varying the driving pressure for the projectile that impacts the piston, the peak pressure of the pressure pulse applied to the bubble was varied from 2.5 to 13 MPa. The peak pressure and impulse of the applied pressure wave are shown in Figs. 6 and 7, respectively, which show the results of a series of trials with bubble diameters ranging from 0.3 to 2±0.1 mm. Explosion of the bubble was detected by monitoring the light emission during the trial. No explosions were observed for bubbles smaller than 0.7 mm. As the peak pressure and impulse of the applied pressure pulse decreased, the minimum diameter for explosion increased. There will exist a minimum impulse, corresponding to the work necessary to compress the bubble to the autoignition temperature, below which no explosions will occur. From Fig. 7, the minimum impulse is less than 0.15 MPa-ms. For a given peak pressure, the scatter in the separation between the *explosion* and *no explosion* regions is a measure of the variation in the shape of the incident pressure pulse from one trial to the next.

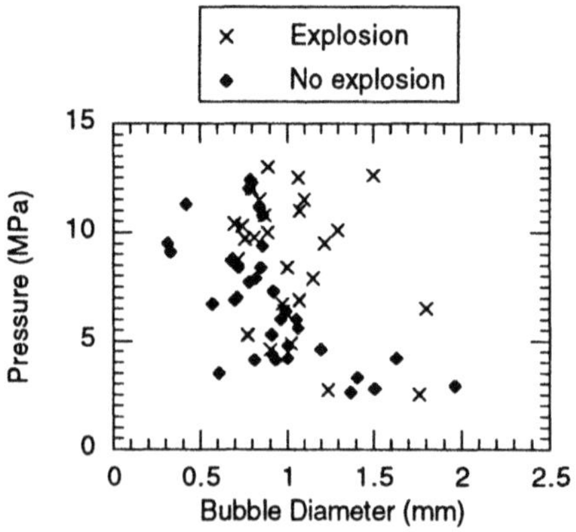

Fig. 6 Effect of peak applied pressure on small bubble explosion limit

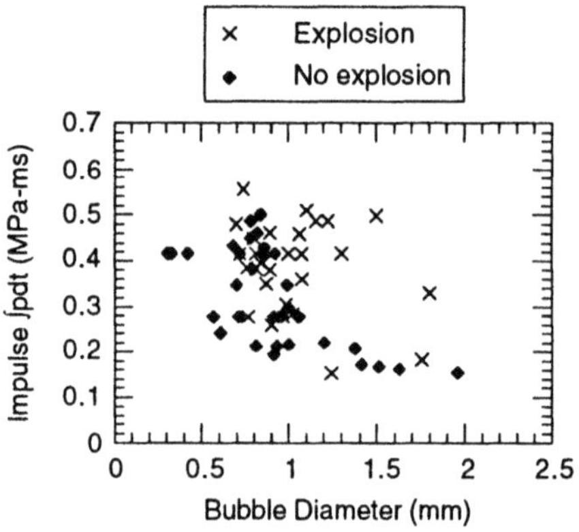

Fig. 7 Effect of applied positive phase pressure impulse on small bubble explosion limit

3.2 Large Bubble Explosion Limit. As the bubble size was increased, asymmetry in the bubble shape was first observed during the *expansion* phase following the initial collapse of the bubbles. An example of this is illustrated in Fig. 8 which shows the explosion of a 6 mm dia bubble. Occasionally, a noticeable asymmetry in the spherical shape of the bubble was observed even during the initial bubble collapse stage. This is illustrated in Fig.9 which shows a digitally enhanced image of a bubble, with an initial diameter of 6.3 mm, near its minimum diameter, highlighting the asymmetric shape. In this case, the bubble expanded to a nondimensional radius of 1.6 which suggests that ignition did not occur in contrast with the bubble of a similar size shown in Fig. 8 which expanded to a nondimensional bubble radius of 2.5, characteristic of ignition (see Fig. 4). Insufficient trials were carried out to determine an accurate value of the large bubble diameter limit for explosion, although it is estimated to be on the order of 1 cm. However, as the bubble diameter approached a value of 1 cm, the rapid rise velocity of the bubble caused departures from a spherical shape making a direct comparison with smaller, spherical bubbles difficult.

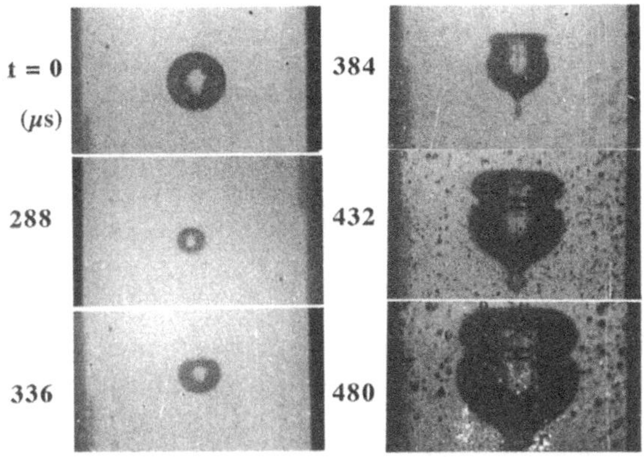

Fig. 8 Single frames taken from the explosion of a 6 mm dia bubble illustrating asymmetric bubble shapes that occur during the expansion of the bubble following ignition

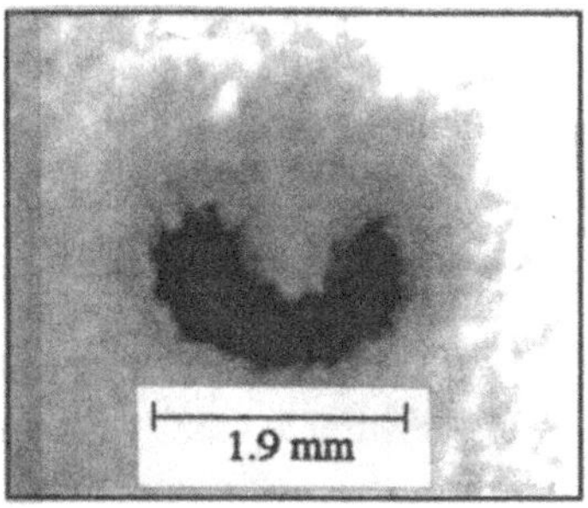

Fig. 9 Digitized image of an initial 6.3 mm bubble illustrating the asymmetric shape of the bubble during collapse

4 Bubble Dynamics Model

A relatively simple model has been developed to see if the general features of the experimental results can be reproduced. The basis of the model and the assumptions are similar to those used by Hasegawa and Fujiwara [2] with some minor modifications to the modelling of the heat transfer and chemical reaction terms. We assume that the bubble remains spherical and the pressure inside the bubble is uniform. The liquid is assumed to be incompressible and its temperature remains constant during the bubble oscillations (Kamath et al. [13] explicitly solved for the temperature field inside and outside of an oscillating bubble and have shown that the heating of the liquid outside the bubble is minimal). Ideal gas behavior is assumed and polynomial expressions for the thermodynamic properties of the gas are used to account for the dependence on temperature. Mass transfer across the bubble interface is neglected. Diffusive heat transfer from the bubble to the liquid is assumed to occur across a thin thermal boundary layer in the gas near the bubble surface and the chemical reaction of the combustible gas is modelled using a two-step Arrhenius reaction model suggested by Korobeinikov [14].

With the above assumptions, the radial bubble history $R(t)$ is governed by the Rayleigh-Plesset equation, i.e.,

$$R\ddot{R} + \frac{3}{2}\dot{R}^2 = \frac{1}{\rho_l}(p - p_\infty - 2\sigma/R - 4\nu\dot{R}/R)\,. \tag{1}$$

Here time derivatives are denoted by dots, ρ_l is the liquid density, p is the pressure in the bubble, p_∞ is the far-field liquid pressure, and σ and ν are the surface tension and viscosity of the liquid, respectively. The conservation of internal energy inside the bubble $E(t)$ is given by

$$\dot{E}(t) = \dot{Q}_r - \dot{Q}_l - \dot{W} \tag{2}$$

where $\dot{Q}_r$ is the rate of chemical energy release, $\dot{Q}_l = k(T - T_\infty)4\pi R^2(\pi\alpha\, t)^{-1/2}$ is the rate of diffusive heat loss, and $\dot{W} = (p - 2\sigma/R - 4\nu\dot{R}/R)\dot{V}$ is the rate of expansion work done by the bubble on the surrounding fluid, where T is the gas temperature, T_∞ is the liquid temperature, V is the bubble volume and k and α are the thermal conductivity and diffusivity of the gas, respectively.

Equations (1) and (2) are solved together with the equation for the conservation of mass within the bubble, the ideal gas law and Korobeinikov's two-step induction-reaction model for the chemical reaction [14]. Since the equations include a fast chemical reaction coupled with relatively slow bubble dynamics, the system of equations is stiff and the implicit Gear method is used to ensure a stable solution procedure.

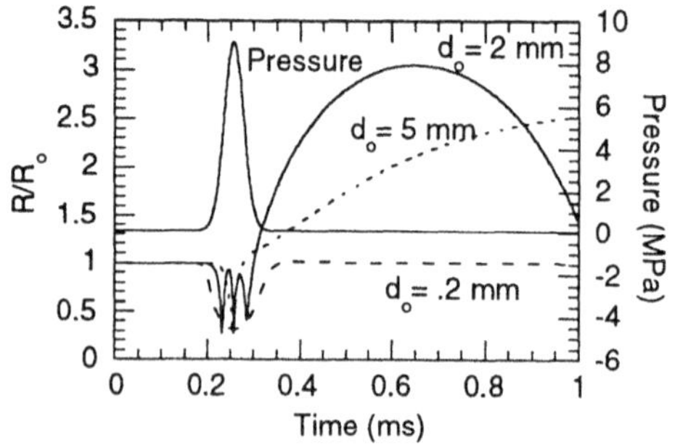

Fig. 10 Bubble dynamics predicted with model using applied gaussian pressure profile for 3 initial bubble diameters

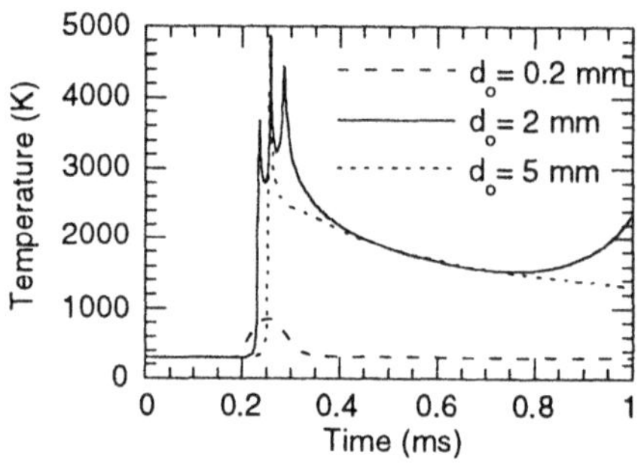

Fig. 11 Temperature history inside bubble for the bubbles shown in Fig. 7. Smallest bubble fails to reach autoignition temperature

4.1 Bubble Dynamics. Figure 10 illustrates the effect of initial bubble size on the bubble dynamics computed using the model described above. The driving pressure imposed, also shown in Fig. 10, has a gaussian profile with a peak pressure (9 MPa) and pulse width (50 μs) characteristic of the pressure pulse obtained experimentally. The temperature history of the gas inside the bubble (see Fig. 11) shows for the bubbles larger than 0.2 mm, that the autoignition temperature is attained during the first bubble collapse and combustion of the gas in the bubble occurs on a timescale of the order of microseconds. For the 2 mm bubble, the bubble collapse occurs sufficiently fast relative to the duration of the applied pressure that the bubble undergoes two small amplitude bubble oscillations prior to expanding to a large size. For the 5 mm bubble, the collapse time is longer and only a single bubble oscillation occurs. Note that since the applied pressure impedes the expansion of the bubble, the maximum nondimensional bubble radius attained for the 5 mm bubble is actually smaller than the corresponding value for the 2 mm bubble. The smallest bubble considered (0.2 mm dia) is just below the explosion limit and Fig. 11

shows that the temperature fails to attain the autoignition temperature (~ 950 K) during the compression stage.

To compare the model predictions with the bubble dynamics observed experimentally, the pressure recorded during the explosion of a 3.4 mm bubble, shown in Fig. 12, was used together with the model to predict the bubble dynamics. Since the pressure recorded also contains a contribution to the pressure field radiated by the exploding bubble itself, the recorded pressure was first used as input to the code to estimate the bubble dynamics. From the computed bubble dynamics, the pressure field generated by the expanding bubble at the transducer location was computed and is shown inset in Fig. 12. The magnitude and timing of this computed pressure pulse suggests that the first of the two pressure peaks recorded may correspond to the pressure generated by the bubble explosion. To get a better estimate of the pressure actually applied to the bubble, the computed pressure pulse was subtracted from the measured pressure profile and the resulting pressure field was used as the input pressure field for the code. The comparison between the predicted and measured bubble dynamics is shown in Fig. 13. The model reproduces the general features of the experimental results although the bubble maxima are not recovered. A variety of factors may account for the differences between the model and the experiments, including the fact that the pressure recorded at the vessel wall may differ somewhat from the pressure field at the bubble location.

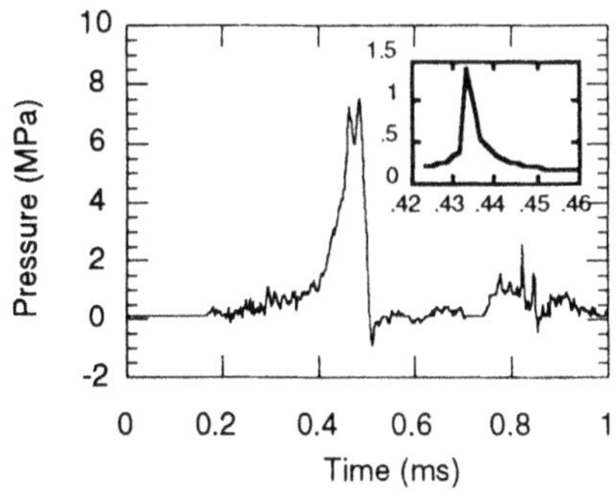

Fig. 12 Pressure pulse recorded during explosion of 3.4 mm bubble. Pressure inset is estimation of pressure field generated by the exploding bubble at the pressure transducer location, calculated using the observed bubble growth rate.

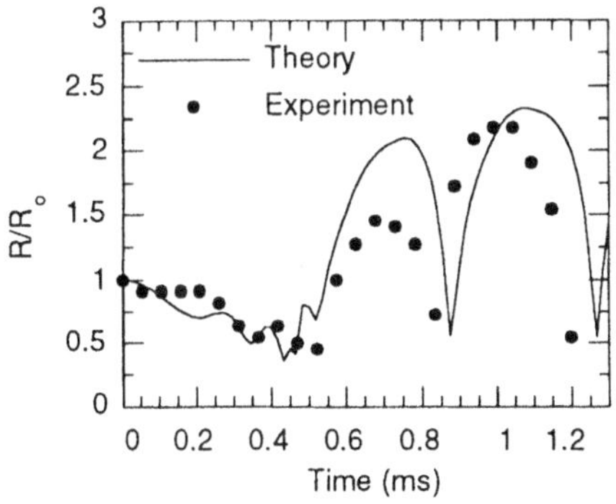

Fig. 13 Comparison between experimental bubble dynamics and that predicted by model using applied pressure shown in Fig. 12, with contribution due to bubble explosion (inset) removed

4.2 Small Bubble Limit. To illustrate which physical processes govern the small bubble explosion limit, the maximum temperature attained within the bubble was computed as a function of initial bubble diameter and is shown in Figs.14 and 15 for peak applied pressures (with gaussian profiles as shown in Fig. 10) of 9 and 5.5 MPa, respectively. From Fig. 14, for bubble diameters greater than about 0.2 mm, the maximum temperature rises sharply to above 4000 K indicating that ignition of the gas has occurred. Therefore a bubble diameter of 0.2 mm represents the minimum bubble diameter for which an explosion is possible subject to the given applied pressure pulse. The maximum temperature attained was calculated using the complete model, then the calculations were repeated with the assumption of no heat transfer and then for an inviscid liquid. Note from Fig. 14 that when the liquid viscosity is neglected, the limiting diameter for explosion of the bubble shifts a small amount whereas when heat transfer is removed, the limiting explosion diameter shifts to an extremely small value. This indicates for a driving pressure of 9 MPa, that the explosion limit is largely governed by heat transfer effects. Figure 15 shows for a gaussian pressure profile with a peak pressure of 5.5 MPa, that the minimum bubble size that explodes increases to about 0.6 mm. In addition, Fig. 15 illustrates that both the viscosity and heat transfer play a relatively equal role in determining the minimum bubble size for explosion.

When a representative experimental pressure trace with a peak pressure of 5.5 MPa, recorded with no bubble present, is used as an input to the model to predict the minimum bubble size that explodes, a value of 1 mm is found. This is comparable with the experimental estimate of 0.9±0.1 mm observed experimentally (see Fig. 6).

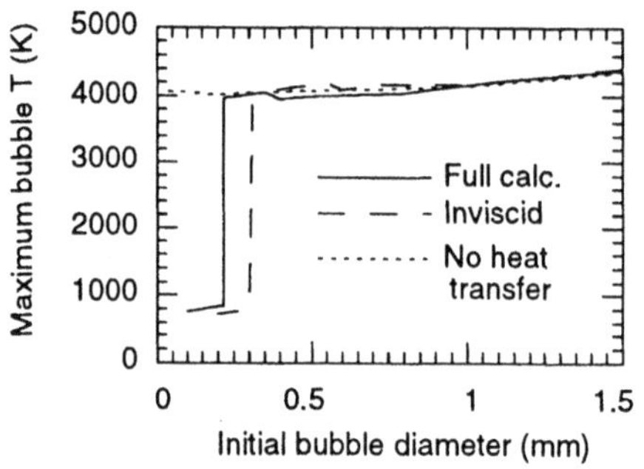

Fig. 14 Maximum bubble temperature attained as a function of initial bubble diameter for a driving pressure of 9 MPa with a Gaussian profile. Sharp rise in temperature indicates ignition.

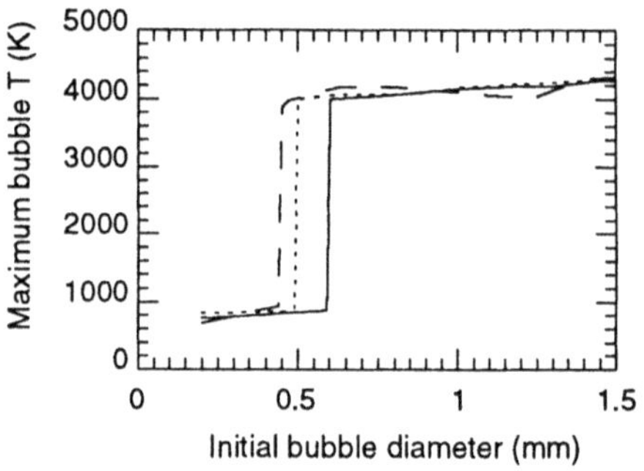

Fig. 15 Maximum bubble temperature attained as a function of initial bubble diameter for a driving pressure of 5.5 MPa with a Gaussian profile

5 Conclusions

The dynamics of a gas bubble consisting of a stoichiometric oxyacetylene mixture immersed in glycerine subject to an impulsive pressure loading has been studied experimentally and theoretically. Compression of the bubbles and ignition of the gas mixture was observed for bubbles ranging in size from 0.3 to 7 mm in diameter for a driving peak pressures ranging from 2.5 to 13 MPa. A simple numerical model was able to reproduce the general features of the observed bubble dynamics. A small bubble limit was identified, both theoretically and experimentally, which depended on the magnitude and form of the applied pressure pulse. The large bubble limit, where asymmetric bubble collapse prevents an explosion, was not determined accurately, but is on the order of 1 cm. To obtain a more accurate estimate of the small bubble explosion limit, higher spatial and temporal resolution photography is required together with a more reproducible method of generating the applied pressure pulse (perhaps the use of milligram quantities of solid explosive may be appropriate). With more highly resolved measurements, refinements to the numerical model, such as the direct computation of the temperature fields in the bubble and liquid, inclusion of evaporation and condensation effects and a more realistic chemical model, may then be justified.

Acknowledgements

The authors gratefully acknowledge the assistance with the construction and operation of the apparatus provided by Jean-Philippe Dionne, Jean-François Bélanger and Leo Grepin. Financial support for this work was provided by the Fonds Pour la Formation de Chercheurs et L'Aide à la Recherche.

References

1. Soloukhin, R. I., "Bubble Mechanism of Shock Ignition in a Liquid," *Dokl. Adad. Nauk. SSSR*, Vol. 136, No. 2, 1961.
2. Hasegawa, T. and Fujiwara, T., "Detonation in Oxyhydrogen Bubbled Liquids," *Proc.19th Symp. (Int.) on Combustion*, The Combustion Institute, 1982, pp. 675-683.
3. Hasegawa, T. and Fujiwara, T., "Propagation Velocity and Mechanism of Bubble Detonation," 9th Int. Colloquium on the Dynamics of Explosions and Reactive Systems, *Progress in Aeronautics and Astronautics*, AIAA, 1983, pp. 309-319.
4. Sychev, A. I., "Detonation Waves in a Liquid-Gas System," *Fizika Goreniya i Vzryva*, Vol. 21, No. 3, 1984, pp. 103-110.
5. Sychev, A. I. and Pinaev, A. V., "Self-Sustaining Detonation in Liquids with Bubbles of Explosive Gas," *Zhurnal Prikladnoi Mekhaniki i Tekhnicheskoi Fiziki*, No. 1, 1986, pp. 133-138.
6. Pinaev, A. V. and Sychev, A. I., "Structure and Properties of Detonation in a Liquid-Gas Bubble System," *Fizika Goreniya i Vzryva*, Vol. 22, No. 3, 1986, pp. 109-118.
7. Kedrinskii, V. K. and Mader, C. L., "Accidental Detonation in Bubble Liquids," in Shock Tubes and Waves, *Proc. 16th Int. Symp. on Shock Tubes and Waves* (editor H. Grönig), Aachen, Germany, 1987, pp. 371-376.
8. Kedrinskii, V. K. and Zamarayev, F. N., "Wave Amplification in Chemically Active Bubble Medium," *Proc. 17th Int. Symp. on Shock Tubes and Waves* (editor Y. W. Kim), Bethlehem, PA, 1989, pp. 51-62.
9. Beylich, A. E. and Gülhan, A., "Waves in Reactive Bubbly Liquids," Adiabatic Waves in Liquid-Vapor Systems, Proc. IUTAM Symp., Göttingen, Germany, 1989, Springer-Verlag, Berlin, pp. 39-48.
10. Borisov, A. A. and Sharypov, O. V., "On the Formation of Bubble Detonation Wave," Izvestia Sibirshogo Otdelenia Akademii Nauk SSSR, Series of Technical Science, Issue 2, 1990, p. 50-59.
11. Scarinci, T., Bassin, X., Lee, J. H. S., and Frost, D. L., "Propagation of a Reactive Wave in a Bubbly Liquid", in *Shock Waves: Proc. 18th Symposium on Shock Waves*, (editor K. Takayama), Sendai, Japan, July, 1991, Springer-Verlag, New York, 1992, pp. 481-484.
12. Beylich, A. E., "Pressure Waves in Bubbly Liquids," *Proc. IUTAM Symp. on Waves in Liquid/Gas and Liquid/Vapor Two-Phase Systems*, Kyoto, Japan, May 9-13, 1994.
13. Kamath, V., Prosperetti, A., and Egolfopoulos, F. N., "A Theoretical Study of Sonoluminescence," *J. Acoust. Soc. Am.*, Vol. 94, No. 1, 1993, pp. 248-260.
14. Korobeinikov, V. A., "Propagation of Blast Waves in a Combustible Gas," *Astronautica Acta.*, Vol. 17, 1972, p. 529.

STABILITY AND NONEQUILIBRIUM PHASE TRANSITIONS IN BOILING-UP SYSTEMS.

V.P.Skripov, V.N.Skokov

Institute of Thermophysics, Urals Division of the Russian Academy of Sciences, 620219, Ekaterinburg, Russia.

Summary.

The interaction of equilibrium (thermodynamic) and nonequilibrium (dynamic) phase transitions in the case when an ordinary phase transition in a nonequilibrium system is overlapped by a change of regime caused by the instability of a flow, is discussed. Several typical examples, such as the thermodynamical aspect of the boiling crisis, bistability in explosive crystallization of amorphous substances and nonlinear phenomena in current-carrying superconductors, cooling by a boiling liquid coolant are considered.

1. Lately the concept of a phase transition has obtained a broadened interpretation. Initially it was associated with thermodynamic systems which can decompose into two different parts (phases) coexisting in equilibrium. At such equilibrium the temperature, the pressure and the chemical potential of the phases are equal. When there is a competition of phases, a lesser chemical potential corresponds to a more stable phase. It is also assumed that the phases possess a recovering reaction to a small local perturbation of density (entropy). The boundary of stability of this kind (spinodal) corresponds to the vanishing of elasticity

$$(dp/d\rho)_T = 0 \qquad (1)$$

In the presence of flows of substance, momentum, energy in a one-phase system an analogue of a phase transition is a change of the dynamic regime of a system with changing governing parameters. The system behaviour is characterized by the hysteresis and the bistability of the dynamic structure. There are many examples of nonequilibrium phase transitions in flow systems: emergence and reconstruction of convective cells, transition from a laminar liquid flow in a pipe to a turbulent flow, transition from noise to coherent (of pure tone) radiation in a laser with increasing pump power, etc.

S. Morioka and L. van Wijngaarden (eds.),
IUTAM Symposium on Waves in Liquid/Gas and Liquid/Vapour Two-Phase Systems, 293-303.

The report draws one's attention to the interaction of an equilibrium (thermodynamic) and a nonequilibrium phase transitions. It is a case when an ordinary phase transition in a nonequilibrium system is overlapped by a change of regime caused by the instability of a flow (heat, hydrodynamic), its dynamic structure [1].

Let us discuss several typical examples of such interacting phase transitions. For convenience they are presented in the form of a table.

Equilibrium (thermodynamic) phase transition.	Nonequilibrium (dynamic) phase transition.	Mechanism of a nonequilibrium phase transition.
Liquid - vapour	⇅ Convection ⇅ Bubble boiling Film boiling	Autowave process Hydrodynamic crisis Thermodynamic crisis
Liquid - crystal	⇅ Slow crystallization Rapid (explosive) crystallization	Positive feedback between the evolution of heat and the crystallization rate. Intensive nucleation.
Superconductor- normal conductor	Bistability of the voltage-current characteristic. Transition to a normal state.	Initiation and development of the thermal instability under Joule self - heating

Together with the liquid - vapour phase transition and the phenomenon of boiling crisis the table reflects the liquid - crystal phase transition, for which a situation of explosive crystallization is possible. Of special interest is the case when not only the liquid undergoes a phase transition, but also the material of the wall that is in contact with the liquid. Such a situation occurs in studying the dynamics of thermal destruction of the superconducting state of current carrying thin films of high-temperature superconductors (HTSC) [2,3].

Now we shall comment on the interacting phase transitions in flow systems presented in the table.

2. The change of heat-exchange regimes of a heater with a boiling liquid can be observed in experiments with a platinum wire placed in a large liquid volume. The wire serves as a heater and at the same time as a resistance thermometer [4,5]. With increasing heat-release power q the convective liquid motion increases, then there originates and develops bubble boiling, and at last a transition to film boiling occurs. Schematically the relationship between q and the excess temperature of the wire surface $\Delta T=T_w-T_f$ is shown in Fig.1, where T_f is the liquid temperature far from the wire. Usually $T_f \lesssim T_s$ was prescribed, where T_s is the saturation temperature at atmospheric pressure. Transitions between regimes reveal a hysteresis and bistability.

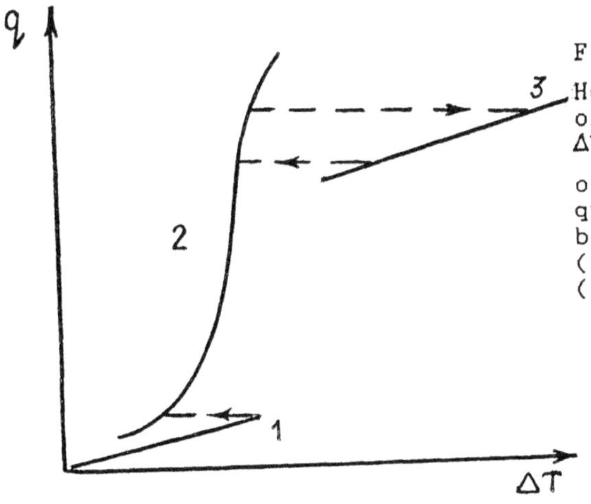

Fig.1
Heat flow q as a function of the temperature head $\Delta T = T_w - T_f$ in the regime of a freely convective liquid motion (1), bubble boiling (2), film boiling (3) near a heated wire (schematically).

For a thin wire (d = 0.03 - 0.08 mm, $l \simeq 60$ mm) one can observe a considerable superheat $\Delta T_{ex}=T_w - T_s$ preceding the beginning of boiling in quasistatic conditions. For diethyl ether, pentane, benzene the value of ΔT_{ex} was as much as 90 K [4,5]. This is caused by the difficulty of origination of boiling centres on a well-wetted clean heater surface. Isolated centres can exist and generate bubbles disturbing but locally the convective heat exchange. But in a superheat $\Delta T_{ex} \simeq 25$ K the origination of one boiling centre resulted in an avalanche - like propagation of bubble boiling for the whole length of the wire. The authors [6] think this process to be autowave.

The inevitability of a transition from bubble to film boiling is evident from the existence of a limiting liquid superheat $\Delta T = T_* - T_s$ [7]. The value of $T_*(p)$ may be identified with the temperature of the spinodal (1) or brought into agreement with a sufficiently high nucleation rate $J(T_*, p)$, for instance, $J \simeq 10^{25}$ $s^{-1}cm^{-3}$. Usually the crisis of bubble boiling regime sets in at a smaller liquid superheat near the wall, $T_{cr} < T_*$ due to the loss of the hydrodynamic and / or thermal stability of a two-phase flow. But during a very rapid heating of a liquid one can realize the shock boiling regime [7,8]. This notion presupposes intensive homogeneous nucleation against a weaker background of ordinary boiling. Experiments [9] with water show that at a heating rate $\dot{T} \geq 10^7$ K/s boiling-up and formation of a vapour film proceed simultaneously over the whole heater. This kind of boiling crisis may be called thermodynamic.

Given below one can find the values that characterize homogeneous nucleation in superheated pentane at atmospheric pressure (T_s = 309.2 K): liquid superheat $\Delta T = T - T_s$, radius of a vapour critical bubble r_*, work of nucleus formation W_* in unit of $k_b T$ (k_b is the Boltzmann constant), nucleation rate J and average time of expectation of a nucleus in the volume $V = 10^{-3}$ cm^3.

ΔT, K	$r_* \cdot 10^7$, cm	$W_*/k_b T$	J, $s^{-1}cm^{-3}$	$\tilde{\tau}$, s
109.1	5.7	81.1	10^{-3}	10^6
110.9	5.3	67.2	10^3	1
113.0	4.8	53.2	10^9	10^{-6}
115.7	4.3	39.2	10^{15}	10^{-12}

These results have been obtained with the help of the classical homogeneous nucleation theory, which has been verified by experiment [7]. A very abrupt increase of J with increasing ΔT ensures the possibility of an explosive boiling regime. Such a regime can be realised not only during a rapid increase in the liquid temperature, but also during a sufficiently abrupt decrease in the pressure in crossing the saturation line. For instance, when a hot liquid ($T/T_c \geq 0.9$, where T_c is the critical temperature) flows through a short nozzle, the mass rate of a two-phase flow can be determined by the model of shock boiling-up [10].

3. Bistability in the crystallization of amorphous solid layers of a substance (water,germanium) is related to the

nonmonotonic temperature dependence of the nucleation rate J and the rate of crystal growth u [11].Fig.2 shows the dependence $J(T)$ for water (1). The linear crystal-growth rate $u(T)$ also changes sharply (curve 2). At $T \geq T_g$, where T_g is the glass transition temperature, the mass crystallization rate on fluctuation nuclei depends on the relative positions of the curves 1 and 2.

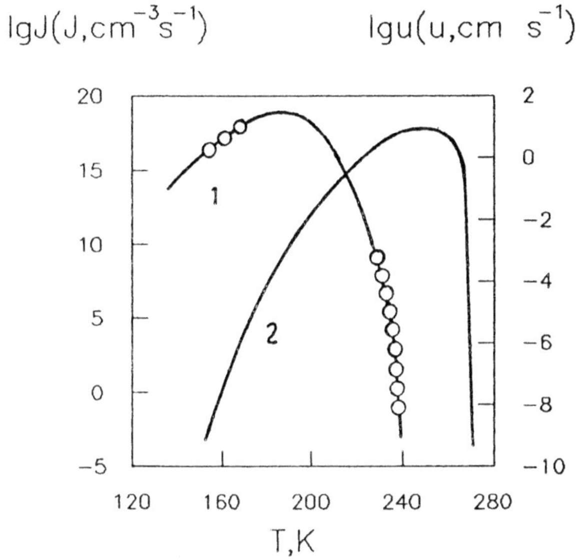

Fig.2
Temperature dependence of the stationary nucleation rate J (1) and the linear rate of growth of ice crystals u (2) in supercooled water.

An amorphous solid layer is obtained by precipitating a substance from a vapour on a cold substrate. The prepared layer contains a certain number of "frozen" crystallization centres. On heating a sample one detects a temperature interval in which crystallization acquires an explosive character. At the dependences $J(T)$ and $u(T)$ shown in Fig.2 for the left portions of the curves there is a positive feedback between the release of crystallization heat and the rates $J(T)$, $u(T)$. This results in self- accelerating mass crystallization in a certain temperature range. (For water the interval is between 160 and 170 K, the glass transition temperature T_g is about 135 K).

Fig.3 shows the bistability which develops during the growth of a crystal of Ge in an amorphous medium [12]. This pattern has been obtained by solving the heat conduction equation at a known dependence $u(T)$.

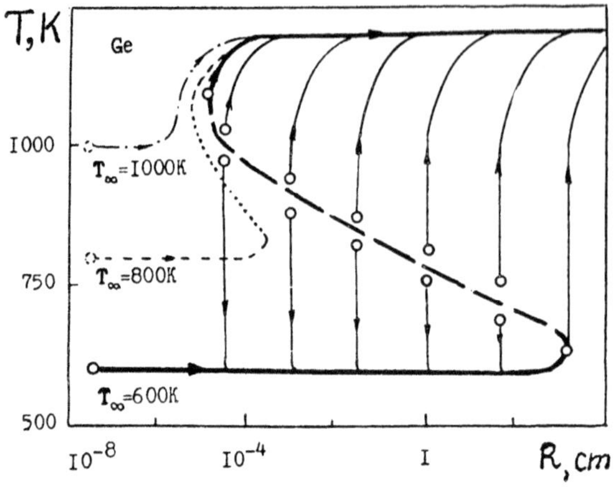

Fig.3

Dependence of the temperature front of a growing germanium crystal on the initial conditions (T_∞, R).

Thus, a transition from slow mass to very rapid crystallization - nonequilibrium phase transition (rearrangement of heat and mass flows) is connected with the characteristics of a liquid - crystal thermodynamic phase transition (supercooling, nucleation rate, viscosity).

4. The thermal instability of superconductors with a current develops when the Joule heat release in resistive or normal zones that have emerged by this or that reason are not compensated by the heat removal into a coolant [13]. The state of thermal bistability is characterised by the presence of two stable states - superconducting (resistive) and normal - at the same value of the transport current. The thermal destruction of a superconducting state may be accompanied by nonlinear thermal and hydrodynamic processes in a coolant. A simultaneous course of phase transitions in a nonlinear heater and a liquid coolant leads to a state of thermal multistability. In thermal multistability several(more than two) stable states correspond to the given value of the governing parameter. The presence of several stable states is connected with the existence of different heat-exchange regimes (heat exchange in one-phase convection, bubble boiling and film boiling). The state of thermal multistability has been realized by experiment with the transport current passing through a thin-film bridge of a high - temperature superconductor $YBa_2Cu_3O_{7-x}$ [2]. Fig.4 gives the voltage-current characteristic of a bridge immersed directly

into liquid nitrogen. At transport currents that exceed the critical one there arises a stable resistive state in a sample (section AB in Fig.4).

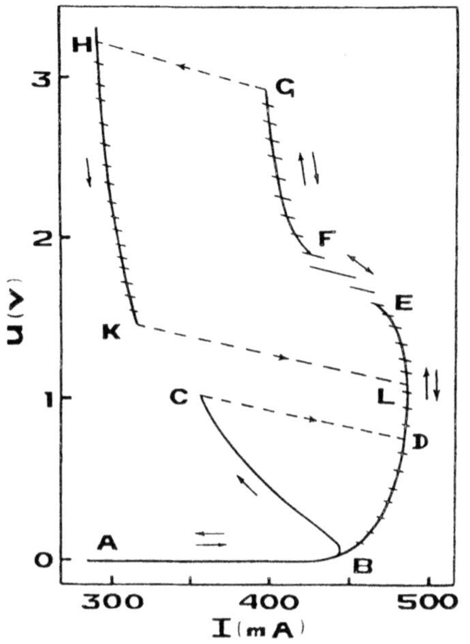

Fig.4
Voltage-current characteristic of a bridge immersed directly into liquid nitrogen (see the text).

With a further increase of the load in the sample there appears a structure unstable in a fixed-current regime, which corresponds to the emergence of a negative differential resistance on the V-I characteristic (section BC in Fig.4). The heat exchange with liquid nitrogen on this section is realized at the expense of one-phase convection - nitrogen near the surface gets superheated. At the point C one observes a local boiling-up of nitrogen, and the system abruptly passes into the point D (along the load line of the circuit diagram). The section DE corresponds to the local boiling of nitrogen at a "weak" place. Oscillations caused by the separation of vapour bubbles were observed on the V-I characteristic. On the section EF there was an abrupt increase of the amplitude of oscillations, and one could observe random jumps from the point E to the point F (along the load line). Visually on this section one observed a certain increase of the separation diameter of bubbles. A transition from the section DE to the section FG of the V-I characteristic may be connected with a local boiling crisis and the formation of a "dry spot" under the vapour bubbles. At a certain load (point G in Fig.4) the vapour-bubble boundary shifted sharply, and on the sample surface there formed a continuous vapour film. A reverse transition from film to bubble boiling occurred at smaller loads (point K in Fig.4), i.e. on the V-I characteristic one could observe a hysteresis.

A more complicated dynamic behaviour is observed if a superconductor is connected to an external electric circuit with an inductance. As a result of taking into account the dynamics of the external circuit the system acquires properties of an autowave medium with restoration in which periodic processes are possible. The temperature profile of a thin superconductor with a sectional area S and a sectional perimeter P is described by the one-dimensional heat equation

$$C \frac{dT}{dt} = \frac{d}{dx} \left(\lambda \frac{dT}{dx}\right) + j^2 \rho(T) - \alpha \, (T-T_0) \frac{P}{S} \qquad (2)$$

where T is the temperature, t is the time, x is a coordinate along the conductor, ρ is the specific resistance, j is the current density, T_0 is the temperature of a coolant, and λ, C and α are the thermal conductivity, the heat capacity and the convective - heat - exchange coefficient, respectively. The current in a superconductor connected to the circuit in series with a load resistance R_0 and a voltage supply u is determined by the equation

$$L \frac{dI}{dt} = u - [\, R(T,I) + R_0 \,]\, I, \qquad (3)$$

where L is the circuit inductance, R is the resistance of a superconducting element. In two limiting cases - a homogeneous transition in a superconductor isothermal along the length and autowave motion of the N-S boundary the heat equation (2) is reduced to an ordinary differential equation. If the time of the current relaxation in the circuit τ_i $L/(R + R_0)$ exceeds the time of the system thermal relaxation τ_h, the system of equations (2), (3) admits periodic solutions of the stable limit-cycle type. The origination of a limit cycle occurs as a result of the Hopf bifurcation.

If the external electric circuit contains a voltage supply

with a harmonic component, i.e. $u=u_0+u_1 \cos \omega t$, the system of equations (2), (3) becomes nonautonomous. In such systems, along with periodic, there may be chaotic solutions [14]. Fig.5 gives experimental oscillograms of the voltage drop on potential contacts at different values of the driving frequency of harmonic oscillations. The same figure gives phase trajectories in the I-U plane.

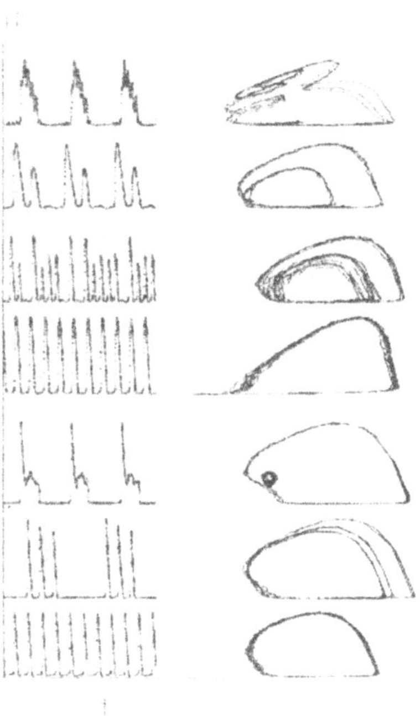

Fig.5

Oscillograms of the voltage drop on potential contacts and phase trajectories in the I-U plane during self-excited oscillations with a harmonic supply (the increase of the driving frequency - from below upwards).

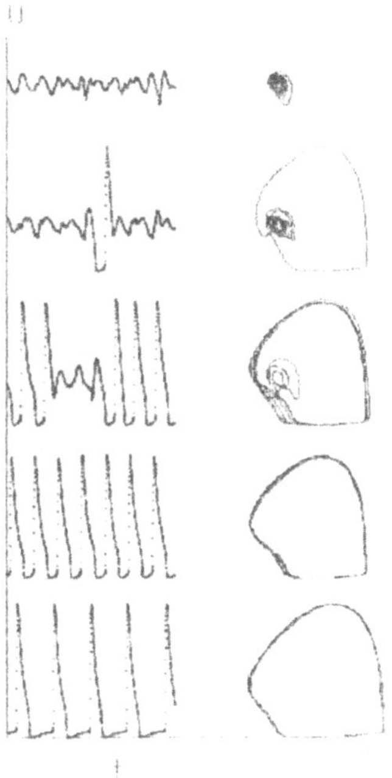

Fig.6

Oscillograms of the voltage drop on potential contacts and phase trajectories in the I-U plane with connection of an HTSC bridge to the circuit of a DC voltage supply with an inductance cooled with boiling nitrogen (the increase of the supply voltage - from below upwards).

In a certain range of frequencies of the harmonic component close to the frequencies of natural relaxation oscillations, there arose effects of entrainment of self-excited oscillations at a driving frequency. With an upset of the frequencies one observed a more complicated behaviour reminiscent of chaos [14]. With a severe upset of the frequencies modulated oscillations were observed.

When the described self-excited oscillations take place during the cooling of a superconductor with boiling nitrogen, the initiation of a chaotic behaviour does not necessarily require the presence of a harmonic driving force. Heat - removal oscillations caused by the bubble boiling of liquid nitrogen play the role of an oscillating driving force. Fig.6 gives the sequence of oscillograms of the voltage drop on the potential contacts of an HTSC bridge connected to the circuit of a DC voltage supply with an inductance and cooled with boiling nitrogen. The same figure gives phase trajectories in the I-U plane. In a wide range of values of the supply voltage there proceeded synchronization of self-excited oscillations of the normal zone in a superconductor with oscillations caused by the separation of vapour bubbles. With increasing supply voltage one observed intermittence and a transition from relaxation oscillations of the limit-cycle type to chaotic ones.

The observed patterns may be interpreted as nonequilibrium phase transitions in a complicated nonlinear system: a superconductor with a current connected to an external circuit with an inductance - boiling nitrogen. Such nonequilibrium phase transitions are accompanied with traditional superconductor - normal conductor and liquid - vapour equilibrium phase transitions.

References

1. Skripov V.P. In: Thermodynamics and Kinetics of Phase Transitions, Ekaterinburg (1992), 312.
2. Skokov V.N. and Koverda V.P. Cryogenics, 33 (1993), 1072-1076.
3. Skokov V.N., Koverda V.P., Skripov V.P., Ivakin V.B. and Semenova N.M. Doklady AN (Russia), 333 (1993), 614-615.
4. Skripov V.P. and Bulanov N.V. In: Hydrodynamics and Heat Transfer, Sverdlovsk (1992), 41-50.
5. Skripov V.P. and Bulanov N.V. Inzhenerno-Fizicheskii Zhurnal, 22 (1972), 614-619.
6. Zhukov S.A. and Barelko V.V. Teplofizika Vysokikh Temperatur, 27 (1989), 920-930.
7. Skripov V.P. Metastable Liquids, Wiley, New-York-Chichester (1974).

8. Pavlov P.A. Boiling Dynamics of Strongly Superheated Liquids, Sverdlovsk (1988).

9. Pavlov P.A. and Nikitin E.D. Teplofizika Vysokikh Temperatur, 18 (1980), 354-358.

10. Skripov V.P. and Isaev O.A. In: Adiabatic Waves in Liquid - Vapor Systems, IUTAM Simposium, Gottingen, Germany, 1989, Springer Verlag (1990), 181-185.

11. Skripov V.P. and Koverda V.P. Spontaneous Crystallization of Supercooled Liquids, Nauka, Moscow (1984).

12. Koverda V.P. and Skripov V.P. Melts (Rasplavy), 1 (1987), 283-304.

13. Altov V.A., Zenkevich V.B., Kremlev N.G. and Sytchev V.V. Stabilization of Superconducting Magnetic Systems, Plenum Press, New-York (1984).

14. Moon F.C. Chaotic Vibration, Wiley Interscience Publ. John Wiley & Sons, New-York, Chichester, Brisbane, Toronto, Singapore (1987).

15. Lvovsky Yu.M. Zhurnal Tekhnicheskoi Fiziki, 58 (1988), 2219-2227.

ON THE PERTURBATIONS OF EVAPORATION WAVES AND LIQUEFACTION SHOCKS

V.M.Teshukov
Lavrentyev Institute of Hydrodynamics
Siberian Division of the Russian Academy
of Sciences, Novosibirsk, 630090, Russia

Perturbation problems related to propagation of liquefaction shocks and evaporation waves are considered. The nonlinear approximate equations governing the evolution of the shear flow in a layer of evaporating superheated liquid are derived. These equations are analyzed and the conditions of nonlinear stability are discussed.
On the background of the stability analysis of liquefaction shocks we discuss the problem of theoretical modelling of the irregular shock front behaviour. It is shown that the conventional van der Waals like equations of state allow the oscillatory behaviour of the shock front distrubances to be simulated. However, the modelling of the shock front instability is impossible when simple equations of state are used.

Introduction

The study of large-heat capacity fluids, composed of molecules with a large number of atoms, exhibited their unconventional properties. Complete evaporation waves and liquefaction shocks were predicted for these complex fluids. The experimental observation of the wave propagation revealed the instabilities in some ranges of the flow parameters [1–5].

Two types of evaporation waves in a superheated liquid were observed. A superheated liquid can result from an equilibrium liquid state through processes of rapid decompression. The following rapid phase change may produce low speed or higher velocity evaporation waves. Taking into account the similarity between the modelling of the phase change propagation and the theory of combustion and detonation of chemical explosives, the low speed wave may be considered as vapor deflagration and higher velocity wave, as vapor detonation. For usual substances the subsonic branch of the adiabat, representing the downstream states of vapor deflagration, terminates inside the liquid-vapor mixture region. However, in fluids of high molar specific heat the complete evaporation wave is possible. Corresponding estimates of the specific heat were given in [6]. In the first part of the paper the propagation of a slow evaporation wave (deflagration) through the layer of a superheated liquid is studied. It is

S. Morioka and L. van Wijngaarden (eds.),
IUTAM Symposium on Waves in Liquid/Gas and Liquid/Vapour Two-Phase Systems, 305-314.

shown that the interaction between the evaporation wave and shear flow may lead to instability. In the second part of the paper the problem of theoretical modelling the instabilities of the liquefaction shocks is discussed. Theoretical models of two-phase flows of complex fluids unclude some approximation of the equation of state. The following question arises: is it possible to simulate the transition to the unstable shock front behaviour for the given approximate equation of state? As far as simple van der Waals like equations of state are concerned, the answer is negative. An oscillatory behaviour of the shocks may be described by conventional equations of state when specific heat of the fluid is large enough.

Approximate model of evaporation wave

Consider an evaporating layer of a superheated liquid lying on a planar solid boundary. We suppose that a fluid in a layer moves in a horizontal direction. In the general case the flow is rotational (sheared). In order to derive the approximate equations for a long wave motion we nondimensionalize the problem. Let H_0 be the mean initial depth of the fluid, g be the gravity acceleration, and L, ρ, R be the scales of wave lenght, vapor and liquid density respectively. We assume that $\delta = H_0 L^{-1} \ll 1$, $\delta_1 = \rho R^{-1} \ll 1$, $\delta_1 = \delta^2$. Introduce the Cartesian coordinates at a plane with a horizontal axis along the bottom. Independent variables such as horizontal, vertical coordinates and time are scaled on L, H_0, $L(gH_0)^{-1/2}$. Horizontal, vertical velocities, pressure and specific internal energy in a liquid domain are scaled on $(gH_0)^{1/2}$, $\delta(gH_0)^{1/2}$, RgH_0, $gL\delta^{-1}$ respectively. Corresponding scales in a vapor domain are as follows: $(gH_0)^{1/2}, \delta^{-1}(gH_0)^{1/2}, RgH_0$, $gL\delta^{-1}$. We nondimensionalize the Euler system of inviscid non heat-conducting flow and pass to the limit $\delta \to 0$. The scaled approximate governing system in a liquid domain is as follows:

$$\begin{gathered}(\rho^\ell u^\ell)_t + (\rho^\ell u^{\ell 2})_x + (\rho^\ell u^\ell v^\ell)_y + p^\ell_x = 0, \quad p^\ell_y = -\rho^\ell, \\ (\rho^\ell)_t + (\rho^\ell u^\ell)_x + (v^\ell \rho^\ell)_y = 0 \\ (\rho^\ell \varepsilon^\ell)_t + (\rho^\ell \varepsilon^\ell u^\ell)_x + (\rho^\ell \varepsilon^\ell u^\ell)_y = 0.\end{gathered} \tag{1}$$

In a vapor domain the following approximate equations are valid:

$$\begin{gathered}(\rho^v u^v v^v)_y + p^v_x = 0, \quad (\rho^v v^{v2})_y + p^v_y = 0, \\ (\rho^v v^v)_y = 0, \quad (\rho^v v^v(\varepsilon^v + p^v(\rho^v)^{-1} + 2^{-1}(v^v)^2))_y = 0\end{gathered} \tag{2}$$

At the liquid-vapor interface $y = h(x,t)$ there are jump conditions:

$$\begin{gathered}u^\ell = u^v + h_x v^v, \quad p^\ell = p^v + \rho^v (v^v)^2, \\ \rho^\ell (h_t + u^\ell h_x - v^\ell) = -\rho^v v^v \\ \varepsilon^\ell = \varepsilon^v + (\rho^v)^{-1} p^v + 2^{-1}(v^v)^2.\end{gathered} \tag{3}$$

Here $u, v, \rho, p, \varepsilon$ are the velocity components, density, pressure, specific internal energy. Superscripts "ℓ" or "v" designate liquid or vapor flow parameters respectively. The vapor domain is bounded above by the material surface $y = h_0(x,t)$,

where the boundary condition $p^v = p_0 = \text{const}$ is satisfied. At the bottom of the liquid layer $y = 0$ the boundary condition is $v^\ell = 0$.
The system (2) posesses the following integrals:

$$\rho^v v^v = J(x,t),\ p^v + \rho^v (v^v)^2 = N(x,t),\ \varepsilon^v + p^v(\rho^v)^{-1} + 2^{-1}(v^v)^2 = E(x,t).$$

As a consequence the equality $p^v = p_0$ is valid in the vapor domain. To complete the mathemetical model we need an additional information on the velocity of deflagration wave.
A) Constant velocity of the deflagration wave.
Let us consider the following condition:

$$h_t + u^\ell h_x - v^\ell = -K, \tag{4}$$

($K = \text{const} > 0$). The jump conditions (3), equation (4), equation of state $\varepsilon^v = \Phi(\rho^v, p^v)$ of the vapour phase allow us to obtain the equality:

$$\varepsilon^\ell = \Phi\left((\rho^\ell K)^2 (p^\ell - p_0)^{-1}, p_0\right) + 2^{-1}(\rho^\ell K)^{-2}(p^{\ell 2} - p_0^2). \tag{5}$$

Relation (5) and the liquid phase equation of state $\rho^\ell = \varphi(p^\ell, \varepsilon^\ell)$ give the boundary condition

$$p^\ell = \chi(\varepsilon^\ell) \tag{6}$$

at the wave front $y = h(x,t)$. Here χ is the monotonically increasing function.
B) Chapman–Jouguet condition. Let the condition $v^v = c^v$ (c^v is scaled value of the sound speed) be fulfilled at the wave front. Then the jump conditions (3) give the boundary conditions:

$$p^\ell = \chi(\varepsilon^\ell), \qquad \rho^\ell(h_t + u^\ell h_x - v^\ell) = -F(\varepsilon^\ell) \tag{7}$$

at the wave front with given functions χ, F. These conditions are derived in the following way. The jump conditions

$$p_0 + \rho^v c^{v2}(\rho^v, p_0) = p^\ell, \quad \Phi(\rho^v, p_0) + p_0(\rho^v)^{-1} + 2^{-1}c^{v2}(\rho^v, p_0) = \varepsilon^\ell$$

allow us to find $\rho^v = \rho^v(\varepsilon^\ell)$ and

$$\chi(\varepsilon^\ell) = p_0 + \rho^v(\varepsilon^\ell) c^{v2}(\rho^v(\varepsilon^\ell), p_0), \quad F(\varepsilon^\ell) = \rho^v(\varepsilon^\ell) c^v(\rho^v(\varepsilon^\ell), p_0)$$

C) Using the formula, suggested in [4]

$$J = \left(\frac{12}{\pi}\right)^{1/2} \eta \frac{\rho^{\ell 2}}{\rho^v} Ja \left[\frac{k}{\tau}\right]^{1/2} \tag{8}$$

for mass flux J we derive the boundary conditions of the form (7) with another functions χ, F. Here η is the constant correction coefficient, k is the thermal diffusivity of the liquid, τ is the bubble life time at the front,

$$Ja = c_p(T^\ell - T_s(p^\ell))(L(p^\ell))^{-1}.$$

$T_s(p)$ is the saturation temperature, L is the latent heat, T^ℓ is the temperature of the liquid.

The boundary conditions (7) are derived in the following way. We represent J in the form (formula (8))

$$J = (\rho^v)^{-1}\psi(\rho^\ell, T^\ell, p^\ell).$$

Jump conditions allow us to find $\rho^v = \psi^{2/3}(p^\ell - p_0)^{-1/3}, \quad v^v = \psi^{-1/3}(p^\ell - p_))^{2/3}$. Another jump condition

$$\Phi(\psi^{2/3}(p^\ell - p_0)^{-1/3}, p_0) + 2^{-1}(p^\ell + p_0)(p^\ell - p_0)^{1/3}\psi^{-2/3} = \varepsilon^\ell$$

and the equations of state $\rho^\ell = \varphi(p^\ell, \varepsilon^\ell)$, $T^\ell = T^\ell(p^\ell, \varepsilon^\ell)$ determine function $\chi(\varepsilon^\ell)$. Function $F(\varepsilon^\ell)$ is given by the following formula:

$$F(\varepsilon^\ell) = \left[\psi(\varphi(\chi(\varepsilon^\ell), \varepsilon^\ell), T^\ell(\chi(\varepsilon^\ell), \varepsilon^\ell), \chi(\varepsilon^\ell))\right]^{1/3} (\chi(\varepsilon^\ell) - p_0)^{1/3}$$

In the general case we suppose that the conditions of the form (7) have to be satisfied at the wave front. As a result we obtain a special free boundary problem in the liquid domain. It is convenient to introduce the new independent variables x', t', λ:

$$x = x', \quad t = t', \quad y = Y(x', t', \lambda). \ (0 \le \lambda \le 1)$$

Here Y is the solution of the problem

$$Y_{t'} + u^\ell(x', Y, t')Y_{x'} = v^\ell(x', Y, t')$$
$$Y(x', 0, \lambda) = \lambda h^0(x').$$

The function $h^0(x')$ defines an initial position of free surface ($h^0(x') = h(x', 0)$). We shall omit the superscript "ℓ" and prime in the following text. The position of free surface is given by the equation $\lambda = \mu(x, t)$, where $\mu(x, t)$ is the function to be determined. The governing system is transformed to the equations:

$$\begin{gathered} u_t + uu_x + (\varphi(p, \varepsilon))^{-1}p_x + \left(\int_0^\lambda (\varphi(p, \varepsilon))^{-1} H d\nu\right)_x = 0, \\ H_t + (uH)_x = 0, \quad \varepsilon_t + u\varepsilon_x = 0, \\ \mu_t + \bar{u}\mu_x = -F(\bar{\varepsilon})\bar{H}^{-1}. \end{gathered} \tag{9}$$

Here $H = \rho Y_\lambda$, $\bar{f}$ designates the boundary value of the function $f(x, t, \lambda) : \bar{f}(x, t) = f(x, t, \mu(x, t))$. The pressure p in a liquid layer $0 \le \lambda \le \mu(x, t)$ may be found in the form:

$$p = \int_\lambda^{\mu(x,t)} H(x, t, \nu)d\nu + \chi(\bar{\varepsilon}). \tag{10}$$

The solution of the Cauchy problem for the system (9), (10) describes the evolution of the flow. The liquid density and the vertical velocity are expressed in terms of p, ε, H:

$$\rho = \varphi(p, \varepsilon), \quad Y = -\int_0^\lambda \rho^{-1} H u d\nu, \quad v = Y_t + uY_x.$$

The stability analysis

In a study of the properties of the system (9) we apply the general approach to the analysis of the integro-differential equations, developed in [7, 8]. We shall consider two simplified statements of the problem.
Let $\varphi(p,\varepsilon) \equiv$ const (incompressible fluid). Assume that $\varepsilon = \varepsilon_0(\lambda)$ when $t = 0$ (an internal energy is constant at material surfaces $\lambda =$ const). The equality $\varepsilon = \varepsilon_0(\lambda)$ is also valid at $t > 0$. The system (9) is reduced to the equations:

$$u_t + uu_x + \left(\int_0^\mu H_1 dv\right)_x + \chi_1'(\varepsilon_0(\mu)) \cdot \varepsilon_0'(\mu)\mu_x = 0,$$

$$H_{1t} + (uH_1)_x = 0, \tag{11}$$

$$\mu_t + \bar{u}\mu_x = -F_1(\varepsilon_0(\mu)) \cdot \bar{H}_1^{-1},$$

$(H_1 = \rho^{-1}H,\ \chi_1 = \rho^{-1}\chi,\ F_1 = \rho^{-1}F)$. Let $k = k(x,t)$ be a root of the equation

$$Q(k) = \int_0^{\mu(x,t)} \frac{H_1 d\nu}{(u-k)^2} - 1 = 0. \tag{12}$$

The equality

$$\left(\frac{\partial}{\partial t} + k\frac{\partial}{\partial x}\right)\left(k - \int_0^\mu \frac{H_1 d\nu}{u-k}\right) - \frac{1}{\bar{u}-k}(\chi_1'(\varepsilon_0)\varepsilon_0'(\mu)(\mu_t + k\mu_x) + \tag{13}$$

$$+(1 + \bar{H}^{-1}\chi_1'(\varepsilon_0)\varepsilon_0'(\mu))F_1(\varepsilon_0(\mu))) = 0$$

is a direct consequence of the system (11). It follows from the equation (13) that the curve $x = x(t),\ x'(t) = k(x,t)$, is the characteristic curve of the system (11). In the case when the equation (12) has complex roots, the Cauchy problem for the system (11) is incorrect. In this case the unstable behaviour of disturbances can be predicted from the consideration of perturbation problem. If the equation (12) has the only real roots the system (11) is hyperbolic one. The regular behaviour of flow disturbances is inherent to this flow regimes. For monotonic velocity profile ($u_\lambda > 0$) we formulate the condition under which the roots of (12) are real:

$$\Delta \arg Q^+(u)/Q^-(u) = 0. \tag{14}$$

Here Δ denotes the increment of the function on the interval $(u(x,t,0),\bar{u})$, $Q^\pm$ are the limiting values of the analytical function $Q(z)$ at this interval from the upper or lower complex half plane.
Consider the case where $\varepsilon =$ const at the initial instant of time. It follows from the system (9) that $\varepsilon = \varepsilon_0 =$ const at $t > 0$ (a barotropic flow). The evolution of the flow is governed by the system:

$$u_t + uu_x + \left(\varphi_0\left(\int_0^\mu H d\nu\right)\right)^{-1}\left(\int_0^\mu H d\nu\right)_x = 0 \tag{15}$$

$$H_t + (uH)_x = 0, \quad \mu_t + \bar{u}\mu_x = -F_0\bar{H}^{-1}.$$

Here $\varphi_0(p) = \varphi(p, \varepsilon_0)$, $F_0 = F(\varepsilon_0)$. We obtain the characteristic equation in a way analogous to that used in the case of an incompressible liquid:

$$Q(k) = \left(\varphi_0\left(\int_0^\mu H d\nu\right)\right)^{-1} \int_0^\mu \frac{H d\nu}{(u-k)^2} - 1 = 0.$$

The condition of the form (14) with the new function $Q(z)$ provides the hyperbolicity of the system (15) and the regular behaviour of the flow. The appearance of the complex roots k leads to an unstable behaviour.
The system (11) admits the solution of the form $u = u_0(\lambda)$, $H_1 = 1$, $\mu = \mu(t)$, where the function μ is determined by the problem

$$\mu'(t) = -F_1(\varepsilon_0(\mu)), \quad \mu\,|_{t=0} = 1.$$

This solution describes the propagation of the evaporation wave through the undisturbed liquid layer. The liquid in a layer flows in the horizontal direction, the velocity field is given by the formulae $u = u(y)$, $v = 0$ when $0 \le y \le h(t)$. Considering the perturbations of the main flow, we have to solve the Cauchy problem for linearized system (11). The characteristic velocities are defined by the equation (12) with u, H_1, μ given by mentioned above formulae. If $u_0'(\lambda) \ne 0$ and condition (14) is fulfilled for the main flow, then the perturbation problem is well posed. If the equation (12) admits complex roots, the instability of the flow will be developed. For some velocity profiles $u = u_0(\lambda)$ the equation (12) admits only real roots as $t \le t_0$. Complex roots appear $t > t_0$. The perturbed flow stability is lost during the evaporation process, the flow is stable for $t \le t_0$ and unstable for $t > t_0$. In the general nonlinear case when k depends on x, t the instability develops locally: for some ranges of x, t the flow is stable and the unstable behaviour may be observed on neigbouring intervals of x.

Stability of the shocks in the van der Waals fluid

Theoretical models of the shock dynamics include some analytical equations of state with a finite number of constant parameters. For an individual fluid these parameters are determined using additional experimental data and theoretical predictions. However, a number of the classical equations of state are inadequate to describe the thermodynamical behaviour in the subcritical region. Experiments on shock waves in fluids with large heat capacities show the evidence for the transverse instability of the shock front near phase transitions [2]. A chaotic behaviour of the disturbances and fracture of the shock front were also observed. The following question arise: is it possible to simulate the transition to the irregular shock front behaviour when the simple analytical equations of state are used? We shall discuss this problem and consider some classical equations of state applied in modelling fluid properties in the subcritical region. We suppose that in the two-phase region the equations of state are determined by the conditions of the liquid-vapor mixture equilibrium. As a

rule, the analytical form of the equation of state in this region cannot be explicitly derived. Some difficulties in the theoretical analysis ensue in this connection.
When the shock moves in a cylindrical tube, the disturbances of the front decay with increasing time if

$$-1 < \Delta < (1 - M^2 - RM^2)(1 - M^2 + RM)^{-1}. \tag{16}$$

The transverse waves appear at the front, if

$$(1 - M^2 - RM^2)(1 - M^2 + RM)^{-1} < \Delta < 1 + 2M. \tag{17}$$

The shock front is unstable, if

$$\Delta < -1 \quad \text{or} \quad \Delta > 1 + 2M. \tag{18}$$

Here $\Delta = (p - p_1)(v_1 - v)^{-1}(\partial v/\partial p)_H$, v is the specific volume, $(\partial v/\partial p)_H$ is the derivative along the Hugoniot curve, M is the Mach number of the downstream state, $R = v_1 v^{-1}$ [9, 10, 11]. Index "1" designates the upstream parameters.
We shall consider the shocks with $M < 1 < M_1$. The second inequality (17) may be reduced to the following:

$$\varepsilon_v(v,p) < -\frac{p_1 + p(p - p_1)^{1/2}(-g_v(v_1 - v))^{-1/2}}{1 + (p - p_1)^{1/2}(-g_v(v_1 - v))^{-1/2}} \tag{19}$$

($g_v = g_v(v,s)$, $p = g(v,s)$ is the equation of state, s is the entropy).
In the mixture region $\varepsilon_v(v,p) = \varepsilon_v(v,T) = Tp_s'(T) - p_s(T)$ where $p_s(T)$ is the saturation pressure. We obtain the following statement as a direct consequence of the inequality (19). Let $(v, p_s(T))$ belong to the mixture region.
Statement 1. If $(v, p_s(T))$ is the downstream state for unstable shock, then $Tp_s'(T) - p_s(T) < 0$.
Consider any analytical equation of state $p = p(v,T)$.
Statement 2. If the function $p(v,T)$ satisfies the inequality $Tp_T(v,T) - p(v,T) \geq 0$, then the function $p_s(T)$ satisfies the analogous inequality $Tp_s'(T) - p(T) \geq 0$.
To prove this statement we differentiate the equation

$$(v_2(T) - v_1(T))p_s(T) = \int_{v_1(T)}^{v_2(T)} p(v',T)dv'. \tag{20}$$

(Maxwell equal area rule) with respect to the variable T and derive the following equality

$$(v_2(T) - v_1(T))p_s'(T) = \int_{v_1(T)}^{v_2(T)} p_T(v',T)dv'. \tag{21}$$

Combination of (20), (21) gives the desired inequality. Here the states $(v_2(T), p_s(T))$, $(v_1(T), p_s(T))$ belong to the mixture-vapor and the mixture-liquid boundaries.
We suppose that in the single phase domains the thermodynamic stability conditions

$p_v(v,T) < 0, \quad \varepsilon_T(v,T) > 0$ are fulfilled.

Statement 3. If (v,T) is the downstream state for the unstable shock in the single phase region, then $Tp_T(v,T) - p(v,T) < 0$.

Statement follows from the equality

$$\varepsilon_v(v,p) = Tp_T(v,T) - p(v,T) - p_v(v,T)\varepsilon_T(v,T)(p_T(v,T))^{-1}$$

and (19).

So, we may conclude that the unstable behaviour of the shock front cannot be simulated if the inequality $Tp_T - p \geq 0$ is valid for the given equation of state. Many well known equations of state satisfy the above mentioned inequality. For example, the inequality may be easily verified for the van der Waals, Redlich and Kwong, Clausius, Abbott equations of state:

$$P = \frac{8T}{3v-1} - \frac{3}{v^2}, \quad P = \frac{3T}{v-B} - \frac{1}{BT^{1/2}v(v+B)},$$

$$P = \frac{4T}{1+4z_0(v-1)} - \frac{27}{T\{3+8z_c(v-1)\}^2},$$

$$P = \frac{12T}{4v-1} - \frac{81\{(59-5\alpha_c)T^{-3/5} + (5\alpha_c - 29)T^{-13/5}\}}{10(8v+1)^2}.$$

Here B, z_c, α_c are the positive constants, $59 > 5\alpha_c > 29$, P, T, v are the pressure, temperature and specific volume scaled on critical values. So, these equations of state are inadequate enough for a simulation of the transition to the unstable shock wave behaviour. However these equations may be used in a simulation of an oscillatory behaviour of the shock front disturbances in the fluids with sufficiently large specific heat capacity. The condition providing the appearance of the tranverse waves [11, 12] may be transformed to the following form:

$$p(v,T) - \frac{Tp_T^2(v,T) - p_v(v,T)\varepsilon_T(v,T)}{\varepsilon_T(v,T) + vp_T(v,T)} \geq p_1 \tag{22}$$

(p_1 is the upstream pressure at the shock front; v, T are the parameters of the downstream state). For any given analytical equation of state $p = p(v,T)$ the specific heat is expressed by the formula:

$$\varepsilon_T(v,T) = c_v(T) - T\int_v^\infty P_{TT}(v',T)dv'$$

($c_v(T)$ is the positive valued function). In a mixture region the inequality (22) is reduced to the following

$$p_s(T) - \frac{T(p_s'(T))^2}{\varepsilon_T^m(v,T) + vp_s'(T)} \geq p_1 \tag{23}$$

Here $\varepsilon_T^m(v,T)$ is the specific heat in a mixture region, $\varepsilon_T^m(v,T) = c_v(T) + \varepsilon_T^{m1}(v,T)$, $\varepsilon_T^{m1}(v,T)$ is independent of $c_v(T)$. Hence the inequality (23) will be valid as the function's $c_v(T)$ values are assumed to be sufficiently large. We can conclude that the transverse wave shock front instability is inherent for the large heat capacity fluids with conventional equations of state in a mixture region.

Some numerical results concerning the van der Waals equation of state are presented below. We assume that the internal energy of the fluid is defined by the expression

$$\varepsilon = \frac{8}{3}\, c_v T - \frac{3}{v} \qquad (c_v = \text{const})$$

(temperature and specific volume are scaled on critical values, the specific heat is scaled on the gas constant R, the internal energy is scaled on $p_c v_c$). In Fig. 1 the domain I bounded by the saturation line S and the line Γ contains the single phase thermodynamic states where the transverse wave instability of the shock front may occur. For any point $(v,p) \in I$ considered as the downstream state for the shock wave one can find the sufficiently large constant c_v and upstream state (v_1, p_1) so that oscillations will develop at the shock front connecting the thermodynamic states (v,p) and (v_1,p_1). When (v,p) approaches Γ then $c_v \to \infty$. In Fig. 2 the graphic of minimal values of c_v, sufficient for the appearance of the shock front oscillatory behaviour is presented for the states $(v,p) \in S$. In Fig. 3 the analogous domain I belonging to the mixture region is presented. If $c_v = 1.5$ then the right boundary of the domain of "oscillatory" states is Γ_1, if $c_v = 2.5$ then the boundary is Γ_2. Presented in Fig. 4 is the family of inverse shock adiabats originating from the liquid-mixture boundary ($c_v = 85$). Every adiabat is divided into two parts by the bold point. For all states (v_1,p_1) belonging to the upper part of the line the liquefaction shock connecting (v_1,p_1) and (v,p) is stable ((v,p) lies on a liquid-mixture boundary). The transverse wave shock front instability appears when the upstream state belongs to the lower part. In particular, the set of upstream states at the vapor-mixture boundary (complete liquefaction shocks) is also divided into two subsets: "stable" and "oscillatory" subsets.

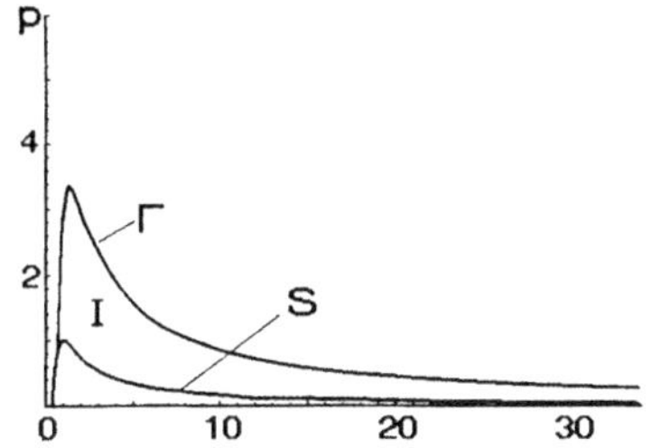

Fig. 1.

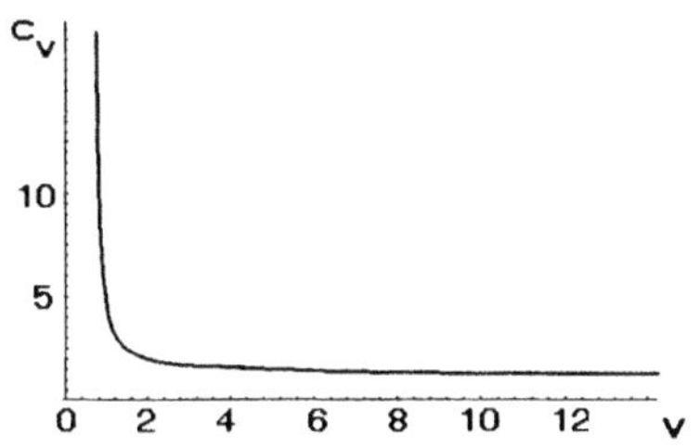

Fig. 2.

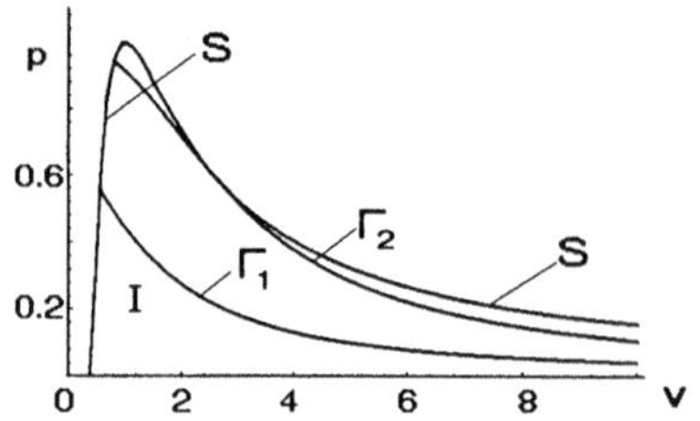

Fig. 3.

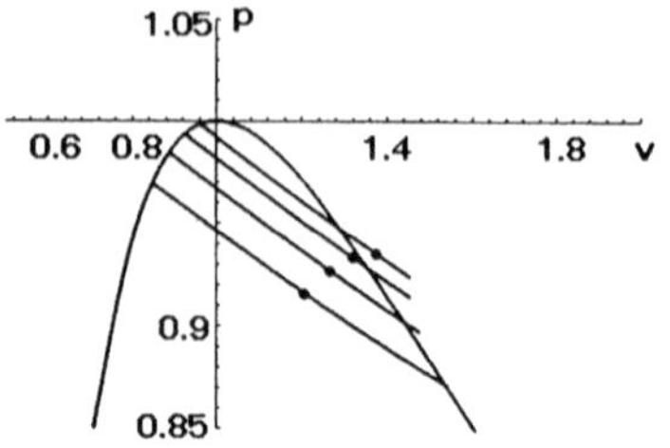

Fig. 4.

Acknowledgments .

This work was supported by the Russian Foundation for Fundamental Research under project 93-013-17621. The author wishes to acknowledge financial support of the attendance at IUTAM Symposium kindly provided by the Organizing Committee and Profs. Shigeki Morioka and Leen van Wijngaarden (Co-chairmen of Symposium).

REFERENCES

1. Thompson P.A., Sillivan D.A.: On the possibility of complete condensation shock waves in retrograde fluid. J. Fluid Mech. 70, N 4 (1975) 639–649.
2. Thompson P.A., Carofano G.C., Kim Y.G.: Shock waves and phase changes in a large-heat capacity fluids emerging from a tube. J. Fluid Mech. 166 (1986) 57–92.
3. Shepherd J.E., Sturtevant B.: Rapid evaporation at the superheat limit. J. Fluid Mech. 121 (1982) 379–402.
4. Hill L.G., Sturtevant B.: An experimental study of evaporation waves in a superheated liquid. In: Adiabatic waves in liquid-vapor systems. G.E.A.Meier, P.A.Thompson, eds., Springer Verlag, Berlin Heidelberg, 1990, 25–37.
5. Chaves H., Kurschat T., Meier G.E.A.: Evaporation waves in fluids of high molar specific heat. In: Adiabatic waves in liquid-vapor systems. G.E.A.Meier, P.A.Thompson, eds., Springer Verlag, Berlin Heidelberg, 1990, 13–24.
6. Shepherd J.E., Mc Cahan, Junhel Cho.: Evaporation wave model for superheated liquids. In: Adiabatic waves in liquid-vapor systems. G.E.A.Meier, P.A.Thompson, eds., Springer Verlag, Berlin Heidelberg, 1990, 3–12.
7. Teshukov V.M.: On the hyperbolicity of long wave equations. Soviet Math. Dokl. 32, N 2 (1985) 469–473.
8. Teshukov V.M.: Long wave approximation for vortex free boundary flows. In: Numerical methods for free boundary problems, ISNM 99, Birkhäuser Verlag, Basel, 1991, 413–421.
9. Jordansky S.V.: On the stabikity of plane stationary shock wave. Prikl. Math. Mech. 21 (1957) 465–472.
10. Kontorovich V.M.: Concerning the stability of shock waves. Sov. Phys.: Tech. Phys. 6 (1957) 1179–1181.
11. Teshukov V.M.: Stability of shock waves and general equations of state. In: Adiabatic waves in liquid-vapor systems. G.E.A.Meier, P.A.Thompson, eds., Springer Verlag, Berlin Heidelberg, 1990, 271–280.
12. Menikoff R., Plohr B.J.: The Riemann problem for fluid flow of real materials. Rev. Mod. Phys. 61, N 1 (1989) 75–129.

Explosive Boiling of a Depressurized Volatile Liquid

R. Barbone, D.L. Frost, A. Makris and J. Nerenberg

Mechanical Engineering Department
McGill University, Montreal, Canada

Summary

The explosive boiling that occurs when a pressure-liquefied gas is suddenly vented to the atmosphere has been investigated experimentally at small scale. The test liquid (refrigerant-22), initially in equilibrium with its vapor, is suddenly depressurized by bursting a foil diaphragm. The competition between the rates of vaporization and venting can lead to substantial repressurization within the vessel. The influence of vent area, liquid fill volume, and initial pressure on the amount of repressurization has been studied in a small (260 ml) steel vessel as well as in a 75 ml glass tube. The time for repressurization is typically more than an order of magnitude larger than the time for the initial pressure drop. The amount of repressurization is found to be proportional to the pressure drop which determines the degree of superheat attained by the liquid. Repressurization within both vessels reaches a maximum value over a range of initial pressures that is independent of the initial liquid fill volume and vessel geometry. High-speed photography shows that the mode of nucleation is dependent on the surface condition of the vessel walls and the initial vapor pressure. With the steel vessel, heterogeneous boiling from the side walls dominates. With the glass tube, for initial pressures less that 1 MPa, an evaporation wave propagates (at a speed ~ 0.15 m/s) from the free surface whereas for higher initial pressures heterogeneous boiling from the tube walls also occurs.

1 Introduction

If a liquid is suddenly superheated, the stored thermal energy can be released explosively during the subsequent rapid phase transition. This rapid boiling process can occur, for example, when a volatile liquid stored under pressure is suddenly depressurized. This phenomenon, when associated with the catastrophic failure of railway tank cars and storage vessels containing pressure-liquefied gases, is often referred to as a BLEVE, or Boiling Liquid Expanding Vapor Explosion [1]. If the vessel contains a combustible liquid such as propane, the dispersion of the tank contents in the form of a two-phase jet can generate a blast wave and a destructive fireball if an ignition source is present. Such accidents have caused extensive damage in the past (e.g., see [2]) and provide the motivation for the fundamental study of the rapid boiling of a depressurized volatile liquid.

If a pressurized volatile liquid contained within a vessel is partially vented (by the formation of a crack or the action of a pressure relief valve), rapid boiling of the liquid may lead to substantial repressurization within the vessel which may then initiate the complete failure of the vessel. The degree of repressurization will depend on the competition between the rates of venting and vapor generation. The dependence of the rate and amount of repressurization on the vessel geometry and the initial thermodynamic state of the liquid has not been studied extensively in the past.

Past experimental work on explosive boiling due to sudden depressurization has ranged from prototypical scale tests involving railway and automotive tanks containing pressure-liquefied gases to small-scale studies investigating the fundamental dynamics of the boiling of a depressurized liquid. For example, Birk et al. [3] have studied the explosion of automotive propane tanks subjected to a pool fire. They observed that the most violent

S. Morioka and L. van Wijngaarden (eds.),
IUTAM Symposium on Waves in Liquid/Gas and Liquid/Vapour Two-Phase Systems, 315-324.

explosions were obtained when the pressure of the tank at rupture was above a threshold value. The difficulty of obtaining reliable rapid pressure measurements within the tank at large scale precluded obtaining detailed dynamic information on the behavior of the liquid inside the tank. At small scale, detailed photographic studies of the boiling of depressurized liquids have been carried out by Hill and Sturtevant [4] and Chaves [5]. They observed the details of the two-phase evaporation wave that propagated into a superheated liquid within a vertical glass tube. In both cases the venting rate was large and no repressurization was observed within the tube.

The rapid depressurization of hot saturated water in a pipe has been studied extensively due to its relevance to nuclear reactor safety although in this case little repressurization is observed (e.g., [6]). For this case, Alamgir and Lienhard [7] have developed a semi-empirical model to predict the pressure undershoot. Their model is based on the assumption that if the rate of depressurization is large enough and the size of any physical heterogeneity present is small enough, then nucleation is initiated stochastically and is governed by homogeneous nucleation theory. To account for nucleation at the wall, they introduce a heterogeneous correction factor (which is determined by fitting to experimental data) to the Gibbs number, which is a measure of the potential barrier to nucleation. Application of the correlation to the present experimental data is described in [8].

Friedel and Purps [9] and Hervieu [10] have studied the blowdown of refrigerant and propane tanks, respectively, with emphasis on the two-phase venting behavior. The pressure behavior observed in [9] during tank venting is qualitatively similar to the present smaller scale tests. However, the tanks were depressurized using fast-opening valves which resulted in rates of depressurization typically several orders of magnitude smaller than those in the present study.

In the present study high-speed photography and fast-response pressure instrumentation are used to investigate the boiling response of a suddenly depressurized volatile liquid. The objective of the present study is to determine how the initial conditions such as initial pressure, fill volume, vent area and vessel wall material influence the boiling process and degree of repressurization within the vessel. In particular, one goal of the study is to determine the conditions that lead to the maximum repressurization within the vessel.

2 Experimental

Refrigerant 22 ($CHClF_2$) is used as the test liquid and has similar thermodynamic properties (e.g., critical conditions, saturation curve) as propane yet is not flammable. Two test vessels were used in the present study and are shown schematically in Fig. 1. Both vessels contain a variable area orifice sealed with a brass foil which is ruptured using a pneumatically-driven plunger. The plunger is located within a 0.12 m^3 dump tank which is placed directly above the test vessel and contains air initially at atmospheric pressure. The vessels are equipped with ports for the delivery of the test liquid and the measurement of pressure and temperature.

The first test vessel shown in Fig. 1 consists of a rectangular steel block containing a vertical slot (2 cm wide x 19 cm high x 7.5 cm deep) which forms a test section with a volume of 260 ml. The vessel walls were coated with Teflon to protect against corrosion. A comparison of preliminary tests with and without the Teflon coating showed that the coating had the effect of slightly reducing the rate of nucleation at the walls. The temperature of the test liquid was controlled by flowing water from a heat exchanger through slots in the exterior walls of the vessel. The second vessel shown in Fig. 1 consists of a 75 ml glass tube (2.5 cm dia., 15.2 cm long) placed within a cylindrical tank.

The glass tube is surrounded by glycerine to reduce diffraction effects and increase the rate of heat transfer from a heat exchanger tube located within the glycerine.

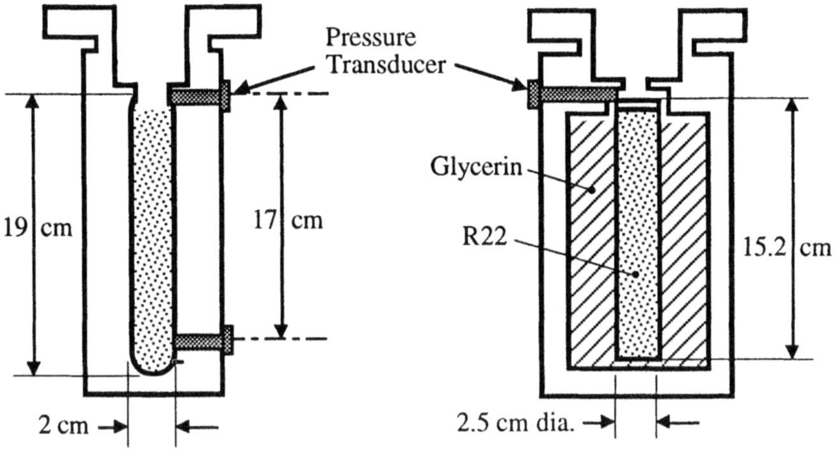

Fig. 1 Cross-section of 260 ml steel vessel (left) and 75 ml glass tube assembly (right)

The pressure within the vessel just below the orifice was recorded using a piezoelectric pressure transducer (PCB 113A24). The boiling process was photographed using a Hycam 16-mm camera (operated typically at 1500 frames/s) through glass windows on the sides of the vessels. Back lighting was provided by a 1000 Watt tungsten flood lamp with a glass diffuser placed directly behind the test section.

3 Results and Discussion

3.1 General Features of Explosive Boiling. Figure 2 shows a comparison between the characteristic pressure history recorded within the steel vessel following depressurization of R-22 (90% liquid fill, initial pressure 1.38 MPa, orifice diameter 1.9 cm) with an experiment at the same initial conditions using ambient temperature water pressurized with air. For the experiment with water, following diaphragm rupture little pressure recovery is observed due to the low vapor pressure of water. In contrast, when

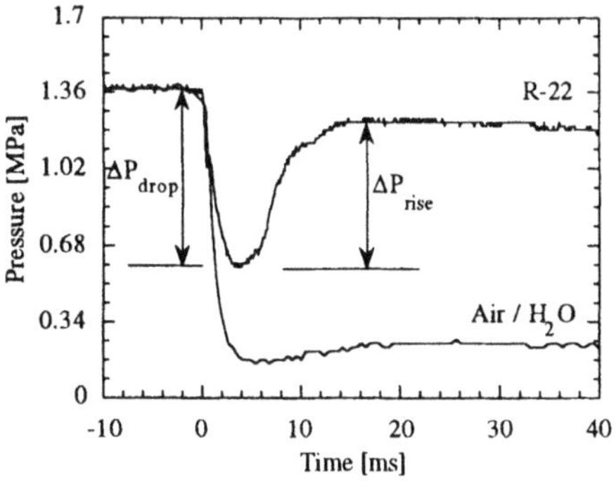

Fig. 2 Characteristic pressure history for venting of R-22 and pressurized water.

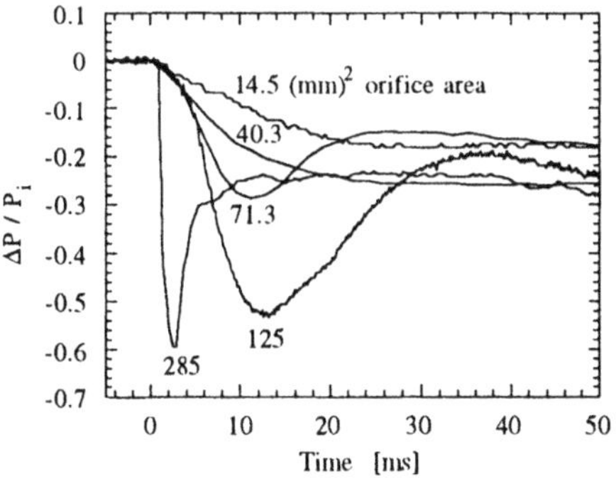

Fig. 3 Effect of orifice area on pressure history for 90% liquid fill.

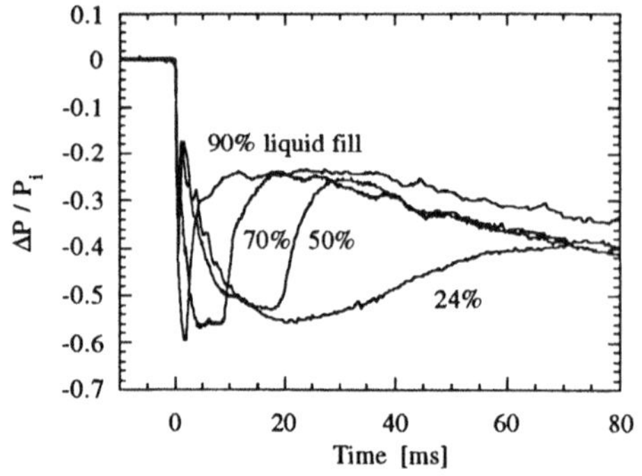

Fig. 4 Effect of initial liquid volume on pressure history. Initial pressure is 3.15 MPa.

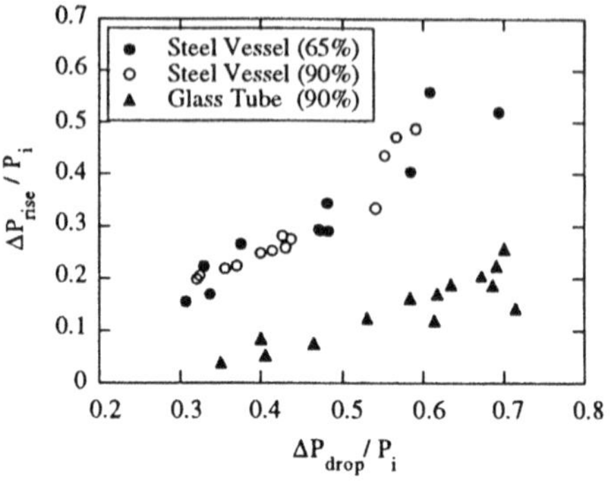

Fig. 5 Relationship between pressure rise and pressure drop (defined in Fig. 2).

depressurized, the liquid R-22 boils rapidly causing a transient repressurization of the vessel, followed by a slow pressure decay to atmospheric pressure on a much longer timescale. The initial rates of depressurization differ due to the larger molecular weight of the refrigerant compared to air.

Figure 3 shows the effect of increasing the orifice area on the normalized pressure history within the vessel for 90% liquid fill and for orifice diameters ranging from 0.4 cm to 1.9 cm. Increasing the orifice area increases the rate of venting and hence increases the rate and amount of depressurization. Figure 4 shows the effect of liquid volume fraction on the pressure history for an initial pressure of 3.15 MPa in each case. The rate of pressure drop and rise are both increased as the amount of liquid increases (corresponding to a smaller initial vapor volume). The relationship between the amount of initial pressure drop within the vessel and the subsequent amount of pressure rise is shown in Fig. 5 for two different fill volumes in the steel vessel as well as for the glass tube apparatus. The data from both vessels indicate that there exists a strong correlation between the degree of superheat attained by the liquid (which is proportional to the amount of depressurization) and the corresponding amount of repressurization for a given liquid fill volume and rupture area.

3.2 Explosive Boiling in 260 ml Vessel. A series of trials was carried out in the steel vessel to investigate the influence of the initial thermodynamic state of liquid R-22 on the explosive boiling characteristics following a sudden depressurization. A sequence of high-speed photographs shown in Fig. 6 illustrates the boiling process that occurs within the vessel, with a orifice diameter of 1.9 cm. Heterogeneous nucleation on the walls dominates and the boiling propagates from the side walls towards the center of the vessel as a wavefront with an average speed of the order of 1 m/s.

The violence of the boiling response may be characterized by the amount of pressure recovery within the vessel which is shown in Fig. 7 as a function of the initial vapor pressure. The experimental data shown in Fig. 7 were obtained with a constant orifice diameter of 1.9 cm and liquid fill volume of 65% and 90%. The degree of repressurization reaches a maximum at about 2 MPa for liquid fill volumes of both 65% and 90%. The initial pressure for which the maximum occurs is close to the initial pressure at which the greatest superheat can be attained following depressurization of the saturated liquid to atmospheric pressure.

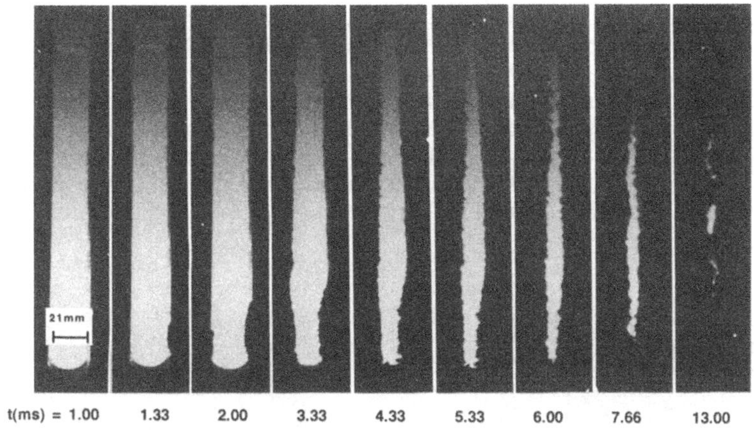

Fig. 6 Sequence of high-speed photographs showing heterogeneous boiling of R-22 from the walls in the 260 ml steel vessel (P_i = 1.03 MPa).

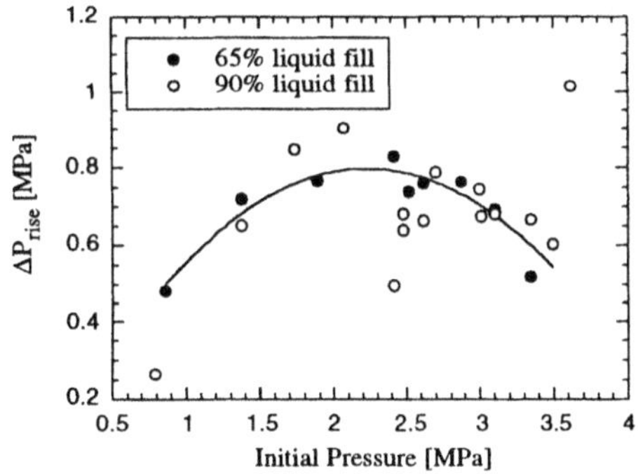

Fig. 7 Degree of repressurization as a function of initial pressure within steel vessel. The curve represents a second order polynomial fit through the 65% fill data.

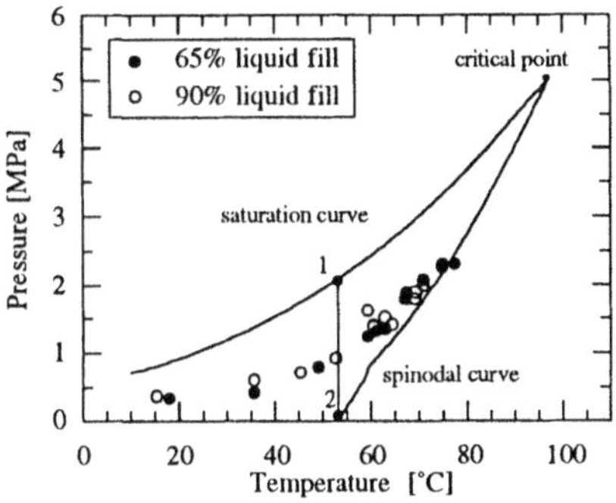

Fig. 8 Locus of thermodynamic end states following pressure drop showing the degree of superheat attained for the steel test vessel.

Figure 8 shows the saturation and spinodal curves for R-22 which bound the metastable region. Path 1-2 corresponds to an isentropic expansion from a pressure of 2.06 MPa (and a saturation temperature of 53°C) to atmospheric pressure which yields the greatest possible degree of superheat at atmospheric pressure. The state of the liquid following depressurization for both the 65% and 90% liquid fill volume trials is also shown in Fig. 8. Although no homogeneous boiling was observed, the locus of data points indicates that there is a limit of superheat attained that is determined by heterogeneous boiling. Even though the degree of superheat attained will depend on the geometry of the vessel as well as the surface properties of the vessel walls, it is interesting to note that the maximum degree of superheat is attained at a liquid temperature that is similar to that predicted by homogeneous nucleation theory.

3.3 Explosive Boiling in 75 ml Glass Tube. The next series of trials was performed in the 75 ml glass tube (25 mm inner diameter) using a 9.5 mm diameter orifice and 90% liquid fill volume. Two modes of boiling were observed which depended on the initial temperature of the liquid. For initial R-22 temperatures less than 25°C (corresponding to a saturation vapor pressure of 1.03 MPa) the boiling occurred as an evaporation wave moving from the free surface of the liquid vertically downwards. Figure 9 shows an enlarged view of the liquid-vapor interface in the early stages following depressurization. After the diaphragm is ruptured, the vapor above the liquid condenses, signaling the passage of an expansion wave, and 7.33 ms later the surface of the liquid erupts violently into a two-phase flow consisting of vapor and fine droplets moving vertically at high speed. A detailed discussion of the mechanism of propagation of the evaporation wave can be found in Hill [11].

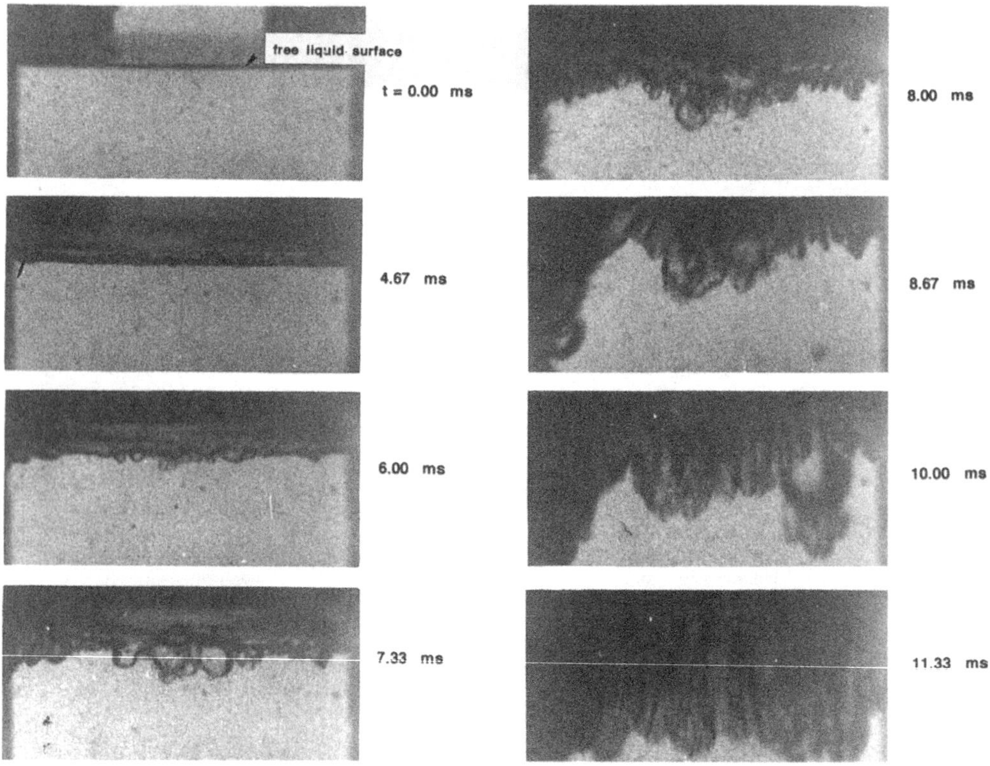

Fig. 9 Breakup of free surface of liquid R-22 [illegible]ned within a glass tube following depressurization (P_i = 1.20 MPa)

Figures 10 and 11 show the average displacement versus time and a schematic of the propagation of the evaporation wavefront within the glass tube, illustrating the random variations in the shape and velocity of the evaporation wave that occur for two trials with the same initial conditions. In both cases the wave initially propagates rapidly downwards from the free surface in a highly asymmetric manner. The wavefront becomes relatively more planar after a short time (denoted *a* on Fig. 10) which corresponds to a decrease in the average wavefront velocity. After about 300 ms the wavefront asymmetry begins to become more pronounced, and the velocity increases beyond the point denoted *c* in the figures.

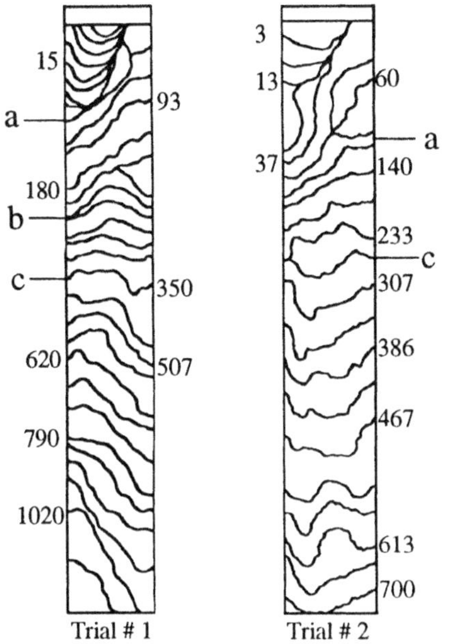

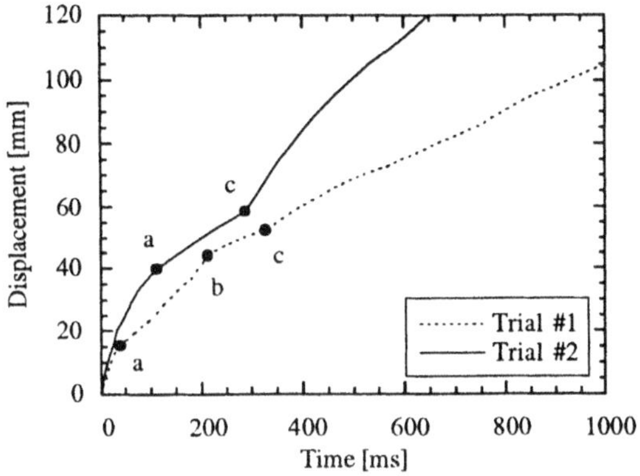

Fig. 10 (above) Evaporation wave displacement versus time (Pi = 1.03 MPa)

Fig. 11 (left) Schematic of the propagation of evaporation wave in liquid R-22. (Pi = 1.03 MPa)

Figure 12 shows single frames from a Hycam film of the propagation of an evaporation wave in the liquid R-22 for an initial vapor pressure of 1.03 MPa. Note that the initial rapid depressurization is sufficient to cause a nucleation site to grow on the right side of the tube wall ahead of the wave front. The vapor bubble growth and pressure buildup within the vessel cause the evaporation wave to virtually stop propagating until the subsequent pressure drop within the vessel restarts the wave propagation.

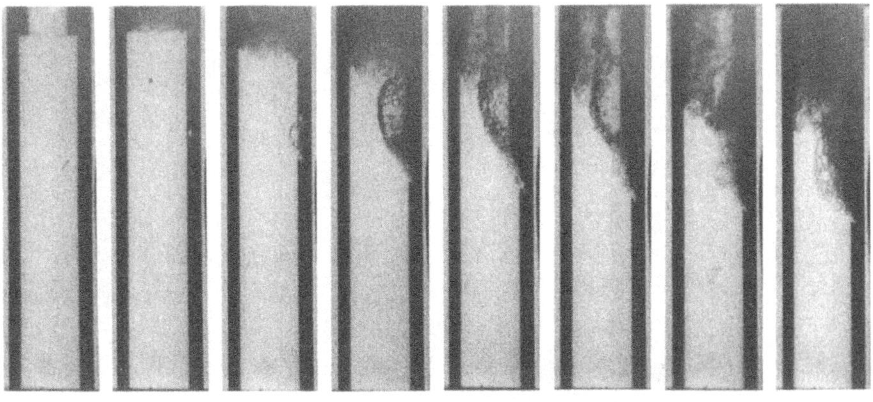

: (ms) = 0.00 12.00 25.33 40.00 70.00 90.00 120.00 146.67

Fig. 12 Sequence of high-speed photographs showing the propagation of the evaporation wave throughout the length of the glass tube. (Pi = 1.03 MPa) (continued on next page)

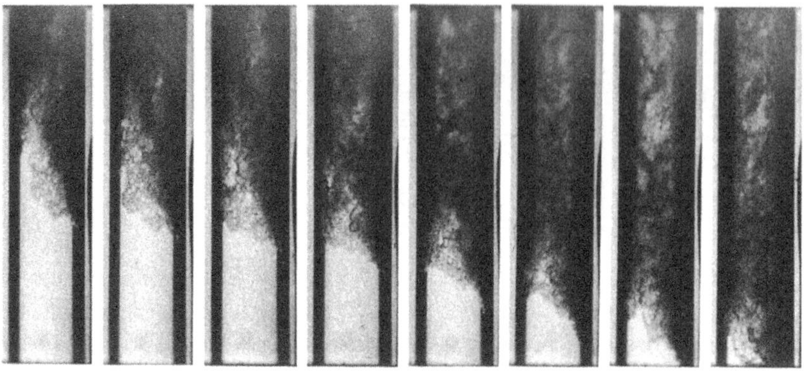

Fig. 12 (Continued)

For initial R-22 temperatures above about 25°C (corresponding to an initial vapor pressure greater than about 1 MPa), the superheat attained by the liquid following depressurization is sufficient to induce heterogeneous nucleation on the tube walls. This is illustrated in Fig. 13 which shows a sequence of photographs for the boiling of R-22 with an initial pressure of 2.24 MPa (corresponding to an initial saturation temperature of 60°C). The growth of the nucleation sites along the length of the tube forces the remaining liquid out the orifice and prevents the propagation of an evaporation wave within the tube. Due to the large increase in liquid-vapor interface area, the liquid R-22 is partially vaporized and expelled from the tube in a time about an order of magnitude faster than when the boiling occurs via an evaporation wave.

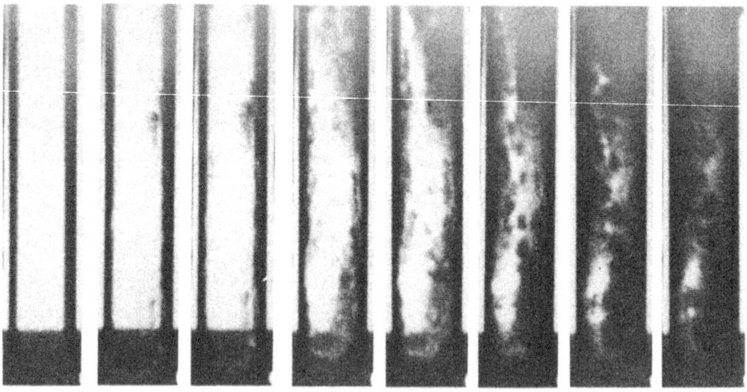

Fig. 13 Sequence of high-speed photographs showing the heterogeneous boiling of R-22 within a glass tube. (Pi = 2.24 MPa).

Figure 14 shows the degree of repressurization as a function of the initial pressure for the experiments in the glass tube. Despite the fact that two modes of boiling were observed (evaporation wave and heterogeneous wall boiling), the repressurization reaches a maximum near an initial pressure of 2 MPa, similar to that observed earlier with the steel vessel. This indicates that although two geometrically different vessels were used, with different wall conditions and vent areas, the maximum pressure recovery is largely a function of the initial R-22 vapor pressure (or temperature), i.e., the initial thermodynamic state of the liquid.

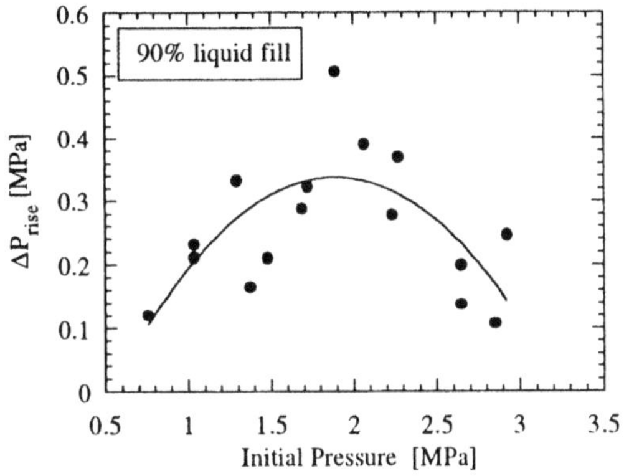

Fig. 14 Degree of repressurization as a function of initial pressure within glass tube. The line is a second order polynomial fit through the data.

5 Conclusions

The explosive boiling response of suddenly depressurized refrigerant 22 has been investigated experimentally in a small-scale steel vessel and a glass tube. In both cases, rapid boiling caused a repressurization within the vessels which was linearly dependent on the pressure undershoot or the degree of superheat attained by the liquid. The degree of repressurization reached a maximum value in both vessels at an initial pressure of about 2 MPa, regardless of the mode of boiling, vessel geometry or liquid fill volume. The mode of nucleation observed depended primarily on the degree of superheat attained by the liquid as well as the surface characteristics of the vessel. For the steel vessel, heterogeneous boiling from the walls dominated. In the glass tube a self-sustained evaporation wave propagated from the free surface into the liquid for initial pressures at and below 1 MPa. For an initial pressure of 1 MPa, after an initial startup phase lasting about 100 ms, the evaporation wave attained an average velocity between 5 and 20 cm/s. Since the wave velocity is relatively slow, fluctuations of the pressure in the vessel downstream of the wave due to the competition between the rates of vaporization and venting can influence the propagation of the wave. Deviations of the wavefront from a planar shape corresponded to an increase in the average propagation velocity and may be caused by spatial variations in the liquid temperature.

For large initial pressures, the boiling in the present experiments was dominated by wall boiling so that the ratio of surface area/liquid volume played an important role in the rate of vapor generation. Therefore the present results cannot be scaled in a quantitative manner easily to larger scale cylindrical or spherical vessels, where boiling on the walls will be relatively less important compared to boiling within the bulk of the liquid. Nevertheless, the trend observed in the present tests, in particular that the violence of the boiling increases with initial pressure or temperature up to a maximum corresponding approximately to the superheat limit temperature at atmospheric pressure (~ 53°C), is expected to be reproduced in larger scale tests. If bubbles or other nucleation sites are present within the liquid prior to depressurization, the "activation energy" for boiling to begin will be lowered limiting the amount of depressurization that can be attained. Regardless of the initial conditions, the most violent boiling will occur if the liquid is brought uniformly to its limit of superheat. Therefore, from a practical point of view regarding the storage of liquefied gases, situations that lead to large rates and degrees of depressurization should be avoided and the initial vessel pressure should be maintained at a level such that the superheat limit cannot be attained, even for rapid venting of the liquid to atmospheric pressure.

Acknowledgments

Financial support for the work was provided by Transport Canada under DSS contract T8200-2-2520/01-XSD

References

1. Reid, R.C., "Possible Mechanism for Pressurized-Liquid Tank Explosions or BLEVE's," *Science*, Vol. 203, No. 23., 1979, pp. 1263-1265.
2. Wormuth, D.W. (ed.), "BLEVE! The Tragedy of San Juanico," *J. Skandia International*, 1985, pp.1-22.
3. Birk, A.M., Ye, Z., Maillette, J. & Cunningham, M., "Summary on Medium Scale Fire Tests Investigating the BLEVE Event", *Internal Report for TDC Transport Canada*, 1991.
4. Hill, L.G. & Sturtevant, B., "An Experimental Study of Evaporation Waves in a Superheated Liquid," IUTAM Symposium, Göttingen, Germany, in *Adiabatic Waves in Liquid-Vapor Systems*, edited by G.E.A. Meier and P.A. Thompson, Springer Berlin, 1989, pp. 25-37.
5. Chaves, H., Lang, H., Meier, G.E.A. & Speckmann, H.-D., *Lecture Notes in Physics: Flow of Real Fluids*, edited by G.E.A. Meier and F. Obermeier, Spinger Berlin, 1985.
6. Bartak, J., "A Study of the Rapid Depressurization of Hot Water and the Dynamics of Vapour Bubble Generation in Superheated Water," *Int. J. of Multiphase Flow*, Vol. 16, No. 5., 1990, pp. 789-798.
7. Alamgir, M. & Lienhard, J. H. "Correlation of Pressure Undershoot during Hot Water Depressurization," *J. Heat Transfer*, Vol. 103, 1981, pp. 52-55.
8. Barbone, R., "Explosive Boiling of a Depressurized Volatile Liquid," M.Eng. Thesis, Department of Mechanical Engineering, McGill University, 1994.
9. Friedel, L. and Purps, S., "Models and Design Methods for Sudden Depressurization of Gas/Vapor-Liquid Reaction Systems," *Int. Chem. Eng.*, Vol. 26, No. 3, 1986, pp. 396-407.
10. Hervieu, E., "Behavior of a Flashing Liquid within a Vessel following Loss of Containment: Application to Propane," *ANS Proc. 1992 National Heat Transfer Conference*, Vol. 6, pp.227-235.
11. Hill, L. G. "An Experimental Study of Evaporation Waves in a Superheated Liquid," Ph.D. Thesis, California Institute of Technology, 1991.

Propagation of Waves in a Vapor-Gas Mixture due to Evaporation and Condensation

Yoshimoto Onishi and Hidekazu Tsuji
Department of Applied Mathematics and Physics
Faculty of Engineering
Tottori University, Tottori 680, Japan

A numerical analysis of the transient behavior of a binary mixture of a vapor and an inert gas due to the evaporation and condensation processes of the vapor occuring at the plane condensed phase boundaries placed in parallel has been carried out, based on the Boltzmann equation of BGK type under diffusive boundary conditions at the interfaces, starting from the initial condition associated with the sudden change in temperature of the condensed phases. Shock and expansion waves are produced due to the onset of these processes. The propagation of these waves and the eventual interaction between them, which occur in the nonuniform flow fields, and also the interaction between the waves and the condensed phases, display intriguing aspects of the transient motions of the gas mixture. The interactions between the waves themselves and the waves and the condensed phases, which are largely responsible for the adjustment of the evaporation and condensation rates at the interfaces in order to bring the flow fields to the possible final states, are quite different from the ones occurring in the pure-vapor case, because of the presence of the inert gas. Especially when the amount of inert gas is small, the condensation process is greatly obstructed by the accumulation of inert gas having been continuously driven toward the condensation surface by the vapor flow, which then reduces the evaporation greatly at the other side to make a balance between the two processes.

1. Introduction

Evaporation and condensation phenomena and flow problems involving these processes are quite common in ordinary circumstances and have aroused interest of scientists in the field of kinetic theory. The reason for this is that the ordinary continuum-based fluid dynamics is not capable of describing qualitatively correctly the behavior of the gas phase within a nonequilibrium region always existent in the close vicinity of the condensed phase. Such a nonequilibrium region has a thickness of the order of the molecular mean free path and is called the Knudsen layer, in which collisions between molecules are not so frequent that the momentum and energy exchanges between the molecules leaving the interface and coming toward it from the outside are not sufficient. It is this nonequilibrium region that is responsible for phase change processes and its existence cannot be neglected in any problems even in the continuum limit as long as these processes are involved in them. The same applies also to a mixture of condensable and noncondensable gases, i.e., to a vapor in the presence of an inert gas. Therefore, the analysis for flow problems of this kind must necessarily be based on kinetic theory or, at least, on the Navier-Stokes equations with appropriate boundary conditions at the interface(s) derivable only from kinetic theory (see e.g., Ref. 1).

Makashev[2] first tried numerical calculations on the evaporation and condensation problem of a vapor-gas mixture between two condensed phases, based on the

S. Morioka and L. van Wijngaarden (eds.),
IUTAM Symposium on Waves in Liquid/Gas and Liquid/Vapour Two-Phase Systems, 325-334.

Boltzmann equation of BGK type[3], and then Gajewski *et al.*[4], based on a slightly different model kinetic equation, also worked out numerically the half-space problem of condensation. Both of their works were concerned with the steady behavior of the gas phase. With the advent of high performance computers, it has become possible to investigate the transient behavior of the gas phase due to the onset of phase change processes and the roles which the various waves play in the transition process of the flow field to its final state. Flow fields of this kind may be of fundamental importance in gas dynamics in the sense that, associated with these phase change processes, shock waves and expansion waves are produced. The propagation of these waves and the eventual interaction between them, together with the interaction between the waves and the condensed phases, give rise to some intriguing aspects of the transient motions of the gas phase, which are surely worth being clarified in detail. In addition, these interactions, which are largely responsible for the adjustment of the evaporation and condensation rates at the interfaces in order to bring the flow fields to their final states, are quite different from the ones occurring in the pure-vapor case (see e.g., Ref. 5) because of the presence of the inert gas, especially when only a small amount is present.

For this reason, motions of a binary mixture of a vapor and an inert gas due to phase change processes at the two plane condensed phase boundaries placed in parallel have been taken up here, in order to investigate what has been stated above. The Boltzmann equation of BGK type[3] was solved numerically under the diffusive boundary conditions at the interfaces, starting from an initial condition corresponding to the sudden change in temperature of the condensed phases, with particular attention being paid to the transient behavior of the mixture and its component gases when the inert gas is present in small amount. The procedure is essentially the same as in Ref. 6, where propagation of shock waves through a vapor-gas mixture of infinite expanse in a half space due to the evaporation process from the plane condensed phase was treated.

2. Kinetic Equations and Initial and Boundary Conditions

The Boltzmann equation of BGK type[3] for a binary gas mixture may be written for the present problem as

$$\frac{\partial f^s}{\partial t} + \xi_x \frac{\partial f^s}{\partial X} = N^s \kappa_{ss}(F_e^s - f^s) + N^t \kappa_{st}(F_e^{st} - f^s), \tag{1}$$

$$\begin{bmatrix} N^s \\ N^s U^s \\ 3N^s R_s T^s \end{bmatrix} = \iiint \begin{bmatrix} 1 \\ \xi_x \\ (\xi_x - U^s)^2 + \xi_y^2 + \xi_z^2 \end{bmatrix} f^s \, d\xi_x d\xi_y d\xi_z, \tag{2}$$

$$P^s = N^s \mathrm{k} T^s, \tag{3}$$

$$F_e^s = \frac{N^s}{(2\pi R_s T^s)^{3/2}} \exp\Big\{-\frac{(\xi_x - U^s)^2 + \xi_y^2 + \xi_z^2}{2R_s T^s}\Big\}, \tag{4}$$

$$F_e^{st} = \frac{N^s}{(2\pi R_s T^{st})^{3/2}} \exp\Big\{-\frac{(\xi_x - U^{st})^2 + \xi_y^2 + \xi_z^2}{2R_s T^{st}}\Big\}, \tag{5}$$

$$U^{st} = U^{ts} = \mu_s U^s + \mu_t U^t, \tag{6}$$

$$T^{st} = T^s + 2\mu_s\mu_t(T^t - T^s) + \frac{m_s}{3\mathrm{k}}\,\mu_t^2\,(U^t - U^s)^2, \tag{7}$$

where $\mu_s = m_s/(m_s + m_t)$. The quantities s and t appearing as superscript and subscript refer to A and B or B and A, i.e., $(s, t) = (A, B)$ or (B, A). Also t is the time; X the space coordinate; (ξ_x, ξ_y, ξ_z) the molecular velocity vector; f^s, N^s, U^s, T^s and $P^s\,(= N^s\mathrm{k}T^s)$ are, respectively, the distribution function, the number density, the mean flow velocity in the X direction, the temperature and the partial pressure of gas S. Further, m_s is the molecular mass of gas S and $R_s = \mathrm{k}/m_s$ the gas constant per unit mass, k being the Boltzmann constant, F_e^s and F_e^{st} are the local Maxwellian distributions for molecules of gas S characterized by the macroscopic quantities (N^s, U^s, T^s) and (N^s, U^{st}, T^{st}), respectively, and $N^s\kappa_{ss}$ and $N^t\kappa_{st}$ represent the number of collisions per unit time of a molecule of gas S due to the self and cross collisions, respectively. Thus $N^s\kappa_{ss} + N^t\kappa_{st}$ is the collision frequency for a molecule of gas S irrespective of its collision partners. The collision frequencies are related to the transport coefficients of the mixture and its component gases and, hence, κ_{ss} and $\kappa_{st}\,(= \kappa_{ts})$, which are assumed to be constant and of the same order of magnitude, can be evaluated from these coefficients at some equilibrium reference state (see e.g., Refs. 1 and 3) as

$$\left.\begin{aligned}
\eta^M &= \frac{P_0^A}{N_0^A\kappa_{AA} + N_0^B\kappa_{AB}} + \frac{P_0^B}{N_0^A\kappa_{AB} + N_0^B\kappa_{BB}}, \\
D_{AB} &= \frac{(m_A + m_B)\,\mathrm{k}\,T_0}{m_A m_B(N_0^A + N_0^B)\kappa_{AB}}, \qquad \eta^s = \frac{P_0^s}{N_0^s\kappa_{ss}}, \qquad \lambda^s = \frac{5}{2}\,R_s\,\eta^s,
\end{aligned}\right\} \tag{8}$$

where the fluid dynamic quantities with subscript 0 such as N_0^s, P_0^s and T_0 refer to those at some equilibrium reference state. In equations (8), η^M and D_{AB} are the viscosity and diffusion coefficients of the mixture, η^s and λ^s being, respectively, the viscosity and thermal conductivity of gas S.

Once the distribution function f^s is obtained, the normal stress τ_{xx}^s, the energy flux E^s and heat flux Q^s for gas S can be calculated from

$$\tau_{xx}^s = -m_s\iiint(\xi_x - U^s)^2\,f^s\,d\xi_x d\xi_y d\xi_z, \tag{9}$$

$$E^s = \frac{1}{2}m_s\iiint\xi_x\,(\xi_x^2 + \xi_y^2 + \xi_z^2)\,f^s\,d\xi_x d\xi_y d\xi_z, \tag{10}$$

$$Q^s = \frac{1}{2}m_s\iiint(\xi_x - U^s)\,[\,(\xi_x - U^s)^2 + \xi_y^2 + \xi_z^2\,]\,f^s\,d\xi_x d\xi_y d\xi_z. \tag{11}$$

Now, let us consider a vapor-gas mixture between the two plane condensed phases placed in parallel, one at $X = 0$ and the other at $X = L$. Initially, the condensed phases and the mixture gas phase are assumed to be in complete equilibrium at temperature T_0, the number density and pressure of the component gas in the mixture then being N_0^s and P_0^s, respectively. The distribution function of gas S at this state is then given by

$$f^s = \frac{N_0^s}{(2\pi R_s T_0)^{3/2}}\,\exp\left\{-\frac{\xi_x^2 + \xi_y^2 + \xi_z^2}{2R_s T_0}\right\} \quad \text{for} \quad 0 < X < L \tag{12}$$

and for all possible values of the molecular velocity vector. At time $t = 0$, the temperature of each of the condensed phases is suddenly changed to a different constant temperature, i.e., from T_0 to T_1 at $X = 0$ and from T_0 to T_2 at $X = L$, respectively. The distribution function f^s then takes

$$f^s = \frac{N_1^s}{(2\pi R_s T_1)^{3/2}} \exp\Big\{-\frac{\xi_x^2+\xi_y^2+\xi_z^2}{2R_sT_1}\Big\} \quad \text{for} \quad \xi_x > 0 \quad \text{at} \quad X = 0, \tag{13}$$

$$f^s = \frac{N_2^s}{(2\pi R_s T_2)^{3/2}} \exp\Big\{-\frac{\xi_x^2+\xi_y^2+\xi_z^2}{2R_sT_2}\Big\} \quad \text{for} \quad \xi_x < 0 \quad \text{at} \quad X = L, \tag{14}$$

for $t \geq 0$. Here N_i^s $(i = 1, 2)$ is the number density of the molecules of gas S leaving the interface after their interaction with the condensed phase. N_i^A is a unique function of T_i and is given as the saturated vapor number density corresponding to T_i, given by the Clapeyron-Clausius relation (see e.g., Ref. 7) as

$$N_i^A = \frac{P_i^A}{\mathrm{k}T_i}, \qquad \frac{P_i^A}{P_0^A} = \exp\Big\{-\frac{h_L}{R_AT_0}\Big(\frac{T_0}{T_i}-1\Big)\Big\} \tag{15}$$

where P_i^A is the saturated vapor pressure corresponding to T_i. h_L is the latent heat of vaporization per unit mass of gas A. N_i^B, on the other hand, is to be determined, as part of the solution, by the condition of no net mass flow of gas B across the interface, i.e., $N^BU^B = 0$ at $X = 0$ and $X = L$. Hence, using Eqs.(13) and (14), we have for N_i^B

$$N_i^B = \mp\frac{2\pi}{(2\pi R_BT_i)^{1/2}} \int_{-\infty}^{\infty}\int_{-\infty}^{\infty}\int_{\pm\xi_x<0} \xi_x f^B \, d\xi_x d\xi_y d\xi_z, \tag{16}$$

where the upper sign applies to N_1^B and the lower to N_2^B.

Now, in order to render the numerical calculation easier, the following new functions have been introduced (cf. Ref. 8)

$$\begin{bmatrix} g^s \\ G^s \\ G^{st} \end{bmatrix} \equiv \iint \begin{bmatrix} f^s \\ F_e^s \\ F_e^{st} \end{bmatrix} d\xi_y d\xi_z, \qquad \begin{bmatrix} h^s \\ H^s \\ H^{st} \end{bmatrix} \equiv \iint (\xi_y^2+\xi_z^2) \begin{bmatrix} f^s \\ F_e^s \\ F_e^{st} \end{bmatrix} d\xi_y d\xi_z, \tag{17}$$

so that

$$\begin{bmatrix} N^s \\ N^sU^s \\ 3N^sR_sT^s \end{bmatrix} = \int \begin{bmatrix} g^s \\ \xi_x g^s \\ (\xi_x - U^s)^2 g^s + h^s \end{bmatrix} d\xi_x. \tag{18}$$

With these new variables, we get a set of simultaneous equations for the four unknown functions (g^A, h^A, g^B, h^B) of three independent variables (t, X, ξ_x) instead of two unknown functions (f^A, f^B) of five independent variables $(t, X, \xi_x, \xi_y, \xi_z)$.

Appropriate nondimensionalization of the set of the governing equations, initial and boundary conditions gives us the following parameters which characterize the motions of the mixture and its component gases:

$$Kn = \frac{l_0^A}{L}, \quad \frac{N_0^B}{N_0^A}, \quad \frac{T_1}{T_0}, \quad \frac{T_2}{T_0}, \qquad \Gamma = \frac{h_L}{R_AT_0}, \quad \frac{m_B}{m_A}, \quad \frac{\kappa_{AB}}{\kappa_{AA}}, \quad \frac{\kappa_{BB}}{\kappa_{AA}}, \tag{19}$$

where Kn is the Knudsen number and l_0^A is the mean free path of the molecules of gas A at the initial state defined by $l_0^A = (8R_AT_0/\pi)^{1/2}/(N_0^A\kappa_{AA} + N_0^B\kappa_{AB})$.

3. Results and discussion

Calculations have been carried out for a number of cases and some of the results are shown in Figs. 1-3. In Fig. 1, theoretical steady state results based on weakly nonlinear analysis[9] are also shown for comparison. Here, the following fluid dynamic quantities of the mixture,

$$\left.\begin{array}{l} N = N^A + N^B, \qquad \rho = \rho^A + \rho^B, \qquad U = (\rho^A U^A + \rho^B U^B)/\rho, \\ P = P^A + P^B + \frac{1}{3}[\rho^A(U^A)^2 + \rho^B(U^B)^2 - \rho(U)^2], \\ T = P/(N\mathrm{k}), \qquad E = E^A + E^B, \end{array}\right\} \tag{20}$$

have also been introduced, where N, ρ, U, P, T and E are the number density, mass density, *mean mass flow* velocity, pressure, temperature and energy flux of the mixture, respectively. Note that P and T are defined in terms of the peculiar velocity of molecules with respect to the *mean mass flow* velocity. In the figures, $k = (\sqrt{\pi}/2)Kn$ is used in place of Kn just for convenience. The time t is measured in terms of a macroscopic time scale τ_0^A defined by $\tau_0^A = L/(2R_AT_0)^{1/2}$. In addition, some other quantities $P_0 = P_0^A + P_0^B$, $N_0 = N_0^A + N_0^B$ and $\rho_0 = \rho_0^A + \rho_0^B$ are introduced together with the sound speed c_0 of the mixture (monatomic gas) at T_0 and its local sound speed c, for the presentation of the results.

The transient behavior of the mixture and its component gases may be summarized as follows: at the initial stage, the shock wave due to the evaporation process at $X/L = 0$ is propagating rightward (toward $X/L = 1$) accompanied by the contact region behind (see Figs. 1-3) and, for strong evaporation cases (see Fig. 3), the expansion waves, which are to propagate in the opposite direction, are being swept backward in the same direction as the shock wave owing to the supersonic evaporation flow behind the shock wave. At the second stage, the shock wave which has reached the condensed phase at $X/L = 1$ is reflected because of the formation of a compressed region caused by the excess in mass flow behind the incident shock wave over the condensation mass flow at the condensed phase. This excess in mass flow becomes larger as the initial evaporation becomes stronger. This is due to the hindrance to the vapor condensation by the accumulation of inert gas near the condensed phase having so far been driven by the vapor flow. It is this hindrance effect of the inert gas that affects the flow fields of the mixture drastically and causes a great difference between the transient behavior of a mixture (with small amount of inert gas) and that of a single component gas (see e.g., Ref. 5). The reflected shock wave, interacting with the approaching contact region and also the swept back expansion waves for initially strong evaporation cases, propagates leftward and is then reflected at $X/L = 0$ slightly, propagating rightward again and then reflected at $X/L = 1$ a second time, going back again to the original phase and being then reflected a second time there. Then the wave-like nature in the flow field disappears almost completely and the diffusion process prevails there, which brings extremely gradually the transient flow field to its final steady state.

References

1) Y. Onishi: "Kinetic theory based nonlinear treatment for motions of a binary gas mixture involving slightly strong evaporation and condensation processes II – A system of macroscopic equations, the boundary conditions and the Knudsen-layer corrections." Z. angew. Math. Phys. **44** (1993) 828-864.
2) N.K. Makashev: "Strong recondensation in a one-component and a two- component rarefied gas for an arbitrary value of Knudsen number." Fluid Dynamics. **5** (1972) 815-821.
3) B.B. Hamel: "Kinetic model for binary gas mixtures." Physics of Fluids. **8** (1965) 418-425.
4) P. Gajewski, A. Kulicki and A. Wisniewski: "A kinetic model of the vapor condensation in the presence of a noncondensable gas." In *Rarefied Gas Dynamics* edited by M. Becker and M. Fiebig, DFVLR - Press, Germany (1974) pp. F.2-1 - F.2-11.
5) Y. Onishi, T. Shoji and H. Tsuji: "Interaction of Shock and Expansion Waves caused in a Vapor between the Two Plane Condensed Phases." In *Proceedings of the International Symposium on Aerospace and Fluid Science*, Institute of Fluid Science, Tohoku University (1993) pp. 517-524.
6) Y. Onishi and H. Miura: "Propagation of shock waves through a gas mixture due to strong evaporation from a plane condensed phase." In *Proceedings of the International Symposium on Shock Waves* (1994) to be published.
7) L.D. Landau and E.M. Lifshitz: *Statistical Physics*. Pergamon Press, New York (1969).
8) C.K. Chu: "Kinetic-theoretic Description of the Formation of a Shock Wave." Physics of Fluids. **8** (1965) 12-22.
9) Y. Onishi: "Nonlinear analysis for evaporation and condensation of a vapor-gas mixture between the two plane condensed phases. Part II: Concentration of inert gas $\sim O(Kn)$." In *Rarefied Gas Dynamics* (*Progress in Astronaut. and Aeronaut.* Vol. 117) edited by E.P. Muntz, D.P. Weaver and D.H. Campbell, AIAA, New York (1989) pp. 492-513.

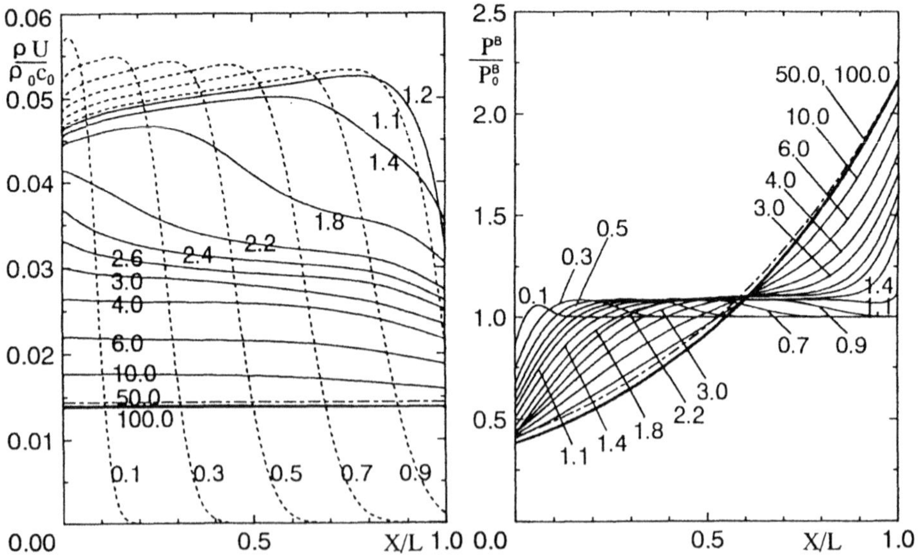

Fig. 1 : see the next page for the caption.

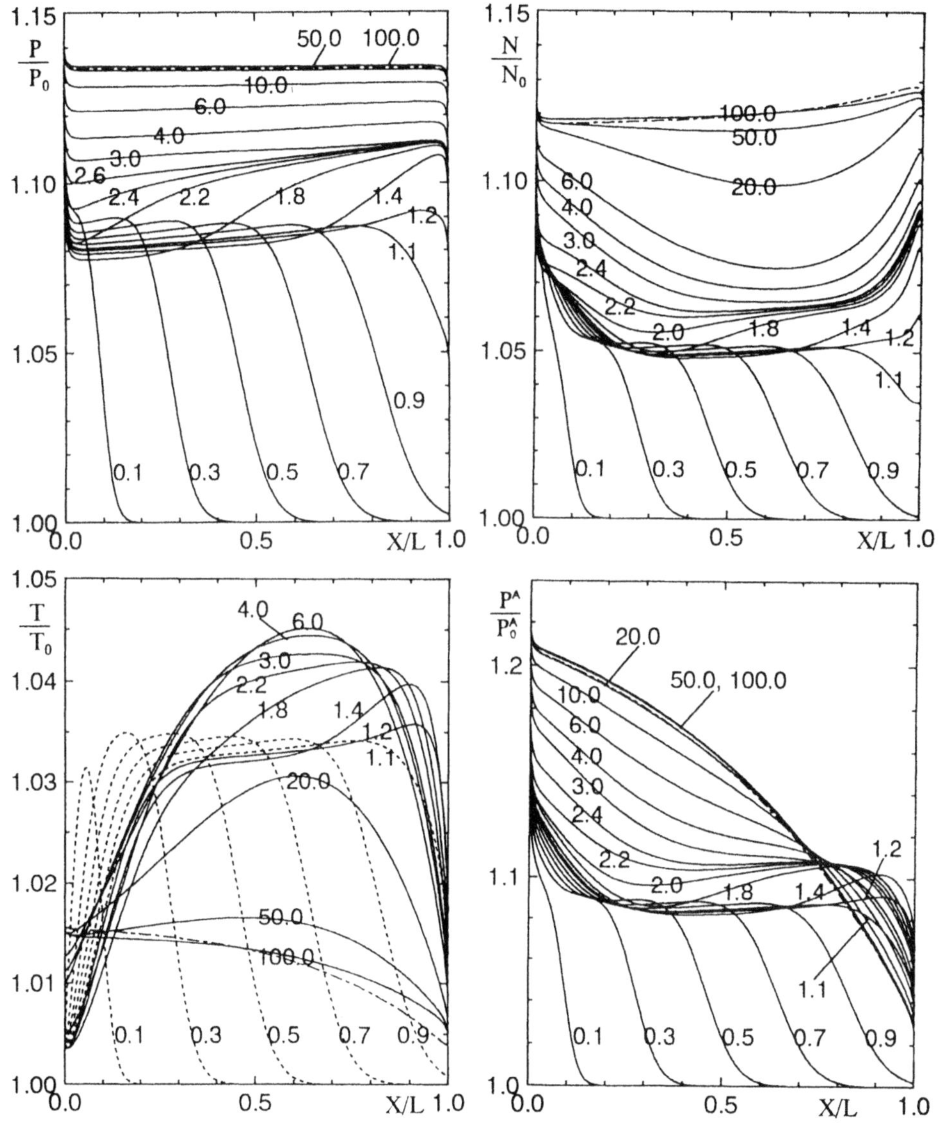

Fig. 1 : Distributions of various fluid dynamic quantities for $k = 0.01$, $N_0^B/N_0^A = 0.1$, $T_1/T_0 = 1.02$, $T_2/T_0 = 1.0$, $\Gamma = 11$ ($P_1^A/P_0^A = 1.24$, $P_2^A/P_0^A = 1.0$), $m_B/m_A = 2$, $\kappa_{AA} = \kappa_{AB} = \kappa_{BB}$. c_0 is the sound speed of the mixture (monatomic) at T_0. Numbers in the graphs indicate the values of t/τ_0^A. Dotted lines are the distributions associated with the incident shock wave. Dashed line (— - —) in each figure indicates the theoretical result for the steady state given by Ref. 9. Fairly good agreement can be seen between the numerical and theoretical results, although the present calculation has not yet reached the complete steady state.

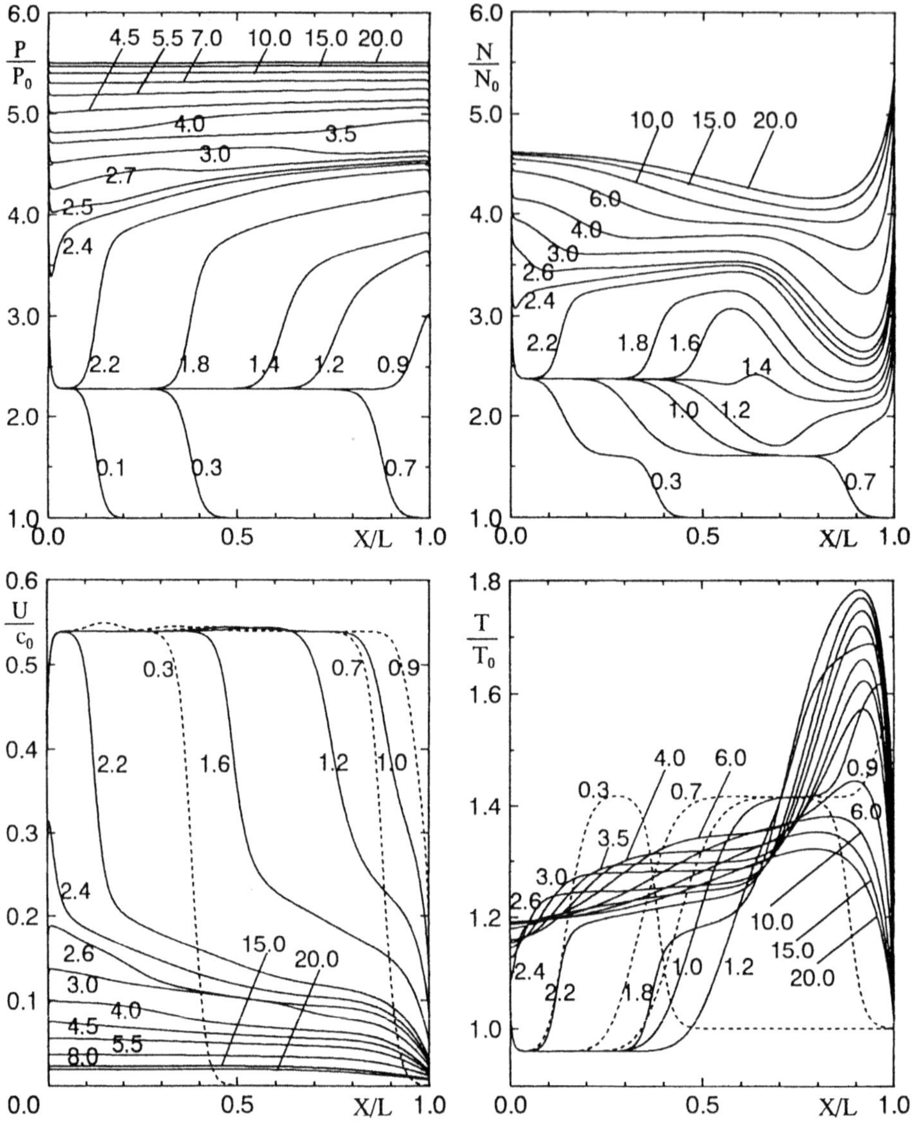

Fig. 2 : Distributions of various fluid dynamic quantities for $k = 0.01$, $N_0^B/N_0^A = 0.1$, $T_1/T_0 = 1.2$, $T_2/T_0 = 1.0$, $\Gamma = 11$ ($P_1^A/P_0^A = 6.25$, $P_2^A/P_0^A = 1.0$), $m_B/m_A = 2$, $\kappa_{AA} = \kappa_{AB} = \kappa_{BB}$. c_0 is the sound speed of the mixture (monatomic) at T_0. Numbers in the graphs indicate the values of t/τ_0^A. Dotted lines are the distributions associated with the incident shock wave.

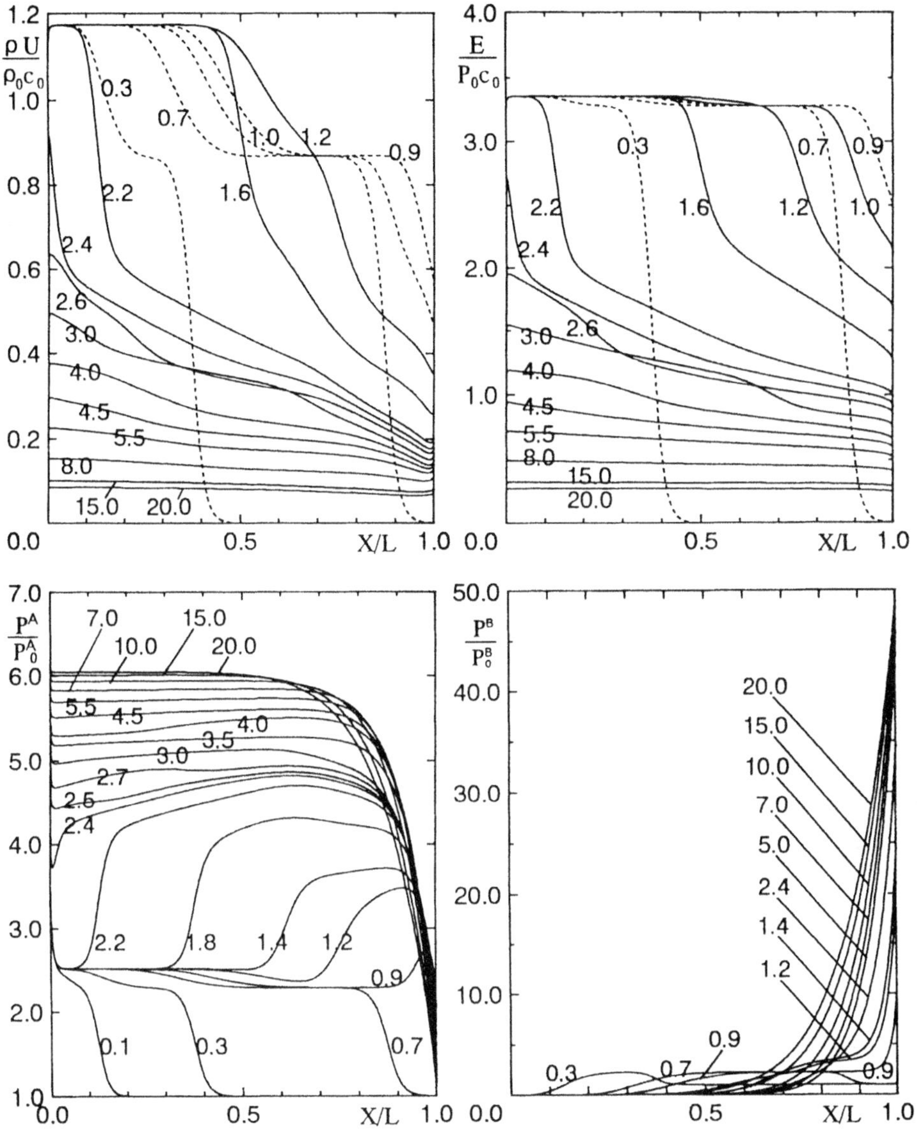

Fig. 2 : see the previous page for the caption.

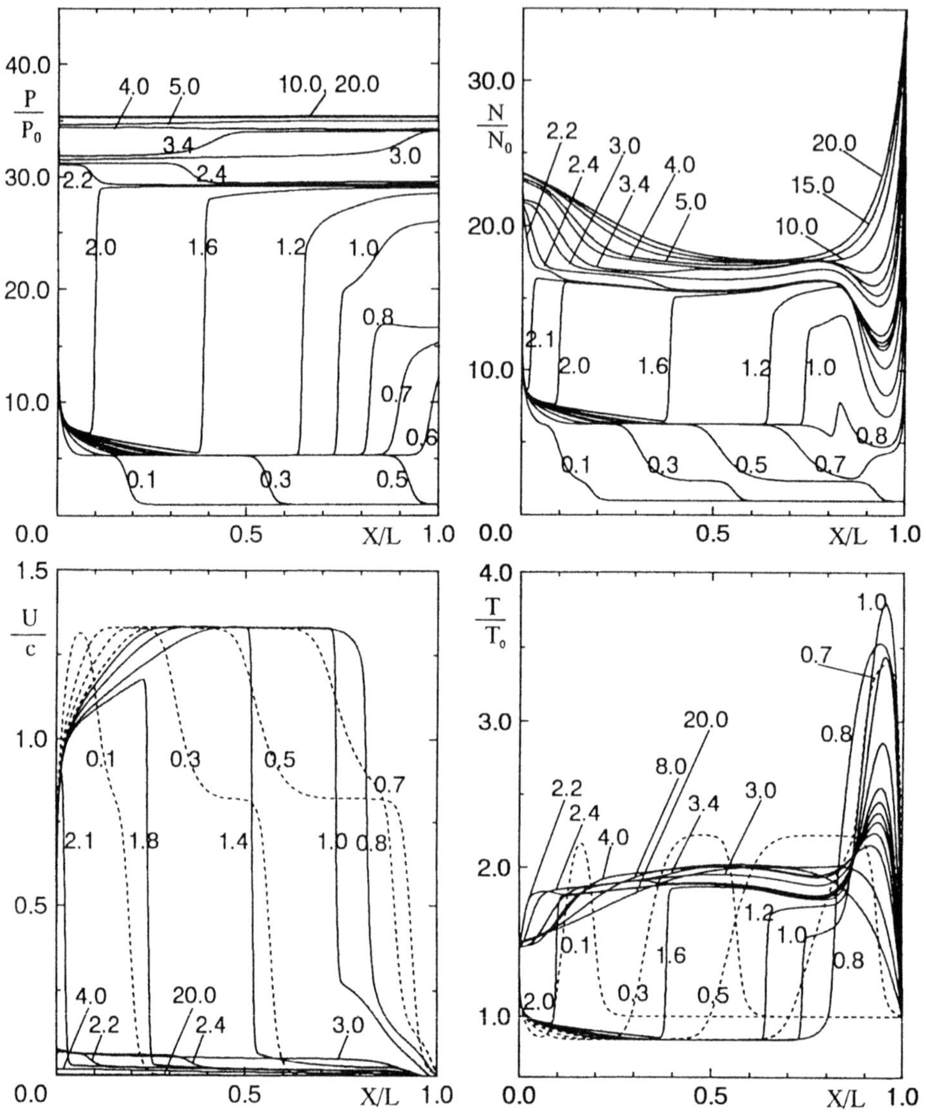

Fig. 3 : Distributions of various fluid dynamic quantities for $k = 0.01$, $N_0^B/N_0^A = 0.1$, $T_1/T_0 = 1.5$, $T_2/T_0 = 1.0$, $\Gamma = 11$ ($P_1^A/P_0^A = 39.12$, $P_2^A/P_0^A = 1.0$), $m_B/m_A = 2$, $\kappa_{AA} = \kappa_{AB} = \kappa_{BB}$. c is the local sound speed of the mixture (monatomic). Numbers in the graphs indicate the values of t/τ_0^A. Dotted lines are the distributions associated with the incident shock wave.

Observations of Explosive Exsolution in Liquids

H. M. Mader*, Y. Zhang†, J. C. Phillips‡, R. S. J. Sparks‡, B. Sturtevant♯ & E. Stolper%

* Institute of Environmental and Biological Sciences, Lancaster University, Lancaster, LA1 4YQ, U.K.

† Department of Geological Sciences, University of Michigan, Ann Arbor, MI 48109-1063, USA.

‡ Department of Geology, University of Bristol, Bristol, BS8 1RJ, U.K.

♯ Graduate Aeronautical Laboratories, California Institute of Technology, Pasadena, CA 91125, USA.

% Division of Geology and Planetary Science, California Institute of Technology, Pasadena, CA 91125, USA.

EXPLOSIVE two-phase (gas-liquid) flows generated by rapid degassing of supersaturated liquids have been investigated experimentally to simulate conditions in violent volcanic eruptions. Two configurations were examined: i) Decompression of CO_2-saturated ($P_{CO_2} \leq 0.7$ MPa) water causes explosive bubble growth and foaming. The large volume increase accelerates the two-phase mixture upward to velocities as large as 14 m/s at nearly constant accelerations up to more than 200 g. Constant acceleration implies that the bubble growth rate is proportional to $t^{2/3}$, *i.e.*, bubble size increases with time more rapidly than the $t^{1/2}$ law expected for simple diffusive growth of spherical bubbles in an infinite medium. ii) Rapid mixing of concentrated K_2CO_3 and HCl solutions generates CO_2 supersaturations up to a few MPa. In this case, reaction and degassing generate an increasingly accelerating flow until the reactants become depleted at peak accelerations of about 100 g and velocities up to 45 m/s. In both experiments the rapid expansion causes the foam to fragment into a heterogeneous spray. The experimental flows are comparable to explosive volcanic flows in terms of decompression ratio, velocities, accelerations and in the large range of scales present. In addition, because the flow is confined to a conduit, the large accelerations cause strong extensional strain and longitudinal deformation of the foam. We speculate that widespread tube pumice in Plinian pyroclastic deposits and ignimbrites could preserve evidence of analogous flow conditions.

We use the term 'explosive volcanic eruption' to describe eruptions powered by the violent release of volatiles, or 'degassing', from magmas in sudden explosions or sustained discharges. The rapid degassing causes violent fragmentation of the magma and acceleration of the two-phase mixture of gas and ejecta to velocities of a few hundred meters per second in volcanic conduits and vents.[1] The interactions among the various processes involved in

S. Morioka and L. van Wijngaarden (eds.),
IUTAM Symposium on Waves in Liquid/Gas and Liquid/Vapour Two-Phase Systems, 335-341.

such an eruption – vesiculation, flow acceleration, fragmentation, *etc.* – are complex and poorly understood. There are, to our knowledge, no complete theoretical treatments of this important geological phenomenon. In view of the incomplete understanding of these processes, we have chosen to carry out laboratory simulations of explosive eruptions with model systems that generate expansion rates and flow velocities of the same magnitude as those in full-scale events. We report two sets of experiments in which explosive flows were generated by rapid degassing of aqueous solutions at room temperature; the experiments differ in the ways in which high, nearly instantaneous supersaturations with respect to gas (CO_2 in both cases) in water at room temperature were generated.

The experiments were carried out in facilities at the University of Bristol and the California Institute of Technology (see Fig. 1) using shock-tube techniques first proposed in this context by Bennett[2] and more recently developed in studies of volcanic jets,[3] explosive vaporization[4] and high-speed dense dusty gases.[5, 6] A key characteristic of the natural systems that we have tried to match closely in our experiments is that the volatile component (largely H_2O or CO_2 in a real volcanic eruption) is in a gaseous state after it has come out of solution, or 'exsolved', but the liquid component (a silicate liquid $\pm$ crystals, in nature), which makes up most of the mass of the system, is essentially entirely condensed, even after the volatile component has nearly completely exsolved. In these respects the systems in our experiments differ significantly from one-component systems[4] and two-component systems that completely or nearly completely evaporate on decompression.

Fig. 2 shows representative single frames taken from high-speed movies of the two experiments. In the experiments of Fig. 2a, velocities up to 14 m/s and accelerations more than 200 g were observed for supersaturation pressures of 0.7 MPa. The flows expand at almost constant acceleration (Figure 3a), which is strongly and positively correlated with the ratio of test cell pressure to reservoir pressure, and weakly negatively correlated with liquid viscosity. Since bubble growth causes expansion and the number of bubbles remains constant following initial nucleation (suggestive of heterogeneous nucleation), constant acceleration implies that the bubble volume increases in proportion to t^2, or the bubble radius increases as $t^{2/3}$. Preliminary measurements on individual bubbles during the time when bubbles are well-separated and roughly spherical confirm this relationship. This growth rate is measurably greater than the $t^{1/2}$ power-law behavior expected for purely diffusive growth of a spherical bubble in an infinite fluid.

In flows generated by chemical reaction (Fig. 2b) peak velocities and accelerations increased with CO_2 yield (Fig. 3b). Velocities up to 45 m/s and accelerations of nearly 100 g were achieved when saturated K_2CO_3 (6 *M*) was injected into concentrated HCl (12 *M*). Faster flows are observed for larger supersaturations. These chemically-generated flows accelerate approximately linearly in time, probably due to continuous mixing of the reactants after injection as the flow expands. The stretching of the accelerating two-phase mixture may also enhance mixing. Eventually the acceleration reaches a maximum and then decreases, possibly when the reactants are depleted. Large gradients of HCl concentration in the neighborhood of the injection jets generate large supersaturation gradients and spatially inhomogeneous liberation of CO_2 vapor. Therefore, the chemically-generated two-phase flows are intrinsically heterogeneous, in contrast to the flows generated by depressurization, which, since nucleation sites are homogeneously distributed, are initially homogeneous when viewed at a scale of several bubble diameters. This difference between the two types of

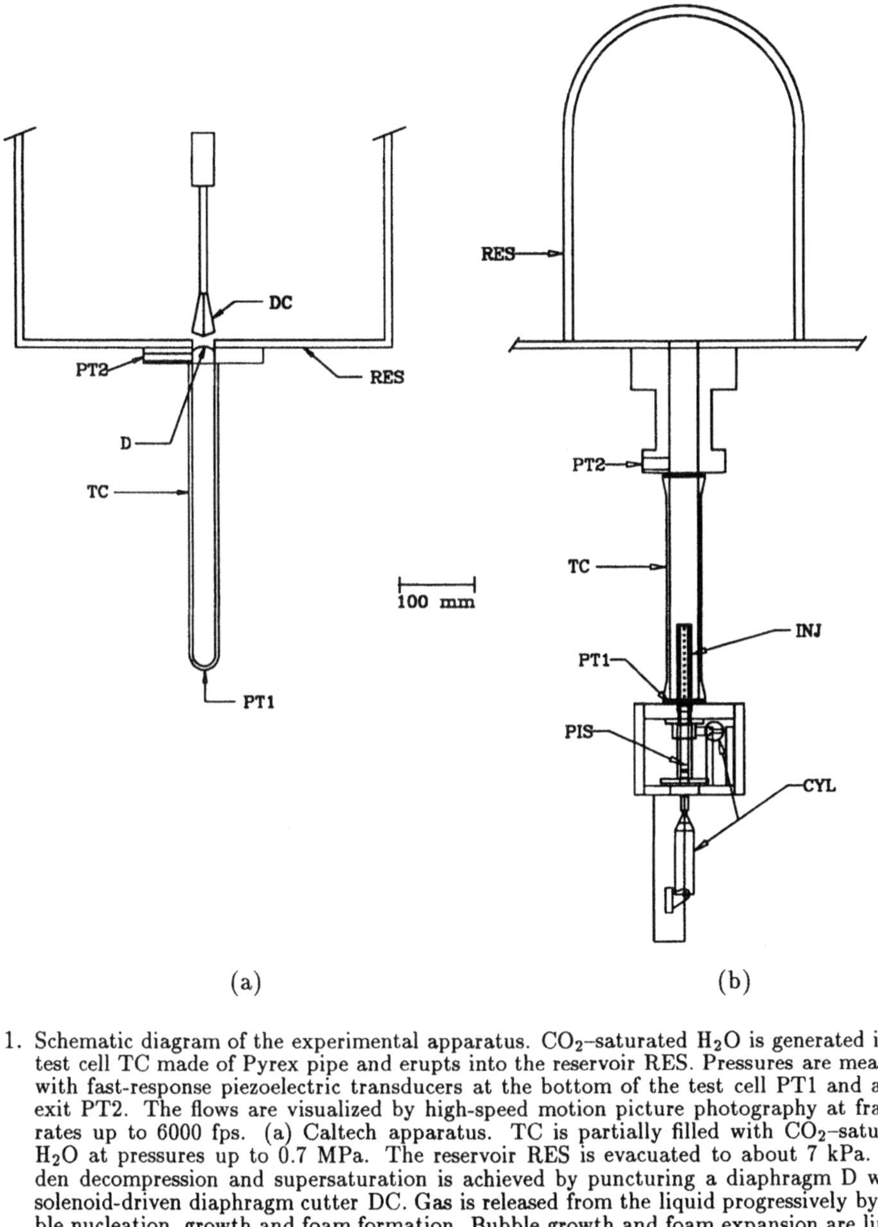

Fig. 1. Schematic diagram of the experimental apparatus. CO_2–saturated H_2O is generated in the test cell TC made of Pyrex pipe and erupts into the reservoir RES. Pressures are measured with fast-response piezoelectric transducers at the bottom of the test cell PT1 and at the exit PT2. The flows are visualized by high-speed motion picture photography at framing rates up to 6000 fps. (a) Caltech apparatus. TC is partially filled with CO_2–saturated H_2O at pressures up to 0.7 MPa. The reservoir RES is evacuated to about 7 kPa. Sudden decompression and supersaturation is achieved by puncturing a diaphragm D with a solenoid-driven diaphragm cutter DC. Gas is released from the liquid progressively by bubble nucleation, growth and foam formation. Bubble growth and foam expansion are limited by mass diffusion of solute to the bubble walls. The viscosity of the test mixture was varied from that of water (10^{-3} Pa s) to 0.7 Pa s by the addition of up to 1% organic polymer. (b) Bristol apparatus. 4 cm^3 of K_2CO_3 solution is injected in a few milliseconds through 96 holes in injector INJ into TC, which initially contains a much larger volume (100 cm^3) of HCl solution filling TC to the top of INJ. Mixing of the reactants initially occurs on the boundaries of the 96 jets and later due to stretching of fluid elements as the gas-liquid mixture accelerates up the test cell. The rate of the ensuing chemical reaction is limited by fluid-mechanical mixing and the rate-limiting step $H_2CO_3 \longleftrightarrow CO_2 + H_2O$, with reaction time scale (the time for the reactant to reach $1/e$ of its initial concentration) about 70 ms. Production of vapor-phase CO_2 is further limited by mass diffusion. By using high concentrations of the reactants very large gas yields are achieved with theoretical supersaturation pressures up to several MPa. The pneumatic cylinders CYL which drive piston PIS to inject the K_2CO_3 and open the 96 holes are computer controlled.

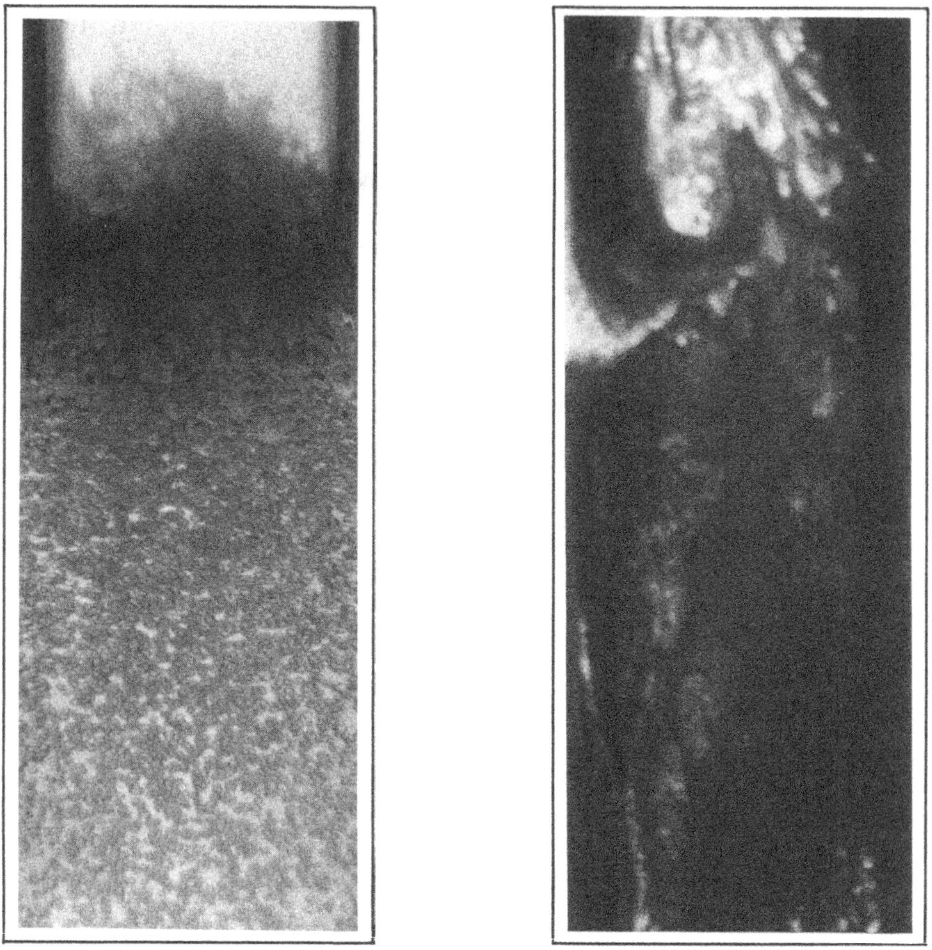

(a) (b)

Fig. 2. Photos of the early development of simulated eruptions. (a) Flow generated by depressurization (Caltech). Initial cell pressure = 565 kPa, initial tank pressure = 6.9 kPa. The photo shows the test solution, initially 10 cm deep, 7.1 ms after depressurization. Of order 10^3 cm^{-3} bubbles nucleate uniformly throughout the solution immediately upon depressurization, and negligible nucleation occurs thereafter. Bubbles near the top of the mixture grow faster than those below. The flow in the photograph has expanded to 12 cm height. Above 10.5 cm the liquid is finely fragmented, while between 9.5 and 10.5 cm the bubbles have merged into a rapidly accelerating foam. The bubbles in the foamed region are stretched in the flow direction. Fragmentation appears to occur by bubbles on a very irregular surface bursting due, perhaps, to a combination of over pressurization, rapid stretching and violent interaction with neighboring bubbles. Run 83. (b) Flow generated by chemical reaction (Bristol). The photograph shows the test solution, initially 9.7cm deep, 34 milliseconds after the initiation of the chemical reaction. After an incubation time of $O(4\text{–}6)$ milliseconds, nucleation and violent expansion form a highly heterogeneous flow. Bubble nucleation occurs over the timescale of the experiment, as the expanding flow produces strong mixing of the reactants. The flow in the photograph has expanded to 40cm in height, with fragmentation appearing to occur in the regions of most violent gas release.

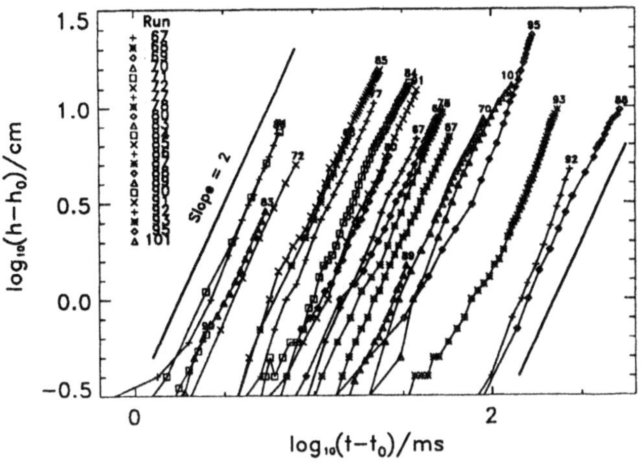

(a)

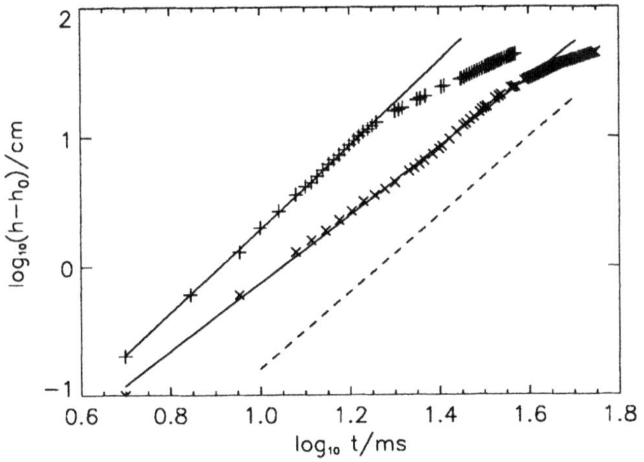

(b)

Fig. 3. Position of the flow head h as a function of time t measured from high-speed motion pictures. (a) Flow generated by depressurization (Caltech). Log–log (base 10) plot of $h - h_0$ in cm *vs* t in ms. Origin of time, $0 \leq t_0 \leq 22.5$ ms, was adjusted for each run to optimize the linearity of the plot. Results of 22 runs at different saturation pressure, ambient pressure, fluid viscosity and fill depth are shown. Lines of slope = 2 (constant acceleration) are indicated at right and left. The acceleration increases monotonically from right to left on the plot from 0.7 m s^{-2} (Run 88) to 2100 m s^{-2} (Run 90). (b) Flow generated by chemical reaction (Bristol). Top curve (6 M $K_2CO_3 \rightarrow$ 12 M HCl), bottom curve (6 M $K_2CO_3 \rightarrow$ 6 M HCl). Each curve is a composite of data from 5 films viewing different parts of the test cell. The origin of time for each film was selected to ensure continuity of the plotted data. A least-squares fit (solid lines) of the initial linear part of the data was made to the function $h - h_0 = At^b$, where h_0 is the initial height of the liquid in the test cell, with the following results: 6 M $K_2CO_3 \rightarrow$ 6 M HCl, $b = 2.69$, $A = 1.49 \times 10^{-3}$; 6 M $K_2CO_3 \rightarrow$ 12 M HCl, $b = 3.26$, $A = 1.07 \times 10^{-3}$. As $b > 2$, these flows experience accelerations that increase with time and with CO_2 yield. Line of slope = 3 is shown on the right (dashed).

flows is visible in Fig. 2. We speculate that heterogeneity in rapidly accelerating magma in volcanic conduits may provide "weak" zones promoting fragmentation.

The two sets of experiments reveal several aspects of the physics of violent degassing. i.) As already stated, the measured bubble growth rate is larger than that expected for diffusion-limited growth, and in the depressurization experiments follows a $t^{2/3}$ law. We speculate that fluid convection due to the large volume changes and the attendant high strain rates occurring in these flows can thin diffusively-depleted layers near bubble walls, increasing concentration gradients and thus increasing bubble growth rate.[7] ii.) Our experiments demonstrate that models[8,1] postulating that fragmentation of magmas occurs at a simple downward-propagating front in almost static foams with vesicularities of 0.7–0.8, and that flow expansion occurs *after* fragmentation, should be revised. Foam acceleration *precedes* fragmentation, rather than the reverse. In addition, in some of the depressurization experiments, vesicularities in excess of 0.7–0.8 are achieved without fragmentation. Silic pumices with high vesicularities are also known.[9] iii.) The precise mechanism of fragmentation in our experiments is not discernible, but it is clear that, as in the industrial and domestic production of aerosols, the foam disintegrates into a spray entirely in the liquid state. Thus, in volcanic systems rapid acceleration in the liquid state *could* be sufficient to cause fragmentation and another mechanism of fragmentation, such as brittle fracture of foam, is not necessary. Nevertheless, the experimental flows accelerate to high velocity over short distances, imposing large extensional strain on the foam. If the strain rate in the volcanic setting were large enough that the material crossed the glass transition,[10] the material could suffer brittle fracture. In the Maxwell model of linear viscoelastic materials the relaxation time τ is given approximately by[11] $\tau = \mu/G$, where μ is the shear viscosity and G is the shear modulus. Taking $\mu \approx 10^8$ Pa s for magma with a few weight percent H_2O at eruption temperature and $G \approx 10^{10}$ Pa, we get $\tau \sim 10^{-2}$ s. If, as suggested by our experiments, acceleration to 100 m/s can occur over a meter or less, then extensional strain rates can be of order 100 s^{-1}, and the ability of magma to flow as a liquid can be exceeded. Thus, key results of our experiments are that magma can rapidly accelerate and expand into a foam, that fragmentation can occur dynamically if the expansion is rapid enough and that fragmentation could be either brittle or in the liquid state.

Application of these experiments to volcanic phenomena requires consideration of scaling and dynamic similarity, but no scaling relationships are known for high-velocity high-density gas-particle flows, where the particles interact with one another fluid-mechanically and by collisions. Thus, the best we can do at present is to try to replicate as closely as possible in laboratory simulations the dynamical processes of the large-scale flows. Velocities in our experiments approach those of volcanic flows (~100 m/s), as do the accelerations (~100 g; note that gravity (1 g) is unimportant in flows that experience such large accelerations). The high velocities and accelerations experienced by these flows mean that inertial rather than viscous forces control the dynamics. The pressure ratio for the experiments ranges from 30 to 300, comparable to magmas with a few percent dissolved water (supersaturation pressures ~ 50–100 MPa) that disrupt explosively at a few MPa.[12,8] Obviously, the length scale of the volcanic system cannot be reproduced in the laboratory. However, using the analogy between dense particulate flows (in which particles interact collectively over a large range of scales) and high Reynolds number turbulent flow, it can be argued that this mismatch in length scale is not likely to undermine the validity of the simulations, since the ratio of the largest scale in the system (*e.g.*, the diameter of the test chamber in the experiments and of the conduit in volcanos) to the size of the smallest particles is much greater than unity in both cases.

Bubble nucleation and growth are ultimately the processes that drive the accelerations observed in our experiments, so scaling of our results to nature requires an evaluation of how they may differ in supersaturated magmas. Molar fractions of volatiles in our experiments (up to 0.004 CO_2 in the depressurization experiments, and perhaps as large as 0.1 in the chemical injection experiments) are comparable to those of intermediate to silicic magmas (0.02–0.06 H_2O), so the ratio of gas to condensed matter is comparable in the experiments and in nature. The diffusivity of CO_2 in H_2O (2×10^{-9} m^2 s^{-1} at 20C) is about 200× larger than that of H_2O in concentrations of about 0.03 in magma at 850C ($\sim 10^{-11}$ m^2 s^{-1}). The diffusively-controlled volume growth rate of voids during degassing is proportional to the product of diffusivity and concentration, which ranges from about 10× larger than in magma in the decompression experiments to locally 200× in the chemical experiments. The viscosity of water is also many orders of magnitude less that that of magma in explosive eruptions. However, numerical calculations[13] (also, J. Barclay & R. S. J. Sparks, unpublished) indicate that, even under conditions of explosive eruption, diffusive bubble growth is not retarded by viscous effects unless magma viscosities exceed 10^8 Pa s. In the fragmentation region, with pressures of a few MPa, magmas are not fully degassed and a few fractions of percent of residual dissolved water are sufficient to keep viscosities at values that would not inhibit explosive expansion rates. Initial bubble densities measured in the decompression experiments are of order 10^9 m^{-3}, while in laboratory experiments on volcanic glasses[14] densities of order 2×10^{14} m^{-3} have been reported, and observations of pumice[15,16] suggest nucleation densities of 3×10^{14} – 3×10^{16} m^{-3}. That bubble densities in the present experiments are up to 5 orders of magnitude smaller than in nature and in decompression experiments on magmatic compositions suggests that natural events could be even more explosive than observed in the laboratory.

1. Wilson, L. *et al.* *Geophys. J. Roy. Astro. Soc.*, **63**, 117-148, 1980.
2. Bennett, F. D. *Amer. J. Sci.*, **274**, 648-661, 1974.
3. Kieffer, S. W. and Sturtevant, B. *J. Geophys. Res.*, **89**, 8253, 1984.
4. Hill, L. G., and Sturtevant, B. in "Adiabatic Waves in Liquid-Vapor Systems" Eds. G. E. A. Meier and P. A. Thompson, 25-37, Springer-Verlag, Berlin, 1990.
5. Anilkumar, A. V., Experimental Studies of High-Speed Dense Dusty Gases, Ph. D. Thesis, California Institute of Technology, Pasadena, CA, 1989.
6. Anilkumar, A. V. *et al.*, *J. Volcanol. Geotherm. Res.*, **56**, 145–160, 1993.
7. Prousevitch, A. A. *et al.*, *J. Geophys. Res.*, **98**, 22283-22307, 1993.
8. Sparks, R. S. J. *J. Volcanol. Geothermal Res.*, **3**, 1-37, 1978.
9. Klug, C. & Cashman, K. V. *EOS* **72**, 312, 1991.
10. Dingwell, D. B. & Webb, S. L. *Phys. Chem. Minerals*, **16**, 508-516, 1989.
11. Scherer, G. W. Relaxation in Glass and Composites. Wiley, N.Y., 331p.
12. Wilson, L. J. *Volcanol. Geotherm. Res.*, **8**, 297-313, 1980.
13. Dobran, F. *J. Volcanol. Geothermal Res.*, **49**, 285-311, 1992.
14. Navon, O. & Hurwitz, S. *Terra abstracts*, **5**, 574, 1993.
15. Sparks, R. S. J. & Brazier, S. *Nature*, **295**, 218-220, 1982.
16. Whitham, A. G. & Sparks, R. S. J. *Jour. Volc.*, **48**, 209-223, 1986.

FILM CONDENSATION OF THE VAPOR FLOW BEHIND A SHOCK WAVE ON THE SHOCK-TUBE SIDE WALL

Yasunori KOBAYASHI, Takahiro WATANABE and Naoki NAGAI
Institute of Engineering Mechanics, University of Tsukuba
Tsukuba, Ibaraki 305, Japan

Summary

Measurements of film condensation were made in the flow field behind the incident shock wave propagating through a vapor-liquid two-phase medium. Major objective of the study is to identify condensing heat transfer rates of the vapor to the shock-tube side wall and to learn condensation in the main flow. Ethanol and E-10 (a heavy liquid named *Afluid* by the manufacturer) were extensively used as working fluid. Steady accumulation of the condensing liquid film was confirmed on the wall surface, as similarly seen in the end-wall experiment conducted elsewhere. A most significant result is that "dual-step" shock pressurization was observed in E-10. The first pressure rise is a normal one created by an incident shock front, whereas the second pressure rise is taken place by some large disturbance in the main flow. The reason for this is not certain yet, but is speculated to be relevant to a long relaxation time or inefficient compressibility of the fluid. The visualized shock front and its vicinity of E-10 is completely different from those of normal gases, implying strong condensation in the main flow.

Key words: Film condensation, Dual-step pressurization, Retrograde fluid

1. Introduction

Shock adiabatic phenomena in the vapor and/or liquid environment have been investigated largely due to their engineering importance in the actual machine designs and safety problems, such as turbo-fluid machinery, chemical and nuclear power plant, mist and rain-drop effect on high-speed vehicles, etc.[1-6]. These problems essentially contain homogeneous and heterogeneous condensation/evaporation phenomena inseparably. Most basic works on homogeneous condensation have been examined by taking advantage of shock expansion flows produced by supersonic nozzle, and it is relatively recently that the significance of condensation by shock compression is realized. Philip A. Thompson and his group proposed an idea of categorizing liquids into "regular" and "retrograde" nature [7-9], in which only "retrograde" liquid can be condensed by compression, although there still remains uncertainties associated with such criteria of regular and retrograde nature and with dynamics of resultant "liquefaction shock waves".

The problem of heterogeneous condensation, on the other hand, appears much common in engineering thermodynamics and heat transfer, and numerous contributions have been made in the past. When the vapor is adiabatically heated by shock compression, the vapor

S. Morioka and L. van Wijngaarden (eds.),
IUTAM Symposium on Waves in Liquid/Gas and Liquid/Vapour Two-Phase Systems, 343-353.

temperature rises above that of the surrounding wall of the shock-tube, and initiates to condense on the relatively cold wall surface which has a huge heat capacity compared with that of vapor. This heterogeneous condensation is well explained as; condensation in the very early stage is proceeded by inter-molecular forces described by kinetic theory, and later by thermal conduction law known as the Stefan problem. However, we need further information on the effects of dominant parameters, such as wall temperature, shock intensity, initial vapor density, purity of vapor, etc. on the condensation rate to understand the phenomena to the extent of practical application to machine designs.

This paper summarizes partial results of a shock tube experiment conducted by using a heavy (large molecular mass) vapor as a driven gas in order to learn such"retrograde" nature be identified, and to measure condensing heat transfer rates of liquid film formation on the solid surface [10-13]. Measurements were also made of pressure changes in the main flow as well as schlieren photography at the observation window of the shock-tube side wall, with emphasis on the E-10 fluid medium.

2. Experimental Setup and Measurement Instrumentation

2.1 Shock Tube Facility

The main body of the shock tube facility used in the experiment is made of stainless steel: it has a lower pressure vessel (LPV) of $5m$ in length with square cross sectional area of $60mm$ by $60mm$. The one end of LPV is connected to a diaphragmless high pressure chamber of $8700cm^3$ in volume which contains two free pistons controlled by electro-magnetic valve to release high pressure gas. To the other end of LVP a cylindrical dump tank of $0.5m$ in diameter with $1.0m$ in length is attached in order to avoid generation of the reflected shock wave as well as to relieve excess pressure rise in LPV. A pair of observation windows for optical measurement are set on the side walls at $4m$ downstream of the high pressure chamber. Pressure changes in LPV are measured at 5 positions by pressure transducers (1 MHz in frequency resolution) mounted flush with the inner surface of the vessel and is connected to multi-channel transient digital memory (1 μs sampling time). A combination of $20mW$ (632.8 nm) CW He-Ne laser light and a photomultiplier detector of 7 mm aperture diameter (P.D.; DC$\sim$15 MHz in frequency range) was used for liquid-film thickness measurement. For flow field visualization by schlieren photography, a conventional magnesium flush light of $2ns$ duration time was used. All the raw data obtained such as for temperature, pressure, beam intensity, etc. are transferred and stored in the memory of personal computer (PC) for further off-line processing and calculations.

The LPV is evacuated to less than 5 Pa prior to discharge of the working liquid in it, and after 30 minutes wait for thermal stabilization in LPV the shock is injected produced by nitrogen gas. The amount of working liquid filled in LPV at each shock-tube run is that corresponding to 120 % of saturation pressure should it be vaporized completely at LPV temperature (almost the same as room temperature). The reason for this is to maintain the LPV in two-phase condition in the entire process of shock wave run.

2.2 Working Liquid

Typical working liquids and a part of their thermophysical properties used in the present

Table 1 Thermo-physical properties of working fluids

Fluid	R-11	Ethanol	E-10
Chemical Constituent	CCl_3F	C_2H_6O	------
Molecular Mass ($kg/kmol$)	137.4	46	445*
Ratio of Specific Heat	1.14	1.13	(1.05)
Boiling Temperature ($℃$)	23.8	77.2	100
Latent Heat (kJ/kg)	180.6	918.1	88.2
Saturated Pressure (kPa)	88.8	5.87	3.0
Surface Tension (N/m)	0.018	0.024	0.012
Density of Liquid (kg/m^3)	1472	783.5	1770
Density of Vapor (kg/m^3)	5.86	0.29	------
Refractive Index in Liquid	1.37	1.362	1.277
$\tilde{C}_v$ value	11.1	12.2	(53.5)

* Measured value, ( ):Best estimation

experiment are tabulated in Table 1, although R-11 has not been used for a last few years. Liquid E-10 is a kind of fluoro-chemical substance with a large molecular mass (we have estimated its value as about 445
through our preparatory experiment, and use this number in data processing), and its thermo-physical properties as well as chemical constituent are not fully disclosed to public. We have used this liquid, however, as it is stated to be nontoxic, thermally stable and above all due to its large molecular mass. This last feature is advantageous to examine the "retrograde" nature of heavy vapor by shock compression. According to the definition proposed by P.A. Thompson et. al., the value of non-dimensional constant-volume heat capacity defined as $\tilde{C}_v$ $(=C_v(T_c)/\mathbf{R})$ is a measure of the liquid nature as;

$\tilde{C}_v < 11.2$ .... "regular" (evaporation by compression)

$11.2 < \tilde{C}_v < 22.4$ .... intermediate

$22.4 < \tilde{C}_v$ .... "retrograde" (condensation by compression)

where T_c and $\mathbf{R}$ stand, respectively, for the critical temperature and the gas constant of the vapor. According to this definition, E-10 apparently belongs to "retrograde" , whereas ethanol and R-11 are rather "regular" fluids (see Table 1).

2.3 Film Thickness Measurement

A multi-reflective interference method is used for film thickness measurement at the shock-tube observation window, which is essentially the same way as Fujikawa, et al. utilized in their end-wall measurement [10], and is commonly explained in the textbooks of optics. The laser beam is injected at 2.5 degrees inclined from the window surface normal in order to reject the reflected beam components other than the one from the inner surface of the window. The multi-reflected beam intensity R detected by P.D. and stored in the PC memory is processed to calculate the liquid-film thickness δ by the use of following optioinal relations as,

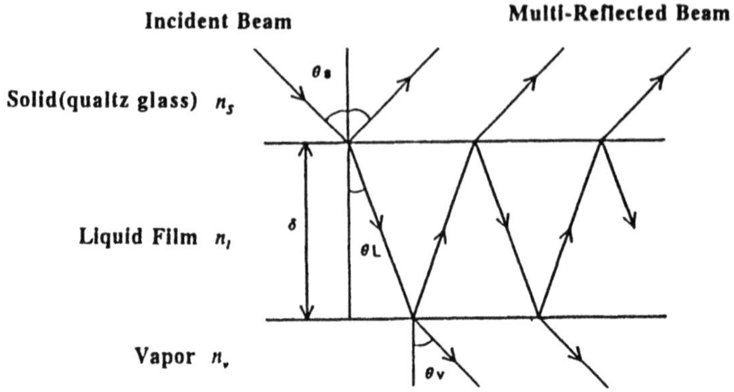

Fig. 1 Schematic optical path for film thickness measurement

$$R = |r|^2 = \frac{r_{sl}^2 + r_{lv}^2 + 2r_{sl}r_{lv}\cos 2\beta}{1 + (r_{sl}r_{lv})^2 + 2r_{sl}r_{lv}\cos 2\beta} \quad , \quad \beta = 2\pi n_l \delta \cos\theta_l / \lambda$$

and
$$r_{sl} = \frac{\tan(\theta_s - \theta_l)}{\tan(\theta_s + \theta_l)} \qquad r_{lv} = \frac{\tan(\theta_s - \theta_v)}{\tan(\theta_s + \theta_v)}$$

where n_l is a refractive index of liquid, and θ_s, θ_l, θ_v are refractive angles for solid (quartz glass), liquid film and vapor, respectively, as schematically shown in Fig. 1. No details for deducing process of δ from the raw R data is shown here, other than the fact that data length for each shock-wave run is $8k$ bites with acquisition sampling time of 1 μs, and that the smoothing process for high-frequency noise in laser intensity signals R is incorporated. Synchronization of beam intensity data with that of pressure rise is also confirmed.

3. Results and Discussions

3.1 Dual-Step Pressure Rise

Figure 2 shows a typical pressure change observed in E-10 vapor, and Fig. 3 shows those of simultaneous measurement of the pressure rise and the multi-reflected laser beam intensity (broken line) representing liquid film formation for ethanol and E-10. The setup conditions and obtained shock intensities for these data are tabulated in Table 2. From Fig. 2 it is apparent that the E-10 flow field is pressurized twice; firstly at the time when the shock front arrives and secondly in rather gradual and random way at a few hundred microseconds (μs) after the first one. Moreover, the second pressure rise coincides approximately with the time when the periodical character of the film growth signal terminates (see Fig. 3, E-10). This coincident position also approximately corresponds to the place where blackening of the main flow field starts as seen in the schlieren photographs shown in Fig. 9. Similar results were obtained for different setup conditions of p_4/p_1 for E-10 vapor. No such dual-step pressure rise nor its obvious correlation with

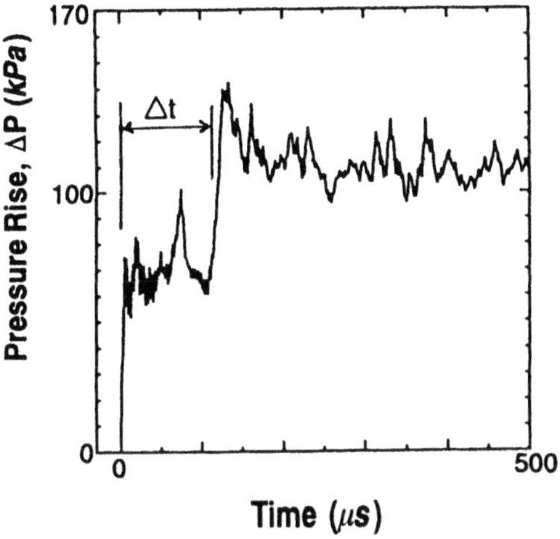

Fig. 2 Typical dual-step pressure rise in E-10 vapor medium
(P_1 = 2.67 kPa, P_4 = 980.7 kPa, T_1 = 24.8 °C, M_s = 5.92, U_s = 469.7 m/s)

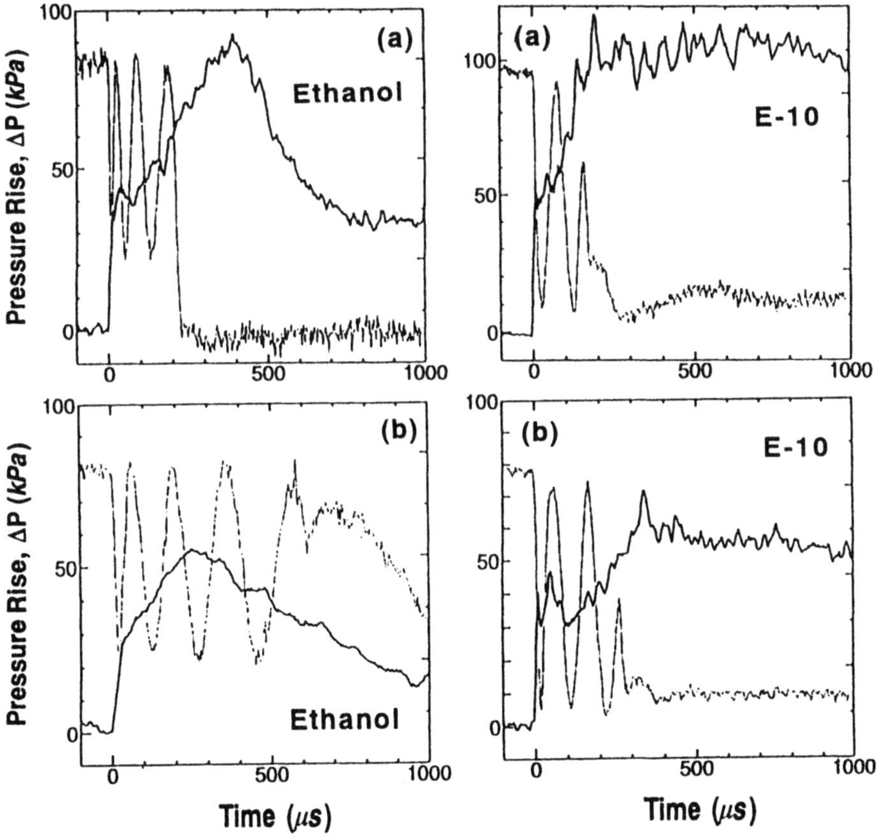

Fig. 3 Output signals of pressure and beam intensity in simultaneous measurements

Table 2 Setup conditions and obtained shock intensities for ethanol and E-10 corresponding to outputs in Figure 3

	Ethanol		E-10	
	(a)	(b)	(a)	(b)
P_4 (kPa)	1176.7	588.2	980.7	196.1
P_1 (kPa)	6.59	6.42	3.61	3.09
T_1 $(^\circ C)$	22.1	21.5	25.2	18.8
U_S (m/s)	650.9	517.4	469.7	338.7
M_S	2.36	1.88	5.89	4.30

a decay of beam intensity signal are observed for ethanol and R-11(data not shown). The time difference Δt (see Fig. 2) between the first and the second pressure rise for different shock Mach number M_s (defined as measured shock velocity divided by the vapor sonic velocity) in E-10 medium is shown in Fig. 4, which indicates that Δt decreases with increasing M_s. This dual-step pressurization implies E-10's insufficient compressibility originating from its large internal energy. In other word, Δt is related to a relaxation time in the main flow to attain thermal equilibrium. It is also suggestive that strong disturbance like boiling of liquid film must take place at the point of second pressure rise, since the contact surface arrives at about 1 *ms* after the shock front, and no reflected shock waves are involved in this experiment.

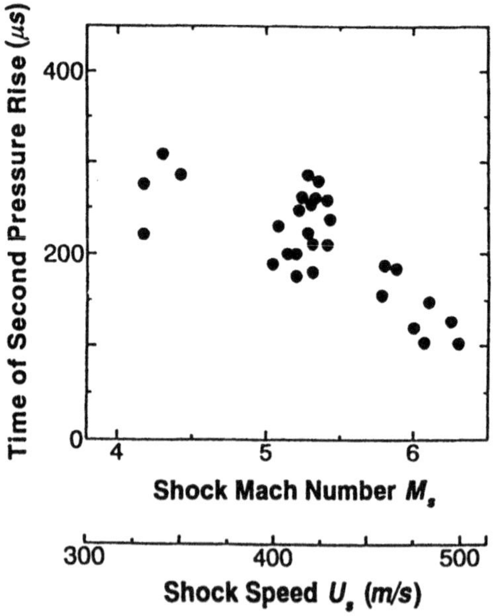

Fig. 4 Time duration between the first and second pressure rise in E-10 vapor

3.2 Film Condensation

Figures 5 shows a growth rate of film thickness at the wall surface after shock wave passing for ethanol and E-10, which is calculated from the same data as shown in Fig. 3. The origin of the graphs represents the time of shock front arrival (similarly used as reference in the other graphs in this paper), and the abscissa is expressed in the square-root of the elapsed time after that. Vapor condensation is confirmed to initiate at the same time of shock front passage within the measuring errors of $\pm 1\ \mu s$, and the film grows relatively steadily with time until the film surface is exposed to some large disturbance in the main flow. The condensation rates (or film growth rates) is promoted with increasing shock intensity, but is terminated sooner. Measured condensation rates in the present experiment is clearly twofold as discussed in the end-wall experiment [11]; initially it increases linearly with the elapsed time as the phenomena be governed by inter-molecular forces, and at a

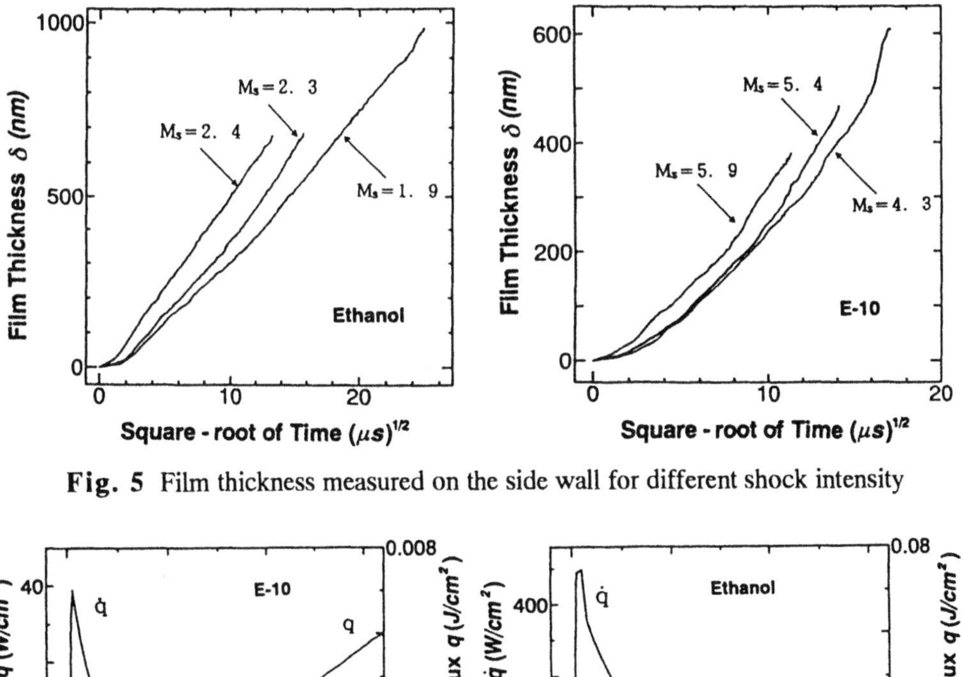

Fig. 5 Film thickness measured on the side wall for different shock intensity

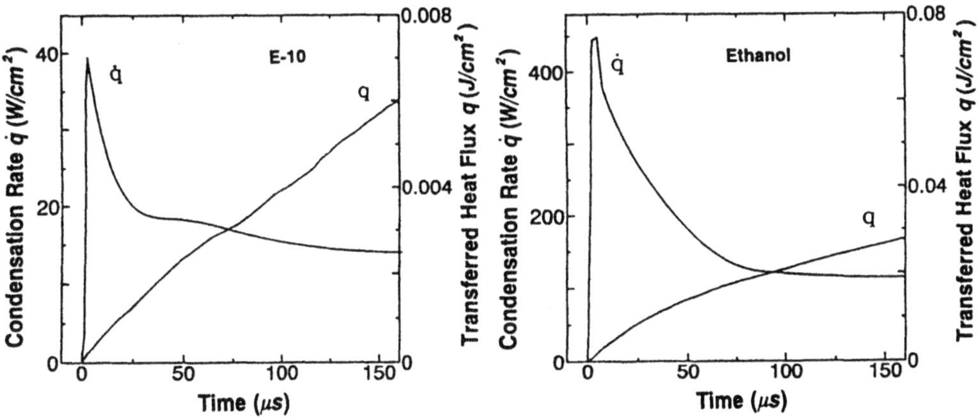

Fig. 6 Condensation rate and transferred heat flux calculated from data in Fig. 5

Table 3 The observed values of δ_c and t_c in the present experiment

	δ_c (nm)	t_c (μs)
R-11*	20	0.3
Ethanol	10 ~ 20	2 ~ 5
E-10	20 ~ 50	8 ~ 10

(*) only a few data are available for R-11

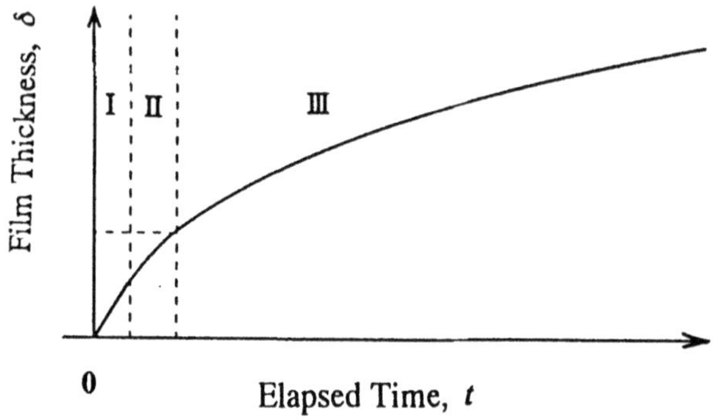

Fig. 7 Schematic expression of three regimes of vapor condensation

later stage it increases approximately in proportion to square-root of the time, controlled by thermal conduction associated with phase change surface transformation. Typical results of condensation rates $\dot{q}$ (W/cm^2) and transferred condensing heat flux $q\,(J/cm^2)$ calculated from Fig. 5 are shown in Fig. 6. These results along with Fig. 4 imply a limit of stable heat transfer condition in this experiment. It is still too rough, however, to derive any empirical equations or correlations for film thickness and/or condensation rates from these rather scattered results. However, the process can be reasonably summarized to proceed in three stages as; the very early stage governed by inter-molecular forces(I), the transient stage (II), and final stage controlled by non-equilibrium thermal conduction (III), as schematically shown in Fig. 7. In the figure simbols of δ_c and t_c represent the limiting film thickness and the time duration, respectively in the region where the molecular motion is dominant (I, II). Numerical values obtained in the present experiment are summarized in Table 3.

3.3 Flow Field Visualization

Some typical shock fronts and its near flow field of E-10 visualized by schlieren photography are shown in Fig. 8, which exhibit a feature far from simple shock front as observed in a pure gas. The setup condition and measured shock intensities for these photographs are summarized in Table 4. In reference to pure gasdynamics, the low pressure P_1 (which corresponds to a saturation pressure) of E-10 in the figure belongs to the range of low density shock waves where a steep pressure rise of the gas becomes difficult and the effect of boundary layer behind a shock front becomes eminent [14-16].

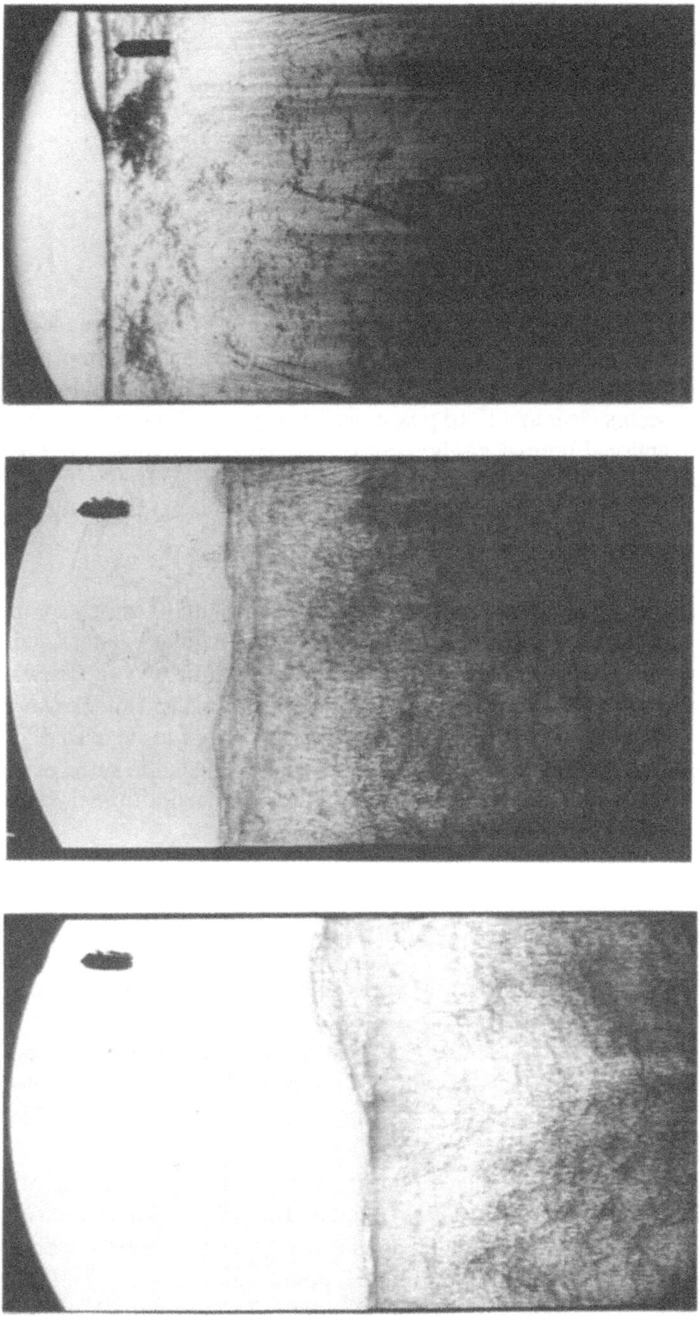

(a)

(b)

(c)

Fig. 8 Typical shock front and its near flow field propagating through E-10 vapor visualized in three different shock intensities tabulated in Table 4

Table 4 Setup conditions and measured shock speeds in E-10 in schlieren photographs

	(a)	(b)	(c)
P_4 (kPa)	196.1	980.7	1372.9
P_1 (kPa)	2.69	3.68	3.93
T_1 $(^{\circ}C)$	18.0	23.7	25.5
U_S (m/s)	364.4	464.1	497.8
M_S	4.63	5.84	6.24

This interpretation does not apply to E-10, as it shows a distinct pressure rise (see Fig. 2). Apparently a strong condensation takes place in the main flow by shock compression, and revealed a rough and irregular shock front with condensed substance, as obviously seen in these photographs. It seems clear that E-10 possesses" retrograde" fluid nature, and cannot be discussed by conventional normal gasdynamics. We have to establish fluid dynamics for heavy-molecules in conjunction with thermodynamics.

4. Concluding Remarks

Condensation phenomena in the vapor-liquid two-phase medium by shock compression was investigated by means of optical technique for ethanol ("regular" vapor) and E-10 (heavy "retrograde" vapor). The significant difference between these two vapors was in the feature of pressure rise and the shock front configuration of the incident shock waves. For E-10, the dual-step compression is taken place which is speculated to be related to a large relaxation energy associated with vapor-liquid phase change on the solid surface as well as to possible homogeneous condensation in the main flow. We need a further examination on these phenomena both in experimental and analytical way.

References

1. P.P. Wegener; Nonequilibrium Flows, Marcel Dekker, Co., Vol.4, (1969), pp.163-243
2. G. Gyarmathy; The Spherical Droplet in Gaseous Carrier Streams, Multiphase Science and Technology, Eds. G.F. Hewitt et.al, McGraw-Hill, (1982), pp.99-279
3. F.F. Abraham; Homogenous Nucleation Theory, Spplememnt 1, Academic Press, (1974), pp.
4. P. Panton and A.K. Oppenheim; Shock Relaxation in a Particle-Gas Mixture with Mass Transfer Between Phases, AIAA J. Vol.6, No.11, (1968), pp.2071-2077
5. D. Barschdroff; Carrier Gas Effects on Homogeneous of Water Vapor in a Shock Tube, Physics of Fluids, Vol.18, No.5, (1975), pp.529-535
6. F.F. Marble; Some Gasdynamic Problems in the Flow of Condensing Vapors, Astronautica Acta, Vol.14, (1969), pp.585-614
7. P.A. Thompson; On Possibility of Complete Condensation Shock Wave in Retrograde Fluids, J.F.M., Vol.70, Part.4, (1975), pp.639-649

8. M. Bratos and G.E.A. Meier; Shock Wave Induced Condensation in the Retrograde Vapor, Archives of Mechanics, Vol.31, No.5, (1979), pp.679-693
9. P.A. Thompson, G.C. Carofano and Yoon-Gon Kim; ShockWaves and Phase Change in a Large-Heat-Capacity Fluid Emerging forma Tube, J.F.M., (1986), pp.57-92
10. S. Fujikawa, M. OKUDA, T. Akamatsu and T. Goto; Non-Equilibrium Vapour Condensation on a Shovk-Tube Endwall Behind a Reflected Shock Waves, J.F.M., Vol.183, (1987), PP.293-324
11. M. Maerafat, S. Fujikawa, T. Akamatsu and T. Goto; An Experimental Study of Non-Equilibrium Vapour Condensan in a Shoch-Tube, Experiments in Fluids, Vol.7, (1989), pp.513-520
12. Y. Kobayashi; Vapor Condensation Behind a Shock Wave Propagating Through Vapor-Liquid Two-Phase Media, Proc. IUTAM Symposium, (1989), pp.59-68
13. H.J. Smolders; Non-Linear Wave Phenomena in a Gas-Vapour Mixture with Phase Transition, Ph.D. Thesis of University of Technology, Einhoven, The Netherlands, (1992)
14. A. Roshko; On Flow Duration in Low-Pressure Shock Tubes, Physics of Fluids, Vol.3, No.6, (1960), pp.835-842
15. G. Dettliff; An Experimetntal Study of Liquefaction Shock Waves, J.F.M., Vol.95, Part.2, (1979), pp.279-304
16. R.E. Duff; Shock-Tube Performance at Low Initial Pressure, Physics of Fluids, Vol.2, No.2, (1959), pp.207-216

EXPERIMENTS OF VAPOR-LIQUID PHASE TRANSITION IN A SHOCK-TUBE

S.Fujikawa(Toyama Prefectural University)
M.Matsumoto(Nagoya University)
M.Kotani(Toyama Prefectural University)
H.Sato(Toyama Prefectural University)

Abstract

This paper is concerned with shock-tube experiments of the phase transition from vapor to liquid and its molecular dynamics computer simulation. The growth of a liquid film formed on the endwall surface of a shock-tube in a shock wave reflection region is measured by an optical method using light sources with two different wavelengths. The condensation coefficient of methanol vapor is obtained from the conformity between the experiment of the liquid film growth and the gasdynamics theory with kinetic boundary conditions. It is clarified that the condensation coefficient of methanol vapor exists between 0.15 and 0.18. This value is in good agreement with a theoretical condensation coefficient(0.1~0.3) which is computed on the basis of molecular dynamics method. The simulation suggests that vapor molecules colliding with the liquid surface apparently drive other molecules out of the liquid, i.e., the fluxes of condensation and evaporation have a strong correlation.

1 Introduction

It is a challenging work to investigate the phase transition of vapor at vapor-liquid interface on a molecular level. The rate of the phase transition of the vapor and the temperature condition at the interface are of special importance, and they have been investigated for a long time but have not yet been made clear. The reason may be attributed to the lack of appropriate experimental method and to the lack of basic understanding of condensation phenomena.

The first author of the present paper and his group have proposed a shock-tube method for the measurement of the rate of the phase transition, and they have determined the condensation coefficient for several kinds of vapor by this method[(1),(2),(3)]. The shock-tube method is based on the following principle. Figure 1 shows a pressure and specific volume diagram for the change of thermodynamic states of a vapor under the action of a shock wave. The symbol O denotes the initial state of the vapor prior to the arrival of the incident shock wave. The state of the main stream changes from the state O to the state 2 through two stages of adiabatic compression by the incident and the subsequent reflected shock wave. However, the state at the inner surface of the endwall changes instantaneously from the state O to the state $2W$ keeping the pressure constant and the state is a supersaturated one under which the phase transition from the vapor to the liquid takes place

S. Morioka and L. van Wijngaarden (eds.),
IUTAM Symposium on Waves in Liquid/Gas and Liquid/Vapour Two-Phase Systems, 355-363.

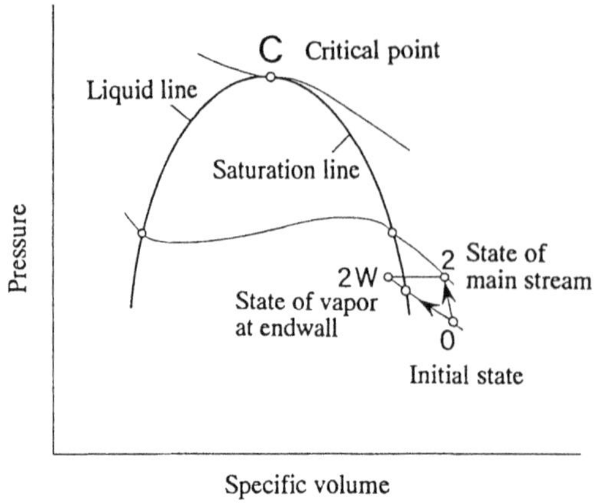

Fig.1 Change of thermodynamic states of vapor under action of shock wave

and a liquid film is formed on the surface of the endwall.

The present shock-tube method has the following characteristics.

(1)A characteristic time of measurement is comparable with a characteristic time of the phase transition.

(2)Content ratio of noncondensable gases to a vapor can be remained at less than 0.01% during measurement.

(3)Vapor-liquid interface is planar.

Owing to the characteristic (1), the shock-tube method makes it possible to measure the rate of the phase transition with sufficient accuracy just from the start of the phase transition. The content ratio of noncondensable gases less than 0.01 % assures that the noncondensable gases in the vapor have no effect at all on the measurement during a measuring time[4]. The characteristic (3) makes it possible to use without any modification kinetic boundary conditions which have been obtained for a planar vapor-liquid interface[5].

In the present paper, the shock-tube method is improved by overcoming several problems which the first author has encountered[1],[2],[3] and the phase transition rate of methanol vapor is measured again with a sufficient accuracy by an optical method using light sources with two different wavelengths.

2 Kinetic Boundary Conditions and Gas Dynamics

Figure 2 shows a physical situation which is produced near the endwall surface of the shock-tube when the shock wave in the vapor is reflected at the surface. Near the vapor-liquid interface a characteristic region is formed, the length scale of which is of the order of the mean free path of molecules and which is called the Knudsen layer. The vapor region may be divided into two regions, the Knudsen layer and the outer gasdynamical region. Sone and Onishi[5] derived equations for the molecular flux j and the vapor temperature T at the vapor-liquid interface as follows:

$$j = -2\pi^{1/2}\sigma_c \cdot \frac{\bar{c}_\infty}{4} \cdot n \cdot \frac{p - p_s}{p_\infty} \tag{1}$$

$$T = T_s + \frac{\sigma_c \sigma_t}{k n_\infty} \cdot (p - p_s) \tag{2}$$

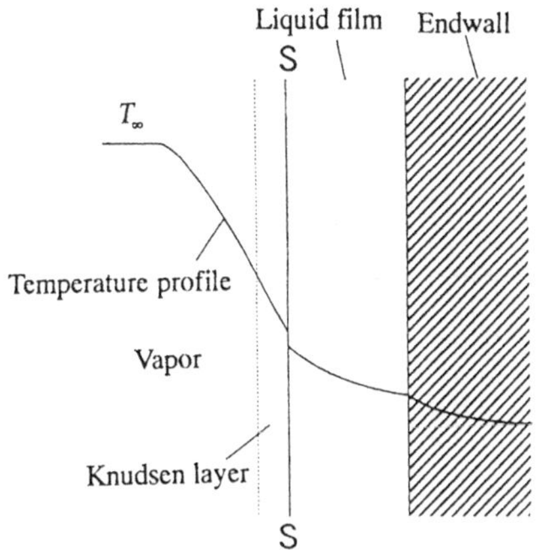

Fig.2 Physical situation near shock–tube endwall

where $\bar{c}_\infty = (8kT_\infty/\pi m)^{1/2}$ in which k denotes the Boltzmann constant, m the mass of a molecule and T_∞ the vapor temperature behind the reflected shock wave, j denotes the net rate of evaporation and condensation, the number of molecules transfered per unit time and unit area at the interface, n and n_∞ denote the number density of vapor molecules at the interface and the number density behind the reflected shock wave, respectively, p, p_∞ and p_s the vapor pressure at the interface, the vapor pressure behind the reflected shock wave and the saturated vapor pressure at the temperature T_s of the liquid surface, T denotes the vapor temperature at the interface, and σ_c and σ_t denote the condensation parameter and the temperature parameter. There is a relation between the condensation parameter σ_c and the condensation coefficient α as follows:

$$\alpha = \frac{2\pi^{1/2}\sigma_c}{1 + 0.798\pi^{1/2}\sigma_c} \qquad (3)$$

Gas flows and the growth of the liquid film in the shock wave reflection region can be analysed by solving the Navier–Stokes equations under the boundary conditions (1) and (2). The growth of the liquid film is greatly influenced by the temperature of the liquid film surface, therefore, the temperature must be obtained correctly by solving simultaneously thermal equations for both the liquid film and the endwall together with the Navier–Stokes equations for the vapor. Details of the analysis are described in a paper of Fujikawa's[6].

3 Experimental Apparatus and Method

3.1 Shock–tube

Figure 3 shows a horizontal type of shock–tube which has been used for the present experiment. The length of the test section is 3 m, while the length of the high pressure section is 2 m. The cross section of the tube is the square of 60 mm x 60 mm. The endwall consists of BK7– optical glass: thickness 16 mm and flatness $\lambda/20(\lambda=632.8\text{nm})$.

The attained vacuum level of the test section is 0.03 Pa, and the leak speed of the test section is about 0.05 Pa/min. It takes about 3 minutes to introduce the vapor into the test section, rearrange and calibrate the optical system, so the final pressure of noncondensable gases in the test section may be 0.25 Pa. Methanol vapor is used in the present experiment. An initial pressure of the vapor is set at about 5 kPa. The percentage of the

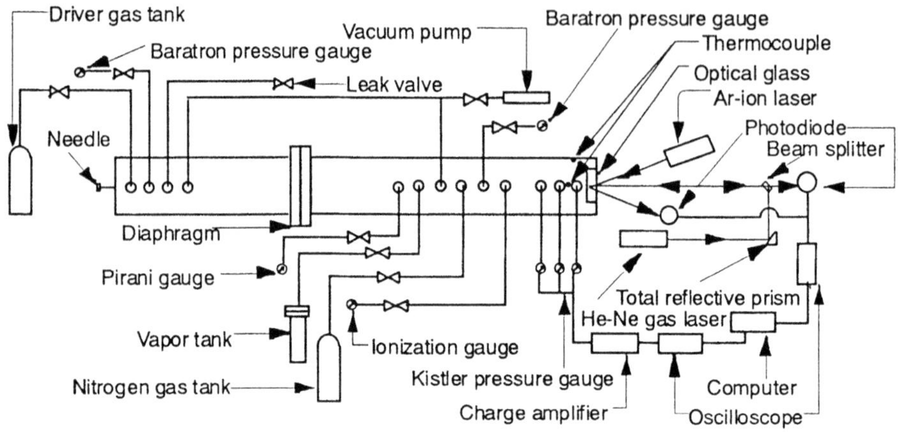

Fig.3 Shock–tube and optical system

noncondensable gases to the vapor is about 5 x 10^{-3} %, and it is greatly improved in comparison with that in the previous work[1] and it assures that the noncondensable gases have no effect at all on the growth of a liquid film[4].

Diaphragm is a thin aluminum foil and it is ruptured by a hand–driven needle after setting a driver gas at a certain pressure. Mach number of an incident shock wave is estimated from the distance(=500mm) between two pressure gauges and the passage time of the shock wave between them. The initial pressure of the vapor is measured by a MKS baratron pressure gauge(Type 122A) which makes it possible to measure the initial vapor pressure with the accuracy of 0.5 %. Shock–compressed pressures are measured by three Kistler pressure gauges(Type 701A). The passage time is measured by a universal counter. Initial temperatures of the vapor and the shock–tube sidewall near the endwall are measured by thermocouples just before the start of each experiment. These data except the temperatures are recorded in digital storage oscilloscopes and then processed by a computer.

3.2 Optical measuring system

The optical system for the measurement of liquid film thickness is also shown in Fig.3. Helium–Neon laser (λ=632.8nm) and Argon ion laser(λ=457.9nm) are used as light sources. Helium–Neon laser is normally projected on the endwall, while Argon ion laser is projected with the incident angle of 5°. The two lasers are used to confirm the uniformity of the liquid film at different places on the endwall surface. Different wavelengths are in particular chosen in order to extend the present method for the simultaneous measurement of two kinds of physical quantities, e.g., liquid film thickness and refractive index of liquid film. Measuring points of the liquid film thickness are different between Helium–Neon laser and Argon ion laser. It is ascertained by numerical simulations that the reflectance of Argon ion laser is the almost same as the case of the normal incidence. Electronic circuits for the measurement of photodiode outputs have the rise time of 0.085 μs. The outputs are recorded in the digital storage oscilloscopes, and then processed by the computer.

The variation in time of the thickness of a liquid film can be estimated by measuring the reflectance of the lights at the optically transparent film system, because the varying film, together with the window, forms a kind of interferometer. Details for data processing method are described in a paper of Maerefat et al.'s[1].

4 Experimental Result and Discussion

4.1 Pressure traces

Figure 4 shows the pressure trace of the incident and the subsequent reflected shock wave in methanol vapor. The initial condition for the experiment is : temperature 297.0 K, pressure 5.706 kPa and incident shock wave Mach number 1.602. This condition is chosen so that condensation may not occur on the sidewall behind the incident shock wave but may take place both on the sidewall and the endwall behind the reflected shock wave. After introducing the vapor into the shock–tube, it takes a few minutes until an initial liquid film is formed by adsorption on the surface of the endwall.

The pressure behind the reflected shock wave coincides with the theoretical pressure evaluated by the Rankine–Hugoniot relation with the accuracy of 4.8 %. This result of the pressure measurement is greatly improved in comparison with the result of the previous work(1) where the pressure behind the reflected shock wave was smaller than the Rankine–Hugoniot value by 15 % and it was dependent on time. The reason for the discrepancy between the present work and the previous one may be due to that the low pressure section of the present shock–tube has a volume larger than that of the previous work by 6 times, therefore condensation on the endwall and the sidewall has little effect on the pressure in the shock wave reflection region.

4.2 Optical output

Figure 5 shows the time histories of light reflectance for Helium–Neon laser(upper) and Argon ion laser(lower) under the same condition as in Fig.4. The abscissa denotes the time. The ordinate denotes the variation of the light reflectance. The histories of the reflectance show that the liquid film grows uniformly on the endwall surface because they vary in ideally regular ways between the maxima and the minima.

4.3 Growth of liquid film and estimation of condensation coefficient

Figure 6 shows the time variation of the thickness of the liquid film under the same condition as in Figs.4 and 5. The abscissa is the time, the origin of which is taken to be just at the start of condensation. The ordinate is the variation of the liquid film thickness. The variation of the liquid film thickness is shown up to 10 μs after the start of condensation. The above time duration is a period while the liquid film grows without being influenced by thermophysical properties such as the thermal diffusivity and thermal conductivity of the endwall and the latent heat of the liquid film(6). The circle symbols denote the thickness obtained from the light source of Nelium–Neon laser and the triangle symbols denote the thickness from Argon ion laser. The both results are in good agreement with each other and they show that the liquid film grows unifromly at different places of the endwall surface.

The results of the theoretical analysis for the condensation parameter σ_c=0.05, 0.055 and 0.06 or the condensation coefficient α=0.16, 0.18 and 0.20 are also shown in Fig.6. The most probable values of the condensation parameter and the condensation coefficient are 0.055 and 0.18, respectively. The magnitude of measurement error of the parameter and the coefficient can be estimated by

$$Z = \frac{p_\infty^{exp} - p_s^0}{p_\infty - p_s^0} \tag{4}$$

where $p_\infty^{\exp}$ and p_s^0 denote the actual vapor pressure behind the reflected shock wave and the initial vapor pressure. For the present experiment, we get Z=1.08 which means that the above values of the condensation parameter are larger than the true values by 8 %. Thus, we can get corrected values of the condensation parameter, 0.046, 0.051 and 0.055 corresponding to the above values in turn(the corrected condensation coefficients = 0.15,

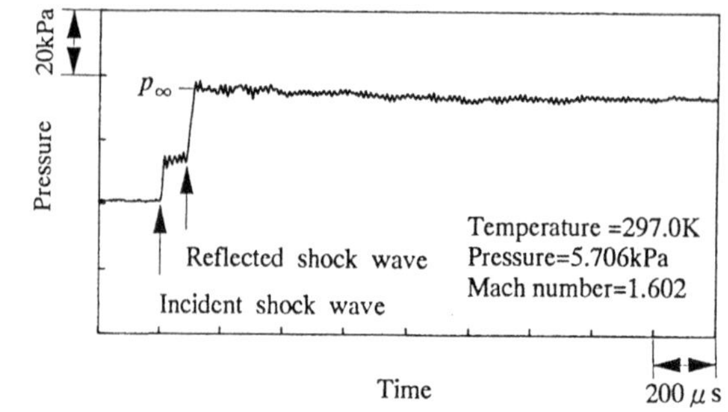

Fig.4 Pressure trace of incident and reflected shock wave

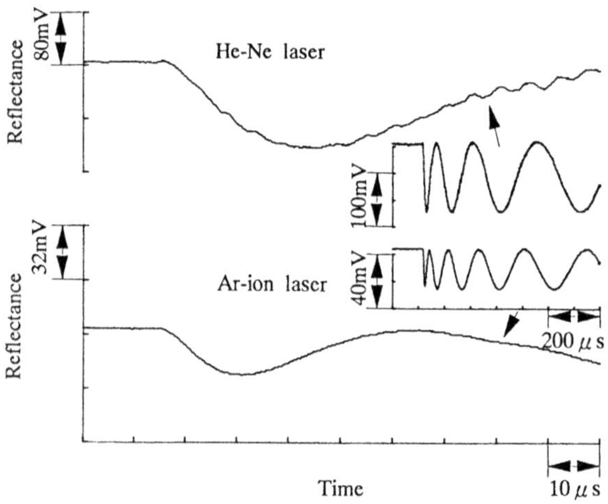

Fig.5 Time histories of light reflectance

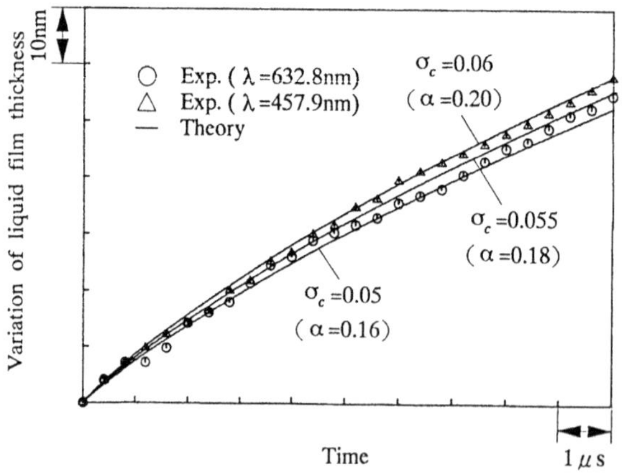

Fig.6 Time variation of liquid film thickness

0.17 and 0.18). Therefore, the most probable corrected condensation parameter and the condensation coefficient are found to be 0.051 and 0.17.

5 Molecular Dynamics Simulation

5.1 Simulation technique

A microcanonical ensemble MD method is adopted. The computer program is similar to what the second author used before to examine static properties of a liquid surface for methanol[8]. Figure 7 shows a rectangular prism with periodic boundary conditions for all three dimensions, used as a unit cell of calculation.

5.2 Dynamic processes of evaporation and condensation

Figure 8 shows superimposed molecular trajectories as the function of time. It is found that most of molecules exist in the central part(roughly 3nm ~ 6.5nm, or the liquid phase), but that several trajectories are in the outer region(the vapor). Evaporated molecules, shown by the trajectories going from the liquid to the vapor, return to the liquid from the outer side of the cell due to the periodic boundary condition. The complexity of the trajectories in the vapor phase suggests the existence of various mechanisms for dynamic processes near the surface.

5.3 Particle exchanges

In visualizing the molecular configurations, we have sometimes observed that vapor molecules colliding with the liquid surface apparently drive other molecules out of the liquid.

Let j_{ent} be the flux of molecules entering the bulk vapor region and j_{coll} be the flux of molecules escaping from this region(or colliding with the surface). The equilibrium condition is

$$\langle J_{ent}\rangle = \langle J_{coll}\rangle \tag{5}$$

where the bracket < > denotes the average with respect to time. The correlation of j_{ent} and j_{coll}, which includes both self-reflection and particle exchange, can be expressed in terms of a memory function $\kappa(t)$:

$$J_{ent}(t) = J_{evap}(t) + \int_0^\infty \kappa(t')\cdot J_{coll}(t-t')dt' \tag{6}$$

where t is the time, j_{evap} is the entering flux independent of j_{coll}, or the flux of true evaporation, which is the number of molecules spontaneously evaporating form the liquid side, not induced by the colliding vapor molecules. Taking the average of Eq.(6) with respect to t gives

$$\langle J_{ent}\rangle = \langle J_{evap}\rangle + \langle J_{coll}\rangle \int_0^\infty \kappa(t')dt' \tag{7}$$

and using Eq.(5), we obtain the following expression for α:

$$\alpha = \frac{\langle J_{evap}\rangle}{\langle J_{coll}\rangle} = 1 - \int_0^\infty \kappa(t')dt' \tag{8}$$

On the other hand, assuming j_{coll} and j_{evap} are independent, the memory function $\kappa(t)$ can be given as follows:

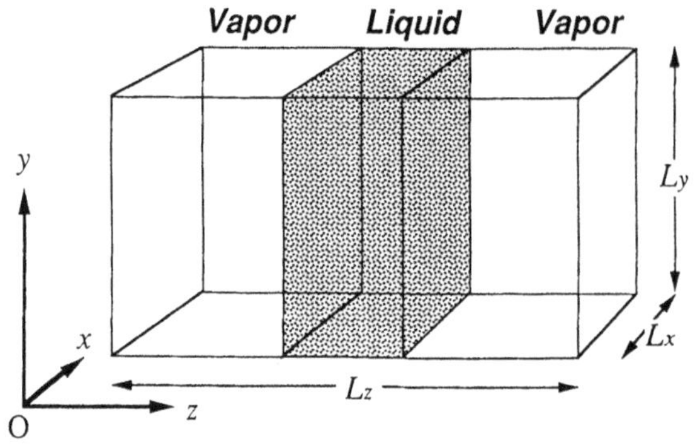

Fig.7 Unit cell for molecular dynamics simulation

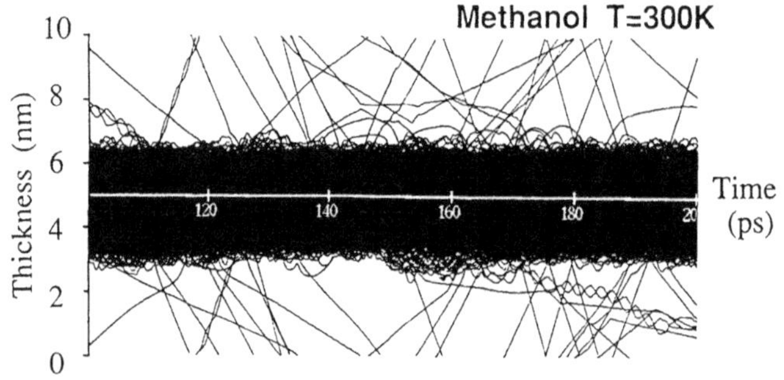

Fig.8 Superimposed molecular trajectories

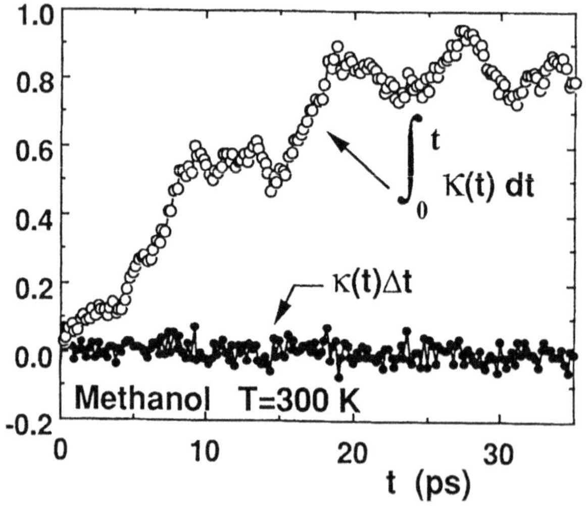

Fig.9 Behavior and time integral of $\kappa(t)$

$$\kappa(t) = \frac{\left\langle \Delta J_{coll}(t_1) \cdot \Delta J_{ent}(t_1 + t) \right\rangle}{\left\langle \Delta J_{coll}(t_1)^2 \right\rangle} \tag{9}$$

The behavior of $\kappa(t)$ is not clear due to the poor statistics, but the time integral has an apparent plateau as shown in Fig.9, from the value of which we can estimate α=0.1~0.3. This result has a remarkable agreement with the experimental result, α=0.15~0.18, of the preceeding chapter.

6 Conclusions

The shock-tube has been applied to measure the rate of the phase transition and the condensation coefficient of methanol vapor. In order to measure them, an optical method using light sources with different wavelengths has been used. It has been clarified that the present shock-tube can provide an ideal physical condition for the investigation of the phase transition and the condensation coefficient. The condensation coefficient of methanol vapor has been found to be between 0.15 and 0.18. The most probable value of the condensation coefficient is 0.17. This value is in good agreement with the value(0.1~0.3) obtained by computer simulation based on molecular dynamics. The simulation has suggested that vapor molecules colliding with the liquid surface apparently drive other molecules out of the liquid, i.e., the fluxes of condensation and evaporation have a strong correlation.

Acknowledgements

This work was partly supported by Grants-in-Aid for Scientific Research on Priority Areas of the Ministry of Education, Science and Culture of Japan in 1993. The authors wish to express their sincere gratitudes for the grants.

References

1. Maerefat,M., Fujikawa,S., Akamatsu,T., Goto,T. and Mizutani,T., An Experimental Study on Nonequilibrium Vapor Condensation in a Shock-Tube, Experiments in Fluids, Vol.7, No.8,(1989), pp.513-520.
2. Maerefat,M., Akamatsu,T. and Fujikawa,S., Nonequilibrium Condensation of Water and Carbontetrachloride Vapor in a Shock-Tube, Expcriments in Fluids, Vol.9, No.6, (1990), pp.345-351.
3. Fujikawa,S.,Maerefat,M., Mizutani,T. and Akamatsu,T., Measurement and Theoretical Consideration of the Condensation Parameter of a Polyatomic Molecule Vapor(The Case of Ethanol), Transactions of JSME, Vol.56, No.530, (1990), pp.3039-3044.
4. Maerefat,M., Fujikawa,S. and Akamatsu,T., Nonequilibrium Condensation of a Vapor-Gas Mixture on a Shock-Tube Endwall behind a Reflected Shock Wave, Fluid Dynamics Research, Vol.6, No.1, (1990), pp.25-42.
5. Sone,Y. and Onishi,Y., Kinetic Theory of Evaporation and Condensation - Hydrodynamic Equation and Slip Boundary Condition-, Journal of the Physical Society of Japan, Vol.44, No.6, (1978), pp.1981-1994.
6. Fujikawa,S., The Theory of Film Condensation on the Shock-Tube Endwall behind the Reflected Shock Wave, Transactions of JSME, (1994), Submitted.
7. Fujikawa,S., Matsumoto,M., Kotani,M. and Sato,H., The Molecular Study of Evaporation and Condensation of an Associating Liquid, Transactions of JSME, (1994), Submitted.
8. Matsumoto,M., Yasuoka,K. and Kataoka,Y., Microscopic Features of Evaporation and Condensation at Liquid Surfaces: Molecular Dynamics Simulation, Thermal Science and Engineering, Vol.2, No.1, (1994), pp.64-69.

Theoretical Investigation of Condensation Dynamics in Shock Tubes

Can F. Delale and Günter H. Schnerr
Institut für Strömungslehre und Strömungsmaschinen
University of Karlsruhe , Germany

Abstract. ***Condensation dynamics is investigated in the rarefaction wave of shock tube flows of a mixture of a condensable vapor and a carrier gas. The asymptotic solution which distinguishes the condensation zones with detailed analytical structure on pathlines for both subcritical and supercritical expansions is presented. An algorithm that utilizes the classical nucleation theory and the Hertz-Knudsen droplet growth formula is developed and the onset conditions for a mixture of water vapor in argon are presented against experimental data. A comparison of steady and unsteady nucleation rates for these data shows that nucleation time-lag effects are more important in the initial nucleation period than near maximum nucleation rates.***

1. Introduction

In this paper we present the asymptotic solution of shock tube flows with homogeneous condensation in the rarefaction wave for both smooth (subcritical) flows and flows with an embedded shock wave (supercritical flows). For subcritical flows the solution up to the onset of condensation along pathlines is obtained analytically and the condensation wave front is identified. The solution in the heat addition zones follows in a semi-analytical fashion. For supercritical flows we identify

S. Morioka and L. van Wijngaarden (eds.),
IUTAM Symposium on Waves in Liquid/Gas and Liquid/Vapour Two-Phase Systems, 365-375.

four distinct flow regimes on pathlines. In this case the solution is discussed only qualitatively due to lack of a shock fitting technique. An algorithm that utilizes the classical nucleation theory and the Hertz-Knudsen droplet growth law is developed from the asymptotic solution of subcritical flows. A comparison of the onset conditions for the expansion of water vapor in argon with measurements show deviations which are reasoned from the qualitative supercritical solution. Finally nucleation time-lag effects are taken into consideration for the same experiments. In this case it is shown that the onset conditions are unaltered in spite of the fact that nucleation time-lag effects may be important in the very initial nucleation period.

2. Basic equations for shock tube flows with homogeneous condensation

We herein present the basic equations of flows homogeneously condensing in the driver section of a shock tube. We neglect the effects due to viscosity and heat conduction and we treat both the condensable vapor and the carrier gas as calorically and thermally perfect. We further assume that the condensed phase is uniformly distributed throughout the gaseous components having the same velocity of the mixture (no velocity slip). Introducing the normalized pressure p , density ρ , temperature T and velocity u of the mixture by

$$p \equiv \frac{p'}{p'_4} \quad , \quad \rho \equiv \frac{\rho'}{\rho'_4} \quad , \quad T \equiv \frac{T'}{T'_4} \quad , \quad u \equiv \frac{u'}{a'_4}$$

where the actual variables are denoted by a prime and subscript 4 is reserved for the initial state of the mixture in the driver section with a'_4 denoting the initial speed of sound in the driver section, together with the normalized coordinates t and x defined by

$$t \equiv \frac{t'}{\Theta'} \quad , \quad x \equiv \frac{x'}{\Theta' a'_4}$$

where Θ' denoting some characteristic flow time, we can cast the classical inviscid flow equations into the characteristic form [1,2] as

$$\left(1-\frac{\mu_m}{\mu_v}g\right)dlnp \pm \frac{\Upsilon a_f}{T}du + \frac{\left[\left(c_{pm}-L_1 g\mu_m/\mu_v\right)-\left(1-g\mu_m/\mu_v\right)L/T\right]}{\left[c_{pm}+\left(1-L_1\right)g\mu_m/\mu_v-1\right]}\frac{\mu_m}{\mu_v}\chi(t)\,dt = 0 \tag{1}$$

on the characteristics $dx\,/\,dt = u \pm a_f$ and

$$dg = \chi(t)\,dt\ , \tag{2}$$

$$\left(1-\frac{\mu_m}{\mu_v}g\right)dlnp - \left(c_{pm}-\frac{\mu_m}{\mu_v}L_1\,g\right)dlnT + \frac{\mu_m}{\mu_v}\frac{L}{T}\chi(t)\,dt = 0 \tag{3}$$

on the pathlines $dx\,/\,dt = u$. In eqs. (1)-(3) μ_m and μ_v are respectively the molecular weights of the mixture and vapor, $L \equiv L'\mu_v\,/\,(\Re T_4')$ with L' denoting the latent heat of condensation and $\Re$ denoting the universal gas constant, g is the condensate mass fraction, $c_{pm} \equiv c_{pm}'\,\mu_m\,/\,\Re$ with c_{pm}' denoting the specific heat of the mixture at constant pressure, Υ is the adiabatic exponent defined by $\Upsilon \equiv c_{pm}\,/\,(c_{pm}-1)$ and a_f is the normalized (with respect to a_4') local speed of sound given by

$$a_f^2 = \frac{[c_{pm}-L_1 g\mu_m/\mu_v]\,[1-g\mu_m/\mu_v]}{\Upsilon\,[c_{pm}-1+(1-L_1)g\mu_m/\mu_v]}\,T \quad .$$

Equations (1)-(3) together with the thermal equation of state of the mixture, namely

$$p = \rho\,T\left(1-\frac{\mu_m}{\mu_v}g\right) \tag{4}$$

are the basic equations of shock tube flows with homogeneous condensation where the function $\chi(t)$ of eq. (2) is to be identified from the kinetic equations of homogeneous condensation. The kinetics of homogeneous condensation occurs in two stages : The initial phase formation in the form of condensation nuclei in the absence of impurities (homogeneous nucleation) and subsequent growth of these nuclei into droplets or crystals. In spite of the fact that different theories of homogeneous nucleation yield nucleation rates which are sometimes off by several orders of magnitude, they can be cast into the common normalized form as

$$J \equiv \frac{J'}{\zeta'} = \Sigma(p,T,g)\, exp\,[-K^{-1}B(p,T,g)] \tag{5}$$

where J' is the actual nucleation rate, ζ' is the normalization constant for nucleation, Σ is the O(1) pre-exponential function of nucleation theories, B is the normalized activation function of nucleation and K is the nucleation parameter, assumed to be much less than unity ($K \ll 1$) signifying a relatively slow process.

The rate of growth of droplets, taken spherical in shape with a normalized radius r , can also be cast into a normalized droplet growth law as [2]

$$\frac{dr}{dt} = \lambda\, \Omega(p,T,g) \tag{6}$$

where it is assumed that the growth rate is independent of droplet radius (this latter assumption is introduced for simplicity only and may be relaxed). In eq. (6) λ is the normalized growth parameter taken to be much greater than unity ($\lambda \gg 1$) signifying rapid droplet growth and Ω is the O(1) droplet growth function. The integral condensation rate equation then assumes the form [2]

$$g(t) = \int_{t_s}^{t} \left[r^*(\tau) + \lambda \int_{\tau}^{t} \Omega(\eta)\, d\eta \right]^3 \frac{\Sigma(\tau)\, exp\,[-K^{-1}B(\tau)]}{\rho(\tau)}\, d\tau \tag{7}$$

where t_s is the normalized time when the condensable vapor reaches saturation and r^* is the normalized critical radius beyond which condensation nuclei grow into droplets. The function $\chi(t)$ of eq. (2) necessary for completion of the basic equations (1)-(4) can now be obtained from eq. (7) by direct differentiation under the integral sign.

3. Asymptotic solution for subcritical and supercritical flows

The asymptotic solution of shock tube flows with homogeneous condensation is constructed by combining the asymptotic solution of the condensation rate equation (7) in the double limit as $K \to 0$ and $\lambda \to \infty$ with the basic equations of the previous section. In this case the behavior of the activation function B along pathlines distinguishes the physically distinct condensation zones. Depending on the amount of heat released by condensation the flow may be subcritical or supercritical with an embedded shock wave.

For subcritical flows where the amount of heat release does not exceed a critical value so that a continuous solution is possible, the condensation zones along a pathline are shown in Figure 1(a). These zones are the initial growth zone (IGZ), further growth zone (FGZ), rapid growth zone (RGZ), onset zone (OZ), nucleation zone with growth (NZ) and droplet growth zone (DGZ). The analytical structure of these zones together with the asymptotic expression of the rate equation (7) in each zone along pathlines are exhibited in [2}. The asymptotic solution for the flow field in each zone is obtained by linking the asymptotic expressions to the basic flow equations. In the interval $t_s \leq t \leq t_\ell$ (i.e. in the zones IGZ, FGZ, RGZ and OZ) there is negligible influence of condensation on the flow field (weak coupling between the flow and rate equation). Thus for these zones we can introduce *nearly frozen flows* as those flows where the flow variables p, ρ, T and u are given by the classical isentropic solution in the first approximation (higher approximations are obtained by iteration) whereas the solution for g is obtained from the corresponding asymptotic expressions. The thermodynamic functions B, Σ, Ω and r^* and their total derivatives, except in IGZ, will deviate from their frozen values and the g dependence is taken into account, especially in the zones RGZ and OZ, in their nearly frozen approximation. With this in mind the asymptotic solution in each of the zones FGZ, RGZ, and OZ can be exhibited along any pathline of our choice. In particular *the relative onset point*, which marks the end of the onset zone with maximum nucleation rate, on any pathline can now be identified as the point where

$$\left(\frac{dB}{dt}\right)_{t=t_\ell} = 0 \ . \tag{8}$$

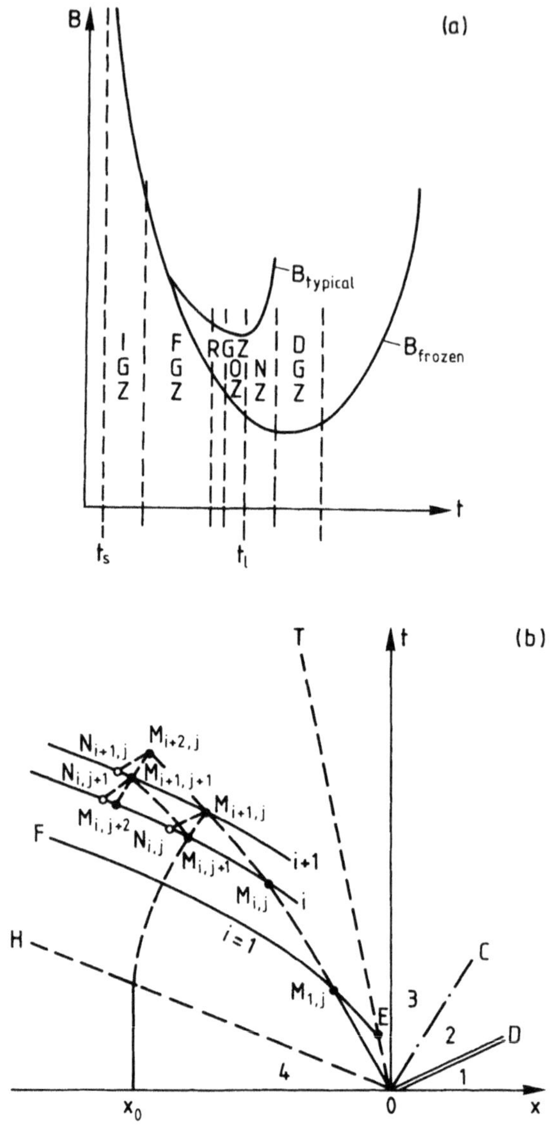

Figure 1 : **(a)** The variation of the normalized activation function $\boldsymbol{B}$ along a typical pathline exhibiting the physically distinct condensation zones in subcritical flows (t_s is the time when saturation is reached and t_ℓ is the time when $\boldsymbol{B}$ exhibits a minimum). **(b)** Construction scheme for the global subcritical solution in the heat addition zones (OH and OT are respectively the frozen rarefaction wave head and tail, EF is the condensation wave front corresponding to a maximum nucleation rate, OC and OD respectively denote the contact surface and the right running shock wave).

Repeating the procedure for different pathlines, we construct the locus of the relative onset points which we call *the condensation wave front* (curve EF in Figure 1 (b)). In particular the condensation wave front in the first approximation of nearly frozen frozen flows, which suffices for all practical purposes, can be constructed analytically independent of any particular nucleation and droplet growth theories to be employed. On the other hand the solution of the flow field in the heat addition zones NZ and DGZ can be obtained semi-analytically by combining the asymptotic expressions of the rate equation in these zones with eqs. (1)-(4). Figure 1 (b) shows such a sketch for the computation of the flow field in these zones. Assuming $M_{i,j}$ i,j=1,2,.. to be points chosen conveniently close to eachother on the i th curve in the heat addition zones (i=1 is taken as the condensation wave front EF), the points $M_{i+1,j}$ on the (i+1)st curve to be constructed are obtained by solving numerically from eq. (1) with (-) sign along $M_{i,j}\,M_{i+1,j}$ where $dx / dt = u - a_f$, from eq. (1) with (+) sign along $N_{i,j}\,M_{i+1,j}$ where $dx / dt = u + a_f$, and from eq. (3) on $M_{i,j+1}\,M_{i+1,j}$ where $dx / dt = u$ together with the appropriate asymptotic expressions (details are given in [2]).

When the amount of heat added to the flow exceeds a critical valuue, the left running characteristics intersect in the heat addition zones and the inclusion of embedded shock waves become necessary (supercritical flows). In this case depending on the location of the embedded shock along different pathlines, four distinct supercritical flow regimes can be distinguished. On pathlines falling in regimes I and II, the embedded shock is located prior to t_ℓ respectively in the zones FGZ and RGZ whereas on pathlines falling in regimes III and IV, the embedded shock proceeds t_ℓ and is located in the zones NZ and DGZ respectively. Figure 2 shows typical supercritical flow fields in shock tubes. The origin K of the embedded shock front in each case can be obtained from the envelope construction of the left running characteristics. Figure 2 (a) shows a weak embedded shock where the condensation wave front EF lies below the shock front. For embedded strong shocks shown in Figure 2 (b) only the portion EG of the curve EF designates the condensation wave front. In this case the onset zone along pathlines intersecting GF precedes the

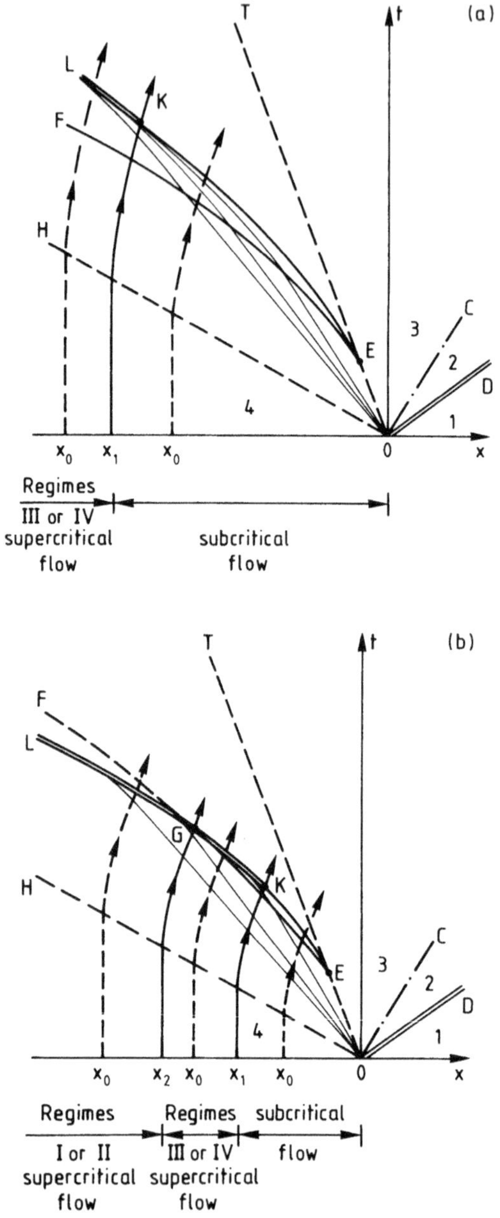

Figure 2 : Global solution of supercritical flows **(a)** with a weak embedded shock wave **(b)** with a strong embedded shock wave (EK is the rarefaction wave tail, KL is the embedded shock front and EF is the condensation wave front. The portion GF of EF in (b) is hypothetical and does not represent the onset of condensation).

shock front GL (supercritical regimes I and II). The asymptotic solution in each regime along pathlines can in principle be obtained by combining the asymptotic expressions in the condensation zones with the basic equations and shock relations (details are given in [2}) provided that the shock location is known (shock fitting). The latter presents a serious obstacle, which has to be overcome, for the computation of such flows.

4. Onset conditions and comparison of steady and unsteady nucleation rates

An algorithm, which locates the condensation wave front EF from the asymptotic solution on different pathlines, for the subcritical expansion of water vapor in argon is developed utilizing a fixed droplet growth law (the Hertz-Knudsen law) and the classical nucleation rate equation. In particular nucleation time-lag effects in the condensation zones along pathlines, as demonstrated recently by the present authors [3], allow a comparison of steady and unsteady rates from the classical theory (details of the algorithm is presented in [2]). This algorithm is tested against experimental data by Barschdorff [4] and Peters [5}. Figure 3 (a) shows the onset of condensation at the observation stations of [4] and [5] in the $p'_v - T'$ diagram. The results show better agreement at low pressures (1-2 Torr) and temperatures (210-225 K) with Peters' experiments as expected since the flow in this case is subcritical. The predicted delay of condensation onset along pathlines for Barschdorff's data is mainly due to the supercritical nature of the flow where the onset of condensation precedes the embedded shock (curve GL in Figure 2 (b)). Taking into account nucleation time-lag effects for the same experiments [4] shows that the onset conditions are unaltered by unsteady nucleation rates. Figure 3 (b) shows a comparison of steady state (J_S) and unsteady (J) nucleation rates from the classical theory along a typical pathline in Barschdorff's experiments. Obviously unsteady nucleation effects seem to be more significant in the initial nucleation period than near maximum nucleation rates.

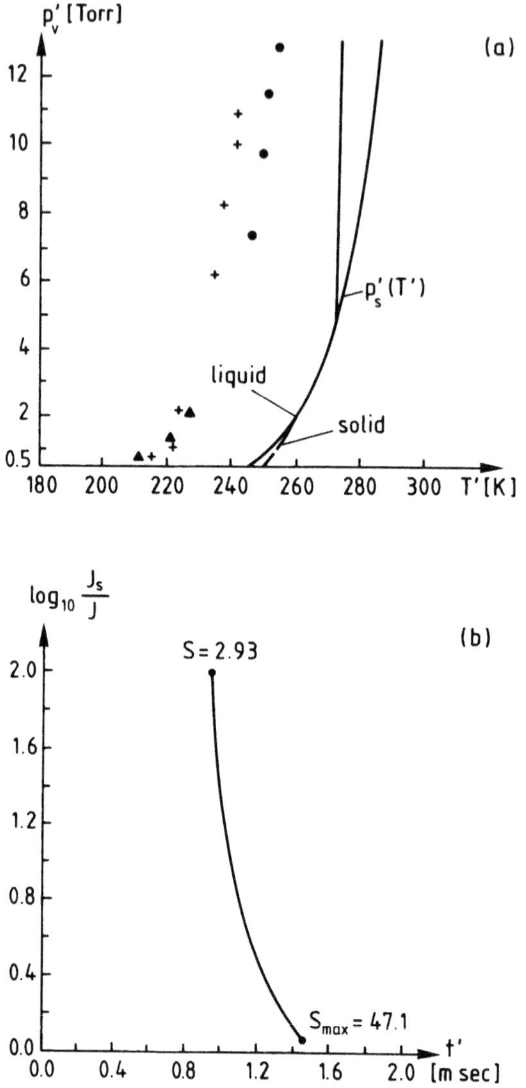

Figure 3 : **(a)** The onset conditions for the expansion of water vapor in argon in a shock tube where $p'_s(T')$ denotes the saturation pressure. • experimental data of Barschdorff [4], ▲ experimental data of Peters [5], + the present asymptotic predictions using the subcritical algorithm. **(b)** Comparison of steady and unsteady nucleation rates along a pathline at $x'_o = 20$ cm in Barschdorff's experiment (S denotes supersaturation with maximum value S_{max} , initial driver conditions are $P'_4 = 290.6$ Torr, $T'_4 = 297.2$ K and partial vapor pressure $(p'_v)_4 = 4.3$ Torr).

5. Future Perspectives

This work demonstrates the precise nature of the asymptotic solution of shock tube flows with homogeneous condensation with detailed structure of the condensation zones in comparison with direct computational methods. The algorithm developed for subcritical flows can be extended for different working substances allowing a comparison for some of the recent nucleation theories (e.g. see [6,7]) in in low density flows. A similar algorithm for supercritical flows can also be developed when a reliable shock fitting technique is available. Condensation phenomena in reflected rarefaction waves can also be discussed using the same ideas.

Acknowledgements

Part of this work was carried out while one of the authors (C.F.D.) held a Research Fellowship by the Alexander von Humboldt Foundation at the University of Karlsruhe. The authors are also grateful to Professor J. Zierep for very valuable suggestions.

References

[1]. J.P.Sislian and I.I.Glass, *AIAA J.* **14** (12), 1731 (1976).

[2]. C.F.Delale, G.H.Schnerr and J.Zierep, *Asymptotic solution of shock tube flows with homogeneous condensation*, preprint, submitted for publication (1993).

[3]. C.F.Delale and G.H.Schnerr, *Nucleation time-lag effects in steady and unsteady expansions*, preprint, submitted for publication (1994).

[4]. D.Barschdorff, *Phys. Fluids* **18** (5), 529 (1975).

[5]. F.Peters, *J. Phys. Chem.* **91**, 2487 (1987).

[6]. A.Dillmann and G.E.A.Meier, *J. Chem. Phys.* **94**, 3872 (1991).

[7]. C.F.Delale and G.E.A.Meier, *J. Chem. Phys.* **98**, 9850 (1993).

New Modes of Periodic Shock Formation in Compressible Two–Phase Flows

G.H. Schnerr, St. Adam, G. Mundinger
Institut für Strömungslehre und Strömungsmaschinen, Universität Karlsruhe (TH), Kaiserstr.12, D - 76128 Karlsruhe, Germany

Abstract
We investigate unsteady Laval nozzle flows of water vapor/carrier gas mixtures with self–excited oscillations induced by heat addition after nonequilibrium condensation. In an atmospheric supersonic wind tunnel the 2–D structure and the frequencies of the instabilities are determined by flow visualisation. Three different modes of periodic shock formation can be distinguished. The numerical solution of the two–dimensional time dependent Euler equations coupled with equations for the nucleation and the droplet growth is used to study the interaction of the shock with the homogeneous condensation process. Depending on the relative humidity of the vapor component we found one minimum of the frequency for slender nozzles, whereas two minima exist when curvature effects become important.

1. Introduction

The work presented concentrates on unsteady Laval nozzle flows with periodic oscillations in the frequency range of about 1 kHz. Instantaneous heat addition during the expansion of a mixture of water vapor and a carrier gas leads to the formation of shocks near the minimum cross sectional area of the nozzle. Self–excitation develops because the upstream moving shock interacts with the nucleation process and the droplets. The temperature increase across the shock reduces momentarily the supersaturation which cuts the maximum nucleation rate and the peak value of the heat release. Then nearly steady state flow conditions accelerate the flow to high velocities, i.e. high supersaturation, and thereby a new cycle starts. This phenomenon is important to many applications in science and technology, e.g. to flows in steam turbines because it influences their efficiency and safety.

The phenomenon of self–excited shock oscillations in nozzle flows of moist air was discovered by Schmidt [1]. For moist air and pure steam Barschdorff [2, 3] conducted quantitative experiments. The first similarity law for a dimensionless frequency has been derived by Zierep and Lin [4] for a constant nozzle throat height and varying supply conditions. Recently White and Young [5] presented the first numerical results of unsteady condensing wet steam flow in a slender (nearly one–dimensional) nozzle. Two–dimensional effects in unsteady nozzle flows of water vapor/carrier gas mixtures and the frequency dependence on the nozzle geometry and on the supply conditions were investigated by Schnerr and Mundinger [6] and Schnerr et al. [7].

Herein, we present experimental and numerical results of new modes of periodic shock formation and oscillation in transonic flows of water vapor/carrier gas mixtures.

S. Morioka and L. van Wijngaarden (eds.),
IUTAM Symposium on Waves in Liquid/Gas and Liquid/Vapour Two-Phase Systems, 377-386.

2. Experiments

2.1 Experimental Setup

Experiments are carried out in an intermittent supersonic wind tunnel with a maximum effective cross sectional area of $200 \times 50\, mm^2$. Atmospheric moist air is sucked through the test section into vacuum tanks ($34\, m^3$) which allow about $10 - 30\, s$ blowing time.

Details of the 2–D flow field are visualized by schlieren photographs using a sparc light source Strobokin (Impulsphysik, Hamburg; exposure time $1\, \mu s$). The schlieren streak method is applied to register the shock motion along the nozzle axis. Here the flow field in the area of a narrow slit at the nozzle axis is projected onto a moving film. The superposition of the shock and the film motions yields the time dependent shock location (rotational speed of the Strobodrum camera $2000\, rpm$, corresponding to a film velocity of $50\, m/s$). Timing marks on the film indicate the frequency of the shock formation and oscillation.

Here we consider two different circular arc nozzles. The more slender nozzle S2 has a total throat height of $2y^* = 30\, mm$ and a radius of wall curvature of $R^* = 400\, mm$. In the strongly curved nozzle S1 ($2y^* = 120\, mm$, $R^* = 100\, mm$) 2-D effects become important. The critical cooling rate ($(-dT/dt)^* \approx 0.3°C/\mu s$) at the nozzle throat area is the same for both nozzles since the characteristic length $\sqrt{y^* R^*}$ is kept constant (one–dimensional theory) [7].

2.2 Modes of Periodic Shock Formation in a Slender Nozzle

Figure 1a shows schlieren photographs of the periodic shock formation in the nozzle S2. The time increases from the top to the bottom, flow from left to right. The reservoir conditions (moist air) are given by the temperature $T_{01} = 291.7\, K$, the pressure of the mixture $p_{01} = 1.006\, bar$ and the relative humidity $\phi_0 = 91.1\%$, corresponding to a mixing ratio $x = 12.2\, g_{H_2O vapor}/kg_{dry\ air}$. Due to the strong latent heat release a shock is formed in front of the condensation zone near the throat of the nozzle. This shock moves upstream through the throat area into the oncoming subsonic flow, where it accelerates and dies out. Simultaneously a new shock develops and the following cycle starts. Obviously the shock does not oscillate, it accelerates monotonically opposite to the main flow direction. This mode of periodic shock formation is well known in the literature [1, 2, 3]; we call it **mode 1**.

If we decrease the vapor pressure in the reservoir, the diabatic compression weakens as well and causes a different mode of instability (**mode 2**, Fig. 1b). Here the shock stops before it reaches the nozzle throat area and then it moves downstream in the main flow direction. Simultaneously the shock strength decreases and it dies out before it returns to the origin, i.e. the shock performs an incomplete oscillation in the divergent nozzle section.

Schlieren streak registrations corresponding to these modes are shown in Fig. 2a (**mode 1**) and 2b (**mode 2**) – time increases from top to bottom, flow from left to right. The vertical black line represents the location of the nozzle throat, the dots indicate the duration of one period. **Mode 1** represents a frequency of 1013 *Hz* whereas for **mode 2** the frequency is significantly lower, only 742 *Hz*. Figure 2c shows a third mode of oscillation for supply conditions just a little bit above the stability limit (**mode 3**). Now the shock oscillates periodically with nearly constant strength in the divergent nozzle section. The frequency of 956 *Hz* typically is as high as that of **mode 1** but the amplitude of the shock motion is much smaller in comparison to the other **modes 1** and **2**.

Modes 2 and **3** are not yet known for condensing flows of vapor/carrier gas mixtures.

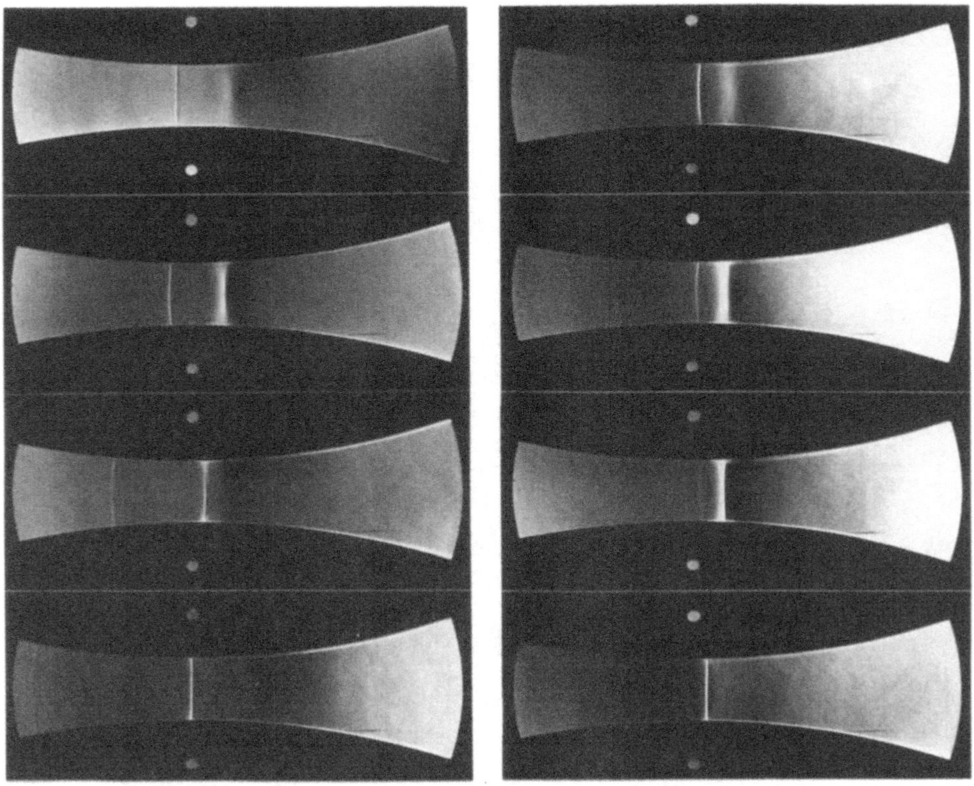

Figure 1: Schlieren pictures of self–excited periodic shock formation in nozzle S2 (time increases from top to bottom, flow from left to right, the holes indicate the location of the nozzle throat). (1a – left) **mode 1** – Reservoir conditions: $T_{01} = 291.7\,K$, $p_{01} = 1.006\,bar$, $\phi_0 = 91.1\,\%$, $x = 12.2\,g/kg$; (1b – right) **mode 2** – $T_{01} = 292.3\,K$, $p_{01} = 1.007\,bar$, $\phi_0 = 79.1\,\%$, $x = 11.0\,g/kg$.

The interesting fact is that under these conditions the oncoming flow ahead of the throat is not affected by the self–excited shock oscillation, i.e. the subsonic flow is steady and there the mass flow is constant. This is a significant difference from **mode 1** and an important aspect for technical applications, e.g. condensing flows in steam turbines.

3. Numerical Simulation

3.1 Computational Method

The inviscid two–phase flow is calculated by solving the 2–D time dependent Euler equations in conservation form. The condensation process is modelled with the classical nucleation theory from Volmer, Frenkel and Zel'dovich and the Hertz–Knudsen law for the microscopic droplet growth. With the surface averaged droplet radius this leads to 4 conservation equations for the condensate mass fraction and 3 auxiliary quantities [7], where the source terms represent the homogeneous nucleation and the droplet growth. The surface tension and the mass accommodation coefficient, which are important for the nucleation rate and the droplet growth, are determined accord-

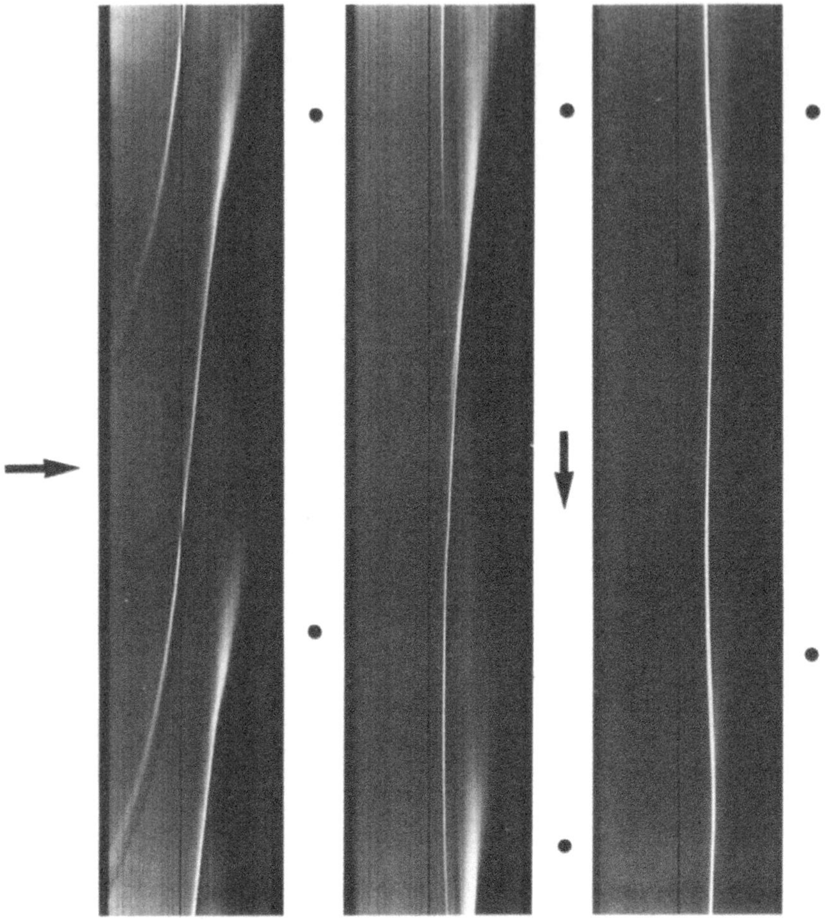

		a) **mode 1**	b) **mode 2**	c) **mode 3**
f	$[Hz]$	1013	742	956
T_{01}	$[K]$	293.2	292.8	297.5
p_{01}	$[bar]$	1.007	1.007	1.005
ϕ_0	$[\%]$	85.3	78.5	75.6
x	$[g/kg]$	12.6	11.3	14.6

Figure 2: Schlieren streak registration of self–excited periodic shock formation/oscillation in circular arc nozzle S2 (time increases from top to bottom, flow from left to right). The vertical black line indicates the throat position in the nozzle, the dots represent the duration of one period.

ing to steady state experiments [9]. Air and water vapor are treated as perfect gases. The hyperbolic system of 8 conservation equations is solved with a MUSCL–type finite volume method on a body fitted grid with explicit, second order accurate time integration. For the calculation of the source terms the fractional step method is applied whereas the cell fluxes are determined with the van Leer flux vector splitting for real gases [8] utilizing the frozen speed of sound [7].

The CPU–time of the vectorized computer code (measured performance ca. 1.8 GFLOPS on SNI S600/20) is in the range of 5 – 10 minutes. Comparison with experiments yields good agreement of all details of the whole 2–D flow field. The difference between the experimentally and the numerically determined frequency is always less than 10 %, and is typically about 2 % – 5 %.

3.2 Frequency Dependence on the Relative Humidity

For the nozzle S2 extensive numerical simulations have been carried out to determine the frequency dependence on the relative humidity under the condition of constant reservoir values of the temperature and the pressure ($T_{01} = 295\,K$, $p_{01} = 1.0\,bar$, Fig. 3). In qualitative agreement with the streak registration we obtain a minimum of the

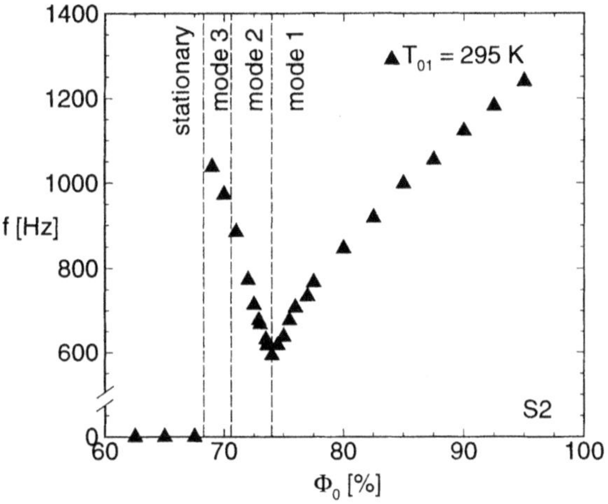

Figure 3: Influence of the relative humidity ϕ_0 on the frequency f of the oscillation for constant reservoir temperature $T_{01} = 295K$ and pressure $p_{01} = 1.0\,bar$, nozzle S2.

frequency as well ($\phi_0 = 74\,\%$, $f = 592\,Hz$). The transition from steady to unsteady flow, from **mode 3** and **mode 2** to **mode 1**, occurs in a narrow range of only 5 % of the relative humidity ϕ_0. In this narrow range the rapid decrease of the frequency leads to a high sensitivity of the flow to small changes of the supply conditions. For the present nozzle S2 the location of the minimum separates the instabilities of **mode 1** and **mode 2**, when the upstream moving shock just reaches the nozzle throat.

The frequency dependence of Fig. 3 depicts a sudden onset of the instability. We assume that at the transition to steady flow the amplitude of the shock oscillation vanishes whereas the frequency rises to a certain maximum value.

3.3 Comparison of Pressure and Nucleation Rate of Modes 1 and 2

In order to analyse the shock dynamics and the interaction of the shock with the nucleation process we compare **mode 1** and **mode 2** at the same frequency $f = 677\,Hz$ and relative humidities of $\phi_0 = 75.5\,\%$ and $\phi_0 = 72.9\,\%$, respectively. Supplementary to the experiments, the numerical calculations make the two–dimensional distribution of all time dependent quantities evident, e.g. the pressure and the nucleation rate. In Figs. 4 and 5 the normalized static pressure disturbance $\Delta p/p_{01}$ with respect to the adiabatic flow and the nucleation rate J are displayed for **mode 1** and **2**, with a time increment of 1/8 period. At the beginning the shock positions are identical. Due to the higher mixing ratio the pressure increase across the shock is stronger for **mode 1** which causes a higher upstream velocity of the shock (shock velocity after the shock formation is 17 m/s for **mode 1**, 12 m/s for **mode 2**). In

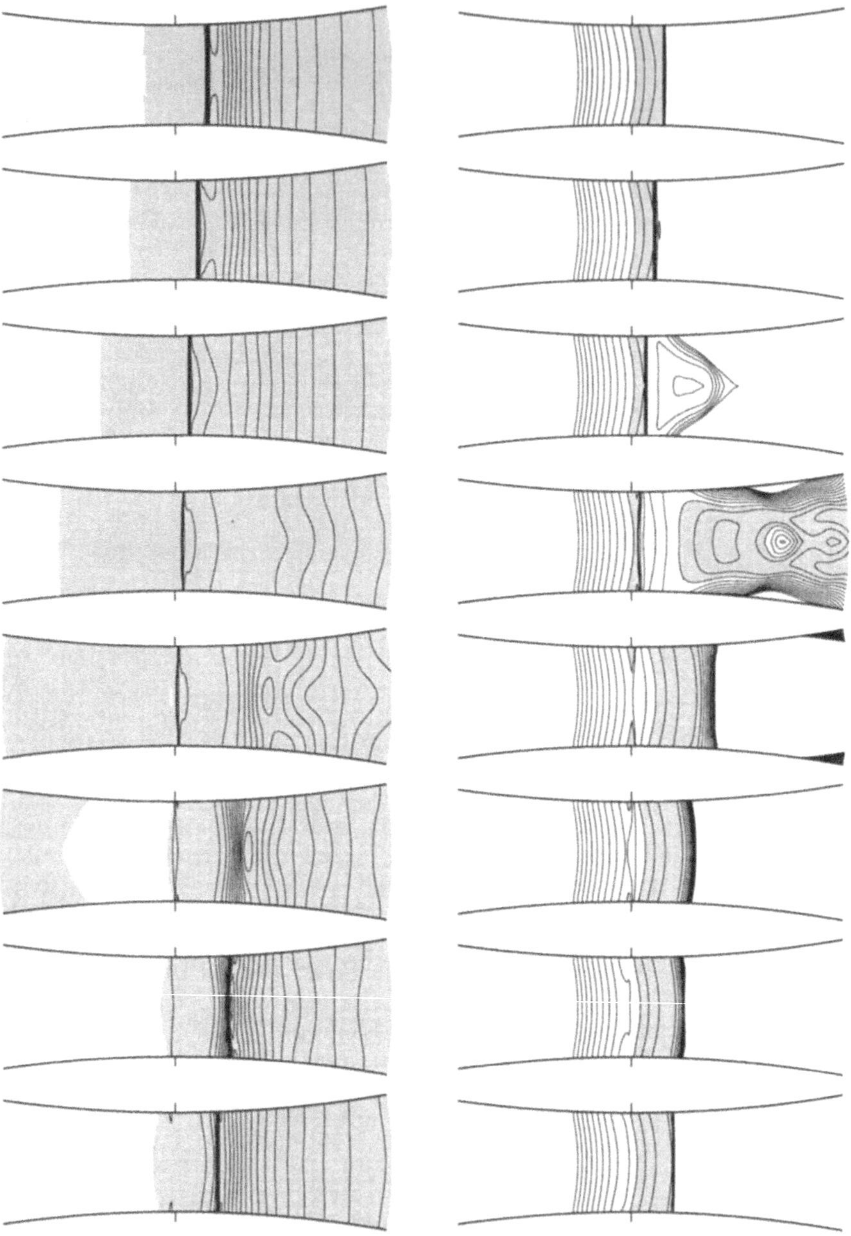

Figure 4: Periodic shock formation – **mode 1**, nozzle S2; time increases from top with increment 1/8 period. Left column: normalized static pressure disturbance with respect to the adiabatic flow $\Delta p/p_{01}$; shaded area: $\Delta p/p_{01} \geq 0.1\,\%$; increment of the isolines: 1%, starting at 1 %. Right column: nucleation rate $\log_{10} J\,[m^{-3}s^{-1}]$; shaded area: $\log_{10} J \geq 19$; increment of the isolines: $\Delta \log_{10} J = 1$, starting at 10. Reservoir state $T_{01} = 295\,K$, $p_{01} = 1.0\,bar$, $\phi_0 = 75.5\,\%$, $x = 12.56\,g/kg$, frequency $f = 677\,Hz$.

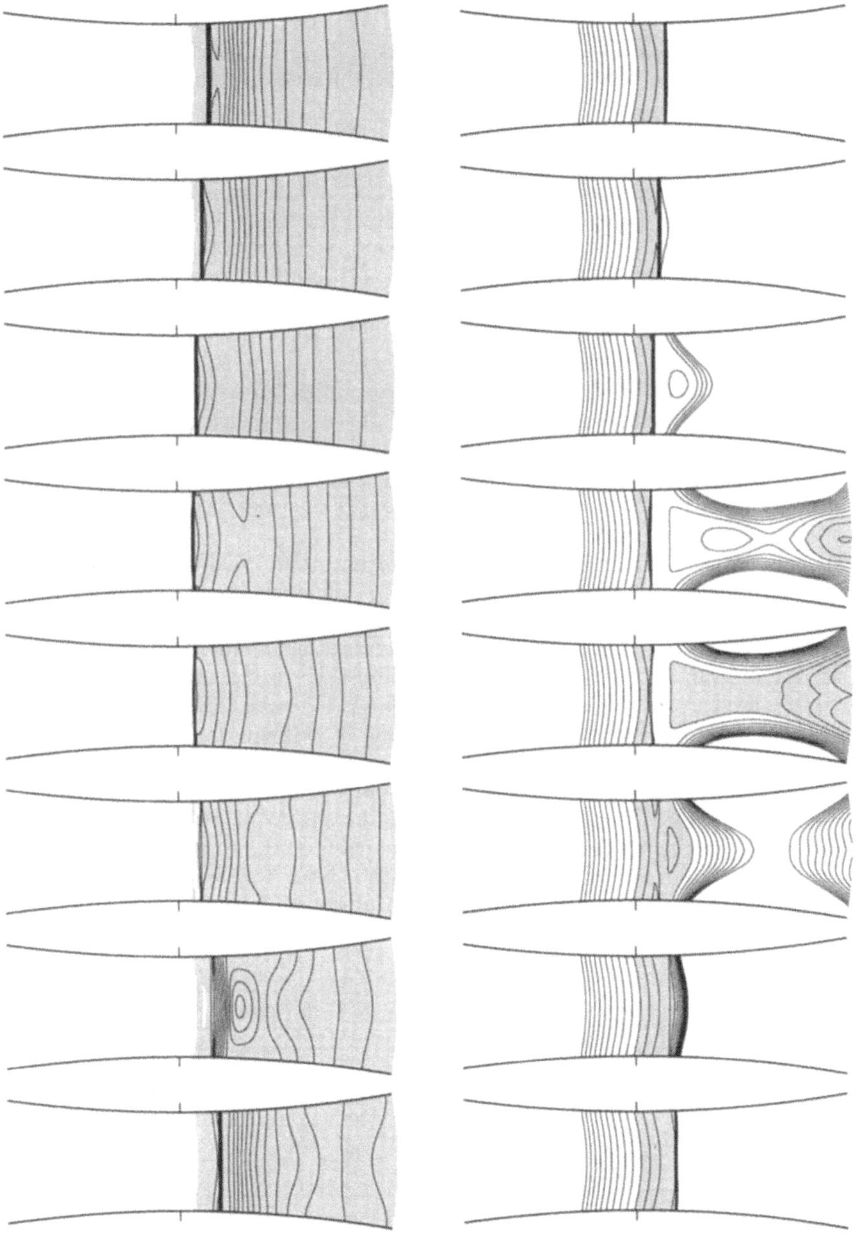

Figure 5: Periodic shock formation - **mode 2**, nozzle S2; time increases from top with increment 1/8 period. Left column: normalized static pressure disturbance with respect to the adiabatic flow $\Delta p/p_{01}$; shaded area: $\Delta p/p_{01} \geq 0.1\,\%$; increment of the isolines: 1%, starting at 1 %. Right column: nucleation rate $\log_{10} J\ [m^{-3}s^{-1}]$; shaded area: $\log_{10} J \geq 19$; increment of the isolines: $\Delta \log_{10} J = 1$, starting at 10. Reservoir state $T_{01} = 295\ K$, $p_{01} = 1.0\ bar$, $\phi_0 = 72.9\,\%$, $x = 12.12\ g/kg$, frequency $f = 677\ Hz$.

the case of **mode 1**, the shock accelerates monotonically and weakens while moving upstream. In the convergent nozzle section the shock reduces to a weak pressure wave approximately propagating with the speed of sound relative to the oncoming flow. The interaction of the moving shock with the nucleation zone can be explained by Fig. 4, right column. The temperature increase across the shock reduces the supersaturation. Obviously, this tends to diminish the homogeneous nucleation process. After the first half period this effect rises to its maximum, when the peak value of the nucleation rate decreases from about $10^{21}\,m^{-3}s^{-1}$ to approximately $10^{19}\,m^{-3}s^{-1}$, i.e. by only about two orders of magnitude. However this results in a strong decrease of the latent heat release to the flow in the zone of interaction. Downstream the nucleation rate rises again to a maximum value of about $10^{21}\,m^{-3}s^{-1}$, which leads to the formation of a new shock. Simultaneously the preceding shock leaves the zone of significant nucleation rates in the nozzle throat (shaded area, $J > 10^{19}\,m^{-3}s^{-1}$).
In contrast, in the case of **mode 2**, the flow ahead of the nozzle throat remains undisturbed (Fig. 5). Since the shock decelerates and stops, there is a much more intensive interaction of the shock and the nucleation process which delays the formation of the following discontinuity while the preceding shock disappears. This is a significant difference from **mode 1**. In the case of **mode 1** the temperature increase across the preceding shock is strong enough to shift the onset of condensation and the formation of a new shock further downstream.
Despite the almost normal shock the distribution of the nucleation rate shows a distinct two–dimensional structure for both modes. Due to the curvature of the isolines of the nucleation rate the upstream moving shock first influences the maximum nucleation rate at the axis. This causes strong gradients perpendicular to the streamlines because of the high sensitivity of the nucleation rate to temperature variations.
Figure 6 shows the calculated time dependent pressure ratio p/p_{01} at two different positions at the nozzle axis. The left part of Fig. 6 confirms that **mode 2** does not affect the oncoming flow ahead of the nozzle throat, whereas **mode 1** definitely causes pressure waves passing this position. The pressure amplitude at the location 0.3 mm upstream of the throat comes up to 1.5 % of the reservoir pressure. At the location 15.7 mm downstream of the throat (Fig. 6 – right), the shock leads to a pressure increase of 14.2 % (**mode 1**). In contrast to the shock formation of **mode 1**, **mode 2** shows a steep increase of static pressure, of about 10.3 %.

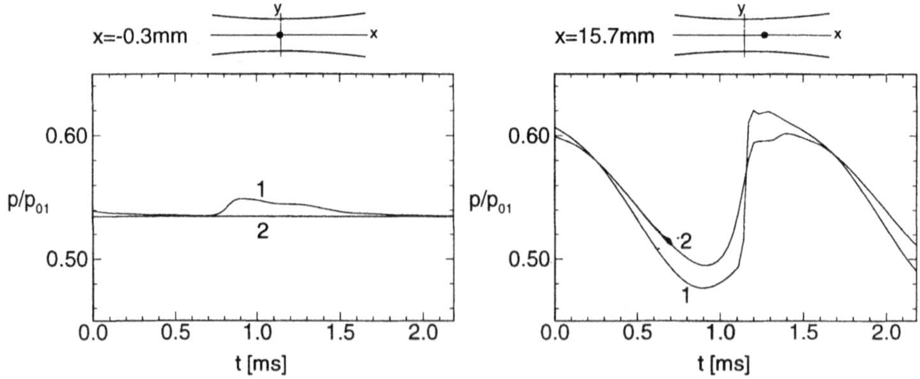

Figure 6: Time dependent pressure distribution of **mode 1** and **mode 2** at two different locations on the axis of the nozzle S2.

3.4 Frequency Dependence in a Strongly Curved Nozzle

From steady flows with homogeneous condensation it is well known that pressure waves due to heat release near curved walls interact with the flow at the nozzle axis [10]. These waves tend to shift the condensation onset at the axis further downstream, which has been observed in unsteady flow too. Because of the stronger acceleration of the flow near the wall, the cooling rate increases from the axis towards the wall. Consequently the condensation zone near the nozzle wall tends to remain in a steady state. In general, 2–D effects lead to lower frequencies, compared with slender nozzles. The frequency dependence on the relative humidity of the nozzles S1 (Fig. 7) and S2 (Fig. 3) shows significant differences. In contrast to the slender nozzle S2, the

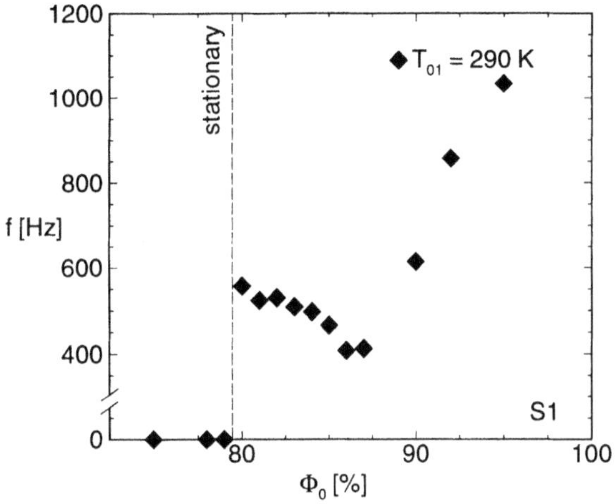

Figure 7: Influence of the relative humidity ϕ_0 on the frequency f of the oscillation for constant reservoir temperature $T_{01} = 290 K$ and pressure $p_{01} = 1.0\,bar$, nozzle S1.

slope of the decreasing branch in Fig. 7 is much smaller than that of the increasing branch. We observe two minima of the frequency. Only for relative humidities below $\phi_0 = 80\ \%$ does the oncoming subsonic flow remain in a steady state. Thus the first minimum corresponds to that of Fig. 3. The second minimum at $\phi_0 = 86\,\%$ is caused by the complex interaction of pressure waves generated in the region near the wall with the condensation zone at the axis. Obviously the frequencies in the nozzle S1 are lower than those in S2.

4. Conclusions

In order to control flows with periodic shock formation due to non–equilibrium condensation, we investigated transonic two–phase flows of atmospheric water vapor/carrier gas mixtures in Laval nozzles. New modes of flow oscillation (**mode 2** and **mode 3**) were found, where the flow ahead of the nozzle throat is steady and thus the oncoming mass flow is constant. The geometry as well as the supply conditions influence the mode and the frequency of the instability. The frequency distribution depicts one minimum for the slender nozzle S2, whereas two minima are observed for the strongly curved nozzle S1. In both cases there exist operating ranges with a high sensitivity of the frequency to small variations of the supply conditions. Further investigations are neccessary to analyse the complex wave phenomena in strongly curved nozzles. It is also not yet clear whether self–excited flow oscillations may occur in external flows or not, e.g. in flows around airfoils. Here the interaction between the shock and the nucleation process in the embedded local supersonic flow field is influenced by the propagation of pressure waves in the surrounding subsonic flow field.

For steady state flows it was found that the boundary layer influence can be reduced to its displacement effect, which does not alter the flow in principle. Neither significant flow separation nor remarkable shock–boundary layer interaction were observed in steady state and unsteady flow experiments [11]. However, it has to be investigated to what extent the frequency of the unsteady flow is influenced by viscous effects, especially in very slender nozzles.

The results presented are of immediate practical importance for transonic flows in steam turbines. Furthermore these results may give some insight into instabilities of flows with supersonic combustion. Nevertheless the interaction of the shock with the nucleation process reduces the supersaturation and therefore the heat release too, whereas the temperature increase across the shock enhances the heat release in flows with combustion.

References

[1] B. Schmidt: Beobachtungen über das Verhalten der durch Wasserdampfkondensation ausgelösten Störungen in einer Überschall–Windkanaldüse. Dissertation, Universität Karlsruhe (TH), 1962.

[2] D. Barschdorff: Kurzzeitfeuchtemessung und ihre Anwendung bei Kondensationserscheinungen in Lavaldüsen. Strömungsmechanik und Strömungsmaschinen, **6**, 18-39, 1967.

[3] D. Barschdorff: Verlauf der Zustandsgrößen und gasdynamische Zusammenhänge bei der spontanen Kondensation reinen Wasserdampfes in Lavaldüsen, Forsch.Ing.-Wes., **37** Nr.5, 146-157, 1971.

[4] J. Zierep and S. Lin: Ein Ähnlichkeitsgesetz für instationäre Kondensationsvorgänge in Lavaldüsen. Forsch. Ing.-Wes., **34**, Nr.4, 97–132, 1968.

[5] A.J. White and J.B. Young: A time–marching method for the prediction of two–dimensional, unsteady flows of condensing steam. Journal of Propulsion and Power, Vol. **9**, pp. 579–587, 1993.

[6] G.H. Schnerr and G. Mundinger: Numerical visualization of self-excited shock oscillations in two–phase flows. In: Proceedings of the 5th Int. Symp. on Comp. Fluid Dynamics, Sendai, (ed. H. Daiguji), Japan Society of CFD, Vol. **3**, pp. 87-92, 1993.

[7] G.H. Schnerr, St. Adam and G. Mundinger: Frequency control of shock oscillations in high speed two–phase flow. In: Proceedings of the Fourth Triennal Intern. Symp. on Fluid Control, Measurement and Visualization (FLUCOME'94), Toulouse, 1994.

[8] M.-S. Liou, B. Van Leer and J.-S. Shuen: Splitting of inviscid fluxes for real gases. J. Comput. Phys., Vol. **87**, pp. 1-24, 1990.

[9] G.H. Schnerr and U. Dohrmann: Drag and Lift in Nonadiabatic Transonic Flow. AIAA Journal, Vol **32**, No. 1, pp 101–107, 1994.

[10] G.H. Schnerr: Transonic Aerodynamics Including Strong Effects from Heat Addition. Computers & Fluids, Vol. **22**, No.2/3, pp. 103–116, 1993.

[11] G.H. Schnerr, R. Bohning, Th. Breitling and H.-A. Jantzen: Compressible Turbulent Boundary Layers with Heat Addition by Homogeneous Condensation. AIAA Journal, Vol. **30**, No. 5, pp. 1284-1289, 1992.

Adiabatic Waves in the Neighbourhood of the Critical Point

A. Kluwick

Institute of Fluid Dynamics and Heat Transfer,
Technical University Vienna

1 Introduction

The past decade has shown a surge in the interest in real fluid effects on wave propagation processes with special emphasis on phenomena caused by large values of the specific heats. These studies have revealed a remarkable number of (antiintuitive) results which are of basic scientific importance and also bear the potential for novel practical applications.

It is well known that the steepening characteristics of unidirectional adiabatic waves are determined by the socalled fundamental derivative (e.g. Thompson (1971))

$$\Gamma = \frac{1}{a}\frac{\partial(\varrho a)}{\partial \varrho}\bigg|_s = \frac{v^3}{2a^2}\frac{\partial^2 p}{\partial v^2}\bigg|_s .$$

Here p, ϱ, v, a and s denote, respectively, the pressure, the density, the specific volume, the speed of sound and the entropy.

Classical gasdynamics is based on the assumption of positive nonlinearity, e.g. that Γ is a strictly positive quantity as in the case of perfect gases. As a consequence, compression shocks are the only discontinuities capable of propagating in the fluid. If, however, Γ changes sign on a transition line in the p,v-plane say, the fluid is said to have mixed nonlinearity and the resulting wave dynamics is much more complicated than according to classical theory. Most important, the formation of expansion shocks can no longer be ruled out. In real fluids Γ may change sign and become negative in the neighbourhood of the critical point for two reasons. The first is associated with the critical point singularity of the isochoric specific heat. The second rests on the fact that isentropes and isotherms differ only slightly if the heat capacities of the fluid under consideration are large in general, e.g. if the fluid consists of complex molecules.

In the limit of equilibrium flow, which will be assumed almost throughout this study, kinematic wave theory provides a unified description of waves with small but finite amplitude. Within the framework of this theory the evolution of a wave is fully determined by a single function, the massflux-density-relationship, which reflects the shape of the isentropes in the p,v-plane in the vicinity of the unperturbed state. In contrast to classical gasdynamics this function is nonconvex if the wave motion causes Γ to change sign. It is this property which leads to a variety of new effects which have no counterpart in classical theory.

S. Morioka and L. van Wijngaarden (eds.),
IUTAM Symposium on Waves in Liquid/Gas and Liquid/Vapour Two-Phase Systems, 387-404.

2 Negative nonlinearity near the critical point

The study of flow phenomena in the near critical region is severely hampered by the fact that the usual analyticity requirements on which classical equations of states are based are not satisfied at the critical liquid vapour point. As a result these equations fail to predict, among others, the asymptotic divergence of the isochoric specific heat and the rate of divergence of the compressiblity. It has recently be shown by Onuki et al (1990), Boukari et al (1990) and Zappoli et al (1990) that the anomalous high compressibility of a nearly critical fluid may substantially enhance the heat transport in closed containers. As far as the properties of adiabatic waves are concerned, however, it is the divergence of the isochoric specific heat which plays a central role, Borisov et al (1983).

Specific equations of state for the critical region have been proposed by various authors. According to Widom (1965) the full range of nonclassical behaviour can be accounted for properly if the free energy $F(\varrho, T)$ in the neighbourhood of the critical point is written as a homogeneous function of its variables. This suggestion is fully consistent with the hypothesis of scale invariance and can, as pointed out by Schofield (1969) be used to derive an especially useful parametric representation

$$\frac{T-T_c}{T_c} = r(1-b^2\Theta^2) \ , \quad \frac{\varrho-\varrho_c}{\varrho_c} = r^{\beta} m(\Theta)$$

$$\varrho_c \frac{\mu(\varrho,T)-\mu(\varrho_c,T)}{p_c} = r^{\beta\delta} a\Theta(1-\Theta^2) \tag{2.1}$$

Here $m(\Theta)$ is a well behaved function from which the free energy function $f(\Theta)$:

$$\frac{F(\varrho,T)-F(\varrho_c,T)}{p_c} = (\varrho-\varrho_c)\frac{\mu(\varrho_c,T)}{p_c} + r^{2-\alpha} f(\Theta) \tag{2.2}$$

may be obtained by quadrature. The function $\mu(\varrho_c, T)$ which is assumed to have a regular expansion for $T \to T_c$ describes the variation of the chemical potential on the critical isochore $\varrho = \varrho_c$. Furthermore, α, β and δ denote the critical exponents $\alpha \approx 0.08$, $\beta \approx 0.33$, $\delta \approx 4.8$ while $a > 0$, $b > 1$ are disposable parameters. Thus $\Theta = 0$, $\Theta = \pm 1$ and $\Theta = \pm b^{-1}$ represent, respectively, the critical isochore, the two sides of the coexistence curve and the two parts of the critical isotherm for $\varrho \gtrless \varrho_c$.

Unfortunately, the validity of equation (2.1) is limited to an extremely small neighbourhood of the critical point. Outside this region $|(\varrho-\varrho_c)/\varrho_c| \lesssim 0.4$, $|(T-T_c)/T_c| \lesssim 0.01$ it fails severely. Here, however, we are interested mainly in the limiting behaviour of the nonlinearity parameter Γ as $\varrho \to \varrho_c$, $T \to T_c$ e.g. as $r \to 0$ where, as shown by Borisov et al (1983), the diverging part of c_v provides the dominant contribution

$$\Gamma \sim -\frac{1}{2a^2\varrho^3}\frac{T^2}{c_v^3}\frac{\partial c_v}{\partial T}\left(\frac{\partial p}{\partial T}\right)^3 .$$

Insertion of the results for a and c_v following from (2.1), (2.2) then yields

$$\Gamma \sim r^{-1+\alpha} L(\Theta) \ , \quad r \to 0 \tag{2.3}$$

where the function $L(\Theta)$ can be expressed in terms of the free energy function $f(\Theta)$. Equations (2.1), (2.2) have been combined by Gulen, Thompson and Cho (1990) with an extended Benedict-Webb-Rubin type equation of state to determine various thermodynamic quantities of SF_6 including Γ in the general neighbourhood of the critical point. In these calculations $m(\Theta)$ was taken to be a linear function of Θ: $m(\Theta) = k\Theta$, $k > 0$ which is supported also by some experimental evidence, Schofield (1969). The numerical results indicate that $\Gamma \to +\infty$ on the critical isochore $\Theta = 0$ as $T - T_c \to +0$ while $\Gamma \to \mp\infty$ on the critical isotherm $\Theta = \mp(1/b)$ as $\varrho - \varrho_c \to \mp 0$. From equation (2.3) we conclude that this is true not only for SF_6 but for the whole class of fluids whose thermodynamic properties can be predicted in terms of linear functions $m(\Theta)$. Furthermore, these results point to the existence of an intermediate value $-(1/b) < \Theta_o < 0$ for wich $\Gamma = 0$. The corresponding curve in the ϱ, T-plane which will be referred to as transition line is sketched in Fig. 1.

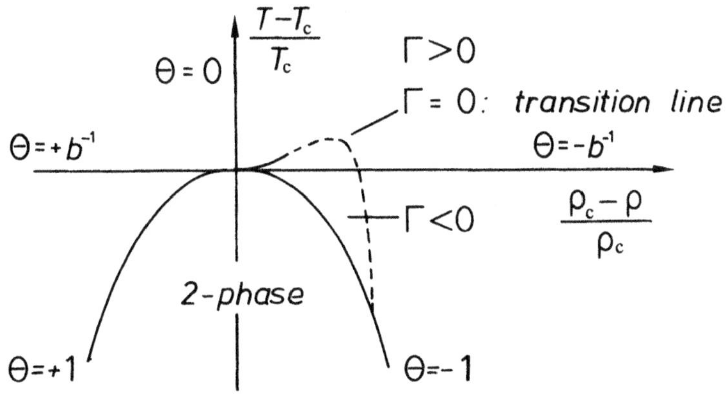

Fig. 1: Schematic representation of the transition line $\Gamma = 0$ near the critical point in the T, ϱ-plane

3 Negative nonlinearity in BZT-fluids

Negative nonlinearity which occurs in fluids consisting of complex molecules has received increasing attention during the past decades. The possibility that Γ may change sign in the dense gas regime of such fluids having large specific heats has been recognized first by Bethe (1942) and independently by Zel'dovich (1946)[1]. The reason for this sign change is seen most easily from a relationship derived by Bethe (1942) which reduces to

$$\Gamma = \frac{v^3}{2a^2}\frac{\partial^2 p}{\partial v^2}\bigg|_s \sim \frac{v^3}{2a^2}\frac{\partial^2 p}{\partial v^2}\bigg|_T + O\left(\frac{R}{c_v}\right) \tag{3.1}$$

[1]According to a recent private information by Prof. Chernyi, Zemplén has reached this conclusion even earlier than Zel'dovich.

in the limit $c_v/R \to \infty$. In geometrical terms this means that the isentropes approach the isotherms in the p, v-plane under this limit. Since the isotherms of all fluids are curved down in the general neighbourhood of the critical point one thus expects that the isentropes there will have negative curvature also provided that c_v/R is sufficiently large. In the case of a *Van der Waals fluid* it is found that a region of negative nonlinearity forms if $c_{v\infty}/R \gtrsim 17.5$ where $c_{v\infty}$ is the ideal gas heat capacity at the critical temperature, Thompson (1971). Using more sophisticated and accurate equations of state several candidates of real negative Γ fluids which include hydrocarbons and fluorocarbons of moderate complexity have been identified by Thompson and coworkers, Thompson (1971), Thompson and Lambrakis (1973). Because of the significance of each of these studies by Bethe, Zel'dovich and Thompson we refer to such fluids as Bethe-Zel'dovich-Thompson (BZT)-fluids. More recently the properties of fluorocarbons have been investigated by Cramer (1989) who appplied the Martin-Hou-equation of state to determine Γ. Each of seven commercially available fluorocarbons were found to have a region of negative Γ large enough to include the critical isotherm.

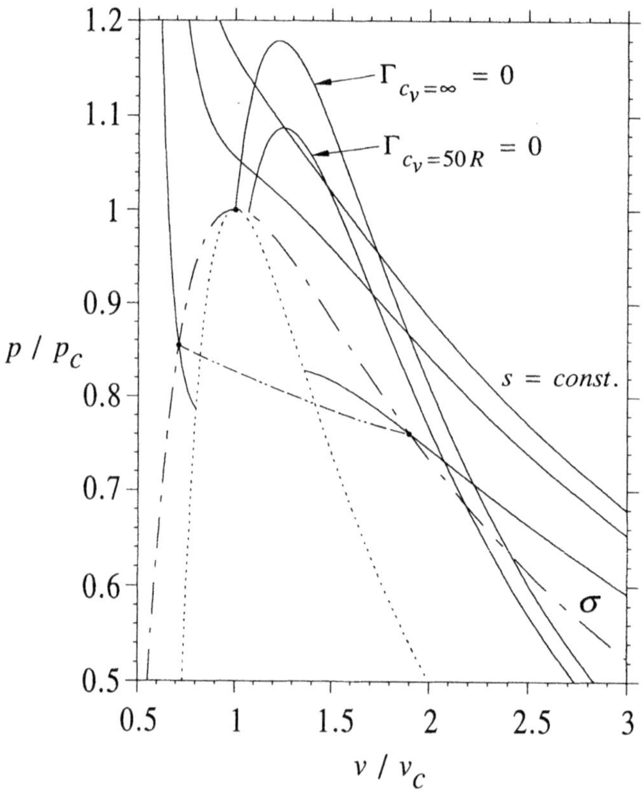

Fig. 2: Transition line $\Gamma = 0$ and isentropes for a *Van der Waals fluid* with $c_v/R = 50 = const.$ $- - \cdot \cdot - -$ equilibrium isentrope in the mixture region, $\cdots\cdot$ spinodals; $- \cdot - \cdot -$ saturation line

To illustrate the region of negative nonlinearity Fig. 2 shows the p, v-diagram of a *Van der Waals fluid* with $c_{v\infty}/R = 50$ which corresponds roughly to normal decane ($n - C_{10}H_{22}$). Displayed in this diagram are the transition line $\Gamma = 0$, various isentropes and the limiting position of the transition line for $c_{v\infty} \rightarrow \infty$. To demonstrate the effect of metastable conditions the mixture part of the region of negative nonlinearity bounded by the vapour spinodal is included also. The dry isentropes in this region are seen to have even stronger negative curvature than those in the single phase regime.

As far as wave motions with phase transition are concerned it is important to note that the slope of the equilibrium isentropes at the saturation line is discontinuous,

$$k(T) = \frac{(\partial p/\partial v)_s - (\partial p/\partial v)_{s,m}}{(\partial p/\partial v)_{s,m}} \ . \tag{3.2}$$

Here the subscript m denotes the equilibrium derivative on the mixture side of the saturation boundary. Geometrically, this socalled 'kink' $k(T)$ can be interpreted as a region of negative Γ which shrinks to a single point thus yielding $\Gamma = -\infty$ at the saturation line. That this interpretation is physically sensible follows from the fact that the properties of weakly nonlinear acoustic waves which will be considered next depend solely on the shape of the equilibrium isentropes as long as nonequilibrium effects can be ignored.

4 Unidirectional planar waves

The basic phenomena associated with negative and mixed nonlinearity are studied best in the most simple case of one-dimensional planar waves. If nonequilibrium effects are neglected the governing equations - in nondimensional form - reduce to

$$\frac{\partial \varrho}{\partial t} + \frac{\partial \varrho u}{\partial x} = 0 \ , \ \frac{\partial u}{\partial t} + u\frac{\partial u}{\partial x} = -\frac{1}{\varrho}\frac{\partial p}{\partial x} \ , \ p = p(\varrho) \ . \tag{4.1}$$

Here u, x, and t denote the velocity, the propagation distance and the time. The dependence of p on s is not displayed explicitely for convenience.

The solution describing right running simple waves is found to satisfy the kinematic wave equation

$$\frac{\partial \varrho}{\partial t} + \frac{\partial j}{\partial x} = 0 \ , \ j(\varrho) = \int_1^{\varrho}\int_1^{\varrho'} \frac{a(\bar\varrho)}{\bar\varrho}\,\Gamma(\bar\varrho)\,d\bar\varrho\,d\varrho' \ . \tag{4.2}$$

The definition of the massflux $j(\varrho)$ which vanishes in the reference state $\varrho = 1$ shows that the wave evolution is traced in a coordinate system moving with the unperturbed speed of sound $a = 1$.

A formal solution to (4.2) is given by

$$\varrho = const. \ \text{on} \ \frac{dx}{dt} = v_w = \int_1^{\varrho} \frac{a(\bar\varrho)}{\bar\varrho}\Gamma(\bar\varrho)\,d\bar\varrho \ . \tag{4.3}$$

Owing to the nonlinear dependence of the convected sound speed v_w on the density initially smooth wave profiles will in general cease to be single valued if t exceeds a critical value t_s. For $t > t_s$ the wave motion then leads to the occurrence of a shock and the simple wave solution ceases to be valid in the strict sense. However, if the wave amplitude is small

$$\varrho = 1 + \bar{\varrho} \ , \ \ |\bar{\varrho}| \ll 1 \tag{4.4}$$

as will be assumed here, this solution remains valid as a first order approximation. Moreover, the jump conditions across weak shocks having the propagation speed

$$\frac{dx}{dt} = v_s = \frac{j_b - j_a}{\bar{\varrho}_b - \bar{\varrho}_a} \tag{4.5}$$

are satisfied to first order also provided that

(a) the Rayleigh line which connects the state before the shock $(j_b, \bar{\varrho}_b)$ and the state after the shock $(j_a, \bar{\varrho}_a)$ does not cut intervening branches of the $(j, \bar{\varrho})$ diagram, and

(b) that the convected sound speeds v_{wb} and v_{wa} are consistent with the wave speed ordering relationship

$$v_{w,b} \leq v_s \leq v_{w,a} \ . \tag{4.6}$$

The first condition is equivalent to the criterion first formulated by Oleinik (1959). According to the second requirement (4.6) shocks may have sonic upstream or sonic downstream conditions (sonic shocks) but also sonic upstream and sonic downstream conditions (double sonic shocks).

Differentation of equation (4.2) with respect to ϱ twice yields

$$\frac{d^2 j}{d\bar{\varrho}^2} = \frac{a}{\varrho}\Gamma \ . \tag{4.7}$$

Taking into account that Γ chracterizes the curvature of isentropes in the p, v-plane we thus infer that the $j, \bar{\varrho}$-diagram for weakly nonlinear waves qualitatively resembles the local shape of the isentrope which passes through the unperturbed state.

If the unperturbed state is sufficiently far from the saturation line and the transition line $\Gamma = 0$ the fundamental derivative in the perturbed fluid is either strictly positive or strictly negative. Consequently, the $j, \bar{\varrho}$-diagram can locally be approximated by a parabola just as in the case of a perfect gas:

$$j \sim \frac{\Gamma_o}{2}\bar{\varrho}^2 \ . \tag{4.8}$$

However, if the unperturbed state is sufficiently close to the transition line with respect to the wave amplitude Γ inside the wave may change sign indicating that the appropriate form of the $j, \bar{\varrho}$-diagram is

$$j \sim \frac{\Gamma_o}{2}\bar{\varrho}^2 + \frac{\Lambda_o}{6}\bar{\varrho}^3 \ , \ \ \Lambda = \left.\frac{\partial \Gamma}{\partial \varrho}\right|_s \ . \tag{4.9}$$

Inspection of Fig. 2 shows that one has also to take into account that the transition line is tangent to the isentrope in some point. If the unperturbed state is in the vicinity of such a point not only Γ but also Λ may change sign inside a wave. Consequently, the $j, \bar{\varrho}$-diagram has to be approximated locally by a quartic

$$j \sim \frac{\Gamma_o}{2}\bar{\varrho}^2 + \frac{\Lambda_o}{6}\bar{\varrho}^3 + \frac{N_o}{24}\bar{\varrho}^4 \ , \quad N = \left.\frac{\partial^2 \Gamma}{\partial \varrho^2}\right|_s . \tag{4.10}$$

Finally, Γ was found to change discontinuously at the saturation line leading to the $j, \bar{\varrho}$-diagram

$$\begin{aligned} j &\sim \frac{\Gamma_o}{2}\bar{\varrho}^2 \qquad\qquad , \ \bar{\varrho} < \bar{\varrho}_s \\ j &\sim \frac{a_{om}\Gamma_{om}}{2}\bar{\varrho}^2 + (a_{om} - 1)(\bar{\varrho} - \bar{\varrho}_s) + \frac{\Gamma_o - a_{om}\Gamma_{om}}{2}\bar{\varrho}_s^2 \ , \ \bar{\varrho} > \bar{\varrho}_s \end{aligned} \tag{4.11}$$

which describes waves with phase transition in the equilibrium limit. Herein $\bar{\varrho}_s$ denotes the density perturbation at the saturation boundary.

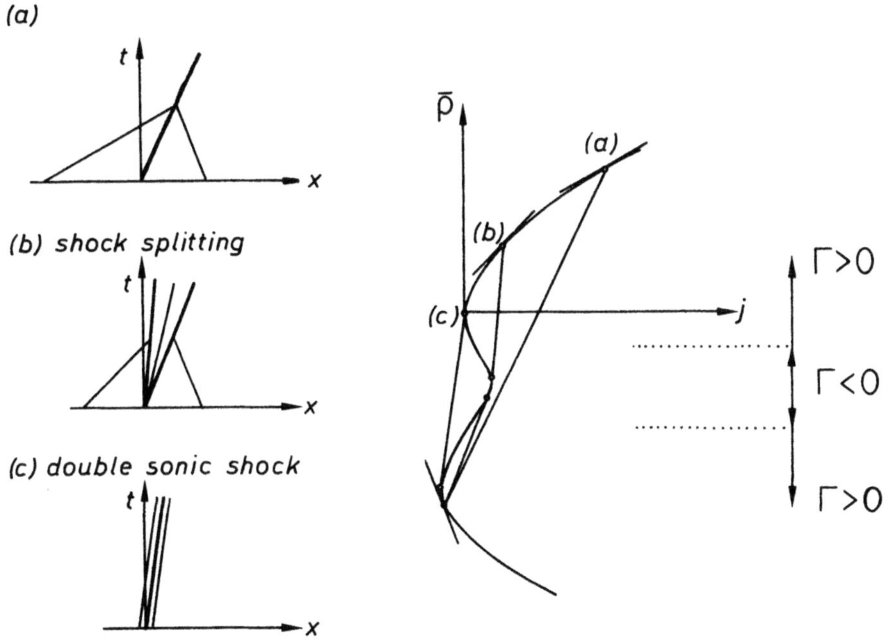

Fig. 3: Example of plot j versus $\bar{\varrho}$ for $B < 1/2$

One important point in which waves in media having both positive and negative nonlinearity differs substantially from waves in perfect gases is the behaviour of shock fronts. Taking into account that the relationship (4.11) qualitatively resembles equation (4.10) and that equations (4.8) and (4.9) can be derived from equation (4.10) by carrying out appropriate limit processes it will be sufficient to consider the case where the function $j(\bar{\varrho})$ is approximated by a fourth order polynomial, Fig. 3.

As a starting point we consider a large density jump (a) which leads to a stable shock which takes the state of the fluid across the whole negative Γ region.

$$\Gamma = \frac{\partial^2 j}{\partial \bar{\varrho}^2} < 0 \ : \ \left|\frac{\Lambda_o}{N_o}\right|\left[1 + \sqrt{1 - \frac{2\Gamma_o N_o}{\Lambda_o^2}}\right] < \bar{\varrho} < \left|\frac{\Lambda_o}{N_o}\right|\left[1 - \sqrt{1 - \frac{2\Gamma_o N_o}{\Lambda_o^2}}\right] . \qquad (4.12)$$

As the strength of the imposed density jump is continuously decreased one eventually reaches the limiting case where the Rayleigh line is tangent to the j versus $\bar{\varrho}$ graph. Discontinuities of even smaller strength (b) violate the shock admissibility criterion (i) and thus cannot propagate as stable shocks but split into two shocks which are separated by a wave fan. The first of these shocks has sonic downstream, the second sonic upstream conditions. However, there exists also a Rayleigh line which is tangent to the j, $\bar{\varrho}$-diagram in both endpoints (c). The corresponding shock discontinuity has sonic upstream and sonic downstream conditions and, therefore, represents a double sonic shock.

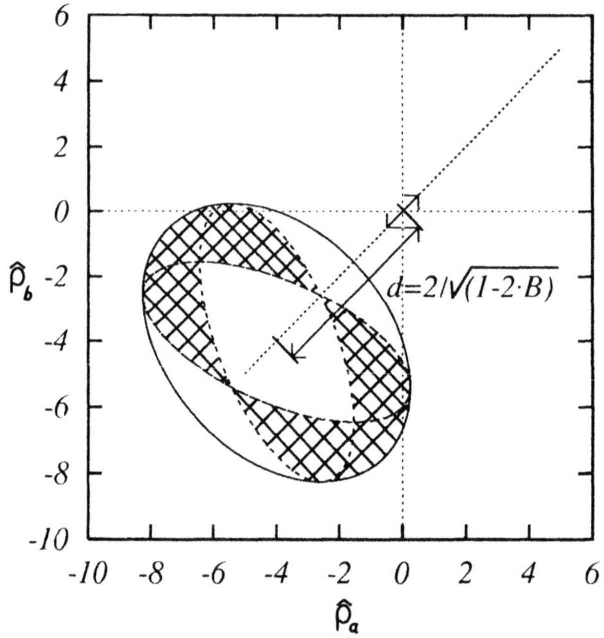

Fig. 4 : Region of inadmissible shocks in $\hat{\varrho}_a$,$\hat{\varrho}_b$-plane

If

$$B = \frac{\Gamma_o N_o}{\Lambda_o^2} > \frac{1}{2} \tag{4.13}$$

the $j, \bar{\varrho}$-relationship (4.10) exhibits a single extremum at $\bar{\varrho} = 0$. As a consequence, shock discontinuities occur in the form of compression shocks ($N_o > 0$) or expansion shocks ($N_o < 0$) only. Effects of mixed nonlinearity come into play if $B < (1/2)$ and as shown in Kluwick and Scheichl (1994) the shock admissibility criterion (i) can then be expressed in general form by introducing the scaled perturbation density

$$\hat{\varrho} = \bar{\varrho}\frac{N_o}{\Lambda_o\sqrt{\frac{1}{2} - B}} \quad , \quad B < \frac{1}{2} \, . \tag{4.14}$$

Inadmissible pairs $\hat{\varrho}_a$, $\hat{\varrho}_b$ lie in the shaded region of Fig. 4 which is bounded by three ellipses characterizing shocks with sonic upstream conditions, sonic downstream conditions and shocks which will split into two shocks if their amplitude is further reduced. It has been pointed out earlier that the $j, \bar{\varrho}$ relationship (4.11) which describes waves with phase transition in the equilibrium limit is qualitatively similar to the result (4.10) insofar as it predicts the occurrence of split shocks and double sonic shocks. The double sonic shock then represents an evaporation discontinuity while split shocks (partial liquefaction shocks) may form inside compression waves leading to the partial condensation of the vapour.

To illustrate how the properties of weakly nonlinear adiabatic waves are influenced by the formation of sonic shocks we briefly discuss the evolution of a square pulse adopting a $j, \bar{\varrho}$-relationship of the form (4.9). It will be assumed that the fundamental derivative is positive in the unperturbed state ($\Gamma_o > 0$) and that Γ decreases during isentropic expansion ($\Lambda_o > 0$). Such a situation may arise in a BZT fluid if the unperturbed state is in the neighbourhood of the high pressure zero of Γ but also in a regular fluid under near critical conditions. Furthermore, it will be assumed that the density decrease caused by the pulse is large enough so that the fluid inside the pulse enters the negative Γ region. As shown in Fig. 5 the resulting wave pattern then is completely different from the classical result, Cramer and Kluwick (1984).

First, the (expansive) discontinuity at $x = 1$ is seen to disintegrate into a wave fan followed by a sonic expansion shock. Second, the amplitude of the initially imposed (compressive) discontinuity at $x = 1$ is too large to lead to an admissible compression shock. It therefore splits into a compression fan which is terminated by a sonic compression shock. The sonic expansion shock is weakened by compression waves merging from behind and in order to stay sonic it has to emanate waves which leave the curved front in the tangential direction. In this way a socalled precursor is formed. Finally, since the compression shock and the expansion shock propagate with different speeds they eventually collide and give rise to yet another interesting phenomenon. According to classical theory of nonlinear acoustics two merging shocks produce a single shock of higher amplitude. In contrast, the collision of an expansion shock and a compression shock is seen to lead to a discontinuous reduction of the shock strength.

Unfortunately, difficulties associated with the required high temperature and pressure levels have prevented so far the direct experimental observation of expansion shocks, sonic shocks, split shocks and double sonic shocks in single phase BZT-fluids. The related phenomena such as evaporation shocks and liquefaction shocks occurring in retrograde fluids, however, have been studied experimentally in detail by Dettleff et al (1979), Thompson et al (1987). Furthermore, expansion shocks inside the region of negative nonlinearity associated with the critical point singularity of the isochoric specific heat have been generated by Borisov et al (1983). In these experiments the flow medium Freon-13 appears to have entered the two phase region. As a result, the experimental observation of a single phase expansion shock still remains a challenging goal for the future.

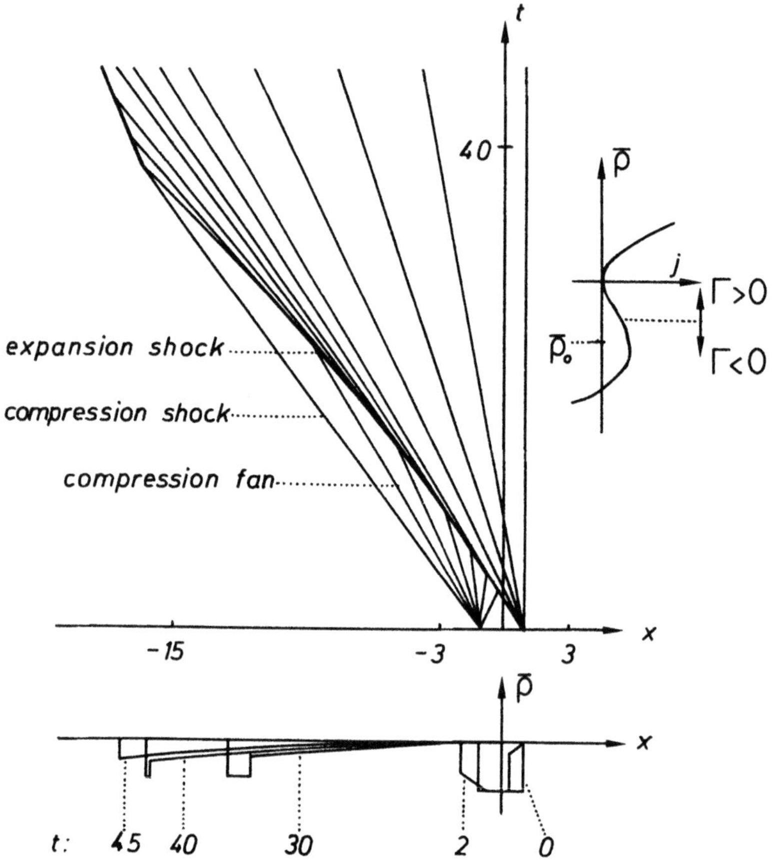

Fig. 5: Evolution of a square pulse (Cramer and Kluwick (1984))

Sofar the considerations have been limited to the case of planar waves in homogeneous media. However, the analysis can easily be generalized to include three-dimensional effects and stratification effects. For example, weakly nonlinear spherical and cylindrical waves have been investigated by Kluwick and Czemetschka (1989), (1990). As in classical acoustics it is found that the variation of the disturbances is proportional to $A(l)^{-1/2}$ where A and l denote, respectively, the ray tube area and the propagation distance. Combination of this classical relationship with the non-classical properties of sonic shocks leads to the new result that sonic shocks will in general terminate at a finite distance from their origin if A increases with l, e.g. in outgoing waves. As shown by Kluwick and Cox (1993) a related phenomenon occurs in stratified media.

5 Transonic nozzle flow

According to classical gasdynamics there exists a close relationship between the perturbation massflux $j(\bar{\varrho})$ carried by weakly nonlinear simple waves and the perturbation massflux $\hat{\jmath}(\bar{\varrho})$ in steady transonic flow:

$$\hat{\jmath}(\bar{\varrho}) = -2j(\bar{\varrho}) \ . \tag{5.1}$$

This is a direct consequence of the fact that the compatibility condition holding along right running waves agrees with the transonic approximation to the energy equation for steady flows. Since this is true also for general fluids equation (5.1) carries over unchanged to media having negative or mixed nonlinearity. As a consequence, equations (4.8), (4.9), (4.10) and (4.11) in combination with the linearized quasi-one-dimensional version of the continuity equation

$$\hat{\jmath}(\bar{\varrho}) + \bar{A}(x) = -2j(\bar{\varrho}) + \bar{A}(x) = Q = const. \tag{5.2}$$

and the well known transonic approximations

$$\bar{\varrho} = \bar{p} = -\bar{u} \ , \tag{5.3}$$

$$M - 1 = \frac{1}{2}\frac{d\hat{\jmath}}{d\bar{\varrho}} = -\frac{dj}{d\bar{\varrho}} \tag{5.4}$$

can immediately be used to study the properties of steady transonic flow in the throat area of slender nozzles, Kluwick (1993). Here M denotes the local Machnumber and $\bar{A}$, $\bar{p}$ and $\bar{u}$ characterize, respectively, the perturbations of the crossectional area, the pressure and the velocity suitably nondimensionalized.

As in the case of one-dimensional waves the following discussion will be based on the $j(\bar{\varrho})$-relationship (4.10). Application of equation (5.4) then yields the important result that the associated Machnumber distribution $M(\bar{\varrho})$ in general exhibits three sonic points $M = 1$ which is in sharp contrast to the classical prediction that sonic conditions are reached only once during isentropic expansion or compression.

Before turning to specific problems we note a simple geometrical interpretation of the mass balance (5.2): the cross-sectional area $\bar{A}$ corresponding to a value of the density $\bar{\varrho}$ is given by the distance between the line $\hat{\jmath} = Q = const.$ and the $\hat{\jmath}$ versus $\bar{\varrho}$ graph.

As a first application we determine the shape of a nozzle leading to a shock free flow as the fluid is accelerated from subsonic to supersonic speeds characterized, for example, by a linear pressure distribution

$$\bar{p} = \bar{\varrho} = -Cx \tag{5.5}$$

Evaluation of equations (5.2) and (4.9) then yields

$$\bar{A} = Q_{max} + \Gamma_o C^2 x^2 - \frac{1}{3}\Lambda_o C^3 x^3 + \frac{1}{12} N_o C^4 x^4 \tag{5.6}$$

where Q_{max} denotes the value of $\hat{\jmath}$ at the maximum of the $\hat{\jmath}(\bar{\varrho})$-relationship.

In agreement with the aforementioned geometrical argument the shape of the nozzle qualitatively resembles the shape of the $(\jmath, \hat{\varrho})$-graph. A linear pressure/density distribution can, therefore, be achieved by a conventional converging-diverging nozzle if $\Gamma_o > 3\Lambda_o^2/(8N_o)$. If $\Gamma_o < 3\Lambda_o^2/(8N_o)$, however, a nozzle having two throats rather than a single throat is required. Of course, these conclusions remain unchanged if the linear pressure distribution (5.5) is replaced by a monotonically decreasing function.

Following this brief discussion of shock-free accelerating flows let us consider the flow through a conventional converging-diverging nozzle

$$\bar{A} = Cx^2 \ , \ C > 0 \ . \tag{5.7}$$

Density/pressure contours corresponding to various value of Q are depicted in Fig. 6 for the choice $0 < \Gamma_o < \Lambda_o^2/(3N_o)$. Using the geometrical interpretation of equation (5.2) it is easily shown that each maximum of the $\hat{\jmath}, \bar{\varrho}$-curve leads to the formation of a saddle point at the throat $x = 0$ while the minimum of the $\jmath$ versus $\bar{\varrho}$ plot leads to the occurrence of a center again at $x = 0$. As a consequence, the resulting density/pressure distributions exhibit regions of multivaluedness. For example, if we follow the density/pressure distribution corresponding to the maximum value of $Q = Q_{max}$ in the flow direction we observe the formation of a fold which has to be eliminated by a shock discontinuity in order to obtain a physically meaningful single valued solution. Application of the shock admissibility criterion yields that the required shock discontinuity has a sonic upstream state. The solution leading from subsonic to supersonic flow then has the following properties: the fluid accelerates in the converging part of the nozzle and reaches a sonic state already upstream of the throat. A sonic expansion shock leads back to subsonic conditions. Downstream of this shock the fluid accelerates again, it reaches a second sonic state at the throat and finally enters the supersonic regime in the divergent part of the nozzle.

Finally, let us discuss briefly a case where all three sonic points of the $\hat{\jmath}, \bar{\varrho}$-diagram come into operation. To this end we consider a solution leading from subsonic to supersonic flow and then back to subsonic flow. As in the case investigated before the fluid is accelerated to sonic conditions upstream of the throat but a sonic expansion shock reduces the Machnumber to a value less than one. The fluid accelerates to a supersonic state as it passes through the throat. An (almost) conventional compression shock reduces the Machnumber to a subsonic value but the flow reaches sonic conditions for a third time before it is finally shocked to a subsonic state by means of a sonic compression shock.

In passing note that (using the same arguments as discussed in the preceeding section) the results described so far can be used also to obtain a qualitative picture of nozzle flows with phase changes. The discontinuities in Fig. 6 then have the meaning of an evaporation shock a single phase compression shock and a partial liquefaction shock, Zauner and Meier (1990).

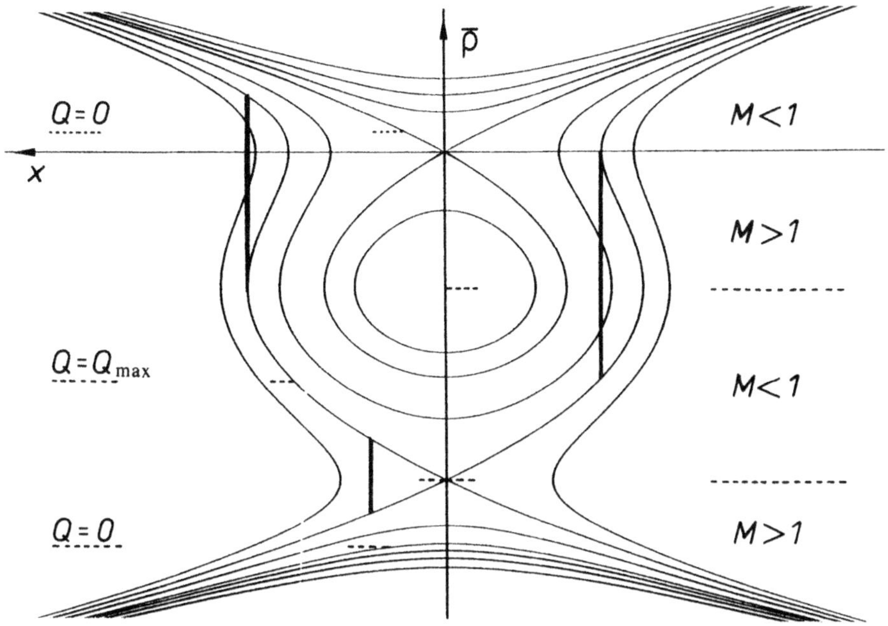

Fig. 6: Laval nozzle: density distributions for various values of Q $((\Lambda_o^2/3N_o) > \Gamma_o > 0$, Kluwick (1993)) —— admissible shocks

Before closing this section let us have a brief look at unsteady transonic nozzle flows. The appropriate generalization of equation (5.2) is (Kluwick and Scheichl (1994))

$$\hat{\jmath}(\bar{\varrho}) + \bar{A}(x) = f(\xi) \tag{5.8}$$

where the characteristic variable ξ is constant on upstream propagating waves

$$\left.\frac{dx}{dt}\right|_{\xi} = \frac{1}{2}\frac{d\hat{\jmath}}{d\bar{\varrho}} \, . \tag{5.9}$$

Equations (5.8) and (5.9) can be used to clarify the important question whether and how a full subsonic supersonic expansion described by equation (5.2) can be achieved starting from a steady subsonic solution by lowering, for example, the exit pressure of the nozzle. The basic conceptual difficulty associated with such a transition process is easily recognized from Fig. 6 if the initial density distribution is identified with the subsonic solution having $M = 1$ at the throat $x = 0$. Consequently, the throat cannot be reached by an upstream propagating characteristic in finite time. This is of no concern if the fluid is a perfect gas as the flow upstream of the throat then remains unchanged. In the case investigated here, however, the density distribution in $x < 0$ has to be modified too.

Numerical results obtained from equations (5.8) and (5.9) are depicted in Fig. 7. Starting at time $t = 0$ the pressure at the nozzle exit is lowered continuously. Eventually an expansion shock forms inside the expansion wave thus generated. It passes through the throat after finite time to approach its steady state position asymptotically in the limit $t \to \infty$. Moreover, since this expansion shock has a sonic upstream state it generates a precursor which in turn causes the necessary modification of the density distribution in the region ahead of the shock.

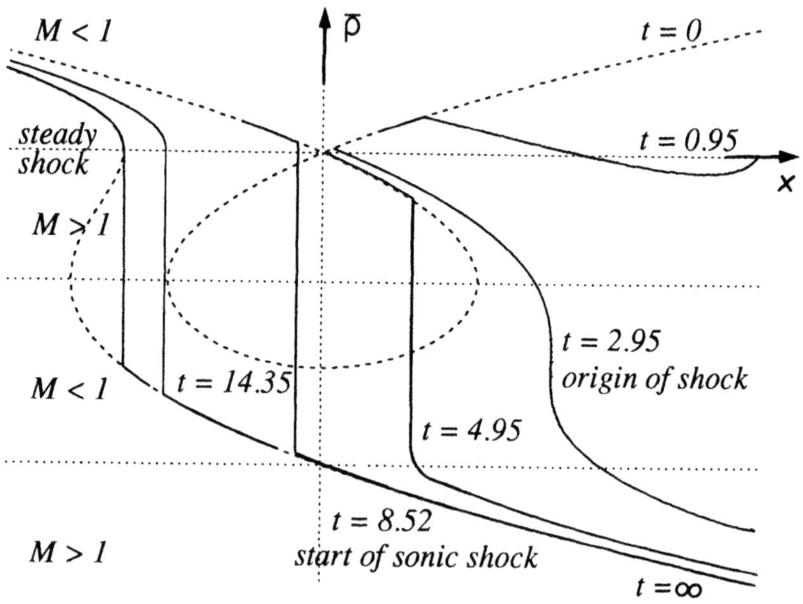

Fig. 7: Start problem for Laval nozzle $((\Lambda_o^2/3N_o) > \Gamma_o > 0)$

Similar calculations have been performed for nozzles with two throats. The results indicate that the shock free expansion cannot be realized via an unsteady process starting from a steady subsonic solution. A more detailed investigation of equation (5.2) shows that in addition to the desired solution (5.5) there exists a second steady state solution containing a shock discontinuity. It is this solution which represents the steady state limit of the unsteady transition process, Kluwick and Scheichl (1994).

6 Nonequilibrium effects

Equilibrium theory has the advantage that it leads to a relatively simple qualitative description of the phenomena under consideration. Nonequilibrium effects may, however, cause significant quantitative modifications. These are especially important if the wave motion is associated with phase changes. For example, it has been shown by Thompson et al (1987) that the overall pressure increase across partial liquefaction split shocks may be substantially higher than predicted by equilibrium theory if the supersaturation of the vapour in the metastable region is taken into account. A more detailed investigation of the internal structure of liquefaction shocks has been carried out recently by Gulen and Thompson (1994). The proposed model applies as long as the downstream states are in the metastable region bounded by the saturated vapour and the spinodal lines.

Thermoviscous nonequilibrium effects on weakly nonlinear unidirectional waves having negative and mixed nonlinearity have been studied by Borisov et al (1983), Cramer and Kluwick (1984) and Cramer and Crickenberger (1991) while thermoviscous transonic nozzle flows have been treated by Kluwick (1993). The disturbances carried by unidirectional waves are found to be governed by the modified Burgers equation

$$\frac{\partial \bar{\varrho}}{\partial t}+\frac{\partial j(\bar{\varrho})}{\partial x}=\frac{\delta}{2}\frac{\partial^2 \bar{\varrho}}{\partial x^2}\,,$$

$$\delta=\left[2\mu_o+\lambda_o+\left(\frac{1}{c_{vo}}-\frac{1}{c_{po}}\right)k_o\right]. \tag{6.1}$$

Here μ_o, λ_o, k_o, c_{vo} and c_{po} denote the first and second viscosity, the thermal conductivity and the heat capacities in the unperturbed state (again suitably nondimensionalized).

Equation (6.1) may be used to obtain the following estimate for the thickness Δ_s of a weak shock relative to the reference length L.

$$\frac{\Delta_s}{L}\sim\frac{\delta}{[p]\Gamma_o}\,. \tag{6.2}$$

Inspection of this result reveals two mechanisms which may lead to an anomalous increase of the shock thickness in the general neighbourhood of the critical point: (i) small values of Γ_o if the unperturbed state is in the vicinity of a transition line (this effect may be pronounced even further by the slow algebraic decay of sonic shocks and the formation of a plateau region in impending split shocks, Kluwick (1993)), (ii) the asymptotic divergence of the second viscosity if the unperturbed state is near critical (for the experiments of Borisov et al (1993) the estimated thickness of the expansion shock generated in Freon-13 is 5.6 cm which is about 600 times larger than the thickness of a compression shock of equal strength far from the critical point).

Parallel to investigations based on first principles the implications following from the more heuristic capillary model of phase transitions proposed by Korteweg 1901 have been explored by Slemrod (e.g. Slemrod (1990)) in a series of papers. In the case of one-dimensional flow these model equations are obtained from equations (4.1) through the substitution

$$p(\varrho)\rightarrow p(\varrho)-\alpha(\varrho)\varrho_x^2-\varepsilon(\varrho)\varrho_{xx} \tag{6.3}$$

where the coefficients α, ε represent long range molecular effects, Serrin (1983). As before the dependence of the various field variables on the entropy s is not displayed explicitely.

The way in which the properties of adiabatic waves are affected by the additional stresses arising in response to density gradients can be studied most easily in the small amplitude limit. If thermoviscous effecs are taken into account also the appropriate generalization of (6.1) is, Kluwick (1994)

$$\frac{\partial \bar{\varrho}}{\partial t} + \frac{\partial j(\bar{\varrho})}{\partial x} = \frac{\delta}{2}\frac{\partial^2 \bar{\varrho}}{\partial x^2} - \gamma \frac{\partial^3 \bar{\varrho}}{\partial x^3} \ , \ \gamma = \frac{\varepsilon_o}{2} \ . \tag{6.4}$$

The Burgers-Korteweg de Vries equation (6.4) with strictly positive nonlinearity arises in a number of physical problems including wave propagation phenomena on liquid films and - more interesting in the present context - wave propagation phenomena in bubbly liquids, e.g. Nakoryakov, Pokusaev and Shreiber (1993). The modified Burgers-Korteweg de Vries equation with $j(\bar{\varrho})$ given by equation (4.9) has been shown by Kluwick (1991) to describe the evolution of long waves in fluidized beds. Here the main point is that the additional capillary force entering equation (6.3) leads to the occurrence of a dispersive term in the transport equation for weakly nonlinear waves. As a consequence shock profiles are no longer monotonous but exhibit oscillations in a certain range of the parameters δ, γ. At present the author is not aware of any experimental data in support of this result.

7 Summary

In the limit of equilibrium flow kinematic wave theory provides a unified description of weakly nonlinear planar waves in fluids having mixed nonlinearity. Moreover, the kinematic wave approach carries over - with slight changes only - to the case of three-dimensional waves and to the cases of steady and unsteady transonic nozzle flows. In fact it is found that the properties of such flows can be derived from the same function $j(\bar{\varrho})$ which determines the evolution of unidirectional planar waves.

The neglect of non-equilibrium effects has the advantage that it leads to a relatively simple qualitative description of the phenomena under consideration. However, this simplification may cause significant quantitative errors, especially if the wave motion is associated with phase changes. Some of these nonequilibrium mechanisms and the resulting modifications of the equilibrium wave properties have been discussed briefly.

References

[1] Bethe, H.A. (1942) The theory of shock waves for an arbitrary equation of state, Office Sci.Res. & Dev.Rep. 545.

[2] Borisov, A.A., Borisov, Al.A. Kutateladze, S.S. and Nakoryakov, V.E. (1983) Rarefaction shock wave near the critical liquid-vapour point. J.Fluid Mech. **126**, pp. 59-73.

[3] Boukari, H., Shaumeyer, J.N., Briggs, M.E., Gammon, R.W. (1990) Critical speeding up in pure fluids, Physical Review A, **41**, 4, pp 2260-2263.

[4] Cramer, M.S. (1989) Negative Nonlinearity in Selected Fluorocarbons, The Physics of Fluids A1, pp 1894-1897.

[5] Cramer, M.S. and Kluwick A. (1984) On the propagation of waves exhibiting both positive and negative nonlinearity. J.Fluid Mech. **142**, pp. 9-37.

[6] Cramer, M.S. and Crickenberger, A.B. (1991) The Dissipative Structure of Shock Waves in Dense Gases. J. Fluid Mech. **223**, pp. 325-355.

[7] Dettleff, G., Thompson, P.A., Meier, G.E.A. and Speckmann, H.-D. (1979) An experimental study of liquefaction shock waves, J. Fluid Mech. **95**, pp. 279-304.

[8] Gulen, S.C., Thompson, P.A. and Cho, H.J. (1990) Rarefaction and Liquefaction Shock Waves in Regular and Retrograde Fluids with Near-Critical End States, in: G.E.A. Meier, P.A. Thompson (Eds): Adiabatic Waves in Liquid-Vapor-Systems, Springer Verlag, pp. 281-290.

[9] Gulen, S.C. and Thompson, P.A. (1994) private communication.

[10] Kluwick, A. (1991) Weakly nonlinear kinematic waves in suspensions of particles in fluids, Acta Mechanica **88**, pp. 205-217.

[11] Kluwick, A. (1993) Transonic nozzle flow of dense gases, J. Fluid Mech. **247**, pp. 661-688.

[12] Kluwick, A. (1994) A note on the small amplitude limit of the Korteweg theory for waves with phase transition, in preparation

[13] Kluwick, A., Czemetschka, E. (1989) Kugel- und Zylinderwellen in Medien mit positiver und negativer Nichtlinearitaet, ZAMM **70**, 4, pp. T207-T208.

[14] Kluwick, A. and Cox, E.A. (1992) Propagation of weakly nonlinear waves in stratified media having mixed nonlinearity, J. Fluid Mech. **244**, pp. 171-185.

[15] Kluwick, A. and Scheichl, St. (1994) Unsteady transonic nozzle flow of dense gases, in preparation

[16] Korteweg, D.J. (1901) Sur la forme que prennent les equations du mouvement des fluids si l'on tient compte des forces capillaires par des variationes de densité, Archives Néerlandaises des Sciences Exactes et Naturelles.

[17] Nakoryakov, V.E., Pokusaev, B.G. and Shreiber, I.R. (1993) Wave Propagation in Gas-Liquid Media, CRC Press, 2nd edition.

[18] Oleinik, O.A. (1959) Uniqueness and stability of the generalized solution of the Cauchy problem for a quasilinear equation, Usp. Mat. Nauk **14**, pp 165-170. (English Translation) Am. Math. Soc. Transl. Ser 2, **33**, pp. 285-290.

[19] Onuki, A., Hong, H., Ferrel, R.A. (1990) Fast adiabatic equilibration in a single-component fluid near the liquid-vapor critical point, Physical Review A, **41**, 4, pp. 2256-2259.

[20] Schofield, P. (1969) Parametric Representation of the Equation of State near a Critical Point, Physical Review Letters **22**, pp. 606-608.

[21] Serrin, J. (1983) The form of interfacial surfaces in Korteweg's theory of phase equilibria, Quarterly of Applied Mathematics, October 1983, pp. 357-364.

[22] Slemrod, M. (1990) Remark on the Traveling Wave Theory of Dynamical Phase Transitions, in: G.E.A. Meier, P.A. Thompson (Eds): Adiabatic Waves in Liquid-Vapor-Systems, Springer Verlag, pp. 325-337.

[23] Thompson, Ph.A. (1971) A Fundamental Derivative in Gasdynamics. The Physics of Fluids **14**, pp. 1843-1849.

[24] Thompson, Ph.A. and Lambrakis, K.C. (1973) Negative Shock waves, J. Fluid Mech. **60**, pp. 187-207.

[25] Thompson, P.A., Chaves, H., Meier, G.E.A., Kim, Y.G. and Speckmann, H.D. (1987) Wave splitting in a fluid of large heat capacity. J. Fluid Mech. **185**, pp. 385 - 414.

[26] Widom, B. (1965) Equation of State in the Neighbourhood of the Critical Point, The Journal of Chemical Physics **43**, pp. 3898-3905.

[27] Zappoli, B., Bailly, D., Garrabos, Y., Le Neindre, B., Guenoun, P., Beysens, D. (1990) Anomalous heat transport by the Piston Effect in supercritical fluids under zero gravity, Physical Review A, **41**, 4, pp. 2264-2267.

[28] Zel'dovich, Ya.B. (1946) On the possibility of rarefaction shock waves, Zh. Eksp. Teor. Fiz. **4**, pp. 363-364.

[29] Zauner, E., Meier, G.E.A. (1990) Phase changes of a large-heat-capacity fluid in transcritical expansion flows, in: G.E.A. Meier, P.A. Thompson (Eds): Adiabatic Waves in Liquid-Vapor-Systems, Springer Verlag, pp. 103-112.

MATERIAL WAVES OF A FLUID IN THE VICINITY OF THE CRITICAL POINT

H. GOUIN
Laboratoire de Mécanique Fondamentale et Appliquée
Faculté des Sciences et Techniques
13397 Marseille Cedex 20
France

J.M. DELHAYE
Commissariat à l'Energie Atomique
Centre d'Etudes Nucléaires de Grenoble
Service de Thermohydraulique pour les Applications Industrielles
38054 Grenoble Cedex 9
France

ABSTRACT

A nonlinear model associated with the mean-field approximation leads to the existence of solitary waves for fluids near the critical point. The model is based on the second gradient theory and yields a Landau-Ginzburg - type behavior.

Above the critical temperature, solitary waves can only be supersonic with respect to the isothermal acoustic speed. Below the critical temperature, all values of the solitary wave velocities are acceptable.

1. INTRODUCTION. - The mean field approximation is convenient for studying an isothermal nonhomogeneous fluid. Isothermal interfacial zones are investigated starting from a local free energy which can be considered as the sum of two terms: a first one corresponding to a medium with a uniform composition equal to the local one and a second one associated with the non-uniformity of the fluid. The additional term may be approximated by a gradient expansion, typically truncated to the second order (Cahn and Hilliard, 1959; Casal, 1972; Rocard, 1967). A representation of the free energy near the critical point therefore allows the study of interfaces of macroscopic size. Capillary layer and bulk phases are not considered independently. This method is similar to the second gradient theory in continuum mechanics (Germain, 1972), and is most likely to be successful and perhaps even quantitatively accurate in the vicinity of the critical point (Malomed and Rumanov, 1984; Rowlinson and Widom, 1984).

Obviously, this model is simpler than models associated with the renormalization group theory (Widom, 1972). Nevertheless, it has the advantage of extending well-known

S. Morioka and L. van Wijngaarden (eds.),
IUTAM Symposium on Waves in Liquid/Gas and Liquid/Vapour Two-Phase Systems, 405-417.

results for equilibrium cases (Rowlinson and Widom, 1984) to the dynamics of interfaces.

De Gennes (1981) pointed out that in some circumstances due to the long range effect of van der Waals forces, the gradient model may be qualitatively incorrect. However, in the vicinity of the critical point, fluid interfaces are not of molecular size and the gradient model is valid.

The idea of studying interface motion as localized travelling waves in a multi-gradient theory is not new and can be traced throughout many problems of condensed matter and phase-transition physics (Poujet, 1980; Slemrod, 1983; Truskinovsky, 1987). The main result of our study is the existence of solitary wave solutions, even for $T > T_c$ and that in this case they are supersonic with respect to the isothermal acoustic speed. Below the critical temperature, the velocity of waves is not lower bounded.

Such waves could be interpreted as the fringes occurring in a Natterer tube experiment (Bruhat, 1968). They are related with the opalescence phenomenon in the vicinity of the critical point. Above the critical temperature, waves look like motions of vibrating strings with velocities depending on temperature. Below the critical temperature, the wave behavior is different and all values of velocity are acceptable.

2. MOTIONS IN CAPILLARY LAYERS IN THE VICINITY OF THE CRITICAL POINT. - Equations of motion are classically given in literature with an additive term of *second gradient* form (Dunn, 1986). The simplest model able to take into account the density and its gradient uses *a unique supplementary quantity, the constant* C *of internal capillarity*. In S.I. units, the value of C for water at 20°C, is of the order of 10^{-16} (Casal and Gouin, 1988). The associated supplementary term is effective only in areas where the density gradient is very important and consequently in interfaces.

The equation of motion reads:

$$(1) \qquad \rho\Gamma + \operatorname{grad} P + \rho \operatorname{grad}\left(\Omega - C\nabla^2\rho\right) - \operatorname{div}\sigma_v = 0$$

where P denotes the thermodynamic pressure corresponding to the reference homogeneous medium, i.e. the pressure entering for example the van der Waals equation of state, Ω is the body force potential, σ_v the viscous stress tensor, Γ the acceleration vector and ∇^2 the Laplacian operator.

In thin interfaces, the viscosity may be neglected (Langevin, 1992) and for an isothermal flow, the equation of motion becomes:

(2) $$\Gamma + \mathrm{grad}\left(\mu - C\nabla^2\rho + \Omega\right) = 0$$

where μ denotes the specific free enthalpy or chemical potential of the homogeneous fluid.

Close to the critical point, the chemical potential of a homogeneous fluid may be written in the following form (Rowlinson and Swinton, 1982):

(3) $$\mu = \mu^c - \mu_{01}^c\, \Delta T - \mu_{11}^c\, \Delta\rho\, \Delta T + \frac{1}{6}\, \mu_{30}^c\, (\Delta\rho)^3$$

where $\Delta\rho \triangleq \rho - \rho_c$ and $\Delta T \triangleq T_c - T$ denote the distance of the density ρ and the Kelvin temperature T to their values at the critical point. The coefficients are the values of the partial derivatives of μ at the critical point:

(4) $$\mu_{ij} \triangleq \frac{\partial^{i+j}\mu}{\partial\rho^i\, \partial T^j}\left(\rho_c, T_c\right)$$

The chemical potential μ has been expanded using a cubic polynomial in terms of density. Defining $A \triangleq \mu_{11}^c > 0$ and $B \triangleq \frac{1}{6}\, \mu_{30}^c > 0$, equation (3) now reads :

$$\mu = \mu^c - \mu_{01}^c\, \Delta T - A\, \Delta T\, \Delta\rho + B(\Delta\rho)^3$$

This is an important point of the study, as it may solve the problem in a universal way, ΔT being an order parameter. Such a chemical potential expansion agrees with the van der Waals equations of state. Only the values A, B and C are representative of the fluid.

3. MATERIAL WAVES IN CAPILLARY LAYERS IN THE VICINITY OF THE CRITICAL POINT. - Near equilibrium conditions, fluid motions can be assumed isothermal. Let us study the one-dimensional problem of isothermal waves where all variables including the velocity V and density ρ are functions only of the variable $\xi \triangleq x - c\,t$ where x is the space variable, t the time and c the wave speed:

(5) $$V = V(x - ct) \qquad \rho = \rho(x - ct)$$

The mass balance equation yields:

(6) $$- c \frac{d\rho}{d\xi} + \frac{d(\rho V)}{d\xi} = 0$$

By integrating Eq. (6) we obtain:

$$\rho(V - c) = q$$

The constant q is the *mass velocity* through every surface of equation ξ = Const. In the case of waves described by Eq. (5), we find:

$$\Gamma = \frac{1}{2}\frac{d}{d\xi}(V - c)^2,$$

or:

(7) $$\Gamma = \frac{1}{2}\frac{d}{d\xi}\left(\frac{q}{\rho}\right)^2$$

From equation (2) we deduce the first integral:

(8) $$C \frac{d^2\rho}{d\xi^2} = \mu - \mu_0 + \frac{q^2}{2\rho^2} + \Omega$$

where μ_0 is an integration constant. Since we know $\mu(P, T)$ from Eq. (3), equation (8) gives the density profile in the wave for each value of q. Then, $V(\xi)$ is deduced from $V = \frac{q}{\rho} + c$, with c as an arbitrary real value. The wave speed with respect to the fluid is $\frac{|q|}{\rho}$.

By neglecting body forces and taking into account equation (3), the equation of waves (8) yields:

(9) $$C \frac{d^2\rho}{d\xi^2} = - A\, \Delta\rho\, \Delta T + B(\Delta\rho)^3 + \frac{q^2}{2\rho^2} + k_0$$

where k_0 is constant at a given temperature.

Equation of waves using nondimensional variables - Nondimensional variables Y, ζ and Q defined by :

(10) $$\rho \triangleq \rho_c\,(1 + \varepsilon Y),\quad \xi \triangleq L\,\zeta,\quad q \triangleq b\,Q$$

are used instead of ρ, ξ and q, where:

(11) $$\varepsilon^2 \triangleq \frac{A}{B\rho_c^2}\,(T_c - T)\ ,\quad L^2 \triangleq \frac{2C}{B\rho_c^2}\,\frac{1}{\varepsilon^2}\ ,\quad b^2 \triangleq 2\,B\,\rho_c^5\,\varepsilon^3\ .$$

Relation (11) assumes that $T < T_c$.

ε is a nondimensional number, whereas L and b have the dimensions of length and mass velocity respectively. Only L involves the capillary constant C.

The values, A and B, depend only on critical values, so that ε can be written in the form:

$$\varepsilon = \left\{\lambda\,\frac{(T_c - T)}{T_c}\right\}^{1/2}$$

If the equation of state is the van der Waals equation, one obtains in the previous expression $\lambda = 4$.

In the same way, b is written in the form $b = \left\{\lambda'\ P_c\,\rho_c\ \varepsilon^3\right\}^{1/2}$. In the case of the van der Waals equation $\lambda' = 3$.

Using these new variables, equation (9) yields:

(12) $$\frac{1}{2}\,\frac{d^2Y}{d\zeta^2} = Y^3 - Y + \frac{Q^2}{(1+\varepsilon Y)^2} + k_1$$

where k_1 is constant.

The term associated with the flow is substituted by a second order expansion in εY :

$$\frac{Q^2}{(1+\varepsilon Y)^2} = Q^2 - 2\varepsilon Q^2 Y + 3\varepsilon^2 Q^2 Y^2 + Q^2\,O\left(\varepsilon^3 Y^3\right)$$

and equation(12) yields:

(13) $$\frac{1}{2}\,\frac{d^2Y}{d\zeta^2} = \left(Y + \varepsilon^2 Q^2\right)^3 - Y\left(1 + 2\,\varepsilon\,Q^2\right) + k_2$$

The new change of variables:

(14) $$Y \triangleq -\varepsilon^2 Q^2 + \alpha Y_1 \; ; \; \zeta \triangleq \frac{\zeta_1}{\alpha} \quad \text{with } \alpha^2 \triangleq 1 + 2\varepsilon Q^2$$

gives the same equation for Y_1 as for Y when $Q = 0$:

$$\frac{1}{2}\frac{d^2 Y_1}{d\zeta_1^2} = Y_1^3 - Y_1 - \frac{1}{4} a$$

and yields the first integral:

(15) $$\left(\frac{dY_1}{d\zeta_1}\right)^2 = \left(1 - Y_1^2\right)^2 - a\, Y_1 - b$$

where a and b are two constants.

4. TWO EXAMPLES OF WAVES IN THE VICINITY OF THE CRITICAL POINT. - The intersections of the straight line $aY_1 + b$ and of the quartic $\left(1 - Y_1^2\right)^2$ yield the density range.

Interface propagation - In the bulk phases, the densities are constant and thus the first and second derivatives of Y_1 are zero. The straight line is tangent to the quartic at the associated points. If the bulk phases are different on both sides of the interface, the straight line has to be bitangential to the quartic which implies that a and b are null (case of two phases). A liquid-vapor interface is similar to the one obtained in the equilibrium case (Q = 0) studied by Rowlinson and Widom (1984) in chapter 9, but mass flows through the interface. Vaporization or condensation phenomena depend on the sign of q. This case corresponds to a "shock wave" in the sense of Slemrod (1983). As in the case of equilibrium conditions studied by Rowlinson and Widom (1984), one has:

$$\lim_{\zeta_1 \to \pm\infty} Y_1 = \pm 1$$

and consequently, $Y_1 = \tanh \zeta_1$ yields the following density variation (fig. 1):

(16) $$\rho = \rho_c \left[1 - \eta + \varepsilon_1 \tanh\left(\frac{\xi}{L_1}\right)\right]$$

where $$L_1 \triangleq \frac{L}{\alpha}\ ,\ \eta \triangleq \varepsilon^3 Q^2 \text{ and } \varepsilon_1 \triangleq \varepsilon\alpha,$$

i.e. $$\eta \triangleq \frac{q^2}{2B\rho_c^5}\ ,\quad \varepsilon_1 \triangleq \left\{\frac{A}{B\rho_c^2}\ (T_c - T) + \frac{q^2}{B\rho_c^5}\right\}^{1/2}$$

or $$\varepsilon_1 \triangleq \left\{\varepsilon^2 + \eta\right\}^{1/2}$$

(17) $$\frac{1}{2}\ (\rho_\ell + \rho_v) = \rho_c\,(1 - \eta)$$

The values ρ_ℓ and ρ_v are not symmetrical with respect to ρ_c except for the case of null flow.

(18) $$\rho_\ell - \rho_v = 2\,\varepsilon_1\,\rho_c$$

Because ε_1 increases when q increases, the difference between the densities increases with q. A characteristic thickness of the interface is $L_1 = \frac{L}{\alpha}$, which is thinner than L, and given by:

(19) $$L_1 = 2\left\{\frac{2C}{B}\right\}^{1/2} \frac{1}{(\rho_\ell - \rho_v)}$$

The surface tension σ is given by (Rowlinson and Widom, 1984):

$$\sigma = \int_{\rho_v}^{\rho_\ell} C\ \frac{d\rho}{d\xi}\ d\rho$$

such that

(20) $$\sigma = \frac{1}{6}\ \left(\frac{BC}{2}\right)^{1/2} (\rho_\ell - \rho_v)^3$$

These quantities only depend on $\rho_\ell - \rho_v$ and not explicitely on q. They are the same as in equilibrium. The surface tension is multiplied by α^3 and is stronger than the one in equilibrium case.

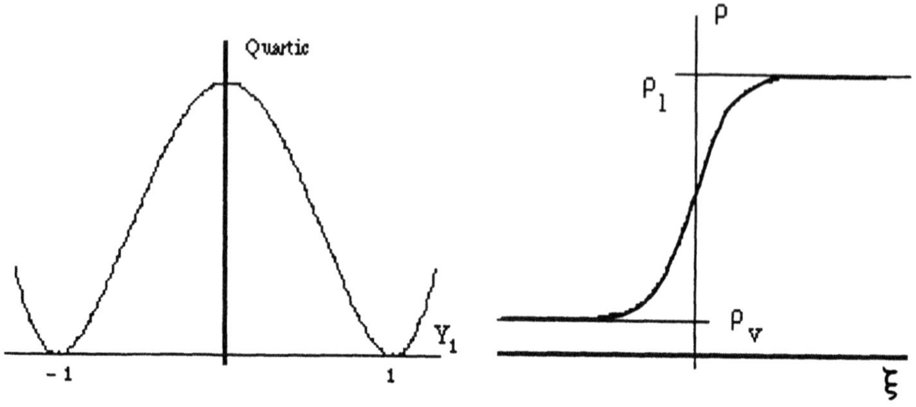

Figure 1

Solitary waves - It is also possible to obtain a solitary wave moving in one of the bulk phases (liquid or vapor). The straight line is tangent to the quartic at the point associated with the bulk $Y_1 = Y_0$ and intersects the quartic at the point associated with the density of the *middle* of the wave $\left(Y_1 = Y_0^{'}\right)$. When Y_1 belongs to the interval between Y_0 and $Y_0^{'}$, the right hand side of equation (15) must be positive. This authorizes only two possibilities:

a) $-1 < Y_0 < -\frac{1}{\sqrt{3}}$ and $Y_0 < Y_0^{'} < 1$: the density increases in the wave (fig. 2).

b) $\frac{1}{\sqrt{3}} < Y_0 < 1$ and $-1 < Y_0^{'} < Y_0$: the density decreases in the wave.

Noting $Y_1 = F(\zeta_1)$ the solution of the differential equation (15). we obtain :

(21) $$\rho = \rho_c \left[1 - \eta + \varepsilon_1 \, F\left(\frac{\xi}{L_1}\right)\right]$$

to be compared with equation (16).
In the bulk phase, the density is $\rho_0 = \rho_c \left[1 - \eta + \varepsilon_1 Y_0\right]$ and the difference between densities in the bulk and the *centre* of the wave is $\rho_0^{'} - \rho_0 = \rho_c \, \varepsilon_1 \, \delta$ with $\delta \triangleq Y_0^{'} - Y_0$ (when Y_0 varies from 1 to $\frac{1}{\sqrt{3}}$, δ varies from 2 to 0).
This solitary wave depends on two arbitrary scalars: the flow q and the value Y_0.

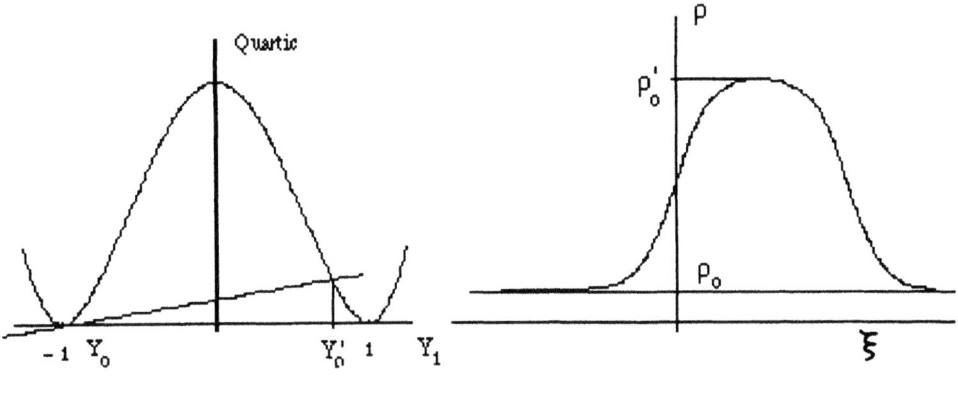

Figure 2

Consequences - The change of variables (11) assumes that $T < T_c$ but obviously, the combination of changes of variables (10), (11) and (14) directly yields equation (15) and only supposes that $(T_c - T) + \dfrac{q^2}{A\,\rho_c^3}$ is positive.

These two kinds of waves exist when the temperature is above T_c *under the restriction that* $(T_c - T) + \dfrac{q^2}{A\,\rho_c^3} > 0$.

For a wave of the first kind and with the van der Waals equation, this was already pointed out in Casal and Gouin (1988). The results obtained here are more general and independent of any type of equation of state.

For $T > T_c$ the flow through an interface must be more important than the limit q_m defined as that:

$$q_m \triangleq \left\{A\,\rho_c^3\,(T - T_c)\right\}^{1/2}$$

As the density in the bulk phase is close to the critical one, the velocity of the wave must be greater than the limiting value $g_m \triangleq \left\{A\,\rho_c\,(T - T_c)\right\}^{1/2}$.

In figure 3, the domain of velocity of material waves is plotted in the vicinity of the critical point of water for a van der Waals fluid. The graph of g_m versus temperature represents a Hopf double-point supercritical bifurcation.

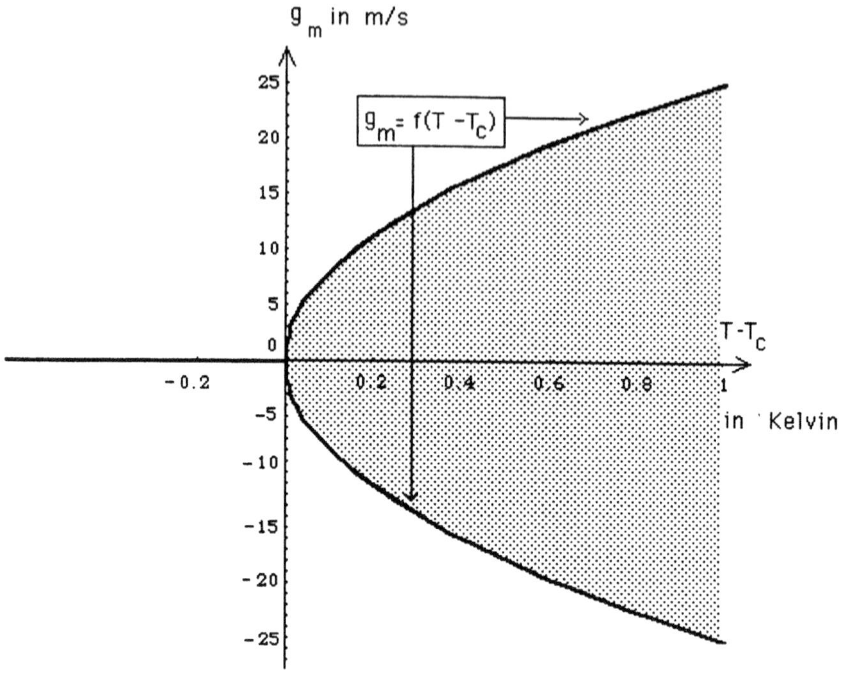

Figure 3

5. FLUID VELOCITY - For a given density, the fluid velocity is deduced at the time $t = 0$ from the mass balance equation (6). An arbitrary constant c is introduced in the velocity field:

$$V(x) = c + \frac{q}{\rho(x)},$$

and at any time t by the relation:

(22) $$V(x - ct) = c + \frac{q}{\rho(x - ct)}$$

The initial conditions yield this arbitrary velocity c.

First case : Interface propagation - By crossing the interface, the fluid changes from liquid to vapor and its volume increases. This phenomenon cannot occur in a closed tube with a constant volume, but only in one with one end closed. It is the *simple* phenomena of vaporization (or condensation).

One imagines that the other side of the tube is closed with a piston whose displacement may be imposed. The fluid velocity at the fixed end is zero (for example in the liquid bulk). This determines the constant c by the condition $c = -\dfrac{q}{\rho_\ell}$.

If we impose a value $\mathcal{V}$ for the velocity of the piston located in the vapor bulk, we obtain

$$\mathcal{V} = q\left(\frac{1}{\rho_v} - \frac{1}{\rho_\ell}\right)$$

By using (17) and (18), it follows a relation that determines the flow q and the velocity c of the "liquid surface."

The velocity $\mathcal{V}$ becomes infinite for a flow q_M associated with a value of η such that $\rho_v = 0$. For $T = T_c$ one deduces $q_M^2 = \left(4-2\sqrt{3}\right)B\,\rho_c^5$ and for the van der Waals equation of state, $q_M^2 = \left(6-3\sqrt{3}\right)P_c\,\rho_c$.

The flow rate can be very high and can reach 10 kg/cm^2s for water. Nevertheless the thermodynamic conditions are far from critical ones in the vicinity of which the calculus is performed.

Second case: Solitary wave . - The *volume of the interface* that moves in only one bulk phase is constant. Such a wave may be moving in a closed tube such as Natterer's one (Bruhat, 1968). At the ends which are assumed far from the interface, the velocity V_0 in the bulk phase is zero and $c = -\dfrac{q}{\rho_0}$. Values of ρ in the bulk phases are imposed on q, except when the temperature is above T_c. In this case, the modulus of q must be above q_m and gives a positive value to ρ_0. It is the case for Y_0 that must verify conditions for solitary waves.

The model of fluid endowed with internal capillarity (Casal and Gouin, 1985) allows travelling waves through a tube where the temperature is close to T_c and depends on two arbitrary parameters.

These waves are different from acoustic waves, as the acoustic waves do not exist for $T < T_c$ in the vicinity of the critical point.

Since the density ρ_0 in the bulk phase is close to the critical value, the velocity c of the wave is close to $\dfrac{q}{\rho_c}$. When the temperature is above the critical point, the velocity must be greater than the limit $c_m = \dfrac{q_m}{\rho_c}$ with

(23) $$c_m = \left[A\,\rho_c\,(T - T_c)\right]^{1/2}$$

In the case of van der Waals fluid, relation (23) yields:

$$c_m = \left[6\frac{P_c}{\rho_c}\frac{(T-T_c)}{T_c}\right]^{1/2}$$

This value corresponds to the velocity of acoustic waves in the vicinity of the critical temperature and above.

6. CONCLUSIONS - The mean field theory is able to predict the existence of solitons in the vicinity of the critical point of a fluid.

The model is useful in interpreting capillary phenomena in both equilibrium and non-equilibrium configurations near the critical point.

Two kinds of material waves were investigated:

(1) liquid-vapor interfaces whose velocity depends on the distance between the temperature and its critical value,

(2) moving solitary wave.

Above the critical temperature, the fluid behaves like an elastic medium and the waves are supersonic with respect to the isothermal sound speed.

Below the critical temperature, the fluid behaves like a nonrigid medium and the wave velocities can take any value.

Thermodynamic functions and parameters in correlation functions vanish or diverge at a critical point proportionally to some power of the distance from that point (often measured as $T - T_c$). Those powers are the critical point exponents, central to any discussion of critical phenomena (Rowlinson and Widom, 1984). It is possible to generalize the form of μ with the critical point exponents and to modify the mean field theory of the near-critical interface, so as to incorporate nonclassical values of critical point exponents of the chemical potential in equation (3).

At least, it would be important if the computations for wave speeds could be shown to correlate favorably or even unfavorably with experimental results for different materials.

REFERENCES

Bruhat, G., 1968, *Thermodynamique*, 6th ed., Masson, Paris, pp.188-192.

Cahn, J.W. and Hilliard, J.E., 1959, Free energy of a non-uniform system III, J. *Chem. Phys.*, vol. 31, pp. 688-699.

Casal, P., 1972, La théorie du second gradient et la capillarité, *Comptes Rendus Acad. Sc. Paris*, vol. 274, pp. 1571-1574.

Casal, P. and Gouin, H., Relation entre l'équation de l'énergie et l'équation du mouvement en théorie de Korteweg de la capillarité, 1985, *Comptes Rendus Acad. Sci. Paris*, vol. 300, pp. 231-234.

Casal, P. and Gouin, H., 1988, Sur les interfaces liquide-vapeur non isothermes, *J. Mécanique Théorique et Appliquée*, vol.7, No 6, pp. 689-718.

De Gennes, P.G., 1981, Some effects of long range forces on interfacial phenomena, *J. de Physique-Lettres*, vol. 42, pp. 377-379.

Dunn, J., 1986, Interstitial working and nonclassical continuum thermodynamics, *New Perspectives in Thermodynamics*, Serrin, J., Ed., Springer-Verlag, pp. 187-221.

Germain, P., 1972, La méthode des puissances virtuelles en mécanique des milieux continus, *J. de Mécanique*, vol. 12, pp. 235-274.

Langevin, D., 1992, *Light scattering by liquid surface*, Marcel Dekker, New York pp. 161-201.

Malomed, B.A. and Rumanov, E.I., 1984, "Intrinsic" velocity of the interfacial boundary, *Dokl. Akad. Nauk SSSR*, vol. 284, pp. 6-12.

Poujet, J., 1990, Non-linear dynamics and instability of twin boundaries in martensitic -ferroelastic materials, *Continuum Models and Discrete Systems*, G.A. Maugin ed. vol 1, pp. 296-312, Longman.

Rocard, Y., 1967, *Thermodynamique*, V, Théorie cinétique, Masson, Paris.

Rowlinson, J.S. and Swinton, F.L., 1982, *Liquid and liquid mixtures*, Butterworth Scientific, London.

Rowlinson, J.S. and Widom, B., 1984, *Molecular theory of capillarity*, Clarendon Press, Oxford.

Slemrod, M., 1983, Admissibility criteria for propagating phase boundaries in a van der Waals fluid, *Arch. Rat. Mech. Anal.*, vol. 81, 4, pp. 301-315.

Truskinovsky, L., 1987, Dynamics of non-equilibrium phase boundaries in a heat conducting non-linearly elastic medium, *P.M.M.*, vol. 51, pp. 777-784.

Widom, B., 1972, Surface tension of fluids, *Phase transitions and Critical Phenomena,* Vol. 2, Domb, C. and Green, M.S., Eds., Academic Press, pp. 79-101.

A MODEL OF A PLUG-CHAIN SYSTEM NEAR THE THERMODYNAMIC CRITICAL POINT: CONNECTION WITH THE KORTEWEG THEORY OF CAPILLARITY AND MODULATION EQUATIONS

Sergey L. GAVRILYUK [1] & Denis SERRE [2]
Lavrentyev Institute of Hydrodynamics, Novosibirsk, 630090, RUSSIA
and
UMPA CNRS UMR 128, Ecole Normale Supérieure de Lyon
69364 LYON Cedex 07, FRANCE

Abstract. We consider a plug flow of a gas-liquid mixture in a pipe of constant cross-section. We assume that the liquid and gas are distinct and non reacting chemicals, and we consider the case when the gas is near the thermodynamic critical point. Pertinent nonlinear models of the flow pattern are derived: a discrete plug-chain model and its long wave approximation. Explicit solutions to both models are obtained.

A mechanical analogy in the discrete case is derived which represents an infinite chain of mass points moving in the transversal direction only.

It is found that the model for the long -wave approximation of a plug-chain system can be considered as a special example of the Korteweg theory of capillarity.

Modulation equations of the Korteweg theory of capillarity along with "pseudothermodynamics" are obtained. It is found that the modulated quantities are described by Godunov's system of quasilinear partial differential equations.

The "pseudotemperature" of the plug flow pattern as a function of other "thermodynamic parameters" is obtained in an explicit form.

Introduction. When a gas-liquid mixture moves in a pipe, a flow pattern occurs in which we have an intermittency of liquid plugs and of big bubbles. This flow pattern is frequently called "plug flow" or "plug-chain" system. We derive a mathematical model of plug flow in section 1, which represents an infinite discrete set of ordinary differential equations.

In section 2 a mechanical analogy of such a flow pattern is found.

Section 3 gives explicit solutions of the discrete model.

Sections 4 and 5 are devoted to long-wave approximation of a plug-flow pattern and its travelling wave solutions.

In section 6 we find that the model for the long-wave approximation of a plug-chain system can be treated as a special example of the Korteweg theory of capillarity.

In section 7 we derive modulation equations for the Korteweg theory of capillarity. We obtain a closed system of four equations in a conservative form and find two additional conservation laws. This fact allows one to introduce "the thermodynamics" of the Korteweg theory of capillarity and transform the modulation equations to a

[1] This work was done while S.L.G. was a visitor at Ecole Normale Supérieure de Lyon, FRANCE
[2] Member of the Institut Universitaire de France

S. Morioka and L. van Wijngaarden (eds.),
IUTAM Symposium on Waves in Liquid/Gas and Liquid/Vapour Two-Phase Systems, 419-428.

symmetrical form of Godunov's type. A sufficient condition for hyperbolicity of the system is obtained in terms of "thermodynamic variables".

In section 8 we introduce new "thermodynamic variables" and obtain an "equation of state" of plug - flow pattern in an explicit form.

1. Governing equations. Let us consider a plug flow of a gas-liquid mixture in a pipe of constant cross-section σ (Fig.1).

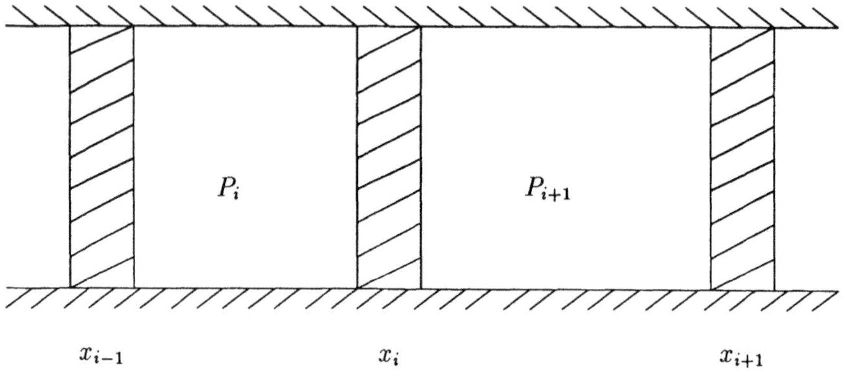

Fig.1.

We assume that the liquid and gas are different and nonreacting chemicals, the liquid plugs are incompressible and have the same mass M. The positions of the mass centre of liquid plug will be denoted by x_i, with p_i and p_{i+1} being the pressures in gas plugs from the left-and right sides of x_i, respectively. The equations of motion take the form:

$$M\frac{d^2x_i}{dt^2} = \sigma(p_i - p_{i+1}), i = 0, \pm1, \pm2, ... \tag{1.1}$$

The equations (1.1) were first proposed by G. Matsui *et al.* [1] as a simple model for a plug regime of a two phase flow. Let us consider the case when the gas is near the critical point. Assuming that the gas flow is isentropic we have:

$$p(V,S) = p_c + \frac{1}{6}p_{VVV}(V_c, S_c)(V - V_c)^3 + O((V - V_c)^4).$$

Here V and S are the volume and entropy of the gas, respectively, and the index "c" denotes the critical state. With $B = -p_{VVV}(V_c, S_c)/6 > 0$ we obtain

$$p_i \approx p_c - B\sigma^3(x_i - x_{i-1} - L_1 - L)^3.$$

Here L_1 is the length of a liquid plug, and L is the equilibrium length of a gas plug. We introduce the new variables q_i by the relation $x_i = q_i + i(L + L_1)$ and write the equations (1.1) as follows

$$\frac{d^2q_i}{dt^2} = \frac{B\sigma^4}{M}\left((q_{i-1} - q_i)^3 - (q_i - q_{i+1})^3\right).$$

In dimensionless variables the equations of motion take the form:

$$\frac{d^2q_i}{dt^2}=(q_{i-1}-q_i)^3-(q_i-q_{i+1})^3, i=0,\pm1,\pm2,... \tag{1.2}$$

The right-hand side of the equations (1.2) contains no linear terms and the situation is quite untypical for mechanical systems.

2. Mechanical analogy. V.F. Nesterenko [2] considered an infinite chain of mass points placed on a massless linear-elastic thread with a cross-section Σ, the Young modulus E, a distance l_0 between the nearest neighbours in an equilibrium state, which move only in a transversal direction (see Fig.2).

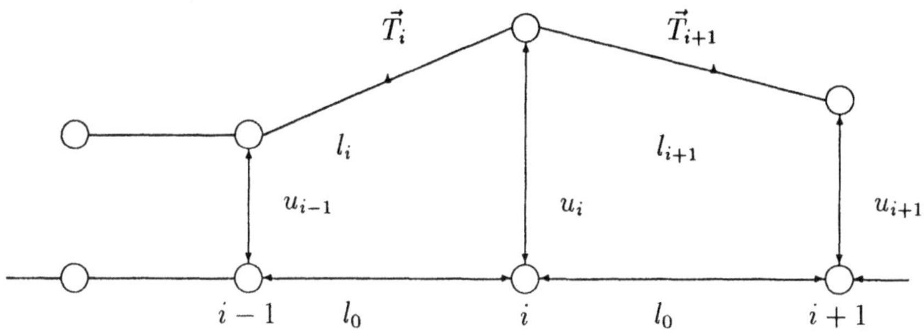

Fig.2.

We shall denote by the displacement from an equilibrium of the i-th mass point by u_i (Fig.2). Let l_i be the distance between the $i-1$ -th and the i-th mass point. Assuming that the tension force T of the thread varies linearly with a change of a lengh we obtain:

$$|\vec{T}_i|=E\Sigma\frac{l_i-l_0}{l_0}\approx\frac{E\Sigma}{2l_0^2}(u_i-u_{i-1})^2,\quad |\vec{T}_{i+1}|\approx\frac{E\Sigma}{2l_0^2}(u_{i+1}-u_i)^2.$$

Hence, the equations of motion take the form:

$$\frac{d^2u_i}{dt^2}=\frac{E\Sigma}{2l_0^3m}\left((u_{i-1}-u_i)^3-(u_i-u_{i+1})^3\right).$$

Performing an obvious change of variables, we obtain the equations (1.2). It is worth noting that the equations (1.2) can be written in canonical form. The Hamiltonian of the system (1.2) is

$$H=\sum_i\left(\frac{\dot{q}_i^2}{2}+\frac{(q_i-q_{i+1})^4}{4}\right). \tag{2.1}$$

3. Explicit solutions. Explicit solutions of an infinite chain can be casted in the form $q_i=g(i)f(t)$. Hence, we get

$$\ddot{f}=cf^3,\quad (g(i-1)-g(i))^3-(g(i)-g(i+1))^3=cg(i) \tag{3.1}$$

where c is a constant. The first equation of the system (3.1) has the integral:

$$\frac{\dot{f}^2}{2} - c\frac{f^4}{4} = h$$

where h is some constant. The functional equation for $g(i)$ has a special nontrivial solution of the form $g(i) = \alpha + i\beta + i^2\gamma$ where constants α, β, γ must satisfy the conditions

$$2\gamma^3 + 6\beta^2\gamma = c\alpha; \quad 24\beta\gamma^2 = c\beta; \quad 24\gamma^3 = c\gamma.$$

If $c = 0$ then $\gamma = 0$ and $f = A_1 + B_1 t$, $g(i) = \alpha + i\beta$ where A_1, B_1 are constants. If $c \neq 0$ then $c > 0$ and

$$\gamma = \pm\left(\frac{c}{24}\right)^{1/2}, \quad \alpha = \gamma\left(\frac{1}{12} + \frac{6\beta^2}{c}\right).$$

Thus, the desired solution is obtained.

Example. If we put $h = 0$, we arrive at the formula

$$f(t) = \pm\frac{f_0}{1 + f_0 t\sqrt{c/2}},$$

where f_0 is a constant. If $f_0 > 0$, we get a solution describing a relaxation to the equilibrium state.

4. Long-wave approximation. If $q_i(t)$ varies slowly with i, we may replace the discrete labeling q_i by a continuously varying parameter $q(x,t) \approx q(il_0, t)$. The potential energy of the particles (see the formula (2.1)) can be written in the form:

$$\frac{(q_{i+1} - q_i)^4}{4} = \frac{l_0^4}{4}\left(q_x^4 + 2l_0 q_x^3 q_{xx} + l_0^2\left(\frac{2}{3}q_x^3 q_{xxx} + \frac{3}{2}q_x^2 q_{xx}^2\right)\right) + O(l_0^7).$$

Hence, the Euler equation for the variational principle

$$\delta \int\int L dt dx = 0, \quad L = \frac{q_t^2}{2} - \frac{l_0^4}{4}\left(q_x^4 + l_0^2\left(\frac{2}{3}q_x^3 q_{xxx} + \frac{3}{2}q_x^2 q_{xx}^2\right)\right).$$

gives

$$q_{tt} = l_0^4\left(q_x^3 + \frac{l_0^2}{4}(q_x q_{xx}^2 + q_x^2 q_{xxx})\right)_x.$$

Changing the variables $x \longrightarrow l_0 x/\mu$, $t \longrightarrow t/\mu^2$ we transform our equation into the final form

$$q_{tt} = \left(q_x^3 + \frac{\mu^2}{4}(q_x q_{xx}^2 + q_x^2 q_{xxx})\right)_x. \tag{4.1}$$

Here a small parameter μ describes the accuracy of the long-wave approximation.

5. Travelling wave solutions. With $y = (x - Dt)/\mu$, $\xi = q_y$ we get from the equation (4.1):

$$\xi^3 + \frac{\xi(\xi^2)''}{8} - D^2\xi = B, \tag{5.1}$$

and $D(= const)$ is the wave velocity, the prime denotes the differentiation with respect to y, B is a constant. Multiplying the equation (5.1) by ξ' we get

$$\frac{\xi^4}{4} + \frac{\xi^2(\xi')^2}{8} - \frac{D^2\xi^2}{2} - \xi B = H \tag{5.2}$$

where H is a constant. The solution of (5.2) can be expressed in terms of elliptic integrals. In some region of parameters D, B, H the equation (5.2) admits periodic or solitary solutions.

6. Korteweg theory of capillarity. The general theory of a continuous medium, wherein the internal energy depends on not only thermodynamic parameters of medium but on their derivatives with respect to a spatial variable as well, is called the Korteweg theory of capillarity. It was further developed in the works by P.Casal & H. Gouin[3],[4]. The representative field of application for this theory is the dynamics of the liquid-vapour phase transitions (M. Slemrod [5], H. Gouin [6] and references therein). The simplest model of a medium with capillarity takes the form

$$v_t - u_x = 0, \quad u_t - \tau_x = 0, \quad \tau \equiv \frac{\delta e}{\delta v} = \frac{\partial e}{\partial v} - \frac{\partial}{\partial x}\left(\frac{\partial e}{\partial v_x}\right). \tag{6.1}$$

Here t is time, x is the mass Lagrangian coordinate, v is the specific volume, u is velocity, $e = e(v, v_x)$ is the specific internal energy. In the theory of liquid-vapour phase transitions, the energy e takes the form (M. Slemrod [5]):

$$e = -\int^v p(w)dw + \frac{C}{2}v_x^2, \quad p(v) = \frac{RT}{v-b} - \frac{a}{v^2}$$

where C, R, T, b, a are positive constants. In our case, the equation (4.1) is reduced to the system (6.1) provided that

$$q_t = u, \quad q_x = v, \quad e(v, v_x) = \frac{v^4}{4} - \frac{\mu^2 v^2 v_x^2}{8}. \tag{6.2}$$

Note that the application field of the Korteweg theory is very wide. For instance, the equations of "shallow water" in the Boussinesq approximation can also be rewritten as (6.1). Indeed, the desired equations have the form (G.B. Whitham [7]):

$$h_t + (hu)_z = 0, \quad u_t + uu_z + gh_z + c_0^2 h_0 h_{zzz}/3 = 0. \tag{6.3}$$

Here t is time, z is the Euler coordinate, h is the depth of the fluid, g is gravity, h_0 is the unperturbed depth, and $c_0^2 = gh_0$. Introducing the mass Lagrangian coordinate x as $z_x = h^{-1} \equiv v$, and the energy as $e(v, v_x) = g/2v - c_0^2 h_0 v_x^2/6v^5$, the equations (6.3) take the form of (6.1).

It is straightforward to check that, *for any function* $e = e(v, v_x)$, the system (6.1) admits two additional conservation laws:

$$\left(\frac{u^2}{2}+e\right)_t - \left(u\frac{\delta e}{\delta v} + e_{v_x}u_x\right)_x = 0, \quad (vu)_t - \left(\frac{u^2}{2} + v\frac{\delta e}{\delta v} + v_x e_{v_x} - e\right)_x = 0. \quad (6.4)$$

The former is the equation for conservation of energy.

7. Modulation equations for the Korteweg theory of capillarity. The equations (6.1) are the Euler equations for a variational principle and the variational approach to the modulation theory can be used (G.B. Whitham [7]). But we shall apply the usual perturbation method substituting suitable expansions directly into the differential equations. The appropriate theory and numerous references can be found in the book by V.P. Maslov [8] (applications to fluid mechanics, nonlinear optics etc.) and in the paper by D. Serre [9] (applications to nonlinear hyperbolic systems).

Let us introduce a small parameter δ measuring the ratio of a wave period to a typical length of modulation and we seek a solution of the equations (6.1) of the form

$$(v,u) = (v_0,u_0)(T,X,\theta) + \delta(v_1,u_1)(T,X,\theta,\delta), \quad (v_1,u_1) = O(1)$$

where $T = \delta t$, $X = \delta x$, $\theta = S(T,X)/\delta$ and (v_i,u_i), $i = 0,1$ depend on θ periodically. The period can be taken as 2π with no loss of generality. A fast variable θ is called the phase, $k = S_X$, $\omega = -S_T$, $D = \omega/k$, are called the local wave number, the local frequency and the phase velocity,respectively. Using the derivative formulae

$$\frac{\partial}{\partial t} \longrightarrow -\omega\frac{\partial}{\partial\theta}+\delta\frac{\partial}{\partial T}, \quad \frac{\partial}{\partial x} \longrightarrow k\frac{\partial}{\partial\theta}+\delta\frac{\partial}{\partial X}, \quad \frac{\partial^2}{\partial x^2} \longrightarrow k^2\frac{\partial^2}{\partial\theta^2}+2k\delta\frac{\partial^2}{\partial X\partial\theta}+\delta^2\frac{\partial^2}{\partial X^2}, \ldots$$

and the periodicity of the solution with respect to θ, we obtain to the lowest order with respect to δ (subscript 0 is dropped on dependent variables):

$$(Dv+u)_\theta = 0, \quad (Du+\tau)_\theta = 0, \quad \tau = e_v - k\frac{d}{d\theta}e_\eta, \quad e = e(v,\eta), \ \eta = kv_\theta,$$

or, equivalently

$$Dv + u = D\langle v\rangle + \langle u\rangle, \quad Du + \tau = D\langle u\rangle + \langle\tau\rangle, \tag{7.1}$$

$$\tau = e_v - k\frac{d}{d\theta}e_\eta, \quad e = e(v,\eta), \quad \eta = kv_\theta. \tag{7.2}$$

Here, for an arbitrary 2π-periodic function $f(T,X,\theta)$ we have defined the average value

$$\langle f\rangle = \frac{1}{2\pi}\int_0^{2\pi} f(T,X,\theta)d\theta.$$

The equations (7.1) and (7.2) admit the first integral

$$e - \eta e_\eta - \langle\tau\rangle(v - \langle v\rangle) - \frac{D^2}{2}(v - \langle v\rangle)^2 \tag{7.3}$$

$$= \langle e + u^2/2\rangle - \frac{\langle u\rangle^2}{2} + D^2(\langle v\rangle^2 - \langle v^2\rangle) - k\langle v_\theta e_\eta\rangle.$$

Since the equations (6.4) follow from the equations (6.1), we obtain no other independent equations to the lowest order with respect to δ. Upon averaging of (6.1) and (6.4), we obtain to the first order with respect to δ the following equations

$$\langle v\rangle_T - \langle u\rangle_X = 0, \tag{7.4}$$

$$\langle u\rangle_T - \langle \tau\rangle_X = 0, \tag{7.5}$$

$$\left\langle \frac{u^2}{2} + e\right\rangle_T - \left\langle u(e_v - k\frac{d}{d\theta}e_\eta) + e_\eta u_\theta k\right\rangle_X = 0, \tag{7.6}$$

$$\langle vu\rangle_T - \left\langle \frac{u^2}{2} + v(e_v - k\frac{d}{d\theta}e_\eta) + e_\eta v_\theta k - e\right\rangle_X = 0. \tag{7.7}$$

The equations (7.6) and (7.7) are now independent of (7.4) and (7.5), and the problem is to obtain the complete system of modulation equations.

We introduce now variables:

$$V = \langle v\rangle, \quad U = \langle u\rangle, \quad \Delta = \langle uv\rangle - UV, \quad \epsilon = \langle e + \frac{u^2}{2}\rangle - \frac{U^2}{2}. \tag{7.8}$$

Recalling (7.1) and (7.2), it is straightforward to show that

$$\langle v_\theta e_\eta\rangle dk = d\epsilon - \langle\tau\rangle dV + Dd\Delta, \tag{7.9}$$

The expression (7.9) is reminiscent of the first law of thermodynamics with temperature $\Theta = \langle v_\theta e_\eta\rangle$, entropy k, the specific energy ϵ, the pressure $-\langle\tau\rangle$, the specific volume V. The term $Dd\Delta$ has no physical interpretation. With $\epsilon = \epsilon(V, k, \Delta)$ we obtain from (7.9):

$$\Theta = \frac{\partial\epsilon}{\partial k}, \quad \langle\tau\rangle = \frac{\partial\epsilon}{\partial V}, \quad D = -\frac{\partial\epsilon}{\partial\Delta}. \tag{7.10}$$

The identity like (7.9) was obtained also by D. Serre [10] (with D = 0) for viscous capillary fluids, and by S.L. Gavrilyuk [11],[12] for bubbly liquids (liquids containing small gas bubbles).

The formulae (7.1), (7.2) and (7.8) give us:

$$\langle uv\rangle = UV + D(V^2 - \langle v^2\rangle), \quad \langle u\tau\rangle = U\langle\tau\rangle + D^2\Delta, \quad \langle u^2\rangle = U^2 - D\Delta, \tag{7.11}$$

$$\langle\tau v\rangle = \langle\tau\rangle V - D\Delta, \quad \langle e_\eta u_\theta\rangle = -D\langle e_\eta v_\theta\rangle \equiv -D\Theta.$$

The final system of governing equations in a conservative form follows from (7.4) - (7.11):

$$V_T - U_X = 0, \quad U_T - \langle\tau\rangle_X = 0, \tag{7.12}$$

$$(\epsilon + \frac{U^2}{2})_T - (U\langle\tau\rangle + D^2\Delta - kD\Theta)_X = 0,$$

$$(\Delta + UV)_T - (V\langle\tau\rangle + \frac{U^2}{2} - 2D\Delta + k\Theta - \epsilon)_X = 0,$$

$$\epsilon = \epsilon(V, k, \Delta), \quad \Theta = \frac{\partial\epsilon}{\partial k}, \quad \langle\tau\rangle = \frac{\partial\epsilon}{\partial V}, \quad D = -\frac{\partial\epsilon}{\partial\Delta}.$$

A nondivergent form with U, V, ϵ and Δ as desired variables reads

$$V_T - U_X = 0, \quad U_T + P_X = 0, \quad \epsilon_T + PU_X - (D^2\Delta - kD\Theta)_X = 0, \tag{7.13}$$

$$\Delta_T + PV_X - (-2D\Delta + k\Theta - \epsilon)_X = 0.$$

Here $P \equiv -\langle\tau\rangle$, D, k, and Θ should be considered as functions of V, ϵ and Δ.

Straightforward calculations show that there are at least two additional conservation laws of the system (7.12) (or (7.13)):

$$k_T - \left(k\frac{\partial\epsilon}{\partial\Delta}\right)_X = 0, \tag{7.14}$$

$$\left(\frac{\Delta}{k}\right)_T - \left(\frac{\Delta}{k}\frac{\partial\epsilon}{\partial\Delta} + \frac{\partial\epsilon}{\partial k}\right)_X = 0 \tag{7.15}$$

The equation (7.14) has a simple sense. It is the consistency relation obtained by eliminating the phase S:

$$k = S_X, \quad \omega = -S_T, \quad D = \frac{\omega}{k} = -\frac{\partial\epsilon}{\partial\Delta}.$$

To obtain a symmetric form of the governing equations, we make the change of variables:

$$(V, k, \Delta) \longrightarrow (V, k, Z) \equiv (V, k, \frac{\Delta}{k}). \tag{7.16}$$

Considering the function

$$\tilde{\epsilon}(V, k, Z) = \epsilon(v, k, Zk) \equiv \epsilon(V, k, \Delta) \tag{7.17}$$

we get

$$\frac{\partial\tilde{\epsilon}}{\partial V} = \frac{\partial\epsilon}{\partial V}, \quad \frac{\partial\tilde{\epsilon}}{\partial k} = \frac{\partial\epsilon}{\partial k} + Z\frac{\partial\epsilon}{\partial\Delta}, \quad \frac{\partial\tilde{\epsilon}}{\partial Z} = k\frac{\partial\epsilon}{\partial\Delta}. \tag{7.18}$$

It follows from (7.13) - (7.18) that the governing equations can be written in the form:

$$V_T - U_X = 0, \quad U_T - \left(\frac{\partial \tilde{\epsilon}}{\partial V}\right)_X = 0, \quad k_T - \left(\frac{\partial \tilde{\epsilon}}{\partial Z}\right)_X = 0, \quad Z_T - \left(\frac{\partial \tilde{\epsilon}}{\partial k}\right)_X = 0. \quad (7.19)$$

If we introduce the variables $q_1 = U, q_2 = V, q_3 = k, q_4 = Z$ and functions $L = q_1 q_2 + q_3 q_4$, $E = q_1^2/2 + \tilde{\epsilon}(q_2, q_3, q_4)$, we obtain Godunov's system (see S.K.Godunov [13])

$$\left(\frac{\partial L}{\partial q_i}\right)_T - \left(\frac{\partial E}{\partial q_i}\right)_X = 0.$$

Hence, a sufficient condition for hyperbolicity of (7.19) is the convexity of the function $\tilde{\epsilon}(q_2, q_3, q_4)$.

It is interesting to note that a system like (7.19) appears also for a special case of nonlinear Maxwell's equations (see D.Serre [14]).

8. An explicit formula for the "temperature". According to (6.2), in the case of plug flow, the energy $e(v, v_x) = v^4/4 - \mu^2 v^2 v_x^2/8$. Hence, taking the definitions (7.8) into account, we obtain from (7.3), (7.9) that

$$\frac{v^4}{4} + \frac{\mu^2 v^2 \eta^2}{8} + P(v - V) - \frac{D^2}{2}(v - V)^2 = \epsilon + D\Delta - k\Theta, \tag{8.1}$$

$$\Theta dk = d\epsilon + PdV + Dd\Delta. \tag{8.2}$$

It is straightforward to check that the identity (8.2) is invariant with respect to transformations

$$\bar{P} = P + D^2 V, \quad \bar{\epsilon} = \epsilon + \frac{D^2 V^2}{2}, \quad \bar{\Delta} = \Delta - DV^2. \tag{8.3}$$

This means that $\Theta dk = d\bar{\epsilon} + \bar{P}dV + Dd\bar{\Delta}$. In terms of (8.3) the equation (8.1) takes the form

$$\frac{\mu^2 v^2 k^2 v_\theta^2}{8} = \bar{G} - \bar{P}v + \frac{D^2 v^2}{2} - \frac{v^4}{4} \equiv \Phi(v; \bar{G}, \bar{P}, D). \tag{8.4}$$

Here we have introduced the "thermodynamic Gibbs potential"

$$\bar{G} \equiv \bar{\epsilon} + \bar{P}V + D\bar{\Delta} - k\Theta, \quad d\bar{G} = Vd\bar{P} + \bar{\Delta}dD - kd\Theta. \tag{8.5}$$

It follows from (8.4) that

$$\frac{\mu k}{\sqrt{2}} \int_{v_1}^{v_2} \frac{v dv}{\sqrt{\Phi}} = 2\pi$$

where $v_i(\bar{G}, \bar{P}, D)$, $i = 1, 2$, $0 < v_1 < v_2$ are the roots of the equation $\Phi = 0$: $\Phi(v_1) = \Phi(v_2) = 0$. It immediately follows that

$$k = \left(\frac{2\pi\sqrt{2}}{\mu}\right) \Big/ \left(\int_{v_1}^{v_2} \frac{v dv}{\sqrt{\Phi}}\right) \equiv -\bar{G}_\Theta. \tag{8.6}$$

Alternatively, in just the same way we find

$$V = V(\bar{G}, \bar{P}, D) \equiv \bar{G}_{\bar{P}}, \quad \bar{\Delta} = \bar{\Delta}(\bar{G}, \bar{P}, D) = \bar{G}_D. \tag{8.7}$$

(8.6) and (8.7) can be considered as an overdetermined system for $\bar{G}$ as a function of Θ, $\bar{P}$, D. It can be checked that compatibility conditions are justified identically. Moreover, we find that the "thermodynamic identity" (8.5) and the equations (8.6), (8.7) are fulfilled with an explicit formula for the "temperature":

$$\Theta = \Theta(\bar{G}, \bar{P}, D) = -\frac{\mu}{\pi\sqrt{2}} \int_{v_1}^{v_2} v\sqrt{\Phi}dv.$$

We stress that the "temperature" of the plug-flow pattern is negative because of concavity of the function $e(v, v_x)$ with respect to v_x (see the formula (6.2)).

Acknowledgments The authors are grateful to V.F.Nesterenko, Lavrentyev Institute of Hydrodynamics for helpful discussions, and H.H. Rugh, Niels Bohr Institute, who read this manuscript. S.L.G. was supported by CNRS under Contract N A69/33/93 and in part by Russian Foundation of Fundamental Research under grant N 93–02–14880.

REFERENCES

[1] G. Matsui, M. Sugihara & S. Arimoto, Propagation characteristics of pressure waves through gas–liquid plug–train systems, Bull. of JSME 22 (1979) 1562 - 1569.

[2] V.F. Nesterenko, Examples of "sonic vacuum", Fiz. Gor. Vzr. 2 (1993) 132 - 134.

[3] P. Casal & H. Gouin, Equations du mouvement des fluides thermocapillaires, C.R. Acad. Sci. Paris,Série II, 306 (1988) 99–104, 1988.

[4] P. Casal & H. Gouin, Invariance properties of inviscid fluids of grade N. In book: PDEs and Continuum Models of Phase Transitions, Lecture Notes in Physics 344, M. Rascle, D. Serre & M. Slemrod, eds.(Springer - Verlag, 1989)

[5] M. Slemrod, The viscosity-capillarity approach to phase transitions. In book: PDEs and Continuum models of Phase Transitions, Lecture Notes in Physics 344, M. Rascle, D.Serre & M. Slemrod, eds. (Springer - Verlag, 1989)

[6] H. Gouin, Ondes matérielles dans les interfaces fluides au voisinage du point critique, C.R. Acad. Sci. Paris, Série II 317 (1993) 1263–1268.

[7] G.B. Whitham, Linear and nonlinear waves (John Wiley & Sons, 1974)

[8] V.P. Maslov, Asymptotic Methods of Solving of Pseudodifferential Equations (Nauka, Moskva, 1987)

[9] D.Serre, Oscillations non linéaires des systèmes hyperboliques: méthodes et résultats qualitatifs, Ann. Inst. Henry Poincaré 8 (1991) 351–417.

[10] D. Serre, Entropie de mélange liquide–vapeur d'un fluide thermocapillaire. Arch. Rat. Mech. Anal. (1994) (to appear).

[11] S.L. Gavrilyuk, Modulation equations for a mixture of gas bubbles in an incompressible liquid, Zh. Prikl. Mekh. Tekh. Fiz. 2 (1989) 86 - 92.

[12] S.L. Gavrilyuk, Large amplitude oscillations and their "thermodynamics" for continua with "memory", Eur. Journal of Mechanics, Fluids B (1994) (to appear).

[13] S.K. Godunov, An interesting class of quasilinear systems, Dokl. Acad. Nauk SSSR 139 (1961) 521 - 523.

[14] D.Serre, Les ondes planes en electromagnetisme non linéare, Physica D 31(1988) 227–251.

Behaviour of a Liquid under Dynamic Loading

V.K.Kedrinskii, A.R.Berngardt, N.N.Chernobaev
Lavrentyev Institute of Hydrodynamics, 630090 Novosibirsk, RUSSIA

Abstract. The paper presents some results of experimental and numerical studies of the late stages of the process of liquid destruction under explosive loading. The problems of relaxation of tensile stresses in a cavitation zone excited by intense rarefaction waves, and the features of the mass velocity distribution, as well as computer processing of the X-ray images of the internal structure of the opaque cavitation zone, are considered.

Key words: Pulse loading, cavitation destruction, two-phase liquid, x-ray pulse technique.

1. Introduction

The dynamics of a real liquid with a free surface under explosive loading is characterized by the essential transformation of its state into the completely new one of a two-phase system, due to the cavitation effects arising behind the front of rarefaction waves. The development of this process leads to the so-called "cavitation destruction" of a medium. Its main features are determined by the intense unlimited development of bubbles in a cavitation zone arising at the nuclei whose role is played by the microinhomogeneities which always exist in real liquids. The different types of microinhomogeneities with their own stabilization mechanisms may provide a wide range of size distribution from nanometers up to tens of microns. Experimental data of state of a real liquid (number density of nuclei N_o , their volume concentration k_o and radius R_0) are given below

Gavrilov L.R.(*)	(1970)	$R_o \simeq 50 - 0.5\mu m$	$N_o \simeq 1\ cm^{-3}$	$k_o \simeq 10^{-8} - 10^{-10}$
Hammitt F.G. et al.	(1974)	6 – 3	1 – 100	
Besov A.S. et al.	(1984)	1.5	$10^3 - 10^4$ ($10^5 - 10^6$)	$< 10^{-7}$

*The range of nuclei size variation in settling a specimen of fresh tap water for several hours is given. The parameters shown include also the experimental data of M.Strasberg.

Practically all results on density N_o concern only gas nuclei. The data in brackets take account of microinhomogeneities of any character, including solid nuclei and their combinations with gas nuclei, at which the vapor-gas bubbles may develop under the action of tensile stresses (Kedrinskii V.K., 1993, review).

At the late stage of cavitation development, a "boiling" liquid during inertial expansion is transformed into a foam structure and then into a gas-drop system. Naturally, in each particular case the duration of one or other stage of the destruction process may differ and depend significantly on the loading dynamics. Nevertheless, on the basis of experimental and numerical studies (Kedrinskii V.K., 1993, review) the following characteristic times of a process may be noted: of the order of microseconds for relaxation of tensile stresses; tens of microseconds for development of dense cavitation clusters; hundreds of microseconds for formation of a foam structure; of the order of milliseconds for its disintegration into cavitating fragments and liquid drops.

Each of these stages represents an independent direction, in principle, in explosion hydrodynamics and at the same time is part of the integral process of cavitation destruction. Therefore, knowledge of the mechanisms responsible for their development is of principal importance. The results of experimental and numerical studies of the late stages of the process will be given below.

S. Morioka and L. van Wijngaarden (eds.),
IUTAM Symposium on Waves in Liquid/Gas and Liquid/Vapour Two-Phase Systems, 429-438.

2. Relaxation of tensile stresses

As it was noted in a previous paper (Kedrinskii V.K., 1993) the problem of destruction on the whole is a new direction in explosion hydrodynamics connected, in particular, with the concepts of the critical tensile stresses and strength of liquid.

It is known that the theoretical value of this parameter is estimated as 10^3 MPa (Frenkel' Ya, 1945). The experimentally obtained values of liquid strength for pulse load vary over a wide range of values, from 1 MPa (Wilson D. et al., 1975, underwater explosions) up to 60 MPa (Carlson G. and Henry K., 1973, pulse electron beam). This scatter turned out to be explained within the framework of the two-phase model of a real liquid (Kedrinskii V.K., 1975) which, as mentioned above, contains the microinhomogeneities playing the role of a second phase.

The suggested approach made it possible to resolve a number of principal questions, and to define more accurately the concept of maximum tensile stresses allowed by a cavitating liquid. It is known that the stress in the rarefaction wave front does not change abruptly and peaks after a finite time interval Δt which may be defined as the front steepness. Solution of the axisymmetric problem of the cavitation zone development in underwater explosion near the free surface shows that this fact is essential in determining the maximum stress (Kedrinskii V.K., 1975). Even during such a short time, the initial value of the volume concentration of the vapor-gas phase k_o was proved to increase by several orders of magnitude, significantly changing the medium state and the applied stress field. As a result, the maximum negative pressure amplitude in cavitating liquid can be up to an order of magnitude lower than that of an ideal one-phase model.

Analytical estimation. In the previous paper (Kedrinskii V.K., 1989) the relaxation time (maximum amplitude decreases by a factor of e) and the influence of front steepness on the amplitude of tensile stresses p_{max} were estimated analytically within the framework of a simulation statement - the development of the cavitation process near the bottom of a tube filled with water and accelerated vertically downward due to an impact :

$$\tau \simeq 14.3(k_o\, R_o^2\, /\, a_o^2)^{1/4}\ , \qquad (1)$$

$$p_{max} \simeq\ -0.77\ (a_o^2\, R_o^3\, /\, k_o\, \Delta t)^{2/5}\ , \qquad (2)$$

where R_o is the initial radius of gas nuclei, a_o is the acceleration, Δt is the steepness of the acceleration profile front. The pressure and acceleration at the tube bottom are connected by the following relationship:

$$p \simeq -\rho_o |a(t)| / \alpha k^{1/6}\ . \qquad (3)$$

Here $\alpha = \sqrt{3k_o/R_o^2}$, $k = (R_o/R)^{-3}$. For example, if $k_o = 10^{-11}$, $R_o = 0.5\ \mu m$, $\Delta t = 0$, $a_o = 3.2810^7\ cm/s^2$, and $a(t) = const$ from (3,1), we have $p_{max} \simeq -30\ MPa$ and $\tau = 0.03\ \mu s$. At $\Delta t = 1\ \mu s$ the expression (2) gives $p_{max} \simeq -3.4\ MPa$.

On the basis of the result obtained, the model of instantaneous relaxation of the stresses in the cavitation zone was developed (Chernobaev N.N, 1989, Kedrinskii V.K., Chernobaev N.N. 1992). According to this model, when the stresses decrease to the saturation vapors pressure p_s, the liquid acquires them during the whole time of a cavitation zone development up to the formation of a foam type structure. In this case the known restrictions of a two-phase model on the value of the volume concentration of gas-vapor phase k are removed, and this value begins to determine the state of the two-phase system: $\rho = (1-k)\rho_l$.

The problem of explosive loading of a liquid volume. Let a charge of HE having an initial radius r_o be surrounded by a liquid shell of radius r_1 with natural content of microinhomogeneities like free gas microbubbles of radius $R_o = 1.5\ \mu m$ with density $N_o = 10^4\ cm^{-3}$ (volume concentration $k_o = 10^{-7}$). It is assumed that when the explosion is initiated the detonation wave achieves the interface "charge-shell" at the instant of time t=0. The value of k_o is too negligible to affect the shock wave propagation before it is reflected from the shell free surface. The cavitation development in the rarefaction wave is calculated following the model of instantaneous relaxation of tensile stress.

The system of equations for calculating the shock wave in a liquid in Lagrangian mass coordinates is of the following form:

$$\frac{\partial r}{\partial t} = u \ ,$$

$$\frac{\partial v}{\partial t} = v_o \xi^{1-\nu} \frac{\partial (r^{\nu-1} u)}{\partial \xi} \ ,$$

$$\frac{\partial u}{\partial t} = -v_o (r/\xi)^{(\nu-1)} \frac{\partial (p+q)}{\partial \xi} \ ,$$

$$\frac{\partial \varepsilon}{\partial t} = (p+q) \frac{\partial v}{\partial t} \ ,$$

$$p = p(\varepsilon, v) \ .$$

Here q is the artificial viscosity, p is the pressure, u is the mass velocity, ε is the specific internal energy, v_o and v are the initial and actual specific volumes, ξ is the Lagrangian coordinate, $\nu = 1, 2, 3$ is the symmetry type, Sternberg H.M., Walker W.A, 1971.

The equation of state for water was taken from the data (Walker W.A., Sternberg H.M, 1967):

$$p_w = f_1/v + f_2/v^3 + f_3/v^5 + f_4/v^7 \ ,$$

where $f_i = f_i(\varepsilon)$ are polynomial functions of the internal energy. For detonation products the equation of state of the type $p = p(\rho)$, where $\rho = 1/v$, was as follows (Kuropatenko V.F.,1989):

$$p_d = [\gamma(\rho) - 1]\rho\varepsilon + \varphi(\rho)) \ .$$

If $x = \delta/\delta_*$ is introduced, where $\delta_* = 0.35$, $\delta = \rho/\rho_*$, ρ_* is the density of the HE charge, D_* is the velocity of its detonation, $A = 0.0139$, and $m = 2.284$, then

$$for \ x > 1 \quad \gamma = 1.667, \quad \varphi = \rho \ D_*^2 \ A \ (x-1)^m$$

$$for \ x \leq 1 \quad \gamma = 1.375 + 0.292 \ x \ (\ 3 \ - \ 3 \ x \ + \ x^2), \quad \varphi = 0.$$

According to the model of instantaneous relaxation behind the front of a rarefaction wave in cavitating liquid, the equations of conservation of momentum, pressure, and specific volume (now for the mean values) have the form:

$$\frac{\partial u}{\partial t} = 0$$

$$p = p_s, \qquad v = v_l + v_g .$$

Here the indexes l and g are appropriated to the liquid and gas phases respectively. The other equations of the system will be the same (also for the mean values of the functions). It follows from the latter that the velocity profile in the cavitation zone is "frozen": $u = F(\xi)$.

The calculations were performed for a charge of TNT/RDX 50/50 with specific volume $v_* = 6.25 \ 10^{-4} \ m^3/kg$, specific explosion energy Q $= 4.88 \ 10^3$ kJ/kg, for detonation parameters $p_j = 2.19 \ 10^4$ MPa, $v_j = 4.7 \ 10^{-4} \ m^3/kg$, and $u_j = 1.85 \ 10^3$ m/s. In solving the problem the balance of the integral energy of the system "detonation products - shell" was checked out.

Let us consider only three characteristics among the calculation results: the dynamics of mean mass velocity distribution u, mean density of two-phase medium $\rho = (1-k)\rho_l$, and characteristic time τ required for the bubbles to achieve the "bulk density" ($k_* \simeq 0.5 - 0.7$), Fig. 1 a, b. The distribution curves are presented for the instants of time 16(1), 20(2) and 24(3) μs (cylindrical symmetry,ν = 2) for a relatively small shell: r_1= 2.83 cm, r_o= 1 cm. Vertical marks show the position of the interface between the detonation products and the cavitating liquid shell.

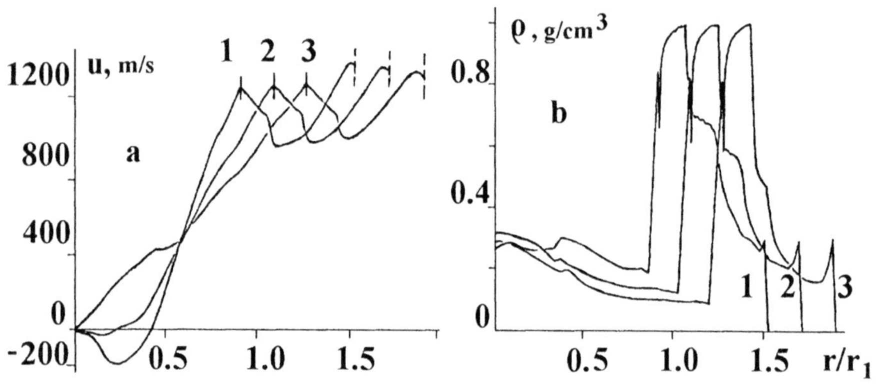

Fig. 1. Dynamics of the state of liquid shell during explosion of TNT/RDX 50/50 charge. The distribution of mean mass velocity (a) and mean density (b) of cavitating liquid for the instants of time : 16 (1), 20 (2), and 24 (3) μs (effect of conservation of mass velocity profile is shown).

It is easily seen that: 1. the velocity profile is practically retained; 2. the shell structure is characterized by two zones - outer (with rapidly developing cavitation) and inner (bordering on the detonation products), with the density of a uniform liquid; 3. the magnitude $k \simeq k_*$ is achieved already when the outer radius of the shell $r \simeq (1.3 - 1.5)r_1$ and $t \simeq (10 - 12)$ μs after reflection of the shock wave. The abrupt increase of the density at the outer boundary of the shell can be explained by its deceleration by the atmospheric pressure.

Experimental simulation of the process and registration methods. In one of the model statements used in studying liquid fracture in intense rarefaction waves, the shock waves were generated by exploding a wire placed along the axis of a liquid cylindrical shell which was bounded at the ends by transparent walls. High-speed optical photography of the process of dynamic loading recorded the complex structure of liquid shell with cavitating outer layer, as in the results of the calculations mentioned above (Chernobaev N.N, 1989).

The energy accumulated in the condenser bank was approximately 150 J, the time of its release was about 4 μs. One may estimate the initial size of the explosive cavity assuming that the pressure in it (as the result of the wire explosion) will be of the same order of magnitude as in instantaneous detonation of the HE charge when the volume is constant. In the experiments the length of the liquid shell was 3 cm, which gave $r_o \simeq 0.4$ mm, and the ratio r_1/r_o= 50 when r_1= 2 cm. Apparently in this case the characteristic time τ required for the bubbles to achieve the "bulk density" will have to significantly increase.

Unfortunately, so far there is no complete understanding of the essentially nonlinear processes of unbounded growth of bubbles in cavitation zones and their hydrodynamic interaction in the formation of dense packing with volume concentration of 0.5 - 0.75, as well as transition through foam structure to spallation and droplet phase separation. Therefore, the stage of accumulation of the experimental information, and development of the techniques allowing one to determine the distinctive features of the process, comes into principal importance.

The most reliable results in this field are obtained by a combination of two techniques which allow one to evaluate to a certain extent the basic elements of the process. The first technique is rather simple: a transducer registering the dynamic structure of pressure when it is flown around by the expanding liquid shell are placed at different distances in the vicinity of the outer boundary of the shell. It is evident that during the evolution process the outer cavitating layer should disintegrate into fragments (the medium stops being continuous), and the transducer registering consequently the dynamic pressure in some vicinity of the free surface should show a smooth transition from one state to another.

Fig.2a shows the "two-pulsed" structure of deceleration pressure in the form of a spatial diagram of pressure as a function of time and distance to free surface. It is visible that the pulse (1) corresponding to the flow of cavitating liquid noticeably weakens as we recede from the registration point. Approximately, by 150-200 μs the pulse practically vanishes, which may be considered as a result of the formation of a foam type structure as an intermediate stage during destruction. The second pulse (2) corresponds to the layer of dense liquid which fractures noticeably later due to the developing instability. The experiment shows that during reversible development of cavitation the fracture fails to occur and after a time both "waves" of pressure flow together.

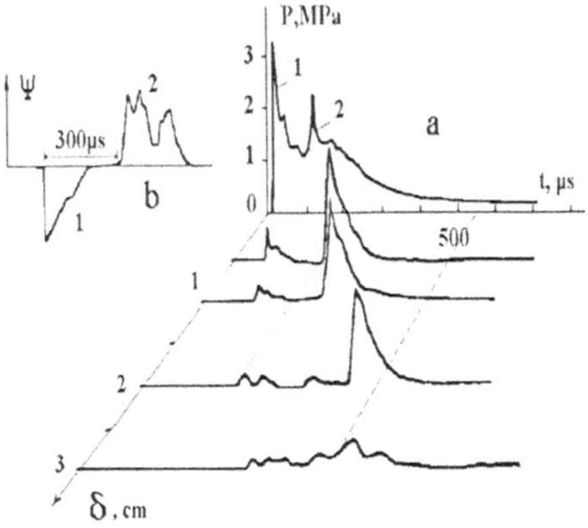

Fig.2. Spatial diagram of two-pulsed structure of deceleration pressure (P) in two-phase flow and dynamics of electrical potential (ψ) as result of two-phase state inversion.

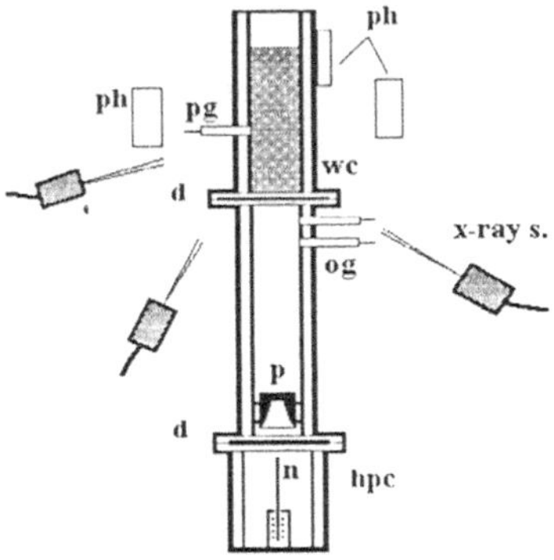

Fig.3. Scheme of two-diaphragm cavitation shock tube with 3 x-ray sources, high pressure (hpc) and working sections (wc).

The fact of inversion may be registered more precisely by measuring the electric potential in the medium. In a dispersed system, as is known, in relative motion of phases there appears a potential difference in the direction of the relative velocity. The reason is the existence of a double electric layer at the phase interface and flow-induced separation of ions adsorbed on the surface of the bubble or drop. Clearly the sign of the ions will depend on the particular two-phase structure. Thus, registering the instant when the potential changes sign, one may determine when inversion of the two-phase state has occurred. Measurements of this kind were carried out using a setup with an explosion wire (Stebnovsky S.V., 1989). They show that after approximately 300 μs the inversion process is completed (Fig. 2b, 1 - cavitating liquid, 2 - the system of "gas - drops").

3. Dynamics of cavitation zone density and mass velocity

The opacity of the intensively developing cavitation zone and screening of its inner structure by a layer of bubbles at the walls of the shock tube makes the use of standard high-speed optical photography unsuitable. The use of the pulse x-ray technique is much more promising (Baykov I.R. et al.,1984).

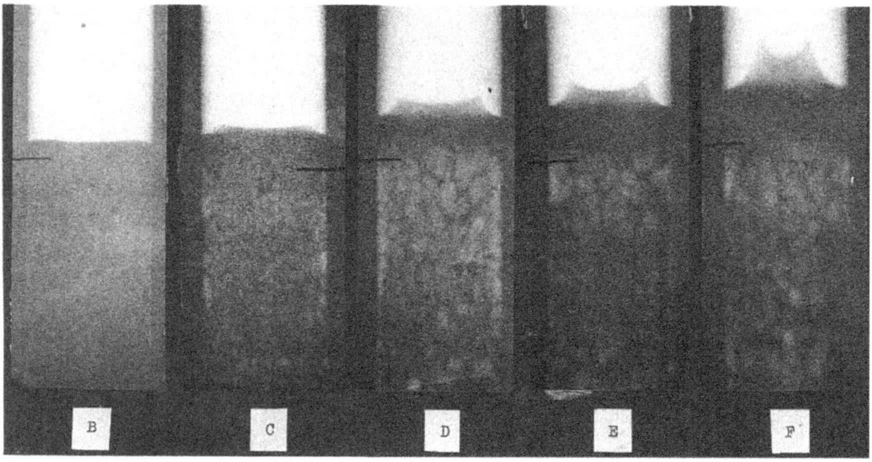

Fig.4. Flash radiograph series of a cavitation zone development (dia - 3 cm): B - 200, C - 400, D - 600, E - 800, F - 1000 μs after a shock wave reflection from free surface.

The later stage of the cavitation process in intense rarefaction waves was studied in a hydrodynamic shock tube by three X-ray apparatuses (Fig. 3, the scheme of the cavitation shock tube). The shown tube consistes of three sections: a high-pressure chamber (hpc), a vacuum acceleration channel (with a driving piston P and two separating diaphragms d on the ends) and the working section (wc) made of duralumin and containing the liquid under study. The radiation sources (x-ray s.) were located along the periphery in the plane normal to the axis of the working section of the shock tube with the interval 60° .The lower diaphragm was ruptured by an electromagnetic system with a needle (n). The piston speed before the impact on the upper diaphragm separating the channel from the liquid was measured by fiber-optical sensors (og). The entire system was synchronized with three X-ray apparatuses which were triggered by the pressure gauge (pg) signal with different assigned lags allowing one to perform ultra-high-speed X-raying. The mean photon energy was about 70 keV, radiation pulse duration - 80 ns.

The shock wave was excited in the liquid by the impact of the piston. Its positive phase lasted several tens of microseconds and the amplitude varied between 20 and 30 MPa. Reflection of the shock wave from the liquid free surface produced intense development of cavitation. Typical X-ray photographs for different instants of time from 0.2 to 1 msec are presented in Fig. 4 (the interval between the frames is 200 μs). It is obvious that already after 600 μs the cavitation zone is "densely packed" with large vapor-gas cells.

Computer processing of the density distribution of the X-ray negatives was necessary to be able to analyze rather freely the dynamics of the process without interference by the transducers. It allows one to obtain easy-to-interpret computer versions of experimental data which grant a unique possibility of reconstructing the structure dynamics of the cavitation zone at any cross-section (the lower resolution limit on concentration is about 2 ÷ 4 percent).

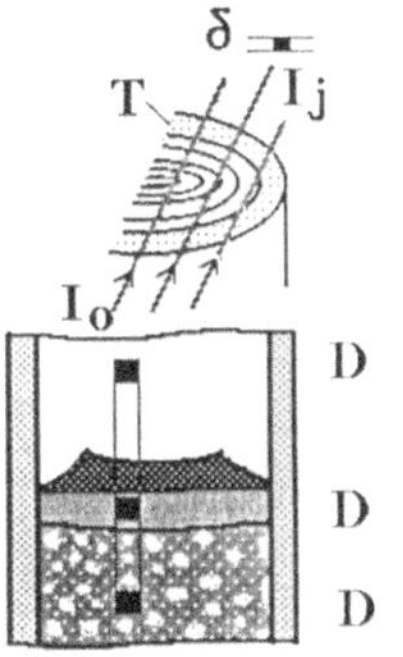

Fig. 5. Calculation diagram of the density of two-phase medium: a path of an elementary beam of intensity I_o through a cylindrical specimen placed in shock tube T.

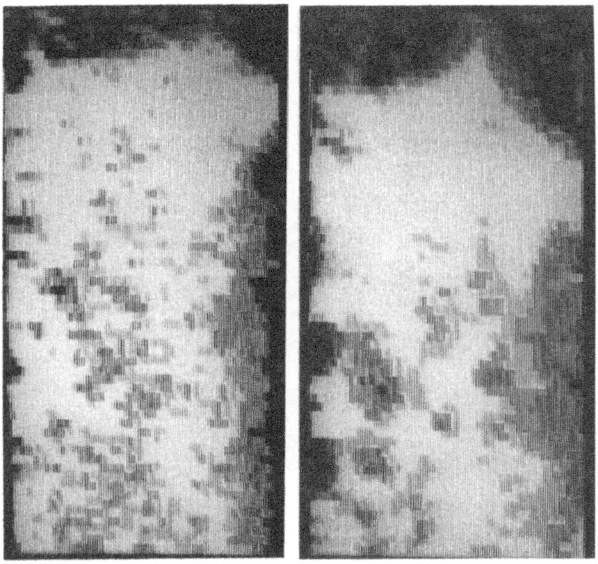

Fig. 6. Characteristic computer version of cavitation zone development: 400 and 800 μs.

The method for calculating the density is presented in Fig. 5. The path of an elementary beam of intensity I_o through the cylindrical specimen placed in the shock tube T consists of sections cut by concentric circles into which the transversal cross-section of the specimen is split. The resulting radiation having passed through the specimen layered structure forms an image at the film (ph, Fig.3). The image is composed of the appropriate element-pixels equal in size to the transversal cross-section of the elementary beam. Intensity of the j-beam at the exit is

$$I_j = I_o\ exp(-2\rho_t\mu_t x_t - 2\Sigma\rho_i\mu_i x_i)$$

where the indexes t and i correspond to the material of the tube and tested medium respectively, ρ is the density, μ is the effective mass absorptivity, x is the beam path in the i-circle. It is assumed that the values of photometric density are within the linear section of the sensitometric curve of the photomaterial, i.e. the density of blackening for two arbitrary points of the photographs is determined by the relationship $D_1 - D_2 = \gamma\ log(I_1/I_2)$, $\gamma = const.$

Preliminary averaging for each group of elements of the (3x3) image matrix and local averaging for each region of (4x8) pixels were performed in calculating the distribution of the medium density. The distributions obtained were monitored at the computer display. The characteristic computer version for two moments of cavitation zone development is presented in Fig. 6 (for 400 and 800 μs) . The average distributions of local density along the symmetry axis are presented in Fig.7. Volume concentration k of vapor-gas phase can vary from zero (heterogeneous liquid) to

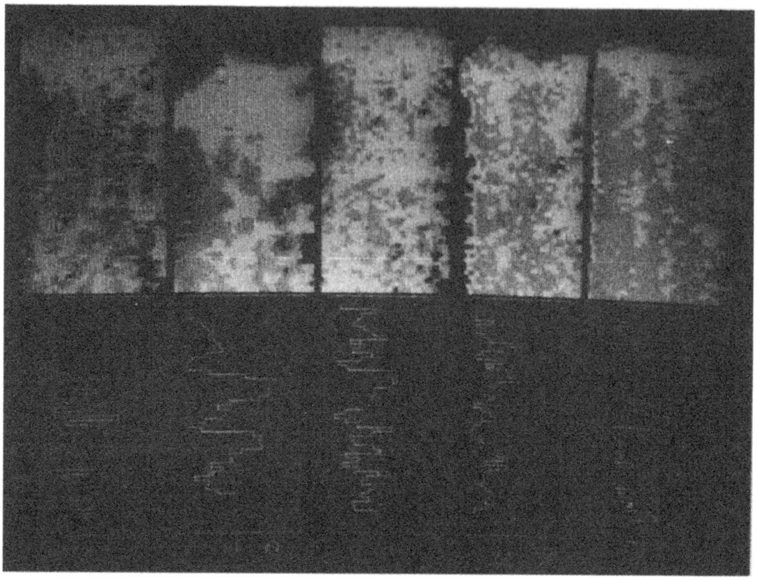

Fig. 7. Computer series of the distributions of an average local density in cavitating liquid (along a symmetry axis) for several instants of time.

several tens of percents, as well as up to the state of system with dense package of macrocavities with average density 0.1 g/cm^3. The experimental data presented in Figs. 6-7 show that macrocavities may change their configuration and unite. Note that the average value of density in the zone decreases monotonously in the experiments with identical loading parameters.

The method of pulse X-ray photography was used to measure the mass velocities in the cavitation zones and study the dynamics of their distribution by means of special traces of lead foil. The traces were produced either as thin bands (thickness of X-ray image - 0.2 mm) attached at different levels to a metal frame by strips of cigarette paper, or as platelets 1x1 mm in size (image - 0.5 mm). The platelets were attached by thin vertical threads. The mean mass velocity was determined by the displacement of the traces subsequently registered X-ray photographs.

The distribution of mass velocities for the instants of time 0.2 , 0.3 and 0.5 ms is presented in Fig. 8. Each point represents the mean value taken from several experiments, solid lines correspond to the calculation results following the mentioned model of instantaneous relaxation of tensile stresses. Here the characteristic profile of the incident shock wave with amplitude about 15 MPa and duration of 60 μs is presented. Owing to the relatively great size of gas-vapor hollows in the cavitation zone and its inhomogeneity, one can conclude that the model is in good agreement with the experimental data and the method of traces in combination with pulse X-ray technique is reliable enough.

4. Conclusion

The studies showed that the relaxation time of tensile stresses in cavitating liquid is much shorter than the characteristic times of evolution of the cavitation process. Numerical model may be essentially simplified by eliminating this parameter. The calculations by the model of instantaneous relaxation made it possible to demonstrate clearly the dynamics of mean density of cavitation

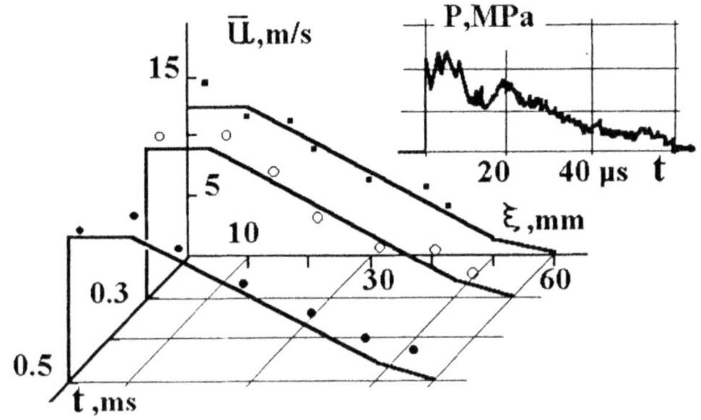

Fig.8. Experimental (points) and theoretical (solid line) profiles of mass velocities into cavitation zone for instants of time : 0.2 , 0.3 and 0.5 ms.

zone and distribution of mass velocities. Experimental data obtained by the method of traces and pulse X-ray photography are in agreement with numerical studies on "freezing" of the profile of mass velocities during evolution of the cavitation zone.

High-speed photography on the basis of X-ray pulse combined with computer processing of the negatives allowed us to obtain the computer analogs of the images of the internal structure of the opaque cavitation zone, as well as unique data on the dynamics of its local and mean density.

Acknowledgement

This work was supported by the Russian Foundation for Fundamental Studies, Grant 93-013-16383.

The authors wish to acknowledge financial support of their attendance at IUTAM Symposium on "Waves in LG and LV two-phase system" kindly provided by the Organizing Committee and Profs. Shigeki Morioka and Leen van Wijngaarden (co-chairmen of Symposium).

References

Baykov IR et al (1984) Experimental methods of investigation of cavitation cluster dynamics. Zh. Prikl. Mekh. Tekh. Fiz. (JAMTP, translated in English), Nauka, Sib.Div., No. 5.

Carlson GA, Henry KW (1973) Technique for studying tension failure in application to glycerol, J. Appl. Phys. V. 42. No. 5.

Chernobaev NN (1989) Modeling of shock-wave loading of liquid volumes. Adiabatic waves in liquid vapor systems: Proc. of IUTAM Symp., Gottingen, 1989. Berlin et al.: Springer.

Frenkel' YaI (1945) Kinetic Theory of Liquids. Izd.AN SSSR.

Hammitt FG, Keller A, Ahmed O, Pyun J, Yilmaz E (1976) Cavitation threshold and superheat in various fluids. In: Cavitation. Proc. of Int. Conf., Mech. Eng. Publ. Ltd London NY, Edinburg, September 3-5

Kedrinskii VK (1976) Negative pressure profile in cavitation zone at underwater explosion near a free surface. Acta Astronautica 3: 623-632

Kedrinskii VK (1989) On relaxation of tensile stresses in cavitating liquid. 13th Int. Congress on Acoustics, Beograd, 1989: Proc. Sabac: Dragan Srnic Press.

Kedrinskii VK, Chernobaev NN (1992) Throwing of liquid shell by HE explosion. Zh. Prikl. Mekh. Tekh. Fiz. (JAMTP, translated in English), Nauka, Sib.Div., n 6.

Kedrinskii VK (1993) Nonlinear problems of cavity liquid fracture at explosive loading (review). J Appl Mechanics and Tech Physics 3:74-91

Kuropatenko VF (1989) State equation of detonation products of HE. Fiz. Goreniya Vzryva (Shock Waves and Combustion, translated in English), n 6.

Stebnovskii SV (1989) On a mechanism of pulsed breakdown of a liquid volume. Zh. Prikl. Mekh. Tekh. Fiz.(JAMTP, translated in English), Nauka, Sib. Div., No. 2.

Sternberg HM, Walker WA (1971) Calculated flow and energy distribution following underwater detonation of a pentolite sphere. Physics of Fluids, 14, No. 9.

Walker WA, Sternberg HM (1967) The Chapman - Jouguet isentrope and the underwater shock wave performance of pentolite. Proc. of 4th Int Symp on Detonation.White Oak USA,1965. Washington.

Wilson DA, Hoyt JM, McKune JW (1975) Measurement of tensile strength of liquid by explosion technique, Nature. V. 253. No.

Mechanics

FLUID MECHANICS AND ITS APPLICATIONS

Series Editor: R. Moreau

Aims and Scope of the Series

The purpose of this series is to focus on subjects in which fluid mechanics plays a fundamental role. As well as the more traditional applications of aeronautics, hydraulics, heat and mass transfer etc., books will be published dealing with topics which are currently in a state of rapid development, such as turbulence, suspensions and multiphase fluids, super and hypersonic flows and numerical modelling techniques. It is a widely held view that it is the interdisciplinary subjects that will receive intense scientific attention, bringing them to the forefront of technological advancement. Fluids have the ability to transport matter and its properties as well as transmit force, therefore fluid mechanics is a subject that is particularly open to cross fertilisation with other sciences and disciplines of engineering. The subject of fluid mechanics will be highly relevant in domains such as chemical, metallurgical, biological and ecological engineering. This series is particularly open to such new multidisciplinary domains.

1. M. Lesieur: *Turbulence in Fluids.* 2nd rev. ed., 1990 ISBN 0-7923-0645-7
2. O. Métais and M. Lesieur (eds.): *Turbulence and Coherent Structures.* 1991 ISBN 0-7923-0646-5
3. R. Moreau: *Magnetohydrodynamics.* 1990 ISBN 0-7923-0937-5
4. E. Coustols (ed.): *Turbulence Control by Passive Means.* 1990 ISBN 0-7923-1020-9
5. A.A. Borissov (ed.): *Dynamic Structure of Detonation in Gaseous and Dispersed Media.* 1991 ISBN 0-7923-1340-2
6. K.-S. Choi (ed.): *Recent Developments in Turbulence Management.* 1991 ISBN 0-7923-1477-8
7. E.P. Evans and B. Coulbeck (eds.): *Pipeline Systems.* 1992 ISBN 0-7923-1668-1
8. B. Nau (ed.): *Fluid Sealing.* 1992 ISBN 0-7923-1669-X
9. T.K.S. Murthy (ed.): *Computational Methods in Hypersonic Aerodynamics.* 1992 ISBN 0-7923-1673-8
10. R. King (ed.): *Fluid Mechanics of Mixing.* Modelling, Operations and Experimental Techniques. 1992 ISBN 0-7923-1720-3
11. Z. Han and X. Yin: *Shock Dynamics.* 1993 ISBN 0-7923-1746-7
12. L. Svarovsky and M.T. Thew (eds.): *Hydroclones.* Analysis and Applications. 1992 ISBN 0-7923-1876-5
13. A. Lichtarowicz (ed.): *Jet Cutting Technology.* 1992 ISBN 0-7923-1979-6
14. F.T.M. Nieuwstadt (ed.): *Flow Visualization and Image Analysis.* 1993 ISBN 0-7923-1994-X
15. A.J. Saul (ed.): *Floods and Flood Management.* 1992 ISBN 0-7923-2078-6
16. D.E. Ashpis, T.B. Gatski and R. Hirsh (eds.): *Instabilities and Turbulence in Engineering Flows.* 1993 ISBN 0-7923-2161-8
17. R.S. Azad: *The Atmospheric Boundary Layer for Engineers.* 1993 ISBN 0-7923-2187-1
18. F.T.M. Nieuwstadt (ed.): *Advances in Turbulence IV.* 1993 ISBN 0-7923-2282-7
19. K.K. Prasad (ed.): *Further Developments in Turbulence Management.* 1993 ISBN 0-7923-2291-6
20. Y.A. Tatarchenko: *Shaped Crystal Growth.* 1993 ISBN 0-7923-2419-6

Kluwer Academic Publishers – Dordrecht / Boston / London

Mechanics

FLUID MECHANICS AND ITS APPLICATIONS

Series Editor: R. Moreau

21. J.P. Bonnet and M.N. Glauser (eds.): *Eddy Structure Identification in Free Turbulent Shear Flows.* 1993 ISBN 0-7923-2449-8
22. R.S. Srivastava: *Interaction of Shock Waves.* 1994 ISBN 0-7923-2920-1
23. J.R. Blake, J.M. Boulton-Stone and N.H. Thomas (eds.): *Bubble Dynamics and Interface Phenomena.* 1994 ISBN 0-7923-3008-0
24. R. Benzi (ed.): *Advances in Turbulence V.* 1995 (forthcoming) ISBN 0-7923-3032-3
25. B.I. Rabinovich, V.G. Lebedev and A.I. Mytarev: *Vortex Processes and Solid Body Dynamics.* The Dynamic Problems of Spacecrafts and Magnetic Levitation Systems. 1994 ISBN 0-7923-3092-7
26. P.R. Voke, L. Kleiser and J.-P. Chollet (eds.): *Direct and Large-Eddy Simulation I.* Selected papers from the First ERCOFTAC Workshop on Direct and Large-Eddy Simulation. 1994 ISBN 0-7923-3106-0
27. J.A. Sparenberg: *Hydrodynamic Propulsion and its Optimization.* Analytic Theory. 1995 ISBN 0-7923-3201-6
28. J.F. Dijksman and G.D.C. Kuiken (eds.): *IUTAM Symposium on Numerical Simulation of Non-Isothermal Flow of Viscoelastic Liquids.* Proceedings of an IUTAM Symposium held in Kerkrade, The Netherlands. 1995 ISBN 0-7923-3262-8
29. B.M. Boubnov and G.S. Golitsyn: *Convection in Rotating Fluids.* 1995 ISBN 0-7923-3371-3
30. S.I. Green (ed.): *Fluid Vortices.* 1995 ISBN 0-7923-3376-4
31. S. Morioka and L. van Wijngaarden (eds.): *IUTAM Symposium on Waves in Liquid/Gas and Liquid/Vapour Two-Phase Systems.* 1995 ISBN 0-7923-3424-8

Kluwer Academic Publishers – Dordrecht / Boston / London

Mechanics

SOLID MECHANICS AND ITS APPLICATIONS

Series Editor: G.M.L. Gladwell

Aims and Scope of the Series

The fundamental questions arising in mechanics are: *Why?*, *How?*, and *How much?* The aim of this series is to provide lucid accounts written by authoritative researchers giving vision and insight in answering these questions on the subject of mechanics as it relates to solids. The scope of the series covers the entire spectrum of solid mechanics. Thus it includes the foundation of mechanics; variational formulations; computational mechanics; statics, kinematics and dynamics of rigid and elastic bodies; vibrations of solids and structures; dynamical systems and chaos; the theories of elasticity, plasticity and viscoelasticity; composite materials; rods, beams, shells and membranes; structural control and stability; soils, rocks and geomechanics; fracture; tribology; experimental mechanics; biomechanics and machine design.

1. R.T. Haftka, Z. Gürdal and M.P. Kamat: *Elements of Structural Optimization.* 2nd rev.ed., 1990 ISBN 0-7923-0608-2
2. J.J. Kalker: *Three-Dimensional Elastic Bodies in Rolling Contact.* 1990 ISBN 0-7923-0712-7
3. P. Karasudhi: *Foundations of Solid Mechanics.* 1991 ISBN 0-7923-0772-0
4. *Not published*
5. *Not published.*
6. J.F. Doyle: *Static and Dynamic Analysis of Structures.* With an Emphasis on Mechanics and Computer Matrix Methods. 1991 ISBN 0-7923-1124-8; Pb 0-7923-1208-2
7. O.O. Ochoa and J.N. Reddy: *Finite Element Analysis of Composite Laminates.* ISBN 0-7923-1125-6
8. M.H. Aliabadi and D.P. Rooke: *Numerical Fracture Mechanics.* ISBN 0-7923-1175-2
9. J. Angeles and C.S. López-Cajún: *Optimization of Cam Mechanisms.* 1991 ISBN 0-7923-1355-0
10. D.E. Grierson, A. Franchi and P. Riva (eds.): *Progress in Structural Engineering.* 1991 ISBN 0-7923-1396-8
11. R.T. Haftka and Z. Gürdal: *Elements of Structural Optimization.* 3rd rev. and exp. ed. 1992 ISBN 0-7923-1504-9; Pb 0-7923-1505-7
12. J.R. Barber: *Elasticity.* 1992 ISBN 0-7923-1609-6; Pb 0-7923-1610-X
13. H.S. Tzou and G.L. Anderson (eds.): *Intelligent Structural Systems.* 1992 ISBN 0-7923-1920-6
14. E.E. Gdoutos: *Fracture Mechanics.* An Introduction. 1993 ISBN 0-7923-1932-X
15. J.P. Ward: *Solid Mechanics.* An Introduction. 1992 ISBN 0-7923-1949-4
16. M. Farshad: *Design and Analysis of Shell Structures.* 1992 ISBN 0-7923-1950-8
17. H.S. Tzou and T. Fukuda (eds.): *Precision Sensors, Actuators and Systems.* 1992 ISBN 0-7923-2015-8
18. J.R. Vinson: *The Behavior of Shells Composed of Isotropic and Composite Materials.* 1993 ISBN 0-7923-2113-8

Kluwer Academic Publishers – Dordrecht / Boston / London

Mechanics

SOLID MECHANICS AND ITS APPLICATIONS

Series Editor: G.M.L. Gladwell

19. H.S. Tzou: *Piezoelectric Shells.* Distributed Sensing and Control of Continua. 1993 ISBN 0-7923-2186-3
20. W. Schiehlen (ed.): *Advanced Multibody System Dynamics.* Simulation and Software Tools. 1993 ISBN 0-7923-2192-8
21. C.-W. Lee: *Vibration Analysis of Rotors.* 1993 ISBN 0-7923-2300-9
22. D.R. Smith: *An Introduction to Continuum Mechanics.* 1993 ISBN 0-7923-2454-4
23. G.M.L. Gladwell: *Inverse Problems in Scattering.* An Introduction. 1993 ISBN 0-7923-2478-1
24. G. Prathap: *The Finite Element Method in Structural Mechanics.* 1993 ISBN 0-7923-2492-7
25. J. Herskovits (ed.): *Advances in Structural Optimization.* 1995 (forthcoming) ISBN 0-7923-2510-9
26. M.A. González-Palacios and J. Angeles: *Cam Synthesis.* 1993 ISBN 0-7923-2536-2
27. W.S. Hall: *The Boundary Element Method.* 1993 ISBN 0-7923-2580-X
28. J. Angeles, G. Hommel and P. Kovács (eds.): *Computational Kinematics.* 1993 ISBN 0-7923-2585-0
29. A. Curnier: *Computational Methods in Solid Mechanics.* 1994 ISBN 0-7923-2761-6
30. D.A. Hills and D. Nowell: *Mechanics of Fretting Fatigue.* 1994 ISBN 0-7923-2866-3
31. B. Tabarrok and F.P.J. Rimrott: *Variational Methods and Complementary Formulations in Dynamics.* 1994 ISBN 0-7923-2923-6
32. E.H. Dowell, E.F. Crawley, H.C. Curtiss Jr., D.A. Peters, R. H. Scanlan and F. Sisto (eds.): *A Modern Course in Aeroelasticity.* Third Revised and Enlarged Edition. (forthcoming) ISBN 0-7923-2788-8; Pb: 0-7923-2789-6
33. A. Preumont: *Random Vibration and Spectral Analysis.* 1994 ISBN 0-7923-3036-6
34. J.N. Reddy (ed.): *Mechanics of Composite Materials.* Selected works of Nicholas J. Pagano. 1994 ISBN 0-7923-3041-2
35. A.P.S. Selvadurai (ed.): *Mechanics of Poroelastic Media.* 1995 (forthcoming) ISBN 0-7923-3329-2
36. Z. Mróz, D. Weichert, S. Dorosz (eds.): *Inelastic Behaviour of Structures under Variable Loads.* 1995 ISBN 0-7923-3397-7
37. R. Pyrz (ed.): *IUTAM Symposium on Microstructure-Property Interactions in Composite Materials.* Proceedings of an IUTAM Symposium held in Aalborg, Denmark. 1995 (forthcoming) ISBN 0-7923-3427-2
38. M.I. Friswell and J.E. Mottershead: *Finite Element Model Updating in Structural Dynamics.* 1995 ISBN 0-7923-3431-0

Kluwer Academic Publishers – Dordrecht / Boston / London

Mechanics

MECHANICS OF FRACTURE

Editors: G.C. Sih

1. G.C. Sih (ed.): *Methods of Analysis and Solutions of Crack Problems.* 1973 ISBN 90-01-79860-8
2. M.K. Kassir and G.C. Sih (eds.): *Three-dimensional Crack Problems.* A New Solution of Crack Solutions in Three-dimensional Elasticity. 1975 ISBN 90-286-0414-6
3. G.C. Sih (ed.): *Plates and Shells with Cracks.* 1977 ISBN 90-286-0146-5
4. G.C. Sih (ed.): *Elastodynamic Crack Problems.* 1977 ISBN 90-286-0156-2
5. G.C. Sih (ed.): *Stress Analysis of Notch Problems.* Stress Solutions to a Variety of Notch Geometries used in Engineering Design. 1978 ISBN 90-286-0166-X
6. G.C. Sih and E.P. Chen (eds.): *Cracks in Composite Materials.* A Compilation of Stress Solutions for Composite System with Cracks. 1981 ISBN 90-247-2559-3
7. G.C. Sih (ed.): *Experimental Evaluation of Stress Concentration and Intensity Factors.* Useful Methods and Solutions to Experimentalists in Fracture Mechanics. 1981 ISBN 90-247-2558-5

MECHANICS OF PLASTIC SOLIDS

Editors: J. Schroeder and G.Æ. Oravas

1. A. Sawczuk (ed.): *Foundations of Plasticity.* 1973 ISBN 90-01-77570-5
2. A. Sawczuk (ed.): *Problems of Plasticity.* 1974 ISBN 90-286-0233-X
3. W. Szczepiński: *Introduction to the Mechanics of Plastic Forming of Metals.* 1979 ISBN 90-286-0126-0
4. D.A. Gokhfeld and O.F. Cherniavsky: *Limit Analysis of Structures at Thermal Cycling.* 1980 ISBN 90-286-0455-3
5. N. Cristescu and I. Suliciu: *Viscoplasticity.* 1982 ISBN 90-247-2777-4

Kluwer Academic Publishers - Dordrecht / Boston / London

Mechanics

3. E.B. Magrab: *Vibrations of Elastic Structural Members.* 1979 ISBN 90-286-0207-0
4. R.T. Haftka and M.P. Kamat: *Elements of Structural Optimization.* 1985
   Revised and enlarged edition see under Solid Mechanics and Its Applications, Volume 1
5. J.R. Vinson and R.L. Sierakowski: *The Behavior of Structures Composed of Composite Materials.* 1986 ISBN Hb 90-247-3125-9; Pb 90-247-3578-5
6. B.E. Gatewood: *Virtual Principles in Aircraft Structures.* Volume 1: Analysis. 1989
   ISBN 90-247-3754-0
7. B.E. Gatewood: *Virtual Principles in Aircraft Structures.* Volume 2: Design, Plates, Finite Elements. 1989 ISBN 90-247-3755-9
   Set (Gatewood 1 + 2) ISBN 90-247-3753-2

MECHANICS OF ELASTIC AND INELASTIC SOLIDS
Editors: S. Nemat-Nasser and G.Æ. Oravas

---

1. G.M.L. Gladwell: *Contact Problems in the Classical Theory of Elasticity.* 1980
   ISBN Hb 90-286-0440-5; Pb 90-286-0760-9
2. G. Wempner: *Mechanics of Solids with Applications to Thin Bodies.* 1981
   ISBN 90-286-0880-X
3. T. Mura: *Micromechanics of Defects in Solids.* 2nd revised edition, 1987
   ISBN 90-247-3343-X
4. R.G. Payton: *Elastic Wave Propagation in Transversely Isotropic Media.* 1983
   ISBN 90-247-2843-6
5. S. Nemat-Nasser, H. Abé and S. Hirakawa (eds.): *Hydraulic Fracturing and Geothermal Energy.* 1983 ISBN 90-247-2855-X
6. S. Nemat-Nasser, R.J. Asaro and G.A. Hegemier (eds.): *Theoretical Foundation for Large-scale Computations of Nonlinear Material Behavior.* 1984 ISBN 90-247-3092-9
7. N. Cristescu: *Rock Rheology.* 1988 ISBN 90-247-3660-9
8. G.I.N. Rozvany: *Structural Design via Optimality Criteria.* The Prager Approach to Structural Optimization. 1989 ISBN 90-247-3613-7

MECHANICS OF SURFACE STRUCTURES
Editors: W.A. Nash and G.Æ. Oravas

---

1. P. Seide: *Small Elastic Deformations of Thin Shells.* 1975 ISBN 90-286-0064-7
2. V. Panc: *Theories of Elastic Plates.* 1975 ISBN 90-286-0104-X
3. J.L. Nowinski: *Theory of Thermoelasticity with Applications.* 1978
   ISBN 90-286-0457-X
4. S. Łukasiewicz: *Local Loads in Plates and Shells.* 1979 ISBN 90-286-0047-7
5. C. Fiřt: *Statics, Formfinding and Dynamics of Air-supported Membrane Structures.* 1983 ISBN 90-247-2672-7
6. Y. Kai-yuan (ed.): *Progress in Applied Mechanics.* The Chien Wei-zang Anniversary Volume. 1987 ISBN 90-247-3249-2
7. R. Negruţiu: *Elastic Analysis of Slab Structures.* 1987 ISBN 90-247-3367-7
8. J.R. Vinson: *The Behavior of Thin Walled Structures.* Beams, Plates, and Shells. 1988
   ISBN Hb 90-247-3663-3; Pb 90-247-3664-1

Mechanics

MECHANICS OF FLUIDS AND TRANSPORT PROCESSES

Editors: R.J. Moreau and G.Æ. Oravas

1. J. Happel and H. Brenner: *Low Reynolds Number Hydrodynamics.* With Special Applications to Particular Media. 1983 ISBN Hb 90-01-37115-9; Pb 90-247-2877-0
2. S. Zahorski: *Mechanics of Viscoelastic Fluids.* 1982 ISBN 90-247-2687-5
3. J.A. Sparenberg: *Elements of Hydrodynamics Propulsion.* 1984 ISBN 90-247-2871-1
4. B.K. Shivamoggi: *Theoretical Fluid Dynamics.* 1984 ISBN 90-247-2999-8
5. R. Timman, A.J. Hermans and G.C. Hsiao: *Water Waves and Ship Hydrodynamics.* An Introduction. 1985 ISBN 90-247-3218-2
6. M. Lesieur: *Turbulence in Fluids.* Stochastic and Numerical Modelling. 1987 ISBN 90-247-3470-3
7. L.A. Lliboutry: *Very Slow Flows of Solids.* Basics of Modeling in Geodynamics and Glaciology. 1987 ISBN 90-247-3482-7
8. B.K. Shivamoggi: *Introduction to Nonlinear Fluid-Plasma Waves.* 1988 ISBN 90-247-3662-5
9. V. Bojarevičs, Ya. Freibergs, E.I. Shilova and E.V. Shcherbinin: *Electrically Induced Vortical Flows.* 1989 ISBN 90-247-3712-5
10. J. Lielpeteris and R. Moreau (eds.): *Liquid Metal Magnetohydrodynamics.* 1989 ISBN 0-7923-0344-X

MECHANICS OF ELASTIC STABILITY

Editors: H. Leipholz and G.Æ. Oravas

1. H. Leipholz: *Theory of Elasticity.* 1974 ISBN 90-286-0193-7
2. L. Librescu: *Elastostatics and Kinetics of Aniosotropic and Heterogeneous Shell-type Structures.* 1975 ISBN 90-286-0035-3
3. C.L. Dym: *Stability Theory and Its Applications to Structural Mechanics.* 1974 ISBN 90-286-0094-9
4. K. Huseyin: *Nonlinear Theory of Elastic Stability.* 1975 ISBN 90-286-0344-1
5. H. Leipholz: *Direct Variational Methods and Eigenvalue Problems in Engineering.* 1977 ISBN 90-286-0106-6
6. K. Huseyin: *Vibrations and Stability of Multiple Parameter Systems.* 1978 ISBN 90-286-0136-8
7. H. Leipholz: *Stability of Elastic Systems.* 1980 ISBN 90-286-0050-7
8. V.V. Bolotin: *Random Vibrations of Elastic Systems.* 1984 ISBN 90-247-2981-5
9. D. Bushnell: *Computerized Buckling Analysis of Shells.* 1985 ISBN 90-247-3099-6
10. L.M. Kachanov: *Introduction to Continuum Damage Mechanics.* 1986 ISBN 90-247-3319-7
11. H.H.E. Leipholz and M. Abdel-Rohman: *Control of Structures.* 1986 ISBN 90-247-3321-9
12. H.E. Lindberg and A.L. Florence: *Dynamic Pulse Buckling.* Theory and Experiment. 1987 ISBN 90-247-3566-1
13. A. Gajewski and M. Zyczkowski: *Optimal Structural Design under Stability Constraints.* 1988 ISBN 90-247-3612-9

Mechanics

MECHANICS: ANALYSIS

Editors: V.J. Mizel and G.Æ. Oravas

1. M.A. Krasnoselskii, P.P. Zabreiko, E.I. Pustylnik and P.E. Sbolevskii: *Integral Operators in Spaces of Summable Functions.* 1976 ISBN 90-286-0294-1
2. V.V. Ivanov: *The Theory of Approximate Methods and Their Application to the Numerical Solution of Singular Integral Equations.* 1976 ISBN 90-286-0036-1
3. A. Kufner, O. John and S. Pučík: *Function Spaces.* 1977 ISBN 90-286-0015-9
4. S.G. Mikhlin: *Approximation on a Rectangular Grid.* With Application to Finite Element Methods and Other Problems. 1979 ISBN 90-286-0008-6
5. D.G.B. Edelen: *Isovector Methods for Equations of Balance.* With Programs for Computer Assistance in Operator Calculations and an Exposition of Practical Topics of the Exterior Calculus. 1980 ISBN 90-286-0420-0
6. R.S. Anderssen, F.R. de Hoog and M.A. Lukas (eds.): *The Application and Numerical Solution of Integral Equations.* 1980 ISBN 90-286-0450-2
7. R.Z. Has'minskiĭ: *Stochastic Stability of Differential Equations.* 1980 ISBN 90-286-0100-7
8. A.I. Vol'pert and S.I. Hudjaev: *Analysis in Classes of Discontinuous Functions and Equations of Mathematical Physics.* 1985 ISBN 90-247-3109-7
9. A. Georgescu: *Hydrodynamic Stability Theory.* 1985 ISBN 90-247-3120-8
10. W. Noll: *Finite-dimensional Spaces.* Algebra, Geometry and Analysis. Volume I. 1987 ISBN Hb 90-247-3581-5; Pb 90-247-3582-3

MECHANICS: COMPUTATIONAL MECHANICS

Editors: M. Stern and G.Æ. Oravas

1. T.A. Cruse: *Boundary Element Analysis in Computational Fracture Mechanics.* 1988 ISBN 90-247-3614-5

MECHANICS: GENESIS AND METHOD

Editor: G.Æ. Oravas

1. P.-M.-M. Duhem: *The Evolution of Mechanics.* 1980 ISBN 90-286-0688-2

MECHANICS OF CONTINUA

Editors: W.O. Williams and G.Æ. Oravas

1. C.-C. Wang and C. Truesdell: *Introduction to Rational Elasticity.* 1973 ISBN 90-01-93710-1
2. P.J. Chen: *Selected Topics in Wave Propagation.* 1976 ISBN 90-286-0515-0
3. P. Villaggio: *Qualitative Methods in Elasticity.* 1977 ISBN 90-286-0007-8

Mechanics

From 1990, books on the subject of *mechanics* will be published under two series:

***FLUID* MECHANICS AND ITS APPLICATIONS**

Series Editor: R.J. Moreau

***SOLID* MECHANICS AND ITS APPLICATIONS**

Series Editor: G.M.L. Gladwell

Prior to 1990, the books listed below were published in the respective series indicated below.

MECHANICS: DYNAMICAL SYSTEMS

Editors: L. Meirovitch and G.Æ. Oravas

1. E.H. Dowell: *Aeroelasticity of Plates and Shells.* 1975 ISBN 90-286-0404-9
2. D.G.B. Edelen: *Lagrangian Mechanics of Nonconservative Nonholonomic Systems.* 1977 ISBN 90-286-0077-9
3. J.L. Junkins: *An Introduction to Optimal Estimation of Dynamical Systems.* 1978 ISBN 90-286-0067-1
4. E.H. Dowell (ed.), H.C. Curtiss Jr., R.H. Scanlan and F. Sisto: *A Modern Course in Aeroelasticity.* *Revised and enlarged edition see under Volume 11*
5. L. Meirovitch: *Computational Methods in Structural Dynamics.* 1980 ISBN 90-286-0580-0
6. B. Skalmierski and A. Tylikowski: *Stochastic Processes in Dynamics.* Revised and enlarged translation. 1982 ISBN 90-247-2686-7
7. P.C. Müller and W.O. Schiehlen: *Linear Vibrations.* A Theoretical Treatment of Multi-degree-of-freedom Vibrating Systems. 1985 ISBN 90-247-2983-1
8. Gh. Buzdugan, E. Mihăilescu and M. Radeş: *Vibration Measurement.* 1986 ISBN 90-247-3111-9
9. G.M.L. Gladwell: *Inverse Problems in Vibration.* 1987 ISBN 90-247-3408-8
10. G.I. Schuëller and M. Shinozuka: *Stochastic Methods in Structural Dynamics.* 1987 ISBN 90-247-3611-0
11. E.H. Dowell (ed.), H.C. Curtiss Jr., R.H. Scanlan and F. Sisto: *A Modern Course in Aeroelasticity.* Second revised and enlarged edition (of Volume 4). 1989 ISBN Hb 0-7923-0062-9; Pb 0-7923-0185-4
12. W. Szemplińska-Stupnicka: *The Behavior of Nonlinear Vibrating Systems.* Volume I: Fundamental Concepts and Methods: Applications to Single-Degree-of-Freedom Systems. 1990 ISBN 0-7923-0368-7
13. W. Szemplińska-Stupnicka: *The Behavior of Nonlinear Vibrating Systems.* Volume II: Advanced Concepts and Applications to Multi-Degree-of-Freedom Systems. 1990 ISBN 0-7923-0369-5
    Set ISBN (Vols. 12–13) 0-7923-0370-9

MECHANICS OF STRUCTURAL SYSTEMS

Editors: J.S. Przemieniecki and G.Æ. Oravas

1. L. Frýba: *Vibration of Solids and Structures under Moving Loads.* 1970 ISBN 90-01-32420-2
2. K. Marguerre and K. Wölfel: *Mechanics of Vibration.* 1979 ISBN 90-286-0086-8

GPSR Compliance

The European Union's (EU) General Product Safety Regulation (GPSR) is a set of rules that requires consumer products to be safe and our obligations to ensure this.

If you have any concerns about our products, you can contact us on ProductSafety@springernature.com

In case Publisher is established outside the EU, the EU authorized representative is:

Springer Nature Customer Service Center GmbH
Europaplatz 3
69115 Heidelberg, Germany

Batch number: 10816188

Printed by Printforce, the Netherlands